Biochemistry of Women: Methods for Clinical Investigation

Biochemistry of Women: Methods for Clinical Investigation

A. S. Curry
M.A., Ph.D., F.R.I.C., F.R.C. Path.
Home Office Central Research Establishment
Aldermaston Nr Reading
Berkshire U.K.

J. V. Hewitt
M.Sc., A.R.I.C.
Area Laboratory
King Edward VII Hospital
Windsor U.K.

published by:

CRC Press
Taylor & Francis Group
Boca Raton London New York

CRC Press is an imprint of the
Taylor & Francis Group, an **informa** business

First published 1974 by CRC Press
Taylor & Francis Group
6000 Broken Sound Parkway NW, Suite 300
Boca Raton, FL 33487-2742

Reissued 2018 by CRC Press

A Library of Congress record exists under LC control number: 74076654

Publisher's Note
The publisher has gone to great lengths to ensure the quality of this reprint but points out that some imperfections in the original copies may be apparent.

Disclaimer
The publisher has made every effort to trace copyright holders and welcomes correspondence from those they have been unable to contact.

ISBN 13: 978-1-315-89108-8 (hbk)
ISBN 13: 978-1-351-07018-8 (ebk)

Visit the Taylor & Francis Web site at http://www.taylorandfrancis.com and the
CRC Press Web site at http://www.crcpress.com

INTRODUCTION

The biochemistry of women is of interest to all – to man for obvious reasons, to women for selfish ones!

However, few books have dealt solely with the analytical methodology of those tests that are performed predominantly or solely to investigate the clinical chemistry of the female.

Several factors in the last decade have created a situation which makes the present time particularly opportune to bring the various aspects of the subject into one volume. Increased opulence, better medicine, and higher standards of living have concentrated thoughts onto better antenatal care, the problems of infertility, and, with the impact of the pill, the problems of fertility. Considerable medical attention has been focused on many other conditions affecting women, such as breast cancer and the menopause, and in all these the laboratory has its role.

At the same time as the increase in rate of growth of medical knowledge have come major technological developments, for example, gas chromatography and radioimmunoassays. These have enabled biochemical investigations to delve into hitherto closed areas.

This book is written by experts who, using the latest techniques, describe laboratory investigations into women from conception to the grave. We asked the authors to pay particular attention to the interpretation of laboratory results, so we hope the book will be of interest to clinicians as well as to medical scientists.

We had problems with the title of this book – at one time it was going to be *Investigations of Women in the Laboratory* – we hope the reader is not disappointed!

THE EDITORS

Dr. Alan S. Curry is Director of the Home Office Central Research Establishment, Aldermaston, U.K., and has long experience of analytical methodology in chemistry, biochemistry, and toxicology. Although a chemist by training, he is a Fellow of the Royal College of Pathologists and consultant in forensic toxicology to the Royal Air Force. He is a graduate of Cambridge University and has written extensively in the field of forensic chemistry and toxicology. He is the author of *Poison Detection in Human Organs* and *Advances in Forensic and Clinical Toxicology*, editor of *Methods of Forensic Science*, and has contributed numerous chapters to other books. He has also had nearly 100 research publications published.

Miss J. V. Hewitt is a clinical biochemist with extensive hospital experience. A graduate of Manchester University, after earning her M.Sc. she joined the staff of the Biochemistry Laboratory, Manchester Royal Infirmary. She worked there under the direction of Mr. H. Varley until moving to King Edward VII Hospital, Windsor, in 1970.

Dr. Curry is a Civil Servant and wishes to make it clear that the editing of this book was in no way connected with his official duties. The views expressed in the book are entirely those of the authors of the chapters.

CONTRIBUTORS

J. D. Acland, M.A., B.Sc., B.M., B.Ch., Ph.D.
Group Pathology Department
Manor Hospital
Walsall WS2 9PS
England

D. Fahmy, M.Sc., Ph.D.
Tenovus Institute for Cancer Research
Welsh National School of Medicine
The Heath
Cardiff CF4 4XX
Wales

B. Furst, M.D.
Department of Endocrinology and Metabolism
St. Luke's Hospital
Cleveland
Ohio 44104

M. C. Hallberg, B.S.
Department of Endocrinology and Metabolism
St. Luke's Hospital
Cleveland
Ohio 44104

P. J. Leonard, M.Sc., Ph.D., M.R.C. Path.
Hormone Assay Laboratory
G.D. Searle Scienfific Services
Lane End Road
High Wycombe
Bucks HP12 4HL
England

M. J. Levell, M.A., Ph.D.
Division of Steroid Endocrinology
The University of Leeds
School of Medicine
26-28 Hyde Terrace
Leeds 2
England

O. J. Lucis, M.D., Ph.D.
Department of National Health and Welfare
Place Vanier
355 River Road
Vanier, Ontario K1A 1B8
Canada

M. G. Metcalf, M.Sc., Ph.D.
The Medical Unit
The Princess Margaret Hospital
Christchurch
New Zealand

D. W. Moss, M.A., M.Sc., Ph.D.
Department of Chemical Pathology
Royal Postgraduate Medical School
Ducane Road
London W12 OHS
England

R. E. Oakey, Ph.D.
Division of Steroid Endocrinology
Department of Chemical Pathology
The University of Leeds
School of Medicine
Leeds
England
now
Visiting Scientist
Pregnancy Research Branch
National Institute of Child Health
and Human Development
Bethesda, Md. 20014

J. F. Pearson, M.D., M.R.C.O.G.
Maternity Department
Welsh National School of Medicine
Heath Hospital
Cardiff
Wales

C. Robyn, M.D.
Human Reproduction Research Unit
Hospital St. Pierre
University of Brussels
Brussels
Belgium

D. P. Rose, M.D., Ph.D.
Division of Clinical Onocology
University Hospitals
University of Wisconsin
Madison, Wisc. 53706

C. Swain, M.Phil., Ph.D.
Department of Clinical Endocrinology
Imperial Cancer Research Fund Laboratories
Lincoln's Inn Fields
London WC2A 3PX
England

A. C. Turnbull, M.D., F.R.C.O.G.
Nuffield Department of Obstetrics and Gynaecology
John Radcliffe Hospital
Headington
Oxford
England

M. Vekemans, M.D.
Human Reproduction Research Unit
Hospital St. Pierre
University of Brussels
Brussels
Belgium

D. Y. Wang, M.Sc., Ph.D.
Department of Clinical Endocrinology
Imperial Cancer Research Fund Laboratories
Lincoln's Inn Fields
London WC2A 3PX
England

R. G. Wieland, M.D.
Department of Endocrinology and Metabolism
St. Luke's Hospital
Cleveland
Ohio 44104

S. Winsten, Ph.D.
Department of Laboratories
Albert Einstein Medical Center
Northern Division
Philadelphia, Pa. 19141

E. M. Zorn, M.S.
Department of Endocrinology and Metabolism
St. Luke's Hospital
Cleveland
Ohio 44104

TABLE OF CONTENTS

Nonpregnancy Estrogens 1
P. J. Leonard

Estrogens in Pregnancy 19
R. E. Oakey

Progesterone and Metabolites 45
O. J. Lucis

Gonadotropins and Prolactin 79
C. Robyn and M. Vekemans

Neutral Steroids 129
M. J. Levell

Gas Chromatographic Procedures for the Measurement of Pregnanediol, Andosterone, Etiocholanolone, and Dehydroepiandrosterone 151
M. G. Metcalf

Testosterone and Metabolites in Women Using Radioimmunoassay Techniques 181
R. G. Wieland, E. M. Zorn, M. C. Hallberg, and B. Furst

Hormones and Breast Cancer 191
D. Y. Wang and M. C. Swain

Enzymes in Carcinoma 219
S. Winsten

The Selection, Performance, and Interpretation of Serum Enzyme Tests in Pregnancy 237
D. W. Moss

Problems of Thyroid Analyses in Pregnancy 257
J. D. Acland

Fetal Monitoring – Current Concepts 299
D. Fahmy, J. F. Pearson, and A. C. Turnbull

Assessment of Tryptophan Metabolism and Vitamin B_6 Nutrition in Pregnancy and Oral Contraceptive Users 317
D. P. Rose

Index 351

NONPREGNANCY ESTROGENS

P. J. Leonard

TABLE OF CONTENTS

I. Introduction . . . 1
  A. Urinary Estrogens . . . 1
    1. Colorimetric Methods . . . 2
    2. Fluorimetric Methods . . . 3
    3. Gas-liquid Chromatographic Methods . . . 3
    4. Radioimmunoassay Techniques . . . 4

II. Recommended Methods . . . 4
  A. Determination of Total Estrogens in Urine . . . 4
    1. Fluorimetric Procedure . . . 4
    2. Gas-liquid Chromatographic Method for the Determination of Estrone, Estradiol, and Estriol in Nonpregnancy Urine . . . 6
  B. Measurement of Estrogens in Blood . . . 7
    1. Introduction . . . 7
    2. Recommended Methods . . . 9

III. Clinical Application of Estrogen Estimation in Nonpregnant Women . . . 9
  A. Introduction . . . 9
  B. Test Procedure Using Pergonal . . . 10
    1. Interpretation of Results . . . 12
  C. Test Procedure Using Clomid . . . 12
    1. Interpretation of Results . . . 12

Acknowledgment . . . 13

References . . . 16

I. INTRODUCTION

Estrogens are C18 steroid hormones with the basic structure outlined in Figure 1. The normal female ovary is the main site of production, although the adrenal cortex is responsible for producing a very small amount. Three estrogens – estriol, estradiol, and estrone – have long been identified as the most important physiologically and are excreted in the urine mainly as the conjugates of glucuronide and sulfate.

A. Urinary Estrogens

Techniques for the assessment of urinary estrogens can largely be classified into four groups: methods dependent on a colorimetric reaction, methods dependent on a fluorimetric reaction, gas-liquid chromatography procedures, and radioimmunoassay techniques. In general, these techniques can be further divided into the various stages of assay procedure.

The first technique, hydrolysis, removes the water-soluble conjugate and liberates a free steroid

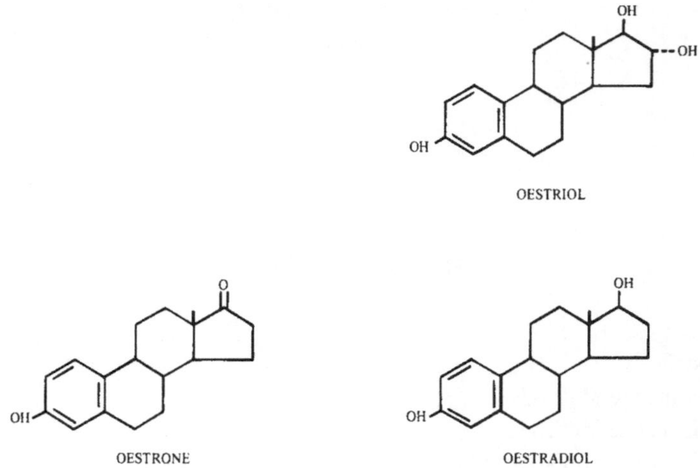

FIGURE 1. Structure of three main classical estrogens.

suitable for extraction into an organic solvent. In general, most methods accomplish hydrolysis by heating with acid either in a boiling water bath or, as recently described by Brown et al.,[1] in an autoclave. Lately, the use of the enzyme β-glucuronidase, obtained from limpet source or helix pomatia, is receiving attention.

The second technique is extraction of the free steroid using a suitable organic solvent. Most procedures use diethyl ether, although dichloromethane is also suitable.

Many purification procedures are in use, but the unique property of estrogen extraction from a carbonate medium at pH 10 to 10.5, leaving behind most interfering materials,[2] is the most widely used method of initial purification. The partitioning of estrogens between water benzene and petroleum ether results in the separation of estriol from estrone and estradiol. The estrone and estradiol can be recovered quantitatively by extraction with sodium hydroxide.[2] Column techniques using alumina or gel filtration have also been used to separate the individual estrogens.

Finally, measurement of the estrogen content of the purified extracts can be accomplished by using one of the procedures to be described.

1. Colorimetric Methods

All methods for estrogen determination using a colorimetric end point utilize the reaction of estrogens with sulfuric acid, as described by Kober[3] and investigated by Brown[4] Bauld,[5] and, more recently, by Brown and co-workers.[6] As originally described, estrogens, when heated with sulfuric acid containing phenol, diluted with water, and reheated, produce a pink color with a maximum absorbance at a wavelength of 520 nm. It was later found that other phenols and reducing agents can be substituted for phenol itself and most methods employing the Kober reaction now use hydroquinone. The reaction has usually been performed at 100°C for 30 to 40 min, but Brown et al.[6] recommend a temperature of 120°C for 5 min. In addition to saving time, the higher temperature has the added advantage that the wavelengths for maximum absorption are the same for all three estrogens, whereas with the lower temperature they differed slightly and the colors produced were more intense.

The procedure of Brown[2] later modified by Brown, Bulbrook, and Greenwood,[7] formed the basis of all techniques that were to follow not only in respect of the absorptiometric stages, but also of the extraction and purification procedures. This method separated the three classical estrogens using alumina chromatography of the methyl ethers. It is an exacting technique that demands a high degree of the technical skill and is limited to the evaluation of about five samples over a 2-day period. Nevertheless, in many respects, this is still the reference method of choice.

Methods for the evaluation of the individual estrogens have been described and the procedure of Brown and Blair[8] for estrone is sensitive to a level of 0.5μg per 24 hr. It is, nevertheless, an exacting and laborious technique. Many methods for the determination of estriol have been published, but the interest in this steroid as a single entity arises mainly from its elevated excretion during pregnancy.

2. Fluorimetric Methods

In 1958, Ittrich[9] examined the factors affecting the Kober reaction and showed that the Kober/estrogen complex could be extracted into chloroform containing *p*-nitrophenol and that this complex emitted a strong yellow fluorescence when exposed to green light. Fluorimetry increases the sensitivity of estrogen determination by a factor of about 50 and is more specific, as demonstrated by Stoa and Thorsen.[10] The sensitivity is limited (approximately 10^{-9} g) by fluorescence of impurities in the extract.[11] Brown et al.[6] examined the effect of heating at 120°C and found a further intensity of the fluorescence of approximately 25% over that obtained by heating at 100°C.

Beling[12] and Eechaute and Demeester,[13] using Sephadex® for the initial purification, published methods based on the Ittrich reaction for the determination of estrogens in pregnancy and nonpregnancy urines, respectively. The Eechaute and Demeester method was shown to be suitable for the determination of total estrogens down to a level of 5 μg per 24 hr. The preparation of the Sephadex and the poor reproducibility of the columns, however, limited the use of this technique.

Corns and James[14] published a modification of the Ittrich procedure in which, following extraction of the conjugates, borohydride reduction was carried out and the estrogens were measured fluorimetrically after phenolic extraction using the Ittrich reaction. This technique and a more recent publication by Goebelsmann[15] appear to offer little advantage over the method of Brown et al.[6]

Recently, a fully automated method for the determination of estriol in nonpregnancy urine has been described by Barnard and Logan.[16] This analysis is performed on the Technicon AutoAnalyzer® using a modification of the pregnancy method of UaConnail and Muir.[17] Samples are manually hydrolyzed with acid and applied to the sampler. A benzene/petroleum ether wash is incorporated into the system to remove estrone and estradiol; after ether extraction the solvent is removed using the digestor helix module. Kober color development is included in this stage and the Kober/estrogen complex extracted into chloroform containing *p*-nitrophenol. Two fluorimeters are incorporated into the flow, the second with wavelengths selected to evaluate interfering background fluorescence. In a short comparison with the method of Brown et al.,[6] good correlation was obtained but the regression line showed values with the automated procedure to be double that of the manual technique.

3. Gas-liquid Chromatographic Methods

In many respects, this form of analysis provides the ideal conditions for the resolution and quantitation of many steroids. However, it has proved essential to purify samples to an even higher degree than was previously required with the colorimetric and fluorimetric methods. The main advantage of using gas chromatography in steroid analysis lies in better specificity and the ability to provide information on a group of compounds from a single extract.

Most of the methods published for estrogen determination utilize extraction and purification stages similar to those described in the previous sections. Free estrogens are not suitable for injection directly onto the gas chromatograph as irreversible absorption or thermal decomposition takes place. Acetates[18] and trimethylsilyl ethers[19-21] are by far the most commonly used derivatives and, in respect of the electron capture detector, heptafluorobutyrates[22] have proved successful.

Gas-liquid chromatographic methods for the determination of estrone, estradiol, and estriol in nonpregnancy urine have been described by many workers and the subject has been reviewed by Adlercreutz and Luukkainen.[23] The method of Cox and Bedford[24] utilizes the extraction system of Brown[2] and gas chromatography of the trimethylsilyl ether derivatives of the estrogen 3-methyl ethers. Menini[25] measured estriol, estrone, and estradiol after borohydride reduction in a procedure similar to the method of Cox and Bedford.[24] Adlercreutz et al.[26] used the Beling purification with Sephadex G25 and after enzymatic hydrolysis and extraction, the free estrogens were applied to a further Sephadex column. The individual estrogen fractions were converted to trimethylsilyl ethers and chromatographed separately.

Kaplan and Hreshchyshyn[27] published a gas chromatographic method for the determination of the three classical estrogens in nonpregnant

FIGURE 2. Semiautomatic extraction apparatus.

women. This method consisted of extraction of the conjugates, enzymatic hydrolysis, phenolic extraction, and TLC separation. An anion-exchange column was used for the final purification and after heptafluorobutyrate derivative formation the samples were chromatographed.

4. Radioimmunoassay Techniques

Recently, with the introduction of small molecule radioimmunoassay,[28-31] there has been much interest in the use of this technique, not only for the estimation of steroids in blood but in other biological fluids. Radioimmunoassay procedures have been described for the estimation of estrogens in urine by Aoki and Hreshchyshyn[32] and Kaplan et al.[33] These workers used enzymatic hydrolysis; purification of the three main estrogens was accomplished using thin-layer chromatography. The estrone, estradiol, and estriol bands were eluted and measured by radioimmunoassay using an antibody to estradiol 17-β succinyl bovine serum albumin (BSA) conjugate.

In summary, although many techniques for the determination of estrogens from nonpregnancy urine are available, the semiautomatic fluorimetric procedure of Brown et al.[6] would, in the routine clinical situation, appear to be the method of choice at the present time. Even this technique demands a high degree of technical skill but, providing the assay is well-controlled, reliable data can be obtained. Where more information is required and fractionation of the individual estrogens is considered important, GLC has many advantages over the older colorimetric techniques, but the potential for radioimmunoassay in this area is encouraging.

II. RECOMMENDED METHODS

The estimation of the total urinary estrogens is all that is usually required in clinical practice, except on rare occasions when it is felt necessary to estimate the individual fractions. Although the selection of any method as the procedure of choice is open to criticism, the following techniques are recommended from experience with large workloads over a period of many years.

A. Determination of Total Estrogens in Urine[1]

1. Fluorimetric Procedure

Extraction is performed using the semiautomatic extractor, as illustrated in Figure 2, equipped with 12 partition tubes. The partition tube is shown in Figure 3 and the complete extraction procedure can be performed with this system.

a. Reagents

1. 15% v/v Hydrochloric acid.
2. Diethyl ether (obtained from MacFarlan Smith, Edinburgh) redistilled over ferrous sulfate immediately before use.

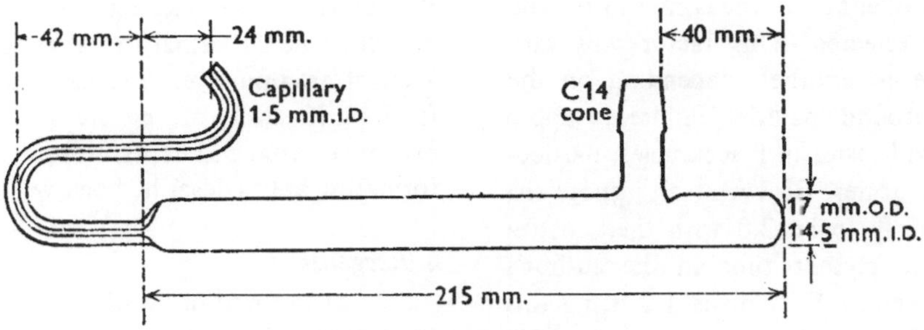

FIGURE 3. Partition tube.

3. Sodium chloride AR.

4. Carbonate buffer, pH 10.5, 21 g sodium hydroxide, and 70 g sodium bicarbonate are dissolved in a liter of distilled water.

5. 1.0 *N* Sodium hydroxide.

6. Petroleum ether – b.p. 40 to 60°C – redistilled.

7. Sodium bicarbonate AR.

8. Kober reagent is prepared by dissolving 20 g quinol (B.D.H. spot test grade) in a liter of 66% sulfuric acid, analytical grade.

9. Ittrich reagent 2 g *p*-nitrophenol is dissolved in 1 ml ethanol with gentle warming and after solution is complete; 100 ml sym-tetrachloroethane is added. This reagent is stored at 4°C and renewed weekly.

10. Estriol standard – 1 μg/ml of estriol in ethanol is prepared from a stock solution of 1 mg/ml.

The estriol can be obtained from Steraloids Ltd. and all other chemicals and solvents from Hopkin & Williams Ltd.

b. Method

Hydrolyze 1 ml of urine with 6 ml of 15% v/v hydrochloric acid by heating in an autoclave at 120°C (15 lb/in.^{2}) for 15 min. The sample is transferred to the automatic extractor and 1 g of sodium chloride added to aid the extraction of estriol. When dissolved, 6 ml diethyl ether is dispensed into the tubes, shaken for at least 2 min and, after removal of the lower aqueous layer, the extract is washed with 2 ml carbonate buffer.

Add 6 ml of petroleum ether 40 to 60° and 6 ml of 1.0 *N* NaOH and extract the phenolic steroids into the alkali by shaking. The upper organic layer containing the neutral steroids is discarded.

Add 0.8 g sodium bicarbonate to the phenolic extract and shake until solution is complete. The estrogens are reextracted into 6 ml of diethyl ether and the carbonate layer discarded. The ether extract is transferred to a 15-ml tube and the solvent evaporated to dryness.

c. Fluorimetry

First, 1 ml of 66% Kober reagent and approximately 1 mg of powdered hydroquinone are added to the dried extracts, to a standard tube containing 100 ng of estriol, and to an empty tube for use as a blank.

The tubes are heated at 120°C in an oil bath for 10 min and cooled in ice water. Dispense 0.75 ml of Ittrich reagent into fluorimeter tubes and cool in the ice water. Add 1.5 ml of ice cold fresh distilled water to the Kober tube, mix, and reimmerse in ice water. The diluted Kober/estrogen complex is transferred to the Ittrich reagent in the fluorimeter tube, shaken vigorously, and the layers separated by centrifugation at 1,500 r.p.m. for 5 min at 0°C.

Fluorescence is evaluated using the Aminco Bowman spectrofluorimeter with mercury/xenon lamp and ellipsoid condensing system. Wavelengths measured are

1. Excitation at 546 nm, fluorescence 565 nm.
2. Excitation at 490 nm, fluorescence 520 nm.

The corrected fluorescence is obtained using the formula

F 546/565 – (K X F 490/590)

K is a factor used to eliminate background fluorescent impurities in the extract and is dependent on the instrument used, the lamp source, and the wavelengths selected. This factor will vary from one sample to another, depending on the amount of background material; failure to apply this correction will result in inaccuracies, particularly in low-titer urines. Brown et al.[1] quote the average value for K to be 2.0 with their instrument, but at the present time in the author's laboratory the average K value is 1.2 using this technique. This factor can best be determined by examination of urine from subjects with very low estrogen levels, e.g., patients with hypopituitarism or patients after adrenalectomy or oophorectomy. The sample is taken through the assay procedure as outlined above. The fluorescence at 546 nm/565 nm (A) provides a measure of the estrogen level plus the impurities while the fluorescence at 490 nm/520 nm (B) provides a measure of the contribution from interfering nonestrogenic substances. K is the ratio of A/B.

d. Calculation

The total estrogens in the sample expressed as μg per 24 hr, can be calculated by

$$\frac{\text{Corrected fluorescence test} - \text{blank}}{\text{Corrected fluorescence standard} - \text{blank}} \times \text{volume} \times \frac{100}{1000}$$

The whole calculation, including fluorescence correction, can readily be applied to many simple desk top calculators.

In over 3 years of daily use, this method has provided reliable and useful data. Quality control samples (freeze-dried aliquots of two separate pools of nonpregnancy urine at concentrations in the region of 30 and 90 μg/l and prepared in our own laboratory) are included. Quality control samples are included in each batch and the between-batch coefficient of variation is less than 10%. Recovery of added estriol to a low-titer urine is also included daily and a mean recovery of 75 ± 3.8% is obtained. A trained technologist can analyze some 30 specimens in a working day.

2. Gas-liquid Chromatographic Method for the Determination of Estrone, Estradiol, and Estriol in Nonpregnancy Urine

This recommended procedure is essentially the method described by Kaplan and Hreshchyshyn.[27] The conjugate extraction with dipyridine sulfate is replaced by the use of Amberlite® XAD-2 resin[34] and after high-temperature enzyme hydrolysis with glucuronidase, samples are extracted using the semiautomatic extractor by the methodology described in the urinary total estrogen procedure. Thereafter, thin-layer chromatography separation, final purification of the steroids with an anion-exchange resin, and heptafluorobutyrate derivative formation are as described below.

a. Reagents

1. Amberlite XAD-2 resin is obtained from British Drug Houses Ltd. and is bulk washed with water. Columns 1 cm in diameter are packed with the resin to a height of 5 cm and are ready for use.

2. β-Glucuronidase is prepared from limpet source. A similar preparation is available from Sigma Chemical Company, extracted from helix pomatia.

3. Thin-layer sheets – aluminum backed silica gel sheets 20 X 20 cm are obtainable from Merck. These are divided into five lanes and used without prior treatment.

4. Anion-exchange resin AG 1 X 2 (200–400 mesh, chloride form) is available from Bio Rad Laboratories and is packed into small columns 0.5 cm in diameter and to a height of 5 cm while suspended in water.

5. One percent QF-1 on Diatomite CQ (80 – 100 mesh) is used as the column stationary phase and is obtained from J & J Chromatography Ltd., Kings Lynn, Norfolk, England. This material is packed into 6-ft coiled glass column under negative pressure and cured in the chromatography oven at 250°C overnight with a flow of nitrogen at 50 ml/min.

6. Heptafluorobutyric anhydride (Pierce Chemical Company) is used as supplied.

7. Tritium labeled steroids, estrone 6-7^3H, estradiol 6-7^3H, and estriol 6-7^3H (Radiochemical Centre, Amersham) are maintained as stock solutions at a concentration of 20 μCi/ml in ethanol and are used without further purification.

8. Standard estriol, estrone, and estradiol (Steraloids Ltd.) are prepared as individual solutions at a concentration of 1 mg/ml in ethanol.

9. Acetate buffer, 1.0 *M*, pH 4.6.

10. Methanol, benzene, chloroform – analytical grade obtained from Hopkin & Williams Ltd.

b. Apparatus

1. Liquid scintillation counting is performed

on the Nuclear Chicago Isocap using a PPO/POPOP counting fluid.

2. The location of the labeled steroids on TLC sheets is performed using the Nuclear Chicago Actigraph, Mark III.

3. Gas-liquid chromatography is evaluated using the Pye 104 Chromatograph equipped with an electron capture detector at an oven temperature of 200°C and with a nitrogen gas flow maintained at 50 ml/min.

c. Method

Conjugate extraction and hydrolysis – 10 ml of filtered urine is applied to the Amberlite XAD-2 column and allowed to enter the resin. The column is washed with 20 ml of water which is permitted to drain. The steroid conjugates are eluted from the column with 20 ml of methanol and the solvent removed by evaporation in a water bath at 65°C under a stream of air. The columns can be regenerated for further use by washing with 50 ml of water.

Hydrolysis is accomplished by dissolving the residue in 5 ml of 1.0 *M* acetate buffer pH 4.6 containing 2,000 Fishman Units of β-glucuronidase per ml incubating at 65°C for 2 hr.

Extraction – 20,000 dpm of 3H-estrone, estradiol, and estriol are added to the extracts and as standards a similar volume dispensed directly into counting vials. The samples are transferred to the Brown semiautomatic extractor and are subject to the purification and extraction methodology as described in the urinary total estrogen procedure.

Thin-layer chromatography – The extracts are transferred to the TLC sheets with three 50-μl aliquots of benzene:chloroform (1:1 v/v). The chromatogram is developed in a solvent of chloroform:acetone (7:3 v/v) in an equilibrated tank lined with filter paper. The radioactive areas are located with the Nuclear Chicago Actigraph and eluted with methanol:ether (1:1 v/v). The extracts are evaporated to dryness at 65°C under a stream of air.

Anion-exchange chromatography – The AG 1 X 2 columns are washed with distilled water, 0.5 *M* sodium bicarbonate, and three times with methanol. The individual estrogen extracts are transferred to the columns with two times 0.5 ml vol of methanol. The columns are washed with 1.0 ml of methanol and the estrogens eluted with 3.0 ml 60% aqueous methanol:ethanol (1:1 v/v). Positive pressure is applied to the columns during chromatography. The extracts are evaporated to dryness at 65°C in a water bath, after the addition of 100 ng estriol-3-methyl ether as an internal standard.

The extracts are dissolved in 200 μl of acetone and 50 μl is removed for scintillation counting to evaluate the recovery of the steroids and compensate for losses incurred.

Derivative formation – A further 50 μl of each extract is removed, taken to dryness, and dissolved in 50 μl of a freshly prepared solution of heptafluorobutyric anhydride in tetrahydrofuran (1:5 v/v). The tubes are incubated at room temperature for 60 min after which the excess reagents are evaporated to dryness in a vacuum desiccator. The derivatives are allowed to remain in the vacuum desiccator overnight. Samples are dissolved in 20 μl of heptane and 5 μl applied to the gas-liquid chromatograph. The retention times in minutes for the heptafluorobutyrates are estradiol 12, estrone 18, and estriol 28.

Standards equivalent to 25 ng of estriol, estrone, and estradiol containing 25 ng estriol-3-methyl ether as internal standard are converted into the heptafluorobutyrate derivative as above.

Although this is a lengthy procedure requiring 3 days for analysis, the precision and accuracy are satisfactory and provide reliable information of the individual fractions.

B. Measurement of Estrogens in Blood

1. Introduction

Estrogens exist in blood plasma in many forms – free, protein-bound, and conjugated. Until the introduction of steroid hapten radioimmunoassay techniques, methods for the determination of estrogens in blood were few and were limited to the evaluation of the total estrogens or the total estrone, estradiol, and estriol fractions, but with radioimmunoassay, the free steroids can now be evaluated.

One of the earliest techniques was that described by Roy and Brown[35] which was essentially very similar to the Brown[2] urine procedure with the use of microcells to increase the sensitivity of the final reaction. Fluorimetry was used by Preedy and Aitken,[36] but the method was complex and lengthy. Even with the introduction of the gas-liquid chromatograph, only a few applications with this instrument for the estimation of estrogens in pregnancy samples have been described. Adlercreutz,[37] Adlercreutz et al.,[26] and Kroman and co-workers[38] used a ^{90}Sr detector for the estimation of plasma estrogens, but the sensitivity was inadequate. The introduction of the electron capture detector increased the sensitivity, particularly when heptafluorobutyrate derivatives were used,[22,39] but even then the sensitivity obtained was such that its use was restricted to the evaluation of total fractions rather than the free steroid.

Recently, Cooper et al.[40] have published details of a plasma estriol procedure using the ditrimethylsilyl ethers of the estriol-3-methyl ether for use with the flame ionization detection. Good resolution is obtained, but 10 ml of sample is required for analysis.

A double isotope derivative procedure for the measurement of estrone and estradiol in human plasma has been described by Baird.[41]

With the introduction of competitive protein binding analysis and radioimmunoassay techniques, many procedures for the evaluation of blood estrogens have been described. In this section, consideration will be given to the methods for determination of plasma estradiol and estriol.

a. Plasma Estradiol

Murphy[42] reported on the binding of testosterone and estradiol with high affinity to a β-globulin termed "the testosterone binding globulin." Mayes and Nugent[43] utilized this binding to develop a competitive protein assay for the evaluation of estradiol in plasma. Similar procedures were described by Dufau et al.[44] and Knox and France.[45]

Korenman[46] described the preparation of rabbit uterine cytosol which had considerable benefits as an assay protein in respect of sensitivity and specificity. Rabbit uterine cytosol was used as the binding protein in a method for the evaluation of estradiol in plasma by Corker and Exley.[47] This technique, which was simple and reliable, used 1 ml of plasma; after extraction with ether, purification was accomplished with alumina TLC. The eluate from the TLC was subjected to competitive protein binding analysis with the cytosol preparation and dextran-coated charcoal was used to separate the free and bound fractions. Similar procedures have been described by Korenman et al.,[48] Concolino and Marocchi,[49] and Mester and co-workers.[50]

Radioimmunoassay for plasma estrogens commenced with antisera to 17β-estradiol coupled to carbon 17 as a hemisuccinate[51] or as an estrogen A ring conjugate.[52] Carbon 17 conjugation resulted in considerable cross-reaction with estrone and 17α-estradiol, while A ring coupled conjugates produced antisera with poor specificity. In an attempt to improve specificity and reduce assay purification stages, Exley et al.,[53] Jeffcoate and Searle,[54] and Lindner and co-workers[55] have described the preparation of antisera coupled to the carbon 6 position as 17β-estradiol-6-(O-carboxymethyl)oxime-BSA. Cross-reaction of this antiserum with estrone and estriol was less than 1%. Using this antiserum, Cameron and Jones[56] have examined samples throughout the normal menstrual cycle with and without chromatography on Sephadex LH 20. Although the chromatographed extracts produced lower values in the cycle samples, the pattern was identical and the significant difference minimal. In the author's department, chromatography is not included for the routine analysis of plasma estradiol during the menstrual cycle and with gonadotrophin stimulation.

b. Plasma Estriol

Several workers have developed methods for the determination of plasma estriol.[57-59] This procedure has recently been simplified by Mathur et al.;[60] using this fluorimetric technique, ten samples can be evaluated in a working day. In general, these methods are unsuitable for routine use. Corker and Naftolin[61] modified their estradiol method using rabbit uterine cytosol as the protein binding reagent to provide an assay for estriol. The samples are hydrolyzed with acid, extracted with ether, and the dried organic phase taken up with water. After washing with benzene/petroleum ether, the estriol in the aqueous phase is evaluated using a system identical to the plasma estradiol method. Tulchinsky et al.[62] have used a similar procedure, but introduced a celite column for purification of the extract. This method offers no advantage over benzene/petroleum ether extraction.

Recently, Gurpide et al.[63] have introduced radioimmunoassay for the determination of estriol in plasma dihemisuccinate and Wilson[64] has used antiserum to estriol (16:17) for the determination of total estriol from pregnancy subjects. Further developments in this area are awaited with interest.

Radioimmunoassay is undoubtedly the method

of choice for the estimation of blood estrogens. It is still too early to say whether the estimation of the individual free estrogens in blood is more valuable clinically than estimating the total urinary estrogen level. The laboratory procedure adopted depends on the specificity of the antiserum used and on whether one wants to estimate the total level or the individual fractions. The method currently used in the author's laboratory is detailed below.

2. Recommended Methods

a. Plasma Estradiol

This is a nonchromatographic procedure for the evaluation of estradiol in plasma using an antisera to 17β-estradiol-6-(O-carboxymethyl) oxime-BSA conjugate prepared in the author's laboratory in rabbits. The cross-reaction with estrone and estriol is less than 1%.

b. Reagents

1. Dichloromethane AR is redistilled just prior to use.
2. Phosphate buffer, 0.05 *M*, pH 7.0, containing 0.1% sodium azide and 0.1% bovine serum albumin.
3. Dextran T2000 (Pharmacia Fine Chemicals) (50 mg) in 100 ml of BSA-phosphate buffer.
4. Activated charcoal (500 mg) in 100 ml of BSA-phosphate buffer (2). Equal parts of dextran and charcoal solutions are mixed as required.
5. Radioactive estradiol 6-7^{3}H is obtained from the Radiochemical Centre, Amersham (specific activity is approximately 40 Ci/nM). For use a solution is prepared in BSA-phosphate buffer to provide a count of 20,000 dpm in 0.1 ml.
6. Antisera to 17β-estradiol-6-(O-carboxymethyl) oxime-BSA conjugate is available from Searle Diagnostic, U.K. For use, the antisera are diluted 1/30,000 in BSA-phosphate buffer.
7. Standard 17β-estradiol is maintained as a stock solution in ethanol at a concentration of 1 μg per ml.

The estradiol can be obtained from Steraloids Ltd. and all routine reagents and solvents from Hopkin & Williams Ltd.

c. Apparatus

All liquid scintillation counting is performed on the Nuclear Chicago Isocap using a PPO/POPOP counting fluid.

d. Method

Plasma (0.5 ml) is extracted with 5 ml dichloromethane by vigorous shaking and the layers allowed to separate. Most of the plasma is removed by aspiration and the extract washed with 1 ml of fresh double-distilled water. The sample is frozen in solid carbon dioxide and the extract transferred to another tube and taken to dryness.

Standard estradiol in the range from 10 to 1,000 pg containing a similar volume of dichloromethane is taken to dryness with the appropriate blanks.

Add 200 μl of antibody solution and 200 μl of 3H-estradiol solution to all tubes and incubate at 37°C for 1 hr. The tubes are placed in an ice water bath at 4°C for 30 min and while still in the ice bath, 200 μl of dextran-coated charcoal is added to the tubes and thoroughly mixed using a vortex mixer. Immediately, the samples are centrifuged at 2,500 r.p.m. for 10 min at 4°C and an aliquot removed for scintillation counting.

The standard curve is plotted on 3 cycle log paper (Chartwell Graph Data Reference 5531) with the percentage bound radioactivity on the vertical axis and the concentration of estrogen on the horizontal axis. The amount of radioactivity bound by the antibody on its own is taken as a 100%. The levels obtained for the different standard values are expressed as a percentage of this and plotted accordingly. The concentration of estrogens in the unknown samples is calculated by reading from the graph.

Using this procedure, some 40 samples can be evaluated in a working day by one trained technologist, with a coefficient of 9% between batches.

A full report on this procedure will be published elsewhere.[66]

III. CLINICAL APPLICATION OF ESTROGEN ESTIMATION IN NONPREGNANT WOMEN

A. Introduction

The main clinical application of estrogen determination outside of pregnancy has been in the investigation of ovarian and pituitary function in cases of primary and secondary amenorrhea. The interrelationship between the urinary excretion of estrogens and the gonadotrophins in a normal

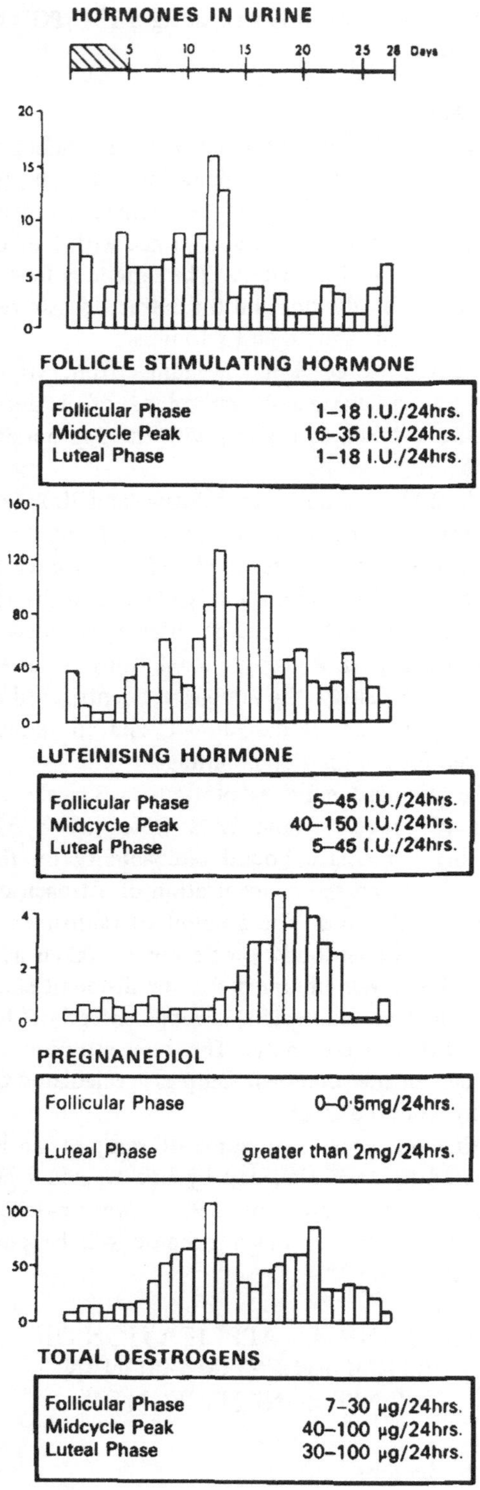

FIGURE 4. Excretion of LH, FSH, pregnanediol, and total estrogens during the normal menstrual cycle.

menstrual cycle is shown in Figure 4, while the control of ovarian function is outlined in Figure 5. It can be seen that during the follicular phase of the cycle, the production of estrogens increases to a maximum at about the time of ovulation. There is a second peak midway in the luteal phase and a rapid decline to basal levels just before menstruation. Measurement of plasma estrogens reveals a similar pattern, as shown in Figure 6. Thus, it can be seen that observing the plasma or urinary estrogen pattern is a very useful tool in evaluating reproductive function in the female. While the use of plasma eliminates the need to collect 24-hr urine samples, it suffers from the disadvantage that the subject may have to come daily to a clinic to have a blood sample taken. Collection of 24-hr urine samples over a period of several days is very tedious, but could be made considerably easier if a spot urine would suffice. Chamberlain et al.[65] investigated the usefulness of collecting untimed overnight urine specimens. This was done during six successive menstrual cycles in one individual; a comparison made with the values obtained for estrogens is shown in Figure 7. It can be seen that the untimed overnight specimen gave the same pattern as the full 24-hr specimen, confirming its usefulness as an outpatient screening technique. Incidentally, there was also a good correlation between the levels of pregnanediol, FSH, and LH between the two specimens.

The use of gonadotrophins in the treatment of infertility due to pituitary insufficiency has aroused considerable interest in recent years. Two methods have been widely adopted. The first is the use of clomiphene citrate (Clomid®) to stimulate the release of FSH and LH from the anterior pituitary; the second is by direct stimulation of the ovary using human menopausal gonadotrophin (Pergonal®). The latter has mainly an FSH-like action on the ovary and on its own is not sufficient to induce ovulation. This is achieved by the added use of human chorionic gonadotrophin (HCG). Adequate monitoring of estrogen production is an essential requirement to minimize the risk of ovarian hyperstimulation and multiple pregnancy. A typical test procedure is given below and is shown graphically in Figure 8.

B. Test Procedure Using Pergonal

The control values of estrogens and pregnanediol should be established immediately prior to the test.

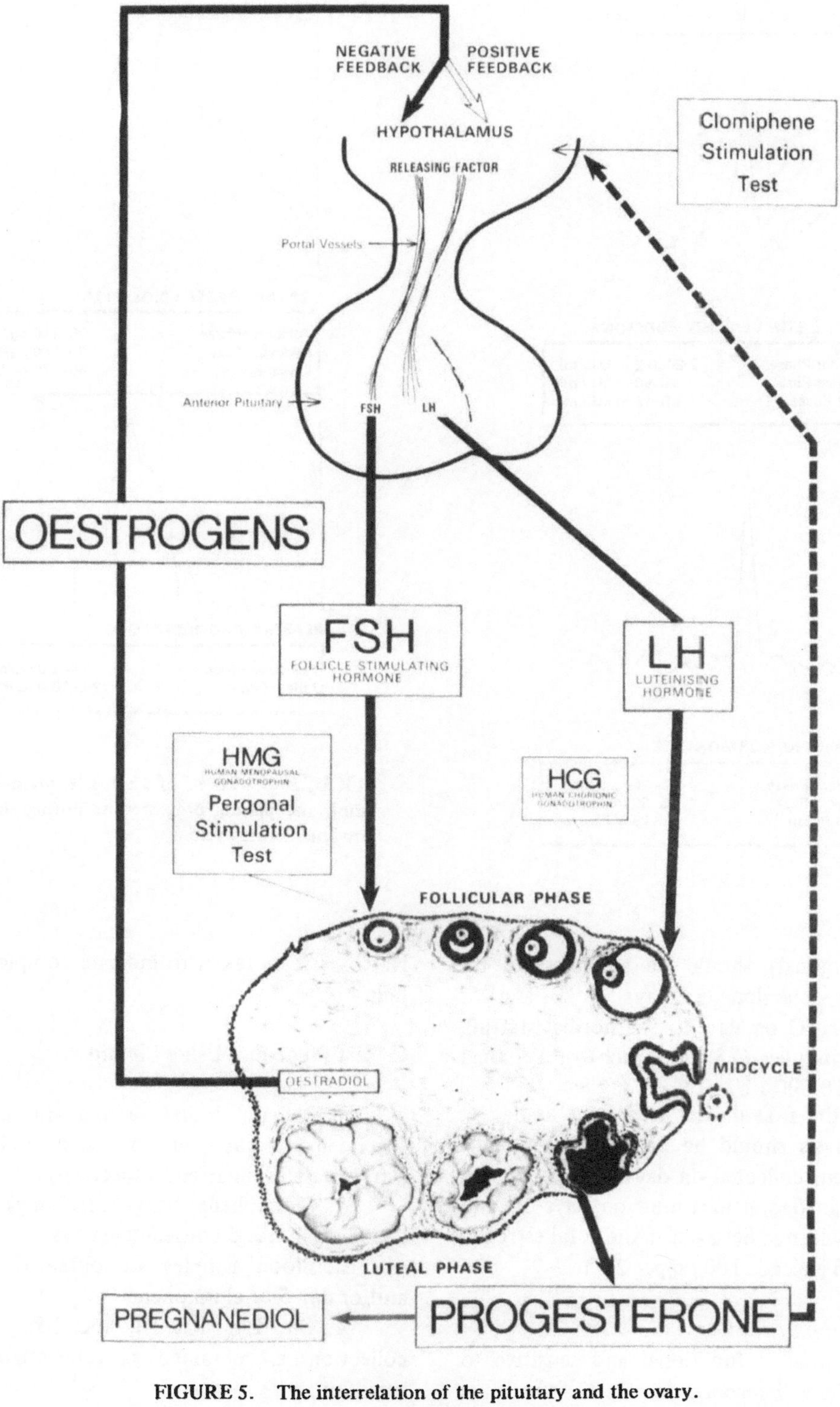

FIGURE 5. The interrelation of the pituitary and the ovary.

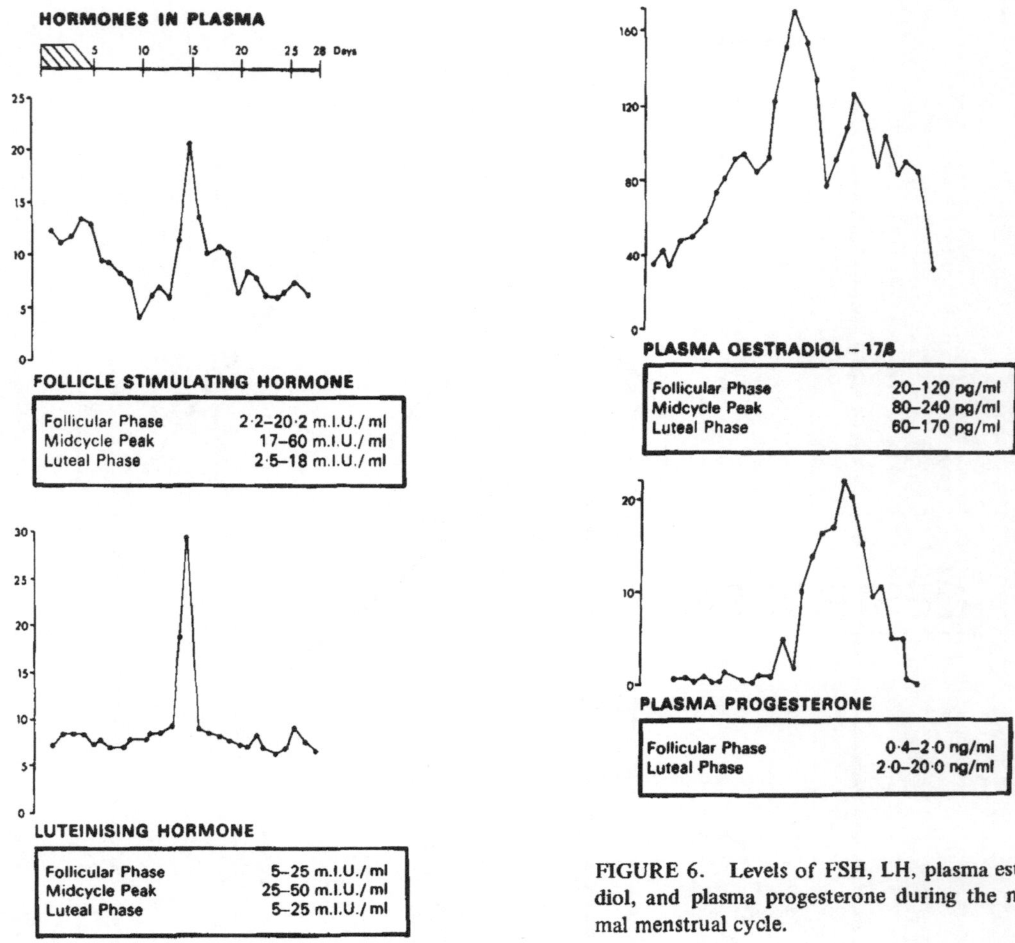

FIGURE 6. Levels of FSH, LH, plasma estradiol, and plasma progesterone during the normal menstrual cycle.

HMG (Pergonal) should be administered by intramuscular injection on days 1, 3, and 5, followed by HCG on day 8. The normal starting doses are 5 ampules (75 units each) of Pergonal followed by 10,000 I.U. HCG on day 8.

Counting the first day of treatment as day 1, estrogen analysis should be performed on 24-hr urine specimens collected on days 4, 6, 8, 10, and 12, with pregnanediol measured on days 16 and 20. HCG should not be given if the total estrogen level on day 6 exceeds 100 μg per 24 hr.

1. Interpretation of Results

An ovary that is functional and sensitive to gonadotrophin will respond to this treatment, as shown by increasing estrogen levels; an increase in pregnanediol levels will indicate that ovulation has probably occurred. In cases where there has been no response, the test should be repeated at a higher dosage. Failure to respond to high levels of HMG must be taken to indicate complete ovarian failure.

C. Test Procedure Using Clomid

1. A basal blood sample and 24-hr urine specimen are obtained for plasma LH and total estrogen determination, respectively.
2. Clomiphene citrate (100 mg) is administered orally for 5 consecutive days.
3. Blood samples are obtained on day 4 and/or day 5 of clomiphene.
4. On days 12, 14, and 16, 24-hr urine collections are obtained for total estrogen determination.

1. Interpretation of Results

The plasma LH should show an increase on days 4 or 5 of clomiphene treatment, the range of the increase with 100 mg of clomiphene being at

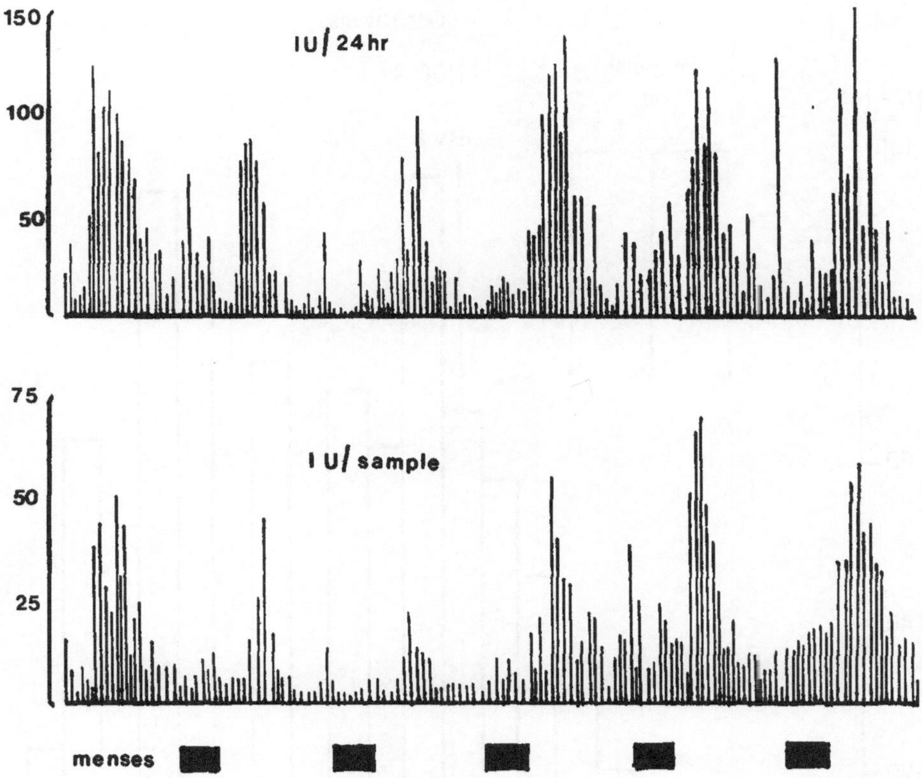

FIGURE 7. The comparison of total estrogen determination in overnight and 24-hr urine collection in one subject over six consecutive menstrual cycles.

least double the basal value. The urinary total estrogen excretion should show a peak value about day 15 of treatment, with a marked increase in the estrogen excretion over the baseline sample. A system for the examination of infertility is outlined in Figure 9.

ACKNOWLEDGMENT

I am indebted to Mr. A. Craig, FIMLT – Manager of the Hormone Assay Services Laboratory – Searle Scientific Services, for his help in the presentation of this manuscript.

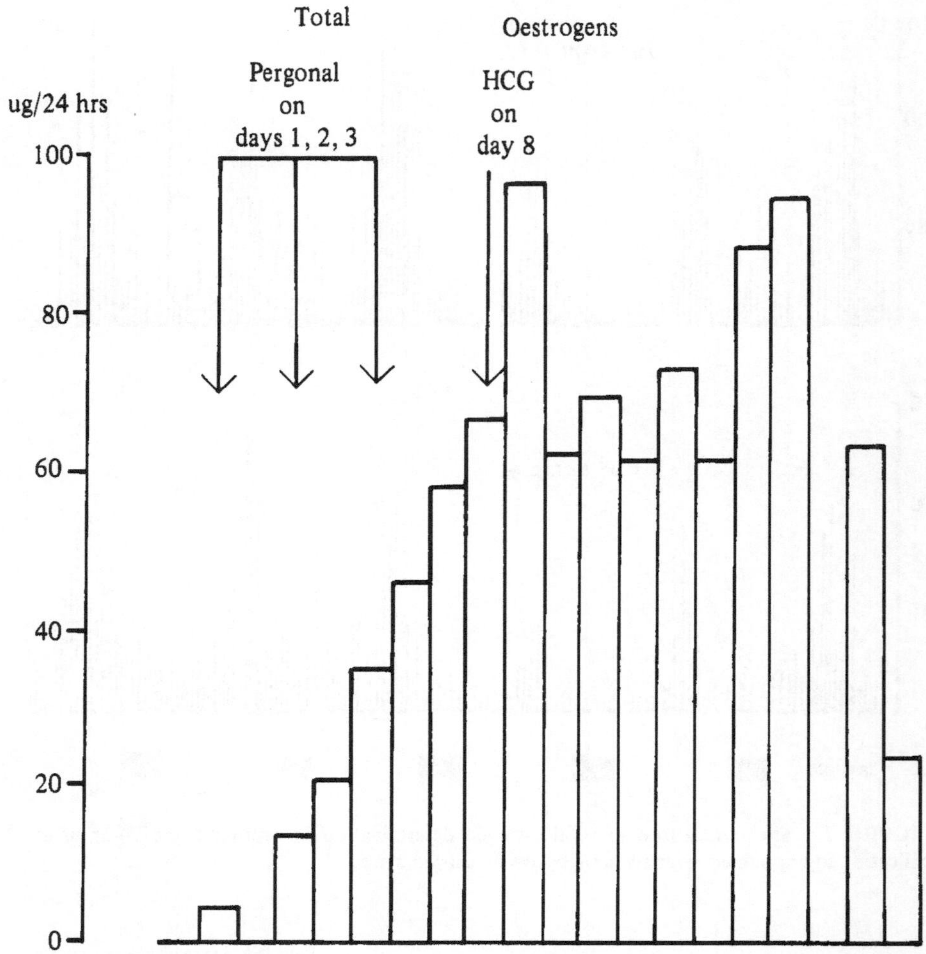

FIGURE 8. Total estrogens and pregnanediol levels in a subject responding to pergonal.

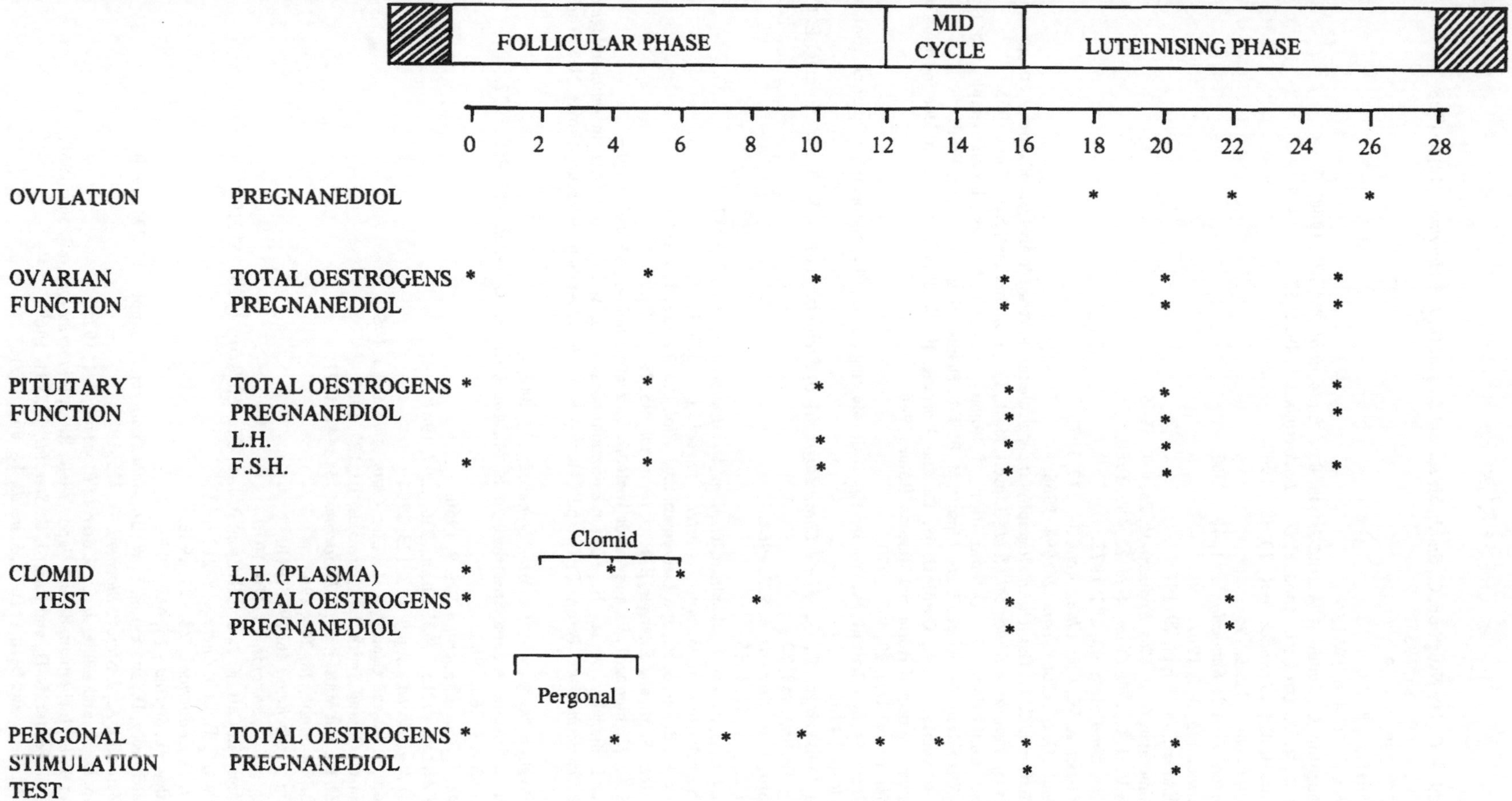

FIGURE 9. The systematic approach for the examination of hormone changes during the menstrual cycle.

REFERENCES

1. Brown, J. B., Macleod, S. C., MacNaughton, C., Smith, M. A., and Smyth, B., *J. Endocrinol.*, 42, 5, 1968.
2. Brown, J. B., *Biochem. J.*, 60, 185, 1955.
3. Kober, S., *Biochem. Z.*, 239, 209, 1931.
4. Brown, J. B., *J. Endocrinol.*, 8, 196, 1952.
5. Bauld, W. S., *Biochem. J.*, 56, 426, 1954.
6. Brown, J. B., MacNaughton, C., Smith, M. A., and Smyth, B., *J. Endocrinol.*, 40, 175, 1968.
7. Brown, J. B., Bulbrook, R. D., and Greenwood, F. C., *J. Endocrinol.*, 16, 49, 1957.
8. Brown, J. B. and Blair, H. A. F., *J. Endocrinol.*, 20, 331, 1960.
9. Ittrich, G., *Z. Physiol. Chem.*, 312, 1, 1958.
10. Stoa, K. F. and Thorsen, T., *Acta Endocrinol.*, 41, 481, 1962.
11. Roy, E. J., *J. Endocrinol.*, 25, 361, 1962.
12. Beling, C. G., *Acta Endocrinol.*, Suppl. 79, 1961.
13. Eechaute, W. and Demeester, G., *J. Clin. Endocrinol.*, 25, 480, 1965.
14. Corns, M. and James, V. H. T., *Clin. Chim. Acta*, 22, 469, 1968.
15. Goebelsmann, U., *Clin. Chim. Acta*, 43, 285, 1973.
16. Barnard, W. P. and Logan, R. W., *Clin. Chim. Acta*, 34, 377, 1971.
17. UaConnail, D. and Muir, G. G., *Clin. Chem.*, 26, 994, 1966.
18. Wotiz, H. H. and Chattoraj, S. C., Gas chromatography and its role in the versatile analysis of urinary estrogens, in *Gas Chromatography of Steroids in Biological Fluids*, Lipsett, M. B., Ed., Plenum Press, New York, 1965.
19. Adlercreutz, H. and Luukkainen, T., Determination of urinary estrogens by gas chromatography, in *Gas Chromatography of Steroids in Biological Fluids*, Lipsett, M. B., Ed., Plenum Press, New York, 1965.
20. Brooks, C. J. W., Chambaz, E. M., Gardiner, W. L., and Horning, E. C., Proceedings of a Meeting on Gas Chromatographic Determination of Hormonal Steroids, Rome, 1966.
21. Richardson, S. J., *Clin. Chim. Acta*, 29, 473, 1970.
22. Wotiz, H. H., Charransol, G., and Smith, I. N., Proceedings of the Meeting on Gas Chromatographic Determination of Hormonal Steroids, Rome, 1966.
23. Adlercreutz, H. and Luukkainen, T., *Gas Phase Chromatography of Steroids*, Eik-Nes, K. B. and Horning, E. C., Eds., Springer-Verlag, Berlin, 1968, 72.
24. Cox, R. I. and Bedford, A. R., *Steroids*, 6, 535, 1964.
25. Menini, E., *Biochem. J.*, 94, 15P, 1965.
26. Adlercreutz, H., Salokangas, A., and Luukkainen, T., *Mem. Soc. Endocrinol.*, 16, 89, 1967.
27. Kaplan, H. G. and Hreshchyshyn, M. M., *Am. J. Obstet. Gynecol.*, 111, 386, 1971.
28. Erlanger, B. F., Borek, F., Beiser, S. M., and Lieberman, S., *J. Biol. Chem.*, 228, 713, 1957.
29. Zimmering, P. E., Beizer, S. M., and Erlanger, B. F., *J. Immunol.*, 95, 262, 1965.
30. Ferin, M., Tempone, A., Zimmering, P. E., and Vande Wiele, R. L., *Endocrinology*, 85, 1070, 1969.
31. Niswender, G. D. and Midgeley, A. R., Jr., Hapten-radioimmunoassay for steroid hormones, in *Immunological Methods in Steroid Determinations*, Peron, F. G. and Caldwell B. V., Eds., Appleton-Century-Crofts, New York, 1970, 149.
32. Aoki, T. and Hreshchyshyn, M. M., *Am. J. Obstet. Gynecol.*, 111, 382, 1971.
33. Kaplan, H. G., Aoki, T., Sansone, A., and Hreshchyshyn, M. M., *Am. J. Obstet. Gynecol.*, 113, 956, 1972.
34. Bradlow, H. L., *Steroids*, 11, 265, 1968.
35. Roy, E. J. and Brown, J. B., *J. Endocrinol.*, 21, 9, 1960.
36. Preedy, J. R. K. and Aitken, E. H., *J. Biol. Chem.*, 236, 1297, 1961.
37. Adlercreutz, H., *Acta Med. Scand.*, Suppl. 412, 123, 1964.
38. Kroman, H. S., Bender, S. R., and Capizzi, R. L., *Clin. Chim. Acta*, 9, 73, 1964.
39. Wotiz, H. H., Charransol, G., and Smith, I. N., *Steroids*, 10, 127, 1967.
40. Cooper, W., Coyle, M. G., and Mills, J. A., *J. Endocrinol.*, 51, 447, 1971.
41. Baird, D. T., *J. Clin. Endocrinol.*, 28, 244, 1968.
42. Murphy, B. E. P., *Can. J. Biochem.*, 46, 299, 1968.
43. Mayes, D. and Nugent, C. A., *Steroids*, 15, 389, 1970.
44. Dufau, M. L., Dulmanis, A., Catt, K. J., and Hudson, B. J., *J. Clin. Endocrinol.*, 30, 351, 1970.
45. Knox, B. S. and France, J. T., *Clin. Chem.*, 18, 212, 1972.
46. Korenman, S. G., *J. Clin. Endocrinol.*, 28, 127, 1968.
47. Corker, C. S. and Exley, D., *Steroids*, 15, 469, 1970.
48. Korenman, S. G., Tulchinsky, D., and Eaton, L. W., Jr., *Acta Endocrinol.*, Suppl. 147, 291, 1970.
49. Concolino, G. and Marocchi, A., *J. Steroid Biochem.*, 3, 725, 1972.
50. Mester, J., Robertson, D. M., and Kellie, A. E., *J. Steroid Biochem.*, 2, 1, 1971.
51. Ferin, M., Zimmering, P. E., Lieberman, S., and Vande Wiele, R. L., *Endocrinology*, 83, 565, 1968.
52. Midgeley, A. R., Jr., Niswender, G. D., and Ram, J. S., *Steroids*, 13, 731, 1969.
53. Exley, D., Johnson, M. W. T., and Dean, P. D. G., *Steroids*, 18, 605, 1971.

54. Jeffcoate, S. L. and Searle, J. E., *Steroids,* 19, 181, 1972.
55. Lindner, H. R., Perel, E., Friedlander, A., and Zeitlin, A., *Steroids,* 19, 357, 1972.
56. Cameron, E. H. D. and Jones, D. A., *Biochem. Soc. Trans.,* 1, 179, 1973.
57. Roy, E. J. and Harkness, R. H., *J. Obstet. Gynaecol. Br. Commonw.,* 70, 1034, 1963.
58. Aitken, E. H., Preedy, J. R. K., Eton, B., and Short, R. V., *Lancet,* 2, 1096, 1958.
59. Nachtigall, L., Bassett, M., Hogsander, V., Slagle, S., and Levitz, M., *J. Clin. Endocrinol.,* 26, 941, 1966.
60. Mathur, R. S., Leaming, A. B., and Williamson, H. O., *Am. J. Obstet Gynecol.,* 113, 1120, 1972.
61. Corker, C. S. and Naftolin, F., *J. Obstet. Gynaecol. Br. Commonw.,* 78, 330, 1971.
62. Tulchinsky, D., Hobel, C. J., and Korenman, S. G., *Am. J. Obstet. Gynecol.,* 111, 311, 1971.
63. Gurpide, E., Giebenhain, M. E., Tseng, L., and Kelly, W. G., *Am. J. Obstet. Gynecol.,* 109, 897, 1971.
64. Wilson, G. R., *J. Endocrinol.,* 54, 15, 1972.
65. Chamberlain, J., Batchelor, A., Craig, A., Gallagher, M. J., and Leonard, P. J., *Ann. Clin. Biochem.,* in press.
66. Horth, C. E. and Palmer R. F., in preparation.

ESTROGENS IN PREGNANCY

R. E. Oakey

TABLE OF CONTENTS

I. Introduction . . . 19

II. Nature and Quantities of Estrogens in Pregnancy . . . 20
- A. Estrogens in Maternal Urine . . . 20
- B. Estrogens in Maternal Plasma . . . 21
- C. Estrogens in Amniotic Fluid . . . 22

III. Estrogen Biosynthesis in Late Pregnancy . . . 22

IV. Methods of Estrogen Analysis . . . 24
- A. End Points . . . 24
- B. Estrogens in Pregnancy Urine . . . 25
- C. Estrogens in Pregnancy Plasma . . . 27
- D. Estrogens in Amniotic Fluid . . . 27

V. Methods in Detail
- A. General Considerations . . . 28
- B. The Method of Oakey et al . . . 28
- C. The Method of Hainsworth and Hall . . . 31
- D. Interfering Substances . . . 34
- E. Compounds Which Suppress Estrogen Excretion . . . 35
- F. Effect of Renal Function . . . 35
- G. Interpretation of Results . . . 36

VI. Quality Control . . . 38

References . . . 39

I. INTRODUCTION

The most common application of the measurement of estrogens in pregnant women is to assist in the clinical management of women whose pregnancies are complicated by any of a group of conditions known from past experience to carry an increased risk of perinatal mortality. These conditions are (1) when the fetus is thought to be "small for dates" (sometimes inaccurately referred to as placental insufficiency); (2) hypertension, especially with proteinuria and edema; (3) previous stillbirth of unknown cause; (4) diabetes; (5) postmaturity; and (6) rhesus isoimmunization. It cannot be emphasized too often that, in the present state of our knowledge, estrogen estimations cannot be used to detect unequivocably any of these conditions, nor can they be used to define accurately the duration of gestation. Attempts to use estrogen estimations for either of these purposes have not been successful. Neither are such attempts fully justified on scientific grounds. For example, "small for dates" fetuses are not always associated with the maternal excretion of subnormal quantities of estrogen.[1,2] However, the risk of intrauterine death in such pregnancies is much greater in those women who are also excreting subnormal quantities of estrogen. In other words, estrogen excretion should be used in the management of the patient once the particular complication has been defined by the obstetrician. A diagnosis of toxemia, for example, is made from measurement of blood pressure, clinical observation of edema, and a urinary protein estimation. Once this condition is recognized, the obstetrician requires information relevant to the status of, and prognosis for, the fetus. This information can be provided by measurement of urinary estrogen excretion.

This viewpoint of the uses of estrogen estimation should not provide a reason to diminish interest in, and research into, estrogen estimation in relation to the onset or development of toxemia or to poor fetal growth, or in relation to the severity of particular complications. Insofar as an increased severity of symptoms may be reflected by increased perinatal mortality, it would be valuable to define such associations. With advances in technology, some causal relationships may be uncovered. This will require careful and accurate classification of the women presently grouped together as, for example, patients with preeclamptic toxemia.

So far, there has been little enthusiasm for measurement of estrogens in pregnant women as a screening procedure, irrespective of the clinical diagnosis of a particular complication. This lack of enthusiasm is thought to be due to difficulties in processing the large numbers of specimens involved and also due to doubt as to the cost effectiveness of the screening procedure. Beischer et al.,[3] however, concluded from a survey of almost 600 patients that routine screening should be carried out in much the same way that tests for rhesus immunization in rhesus-positive women are undertaken at present.

Estrogen estimations in pregnancy have been made on maternal urine, maternal peripheral plasma, and amniotic fluid in efforts to monitor the fetus and predict the likelihood of imminent fetal death. Measurement of estrogen in each of these compartments has been in vogue at different periods of time. Maternal urine has been the choice for many years and continues to be so. This is due in part to the ready availability of urine (and to the erroneous belief that skilled personnel are not required for urine collection) and partly because early analytical techniques were insufficiently sensitive for measurement of estrogens in small volumes of peripheral plasma. Amniotic fluid was considered to be 'nearer to the fetus' and estimation of estrogen in this compartment was proposed as possibly offering a more sensitive index of fetal function than would analysis of maternal urine.[4] Such hopes have not yet been realized. The relative ease of obtaining specimens of urine, blood, and amniotic fluid plays a considerable role in the choice open to the analyst.

This review will deal with methods for the estimation of estrogens in all three fluids. It is necessary first to consider the nature of the estrogens in each compartment and the pathways by which they are synthesized, some of which are peculiar to human pregnancy.

II. NATURE AND QUANTITIES OF ESTROGENS IN PREGNANCY

It must be emphasized that the quantities of each form of estrogen in each of the compartments to be described vary widely from individual to individual. Consequently, the range of values encountered is wide. Mean values will be quoted, but the original papers should be consulted for detailed figures.

A. Estrogens in Maternal Urine

Estrogens are found in maternal urine conjugated with glucosiduronic acid or as esters of sulfuric acid. It is doubtful whether a complete analysis of the estrogens in pregnancy urine has ever been achieved. As techniques improve, more estrogens are added to the long list of compounds known to be present. Breuer[5] isolated 15 different estrogens. Others, including estetrol (1,3,5(10)-estratriene-3,15α,16α,17β-tetrol),[6,7] 15α-hydroxyestrone (17-oxo-1,3,5(10)-estratriene-3,15α-diol),[8] 15β-hydroxyestrone (17-oxo-1,3,5(10)-estratriene-3,15β-diol), and 15β-hydroxyestradiol (1,3,5(10)-estratriene-3,15β,17β-triol)[9] have been detected since then. Interpretation of the relative quantities of the different estrogens present in urine is made difficult by the sensitivity of certain estrogens to acid hydrolysis, a procedure often used to release the free steroid from the water-soluble conjugated form. This sensitivity is particularly marked in estrogens with an α-ketol group in ring D, for example, 16α-hydroxyestrone (17-oxo-1,3,5(10)-estratriene-3,16α-diol) or in those with a catechol group in ring A, as in 2-hydroxyestradiol-17β (1,3,5(10)-estratriene-2,3,17β-triol). Hydrolysis with enzymes such as β-glucuronidase is less destructive, but may not be as effective due to incomplete hydrolysis. Isolation of conjugated estrogens has been made easier by the employment of gel filtration techniques on columns of Sephadex®.[10-12] By this technique, Cohen and Oran[13] recently obtained estriol monoglucosiduronate in crystalline form from pregnancy urine.

Estriol (1,3,5(10)-estratriene-3,16α,17β-triol) appears to be the major estrogen in urine collected in the third trimester of human pregnancy. It is largely excreted conjugated to glucosiduronic acid at C-16 and, to a lesser extent, as the C-3 glucosiduronate.[12] According to Ahmed and Kellie,[14] estriol-3-glucosiduronate and estriol-16α-glucosiduronate account for 80% of the conjugated estrogens in late pregnancy urine. 16α-Hydroxyestrone-3-glucosiduronate (9%) was the next most common estrogen. Other ring D α-ketols accounted for 1.5% of the total estrogen measured. Hobkirk et al.[15] hydrolyzed conjugates with enzymes and concluded that estriol conjugates formed 65 to 90% (mean 76%) of five estrogens measured in late pregnancy urine. Ring D α-ketols accounted for approximately 5% and conjugated estrone about 1%. Cohen[16] suggested that an even higher proportion (up to 33%) of total estrogens may be 'labile' estrogens near term. This view is based on a procedure of precipitation of the conjugates by addition of ammonium sulfate to the urine.[17] This enables certain estrogens, usually destroyed by acid hydrolysis, to be measured.

Despite evidence of this kind, most laboratories continue to use methods involving acid hydrolysis and neglect, or discount, the contribution made by the labile estrogens. In the most recent methods, to be described in detail later, development of a suitable chromogen may occur without hydrolysis of the conjugates and with little destruction of the labile estrogens. These features may contribute to the higher values noted with these methods and with that reported by Cohen.[16]

It must be concluded from this brief survey that conjugated estriol makes the largest contribution to total estrogen measurements. It is not possible to be dogmatic about the exact proportion of estriol in the total quantity of estrogens present in late pregnancy urine.

B. Estrogens in Maternal Plasma

With the aid of methods of improved sensitivity, some progress has been made in the analysis of the different forms of estrogen in maternal plasma. Estrogens in this compartment are found in the free form or as conjugates of glucosiduronic or sulfuric acid or both. Smith found that the proportions of free, glucosiduronate, sulfate, and double conjugate forms of estrogen varied widely among individuals.[18] Near term, estriol was found mainly as the glucosiduronate and estrone as the sulfate, whereas unconjugated estradiol-17β was the predominant form of that hormone.

Generally, acid hydrolysis of plasma or whole blood has been used and the relative and absolute concentrations of three estrogens (estrone, estradiol-17β, and estriol) after this procedure have been recorded. Roy found the mean concentrations of estriol, estrone, and estradiol-17β in acid-hydrolyzed blood from women in hospital to be 6.7, 2.0, and 0.9 μg/100 ml, respectively.[19] Using an essentially similar method, but making a correction for losses in hydrolysis and purification, Rado et al. reported mean values of 18.8, 4.8, and 1.9 μg/100 ml plasma.[20] Similar values were reported by Brown and co-workers.[21]

Since the conjugated forms of estrogen appear to lack biological activity associated with the free hormone, interest has been transferred, to some extent, to estimation of nonconjugated estrogens. Application of a double isotope derivative technique to the problem gave values of 1.5 μg/100 ml plasma for nonconjugated estrone and 3.0 μg/100 ml plasma for nonconjugated estradiol-17β.[22] Sybulski, using competitive protein binding, found the concentration of nonconjugated estradiol-17β in plasma near term to be approximately 2 μg/100 ml.[23] The relative concentrations of the three estrogens were measured throughout pregnancy by Tulchinsky, et al.[24] At term, the concentration of nonconjugated estradiol-17β (1.5 μg/100 ml) exceeded that of estriol (1.3 μg/100 ml) and of estrone (0.8 μg/100 ml). Other workers found concentration of nonconjugated estetrol (0.2 μg/100 ml) to be approximately 12% that of estriol (1.6 μg/100 ml).[25] The concentration of estrone sulfate in peripheral plasma has also been reported.[26] Comparison of the relative concentrations of estrone, estradiol-17β, and estriol in acid-hydrolyzed plasma and urine reveals striking differences. In acid-hydrolyzed plasma, the ratio of estriol:estrone:estradiol-17β is approximately

12:7:1, whereas in urine, the ratio is 60:3:1. The greater preponderance of estriol in the urine reflects the relative renal clearance of the conjugates. The renal tubule actively secretes conjugated estriol.[21]

C. Estrogens in Amniotic Fluid

Estrogens have been measured in amniotic fluid after acid hydrolysis, when conjugated plus free estrogens are included in the results, and after enzyme hydrolysis when the relative proportions of the different conjugates are reported. Schindler and Siiteri hydrolyzed amniotic fluid with β-glucuronidase and sulfatase together.[27] Eight steroids were separated and estimated by gas-liquid chromatography. Estrogen concentrations were (μg/100 ml corrected for manipulative loss) estriol 157.2, estetrol 8.38. After acid hydrolysis of amniotic fluid and estimation colorimetrically, Michie and Livingstone found a mean estriol concentration of 91 μg/100 ml at term.[28] Similar values were reported by Berman et al. using acid hydrolysis and gas-liquid chromatography.[29] Pinkus and Pinkus precipitated the conjugates with ammonium sulfate and measured the estrogens without separation or hydrolysis.[30] Their results, expressed as estriol, are little different from those of workers who isolated estriol specifically. Young and co-workers estimated the estriol released after enzyme hydrolysis of partly purified conjugates.[31] At term, estriol-16-glucuronoside was present in the largest amount, whereas at midterm, estriol-3-sulfate-16-glucuronoside was quantitatively more important.

III. ESTROGEN BIOSYNTHESIS IN LATE PREGNANCY

Although not strictly relevant to a discussion of methodology, some understanding of the complex processes of estrogen production and of estriol production in particular is essential. Only with this information can a proper assessment be made of the possible causes of low or high estrogen production.

In the nonpregnant woman, estrone and estradiol-17β are secreted by the ovary; estriol arises as a peripheral metabolite. There is little evidence for ovarian secretion of estriol. In late pregnancy, after 30 weeks of gestation, estrone, estradiol-17β, and estriol all appear to be direct secretory products of the placenta. Indeed, placental secretion is the major source of estriol. Only a small proportion is formed by peripheral metabolism of estrone and estradiol-17β.

Views of estrogen biosynthesis have changed many times during the past 15 years. The following description probably represents the presently accepted and coherent picture (for reviews, see Oakey,[32] Liggins,[33] and Figure 1). Little estrogen secretion by the placenta occurs in the absence of a supply of precursors from the fetus. The fetal adrenal converts acetate and plasma cholesterol to dehydroepiandrosterone sulfate which is secreted into the fetal circulation.

Some of this androgen sulfate is hydroxylated at C-16α by the fetal liver to form 16α-hydroxydehydroepiandrosterone sulfate. Both these androgen sulfates are secreted to the placenta where they are taken up and converted into estrone and estriol, respectively. From estimation of the precursors in plasma from the cord artery and vein, it appears that 16α-hydroxydehydroepiandrosterone sulfate is utilized more efficiently than is dehydroepiandrosterone sulfate. Consequently, estriol, rather than estrone, is secreted in greater amounts by the placenta. The conversion of androgen sulfates to estrogens in the placenta is itself a complex procedure involving (1) sulfatase; (2) 3β-hydroxysteroid dehydrogenase – isomerase; (3) aromatase complex; and (4) 17β-hydroxysteroid dehydrogenase enzymes (Figure 2). Continued estrogen production during pregnancy requires an adequate supply of precursors from the fetus and maintenance of appropriate enzymes in the placenta. The case for the importance of the fetus, and in particular of the fetal adrenal, for estrogen production in late pregnancy is based largely on the work of Cassmer[34] and Frandsen and Stakemann.[35] The concept that the fetus and placenta work in concert to produce the estrogens is due in large part to the investigations of Diczfalusy and Mancuso.[36] The sensitivity of estrogen production to inadequate function of either the fetus or placenta, or both, can be explained readily on this basis. However, it should be emphasized that, at present, no clear biological reason for the production of these large quantities of estrogen, particularly estriol, has been uncovered. Suggestions that their formation is a means of protecting the fetus from androgens,[37] possibly arising as by-products as the fetus' attempts to produce sufficient quantities of cortisol,[32] have been made. No complete understanding has yet emerged. The major 'use' of

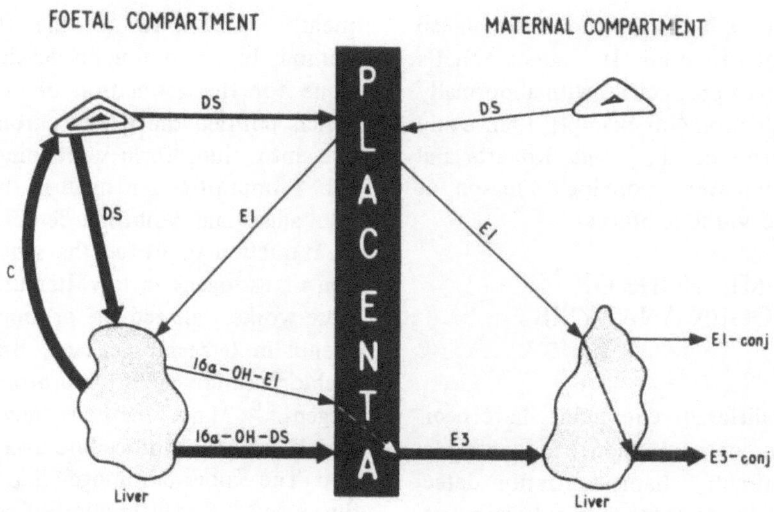

FIGURE 1. Major pathways of estrogen biosynthesis in the fetoplacental unit in late pregnancy. Note secretion of dehydroepiandrosterone sulfate from fetal and maternal adrenal glands, with the major pathway (broad arrow) producing estriol. Abbreviations: DS – dehydroepiandrosterone sulfate; 16α-OH-DS – 16α-hydroxydehydroepiandrosterone sulfate; E_1 – estrone; 16α-OH- E_1 – 16α-hydroxyestrone; E_3 – estriol; E_1-conj – estrone conjugates; E_3-conj – estriol conjugates. (From Oakey, R. E., *Vitam. Horm.*, 28, 1, 1970. With permission.)

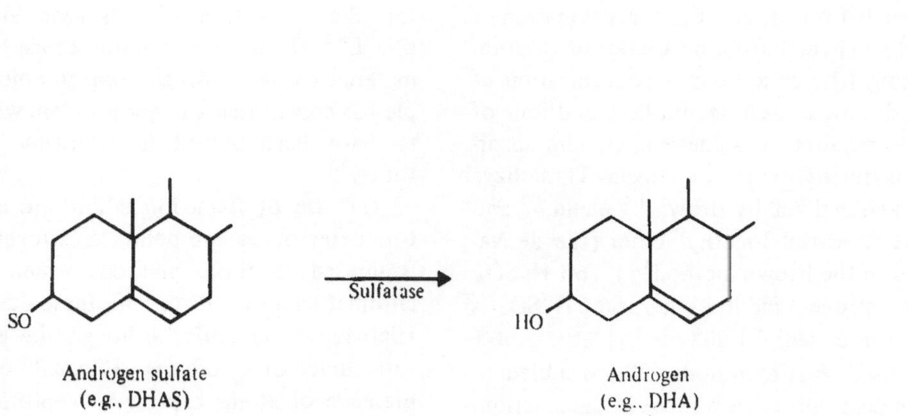

FIGURE 2. The conversion of androgen sulfates to estrogens in the placenta.

the estrogens produced is to provide a biochemical means to assess fetal function. It is possible that a closer examination of pregnancies with abnormally low estrogen production (for example, France and Liggins,[38] Anderson et al.,[39] and Roberts and Cawdery[40]) may uncover a more logical reason for this fascinating biosynthetic process.

IV. METHODS OF ESTROGEN ANALYSIS

A. End Points

A number of different end points have been used in estrogen determination, for example, colorimetry, fluorimetry, flame ionization detector, electron capture detector, and radioimmunoassay. The Kober reaction is probably the most widely used of the colorimetric assays; it is worthwhile to make some comments regarding this reaction. Marrian[41] has given a fascinating account of the original observation by Kober, which resulted in the development of the reaction which bears his name. Essentially, the reaction consists of heating the estrogen with sulfuric acid until a yellow color is formed, and the transformation of this color to a characteristic pink color on dilution and reheating. The conditions – concentration of sulfuric acid, time of heating, dilution, and time of reheating – required to achieve maximum absorbance are different for each estrogen. These have been discussed in detail by Brown,[42] Bauld,[43] and Nocke.[44] For estriol-3-methyl ether (the derivative formed in the Brown method[45]), 76% H_2SO_4 is used; for estrone-3-methyl ether, 66% H_2SO_4 is used; and for estradiol-17β-3-methyl ether, 60% H_2SO_4 is used. Hydroquinone (2%) is added to the sulfuric acid, but takes no part in the reaction. The reaction is highly specific for natural estrogens, their conjugates, and 3-methyl ethers. Diethylstilbestrol does not react. A useful list of extinction coefficients is provided by Breuer.[46]

When estrone, estradiol-17β, and estriol in pregnancy urine are measured without separation into the individual estrogens, the Kober reagent chosen is generally that for estriol, since this is the estrogen present in the largest quantity. Another modification, introduced by Oakey et al.[47] is to discard the heating stage after dilution with water. This serves to simplify the manipulations without loss of sensitivity.

An important addition to the Kober reaction methodology was made by Ittrich and, consequently, is known as the Ittrich extraction method. In attempting to develop a simple technique for the estimation of estrogens in urine, Ittrich purified the Kober chromogen by extraction into chloroform containing 2% *p*-nitrophenol.[48] Impurities remained in the aqueous phase. Salokangas and Bulbrook demonstrated that such an extraction improved the sensitivity of estimation of estrogens in low-titer urines.[49] However, these workers altered the organic phase to *p*-nitrophenol in tetrachloroethane. Bradshaw proposed trichloroacetic acid in chloroform as a less toxic reagent.[50] Many methods now incorporate an Ittrich extraction procedure as a final purification step. The Kober chromogen in organic solvent will fluoresce.[48] For estimation of estrogens in urines of low titer, fluorimetry of the product of the Kober reaction when extracted into organic solvent is now the rule. In estimation of estrogens in pregnancy urine, the Ittrich extraction procedure is generally used to compensate for inadequate purification steps prior to the Kober reaction.

Direct fluorimetry of purified estrogens after heating with sulfuric acid[51] has been advocated for the estimation of estrogens in urine and blood.[52] Quenching of fluorescence by unknown material carried through from the biological sample has constituted a major problem which appears to have been solved in relatively few laboratories.[21]

The use of flame-ionization and electron-capture detectors as end points in estrogen estimation is limited to those methods which involve gas chromatography. Both of these detectors are relatively nonspecific, although the electron-capture detector produces a signal only in the presence of atoms capable of capturing electrons generated in the detector, for example, the halogens. It is essential, therefore, to form a derivative, such as an ester of polyfluorobutyric acid, which contains the necessary halogen atoms.[53-56] Derivative formation for this purpose represents no additional problem, since it is usual to form a derivative such as acetate, methyl ether, or trimethyl silyl ether in order to be able to reduce the temperature at which the chromatographic process is carried out and thereby to reduce destruction of the steroid.[57-63] Other workers have chromatographed the free estrogens.[64-66]

Whichever approach is used, extensive purification procedures are required to reduce signals generated by impurities.

In competitive protein binding and radioimmunoassay techniques, the end point involves the competition between the steroid to be measured and a radioactive steroid or steroid derivative for a limited number of binding sites on a selected protein. In competitive protein binding assays of estrogens, the protein is usually obtained either from peripheral plasma (sex hormone binding globulin) or from a tissue sensitive to estrogens (rabbit uterus). Radioimmunoassays use as binding protein antisera raised by injection of antibodies comprising the steroid linked covalently to a protein, for example, bovine serum albumin. Final estimation is by assay of ^3H (liquid scintillation counter) or ^{131}I or ^{125}I (gamma counter). The techniques are highly sensitive and can be developed to provide a known degree of specificity, either by including chromatographic separation techniques or by using highly specific antisera. A full discussion of the techniques would be out of place in this chapter. The interested reader is referred to two excellent books on the subject.[67,68]

B. Estrogens in Pregnancy Urine

The first reliable chemical methods for the estimation of estrogen in urine from nonpregnant women were devised by Brown[45] and by Bauld.[69] Further refinements were required before the methods were completely satisfactory for estimation of urine from ovariectomized and ovariectomized-adrenalectomized women. However, these methods could also be used for estimation of estrogens in urine from pregnant women. Brown[70] reported the quantities of estriol, estrone, and estradiol-17β in acid-hydrolyzed pregnancy urine. Excretion of each estrogen fraction was found to increase during gestation. By term, the relative quantities of estriol, estrone, and estradiol-17β were approximately 40:2:1 and estriol excretion reached 40 mg/day.

As experience in measuring estrogens in urine from pregnant women accumulated, it became recognized that low levels of estrogen were encountered after delivery or after intrauterine fetal death. Consequently, interest was roused in estrogen estimation as an index of fetal well-being and faster, less complicated methods were produced.

Brown and Coyle[71] introduced a shortened form of the Brown[45] method which measured estriol, but not estrone and estradiol-17β. Essentially, the method involved acid hydrolysis of the urine (50 ml from a 24-hr collection), extraction of free estrogens into ether, alkali wash and partition to remove acidic substances, extraction of phenols, methylation and oxidation of impurities with hydrogen peroxide, and chromatography on alumina columns. Final estimation was by the Kober reaction.

Klopper and Wilson[72] also described an assay with similar steps. It was pointed out that three urines could be analyzed in duplicate each working day. With both these methods, chromatography on alumina, together with derivative formation, contributed to the specificity achieved. A method for routine service work was devised by Frandsen. The estriol in acid-hydrolyzed urine was purified by solvent-solvent partition, but neither methylation nor chromatography was included in the scheme. Final estimation utilized the Kober reaction. In the U.S. at this time, fluorimetry was favored as the end point. Eberlein et al.[74] hydrolyzed urinary estrogens with β-glucuronidase, extracted the free estrogens into ether, treated them with NaOH, and chromatographed estriol on columns of alumina. A modification of this method was used by Green and Touchstone in their pioneering investigation of the value of urinary estrogen estimations in the management of complicated pregnancies.[75] Acid hydrolysis replaced hydrolysis with enzymes. The fluorimetric end point was replaced by use of Bachman's reagent.[76]

The next stage in the process to routine estrogen assays was marked by a method introduced by Montagu.[77] Urine (0.5 ml) was hydrolyzed with hot acid and extracted with ether. After evaporation of the ether, a Kober reaction was performed on the residue, with 80% H_2SO_4. Final estimation was made after extraction of the Kober chromogen into *p*-nitrophenol in chloroform. This method was one of the first in which estrogens were measured without separation. Such assays reduced the number of manipulations required and, therefore, enabled urines from more patients to be assayed in a shorter time. Prompted by similar considerations, Oakey et al. also devised a rapid method.[47] The estrogens were not separated into the individual components; neither was the Ittrich extraction procedure used. The main purification step was merely an alkali wash (pH 10.5) of the ether extract of acid-hydrolyzed urine. The optical density of the Kober chromogen was corrected for interference by use of the Allen correction. This method will be described in detail in a subsequent section. It offered the opportunity of estimating urines from 15 subjects in duplicate and included an internal standard to monitor manipulative losses after the hydrolysis step. Modifications have been introduced, for example, a reduced time of hydrolysis and a shorter time for developing the Kober color.[78] This method remains a popular method in the United Kingdom, partly because of its simplicity and use of inexpensive equipment.

A further modification of the early Brown method was introduced by Brown and coworkers.[79] Acid-hydrolyzed urine was extracted with ether in a series of tubes which were shaken automatically. Purification was by alkali partition and methylation, with final estimation by Kober reaction and colorimetry. Estrogen methyl ethers were measured without separation. An Allen correction was applied. Introduction of this method again increased the number of specimens that could be handled daily.

The literature now contains many methods for the estimation of estrogens in pregnancy urine. Some of these methods appear to be little more than local modifications of methods already available. Others introduced new ideas in efforts to achieve the necessary purification. Huang carried out the usual acid hydrolysis and ether extraction before isolating an 'estriol' fraction from an 'estrone + estradiol' fraction.[80] Final measurement was by Kober reaction. However, there appears to be no advantage in measuring an 'estriol' fraction rather than a 'total estrogen' fraction for the purpose of predicting the likelihood of intrauterine death. Cohen devised a method in which the conjugated estrogens were precipitated by addition of $(NH_4)_2SO_4$ to the urine and were estimated by the Kober reaction.[17] The method gave results higher than conventional techniques, which Cohen considers due to protection of labile estrogens. Osawa and Slaunwhite isolated estrogen conjugates from urine by ion-exchange chromatography and performed a Kober reaction and Ittrich extraction on this conjugate fraction.[81] Again the results appeared to be higher than those obtained by Oakey et al.[47] Ion-exchange chromatography was also used in a similar manner by Lee and Hahnel.[82] Estimation of conjugated estrogens was also used by Rourke et al., who salted out these compounds from urine into ethyl acetate.[83] Final estimation was by the Kober reaction with Ittrich extraction and colorimetry.

A parallel development has been the introduction of methods in which some form of automation is incorporated to increase the throughput of specimens. This can vary from fairly obvious pieces of work simplification to fully automated procedures.

Strickler et al. described equipment in which conjugated estrogens (precipitated as described by Cohen[17]) were estimated by the Kober reaction with Ittrich extraction and a fluorimetric end point.[84] Technicon® modules were used, but two fluorimeters were necessary to adjust for background. A similar method was reported by Thysen and his colleagues.[85] Between 10 and 15 samples could be processed each hour.

In the U.K., Ua Conaill and Muir devised an automated method which processed urine samples that had been hydrolyzed with acid.[86] By means of Technicon modules, the acid-hydrolyzed urine was extracted with ether in a continuous flow system. The ether phase was separated and evaporated at a high temperature while development of the Kober reaction took place. The chromogen was extracted into *p*-nitrophenol-chloroform and estimated colorimetrically by absorbance at 540 nm. Subsequent development of this method included introduction of fluorimetry for final estimation,[87] and automation of the hydrolysis stage.[88] A similar but more specific technique was devised by Barnard and Logan which included extra partition steps to achieve a fraction rich in estriol, the recovery of which was increased to 94%.[89] Final estimation was by fluorimetry of a Kober-Ittrich chromogen. Possibly because of the need for hydrolysis of the urine or precipitation of the conjugates, in the early forms, these techniques, although capable of handling 100 or more samples daily, failed to gain wide acceptance.

A more promising development has followed from an earlier procedure of Ittrich, who suggested applying the Kober reaction to diluted urine.[90] This was developed into a workable manual procedure by Brombacher et al.[91] and by Howarth and Robertshaw.[92] Basically, urine was diluted with water, heated with Kober reagent, diluted, cooled, and extracted with Ittrich reagent. Final estimation was carried out by fluorimetry. The sensitivity of the end point is such that small portions of urine are required and the background or quenching is kept to a minimum. This procedure was adapted for Technicon modules by Dixon (King's College, London), Braunsberg (Medical Research Council Clinical Endocrinology Research Unit, Edinburgh), Hall (Royal Infirmary, Glasgow), Howarth (Royal Infirmary, Bradford), Nunn (Hurstwood Park Hospital, Sussex), and Oakey (University of Leeds), and was published by Hainsworth and Hall.[93] Full details will be given in a later section. The important step forward in this technique is the avoidance of a discrete hydrolysis stage. Quite independently, Campbell and Gardner[94] described a similar method but without the extensive dilution, as described by Hainsworth and Hall.[93] This method has been criticized.[95] From experience with the continuous flow methods, however, it is clear that properly carried out, they have great potential.

The use of radioimmunoassay techniques for estimation of estrogens in pregnancy urine has been reported.[96] However, hydrolysis and purification by solvent partition were required before final assay. This application is essentially, therefore, a modification only of the end point of the assay.

Urinary estrogen estimation by gas-liquid chromatography has been reported by many workers.[57,60,62-66,97,98] All these methods suffer from the defect that lengthy purification procedures, coupled with the formation of derivatives, are required before the chromatographic procedure proper is commenced. Indeed, the partition columns and detectors serve essentially as an end point. The full potential of the gas chromatography method for resolution of compounds is rarely applied. However, gas-liquid chromatography does offer the opportunity, not afforded by the more rapid colorimetric or fluorimetric methods, of measuring estriol and other estrogens specifically. Whether this is of any value in the context of monitoring fetal function is debatable, to say the least. Samples, once purified, can be applied to the column automatically. An ingenious technique for monitoring manipulative loss was devised by Smith and Stitch.[63]

C. Estrogens in Pregnancy Plasma

Methods for the estimation of estrogens in peripheral plasma from pregnant women in general evolved by application of the Brown method.[45] Conjugated estrone, estradiol-17β, and estriol, together with any nonconjugated estrogens, were assayed by the Kober reaction after lengthy purification.[19,20,99] These methods were too involved for purposes of clinical management. Nachtigall et al.[100] devised a method for this purpose involving acid hydrolysis of conjugates, purification by solvent partition during which estrone and estradiol-17β were separated from estriol which was estimated by fluorimetry of the Kober chromogen after extraction into chloroform containing 2% *p*-nitrophenol. A very similar method was recently published by Mathur et al.[101]

In contrast, gas-liquid chromatography techniques for plasma estrogens have been widely reported. This approach has been favored because of the extra sensitivity offered by the detector, rather than for any improvement in speed or precision. Touchstone and Murawec measured free and conjugated estrogens (after hydrolysis), but the purification required chromatography on alumina columns prior to gas-liquid chromatography.[102] Others also reported methods for conjugated estrogens.[54,58,61,103] Nonconjugated estradiol-17β was measured by Attal and Eik-Nes,[56] by Munson et al.,[53] and by Mead and co-workers.[55] It is clear from the literature that none of these have been widely adapted for service work.

A more interesting development is the application of radioimmunoassay (RIA) or competitive protein binding (CPB) techniques to estimation of estrogens in plasma from pregnant women. Corker and Naftolin[104] hydrolyzed conjugated estrogens in plasma, obtained an estriol-rich fraction by solvent partition, and estimated the estrogen present by competitive protein binding with rabbit uterine cytosol. Others have measured nonconjugated estriol,[105,106] nonconjugated estradiol-17β[23,107] or both[24,108] nonconjugated estetrol[25,109] or estrone sulfate.[26] All these methods have only recently been introduced and their relative value in providing an accurate index of the likelihood of fetal death is so far undecided. This must be clarified before widespread introduction of these methods for service work can be recommended.

D. Estrogens in Amniotic Fluid

Amniotic fluid contains conjugated estrogens, particularly conjugates of estriol. Methods of analysis have been devised by Schindler and Herrmann,[110] who used a sequence of acid hydrolysis, solvent partition, formation of estriol trimethyl silyl ether and gas-liquid chromatography; by Michie and Livingstone,[28] who used the Brown[45] method with the additional purification steps advised by Brown et al;[111] by Biggs and co-workers,[112] who used a similar procedure but included, in addition, formation of a diacetate derivative and estimation after gas-liquid chromatography; by Aleem et al.,[59] who used enzyme hydrolysis and formation of a methyl ether and diacetate derivative but without chromatography on alumina of the estriol methyl ether; and by Pinkus and Pinkus.[30] A wide range of estriol concentrations in fluid from individual patients was noted. This range of values was due to variation in individual patients and also in part to the different methods used. Klopper et al.[113] analyzed portions of the same sample of fluid and found widely differing values by the methods used in Aberdeen, Edinburgh, and Belfast. Although amniotic fluid has been shown to be a rich source of estrogens, their measurement is rarely undertaken to provide an index of fetal function. This must be due in part to problems associated with sampling this compartment, especially at an outpatient clinic.

V. METHODS IN DETAIL

A. General Considerations

Two methods for the estimation of estrogens in urine have been selected for detailed description. These are the methods of Oakey et al.[47] and Hainsworth and Hall.[93]

It is clear that there are many methods which might have been selected for detailed description. It is, therefore, appropriate that the reasons for making this particular choice are stated, since these may be relevant in deciding which assay to select for laboratory use. First of all, one must consider the use to which the assay is put. For purposes of assisting in clinical management and detecting those fetuses which are likely to die, the method of choice must (1) provide a good index of likely fetal death, (2) be precise, and (3) be robust for daily use. Other considerations are those of simplicity and cost. The method of Oakey et al.[47] provides a good index of imminent fetal death.[114] It is one of the few estrogen assays which has been tested rigorously for this purpose. It is of known precision; a coefficient of variation of 10% on daily estimations was achieved. It is robust and was in regular daily use from March 1965 until February 1972 in the author's laboratories. The simplicity of the assay can be evaluated from the detailed description, but only test tubes, water baths, and a spectrophotometer are required. The second method[93] has not yet been tested rigorously as an indicator of likely fetal death. Nevertheless, from comparisons in the author's laboratory, good correlations were found in results from individual subjects obtained by the two methods. In our hands, the method of Hainsworth and Hall[93] is more precise than that of Oakey et al.[47] It is robust, but requires a greater initial capital outlay than that of Oakey et al.[47]

The methods selected both relate to estrogens in urine. It was considered advisable, in the light of the present state of our understanding, to describe urinary assays rather than blood assays, because of the experience in interpretation which has been accumulated. Until similar experience has been obtained for the interpretation of blood estrogen values, it is believed that most obstetricians will prefer to act on urinary measurements.

Both methods measure estrogens in urine without separation into individual fractions. It has been argued that so-called total estrogen assays are less precise than those that measure a single estrogen, e.g., estriol only.[115] Evidence to support this statement is lacking. Indeed, the continuous flow methods, e.g., Hainsworth and Hall,[93] have much higher precision than that reported for the older, more complicated methods. No clinical trial appears to have been carried out to assess the relative effectiveness, in terms of perinatal mortality figures, of these two approaches. Until a significant improvement in these figures can be shown by use of specific methods for estriol, total estrogen methods, properly carried out, have much to recommend them.

The method of Brown et al.[79] might have been included in this section. Brown and Beischer[116] have provided a detailed description of their approach and have carefully listed the problems which have arisen with this method in practice and the solutions reached. It was considered inappropriate to present another description of the method here.

B. The Method of Oakey et al.[47]

1. Principle

Free estrogens, together with other compounds in acid-hydrolyzed urine are extracted into ether.

The ether extract is washed with an aqueous solution of sodium bicarbonate (pH 10.5) which removes unwanted materials, but leaves the estrogens in the ether. The residue remaining after evaporation of the ether is heated with 66% H_2SO_4 containing 2% hydroquinone to form a Kober chromogen. The optical density of this chromogen is compared with that from a standard solution of estriol and the quantity of estrogen in the urine is calculated by proportion.

2. Apparatus

Test tubes – (Kober tubes) 150 mm x 25 mm with B24 joints and stoppers (Quickfit, J. A. Jobling & Co. Ltd., Stone, England)

Water baths – to be maintained at 55°C and 100°C

Spectrophotometer – with glass cells 1 cm in length – Unicam® SP600 (Pye Unicam Ltd., Cambridge, England)

All glassware must be washed in detergent. Either Neodisher LA (Scientific Instrument Centre, London N W 1) or Comprox A (Griffin & George Ltd., Wembley, England) is suitable. After washing, the glassware must be rinsed thoroughly and dried before use.

3. Reagents

AR grade chemicals were purchased from BDH Chemicals Ltd., Poole, Dorset, England.

Hydrochloric acid (36.5%) – AR grade

Sulfuric acid (98%) – AR grade

Ether – AR grade supplied and stored in a dark glass bottle containing copper gauze.

Ethyl alcohol – AR grade

Sodium sulfate, anhydrous – AR grade

Sodium hydroxide (20% w/v) – AR grade

Sodium hydrogen carbonate (8% w/v) – AR grade

Hydroquinone – laboratory reagent grade (Griffin & George Ltd., Wembley, England)

Estriol – Steraloids Ltd., Croydon, England

Bauxite chips (no. 16 grade) – Universal Grinding Wheel Co., Stafford, England)

Kober reagent – prepare 2% w/v hydroquinone in 66% H_2SO_4 (made by adding *carefully* 250 ml 98% H_2SO_4 to 136 ml water)

Sodium carbonate solution – mix 1,000 ml 8% w/v $NaHCO_3$ with 150 ml 20% w/v NaOH

Estriol standard solutions in ethanol – 100 μg/ml for internal standard; 5 μg/ml for external standard

4. Procedure

Dilute the 24-hr urine specimen to 2,000 ml. Pipette 2 ml of diluted specimen into each of 3 Kober tubes. Add 0.3 ml HCl to each tube. Stopper the tubes and place in a *boiling* water bath for 1 hr. Cool. (If the volume of the 24-hr urine specimen is more than 2 l., dilute to the next liter and take 1/1,000th of the diluted specimen for assay. Increase the quantity of HCl in proportion.) Add 0.2 ml of 100 μg/ml estriol standard solution to the third tube to serve as an internal standard. To each tube add 10 ml ether, stopper (moisten the stoppers with water), shake for 30 sec, allow the layers to separate, and suck off the lower layer. (Suction from a water pump can be used.) Add 0.5 ml sodium carbonate solution to each tube, stopper, and shake. Add 1 to 2 g anhydrous sodium sulfate to each tube, stopper, and shake. Take one 3-ml aliquot from the first tube, one 3-ml aliquot from the second tube, and two 3-ml aliquots from the third (internal standard) tube and transfer to separate Kober tubes. Pipette 1 ml of estriol standard solution (5 μg/ml) and 3 ml ether into a clean Kober tube to serve as the external standard. Add 0.2 ml 2% hydroquinone in ether to all tubes and to a sixth tube to serve as a blank. Add 2 to 3 bauxite chips to each tube and evaporate to dryness on a water bath at 55°C. Add 2 ml 2% hydroquinone in 66% H_2SO_4 to all tubes. Place in a boiling water bath for 45 min. Cool. Add 1.7 ml distilled water to each tube. Shake. Transfer to a 1 cm spectrophotometer cell. Measure the optical densities at 472, 514, and 556 nm against the blank.

5. Calculation

(1) Allen correction – corrected optical density at 514 nm = $2 \times R_{514} - (R_{472} + R_{556})$ where R_{514}, R_{472}, and R_{556} are the observed readings at 514, 472, and 556 nm, respectively.

(2) Total estrogen – total estrogen (as estriol) = $\frac{20 \times A}{B - A}$ mg/24 hr

where

A = corrected optical density at 514 mμ from urine

B = corrected optical density at 514 mμ from urine with added internal standard.

A set of calculations, taken from routine work, is shown in Table 1. In our practice a calculation of the recovery of the internal standard was always made as follows:

Internal standard 0.2 ml of 100 μg/ml estriol
Quantity of estriol added = 20 μg
3 ml ether taken from 10 ml for color development
$\therefore \frac{3}{10} \times 20$ μg estriol = 6 μg estriol from internal standard
$\therefore$ for 100% recovery $D_{(u+s)} - D_u$ = 6 μg estriol

The external standard color (D_E) is derived from 5 μg estriol

$$\therefore 5[D_{(u+s)} - D_u] = 6 \times D_E$$

$$\therefore \% \text{ recovery} = \frac{5[D_{(u+s)} - D_u] \times 100}{6 \times D_E}$$

Assays in which recoveries of less than 70% or more than 110% are produced should be repeated. The recovery of internal standard provides a good index of technician performance. Once the operator is properly trained, the recovery of internal standard is found to be high and reproducible (see Table 1, column 10). It is important that the ether extract should be evaporated completely to dryness in the 55°C water bath. This is accomplished by application of reduced pressure. That generated by a water pump is adequate.

6. Accuracy

Since the method measures a mixture of estrogens, accuracy in absolute terms cannot be established properly. Accuracy has been assessed from measurement of acid-hydrolyzed male urine to which known amounts of estriol were added. Accuracy ranged from 74 to 104% over the range 0 to 80 mg/24-hr specimen.

7. Sensitivity

Urine from males, assayed by this method, appears to contain 1 to 2 mg total estrogen/24-hr specimen, whereas more specific methods indicate values of 20 μg/24 hr. Consequently, the method is not applicable to estimation of estrogen in urine from men or nonpregnant women and should be used only for specimens from women in the third trimester.

8. Precision

Precision was assessed by multiple estimations

TABLE 1

Typical Optical Densities and Calculations for the Method of Oakey et al.[47]

Patient and urine volume	Tube	R_{472}	R_{514}	R_{556}	$2xR_{514}$ (a)	$R_{472}+R_{556}$ (b)	a − b		(u+s)−u	mg/24 hr
Seymour	u	0.230	0.250	0.120	0.500	0.350	0.150	0.152	–	8.6
850 ml	u	0.215	0.240	0.110	0.480	0.325	0.155			
	u+s	0.420	0.570	0.180	1.140	0.600	0.540	0.502	0.350	
	u+s	0.360	0.490	0.155	0.980	0.515	0.465			
Ramsden	u	0.190	0.240	0.080	0.480	0.270	0.210	0.217	–	11.6
1,700 ml	u	0.195	0.250	0.080	0.500	0.275	0.225			
	u+s	0.355	0.530	0.115	1.060	0.470	0.590	0.590	0.372	
	u+s	0.365	0.530	0.105	1.060	0.470	0.590			
Smith	u	0.495	0.650	0.205	1.300	0.700	0.600	0.605	–	34.4
1,450 ml	u	0.510	0.660	0.200	1.320	0.710	0.610			
	u+s	0.690	0.930	0.235	1.860	0.925	0.935	0.955	0.350	
	u+s	0.730	0.980	0.255	1.960	0.985	0.975			
Friend	u	0.205	0.270	0.095	0.540	0.300	0.240	0.232	–	
2,380 ml	u	0.220	0.270	0.095	0.540	0.315	0.225			
	u+s	0.340	0.540	0.120	1.080	0.460	0.620	0.617	0.385	$12.0 \times \frac{3}{2} = 18.0$
	u+s	0.360	0.550	0.125	1.100	0.485	0.615			
External standard		0.095	0.215	0.015	0.430	0.110	0.320			

of a urine specimen, adjusted to pH 3 with glacial acetic acid, and stored at 4°C. During January and February 1972, 35 measurements were made on a single specimen during the course of normal service work. The mean value recorded was 15.9 mg/24 hr; the standard deviation was 1.6 mg and the coefficient of variation was 10%.

9. Values in Normal Pregnancy

Estrogen excretion in 168 patients (431 urines) with uncomplicated pregnancies varies from 7.5 to 26 mg/24 hr at 30 weeks' gestation to 14.5 to 58 mg/24 hr at term.[47]

C. The Method of Hainsworth and Hall[93]

1. Principle

Diluted urine is heated with 66% H_2SO_4. The Kober chromogen formed after dilution is extracted into chloroform containing 2.5% trichloroacetic acid. Estimation is by fluorimetry. The entire process is carried out on a continuous flow principle.

2. Apparatus

Sampler Mk® II proportioning pump and oil bath (120°C) with glass coil 12 m long, 1.6 mm internal diameter.

Pump tubing – in Tygon: 3.90 ml/min (purple/white); 0.235 ml/min (orange/white); 2.90 ml/min (purple/black); 2.0 ml/min (green/green); 0.10 ml/min (orange/green); 0.8 ml/min (red/red); 1.60 ml/min (blue/blue)

Pump tubing – in Acidflex: 1.44 ml/min (green/green); 1.71 ml/min (purple/purple)

Glass connections: HO, H3, A1, A6

Glass debubblers: Co

Glass separating cell: Bo

Glass pulse suppressor: PCl

Glass coils: double mixing coil, junction coil (116-103-7), single mixing coil, water cooled coil. All from Technicon.

Fluorimeter – Mk V (Locarte Co. Ltd., London SW7) fitted with a thallium lamp (AEG Ltd., London WC2) and a filter λ_{max} 562 nm for emission (Balzers, Liechtenstein)

Recorder – Speedomax® XL680 (Leeds and Northrup Ltd., Tyseley, Birmingham 11) The construction of the continuous flow system is shown in Figure 3. It should be noted that the sizes of the pump tubing used are different from those selected by Hainsworth and Hall.[93]

3. Reagents

Sulfuric acid (98%) – AR grade Chloroform – Laboratory reagent grade (Chromapak-Vickers, Burley in Wharfedale, Yorkshire)

Trichloroacetic acid – Laboratory reagent grade (Chromapak-Vickers, Burley in Wharfedale)

Hydroquinone – Laboratory reagent grade (Griffin & George Ltd., Wembley, England)

Estriol – (Steraloids Ltd., Croydon, Surrey)

Kober reagent – prepare 2% w/v hydroquinone in 66% H_2SO_4 (made by adding carefully 250 ml 98% H_2SO_4 to 136 ml water)

Ittrich reagent – dissolve 50 g trichloroacetic acid in chloroform (2 l.)

Estriol standard solutions – dissolve estriol in the minimum quantity of ethanol and dilute with water to give the following concentrations: 5, 7.5, 10, 15, and 20 mg/l. (equivalent to 10, 15, 20, 30 and 40 mg/2 l.).

4. Procedure

Measure and record the volume of each 24-hr urine specimen and dilute to 2 l. or to the next liter. Arrange in order on the sample plate interspersed with estriol standard solutions and quality control specimens. Table 2 lists the sequence used in the author's laboratory. The procedure is best described by reference to Figure 3, which shows the flow rate, color code, and type (Acidflex or Tygon) of the tubing used. Pulse suppressors of Tygon are shown PS, those of glass are shown PS-G. The other junctions are standard Technicon equipment. To start an analysis, stretch the manifold over the pegs of the peristaltic pump and fit the platen. Place the sampling lines in the water, Kober reagent and Ittrich reagent supplies. Switch on the pump, water supply to water-cooled coil, transformer, fluorimeter lamp, and photomultiplier. Clean the flow cell with methanol, fill the ice bath with ice, and switch on the recorder. When the first sample reaches the flow cell (18 min after initial sampling) note the recorder height and adjust to give 85 units deflection. A typical plot from service work in the author's laboratory is shown in Figure 4. From the recorder chart plot peak heights derived from the standard solutions as a function of concentration. Interpolate peak heights derived from the urine samples to obtain total estrogens (mg/24-hr specimen). At the end of a run, clamp off the flow cell line. Switch off the recorder, photomultiplier, lamp, and transformer. Remove the sample lines from water, Ittrich

TABLE 2

Order of Samples for Analysis by Continuous Flow System*

Chart no	Specimen	Estrogen mg/2 l.
1	40 mg/2 l. estriol standard	Not shown
2	40 mg/2 l. estriol standard	Not shown
3	40 mg/2 l. estriol standard	–
4	10 mg/2 l. estriol standard	–
5	10 mg/2 l. estriol standard	–
6	15 mg/2 l. estriol standard	–
7	20 mg/2 l. estriol standard	–
8	30 mg/2 l. estriol standard	–
9	40 mg/2 l. estriol standard	–
10	Quality control urine no. 1	13.3
11	Quality control urine no. 2	20.7
12	Quality control urine no. 3	30.3
13	40 mg/2 l. estriol standard	–
14	Urine	12.0
15	Urine	25.8
16	Urine	23.5
17	Urine	17.5
18	40 mg/2 l. estriol standard	–
19	Urine	12.0
20	Urine	40.0
21	Urine	21.5
22	Urine	23.5
23	40 mg/2 l. estriol standard	–
24	20 mg/2 l. estriol standard	–
25	Urine	24.0
26	Urine	22.0
27	Urine	16.3
28	Urine	22.8
29	40 mg/2 l. estriol standard	–
30	10 mg/2 l. estriol standard	–
31	Urine	12.3
32	Urine (repeat of 31)	12.3
33	Urine	31.0
34	Urine	>40.0
35	Urine	32.0
36	40 mg/2 l. estriol standard	–
37	Urine	34.0
38	Urine	19.0
39	15 mg/2 l. estriol standard	–
40	Urine (repeat of 14)	11.5
41	Urine	11.0
42	Urine	20.0
43	Quality control urine no. 1	13.0
44	Quality control urine no. 3	29.3
45	Quality control urine no. 2	20.3
46	40 mg/2 l. estriol standard	–
47	30 mg/2 l. estriol standard	–
48	20 mg/2 l. estriol standard	–
49	15 mg/2 l. estriol standard	–
50	10 mg/2 l. estriol standard	–

*For results of one batch, see Figure 4.

reagent, Kober reagent, and urine samples. Allow the pump to pump the tubes dry. Remove the platen and loosen the manifold. Turn off the water-cooled coil.

5. *Trouble Shooting*

The continuous flow system requires at least as much attention to detail as does any manual system, if acceptable results are to be achieved. Continuous flow systems are not common in laboratories concerned only with steroid analyses, so that teething troubles are bound to arise. It is essential that all tubing on the manifold is in good condition, is flexible, and that the cross-sectional area is not distorted. Tubing should be renewed each week. All connections of tubing to glass must be firm and these must not allow air to leak into the system. Glass coils, phase separators, and tubing must be clean and free from grease. Key points to watch are

a. The pattern of the fluid train between the final mixing coil and the phase separator. This fluid train should be made up of alternate segments of chloroform and aqueous phases. Each segment should be about 4 mm in length. Deviations from this pattern lead to poor performance and are usually due to an accumulation of dirt in the mixing or heating coils.

b. The separation of chloroform and aqueous phases in the phase separator. This should proceed almost imperceptibly, with the aqueous phase passing continuously to waste. An accumulation of discrete bubbles points to poor performance.

c. The peak heights of standard estriol solutions on the recorder chart (Figure 4). Variations are usually due to inaccurate sampling or dispensing of reagents. The tubing on the manifold should, therefore, be checked. Regular variation in the height of standard solutions is often indicative, in our experience, of poor temperature control. This may be due to a faulty thermostat control or to degeneration of the quality of the oil in the heating bath. Both these faults can be remedied simply.

d. The shape of the recorder trace. Spiky traces reflect either faulty separation or air bubbles in the fluid line.

6. *Accuracy*

The accuracy of the method cannot

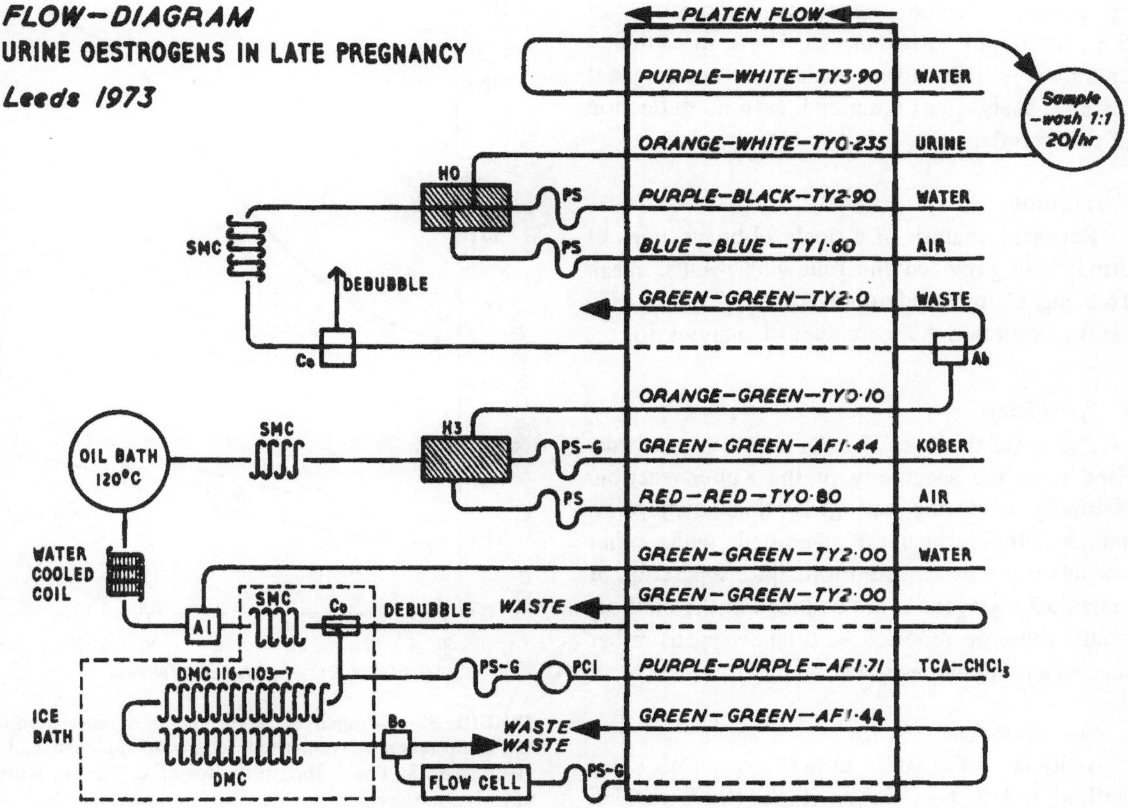

FIGURE 3. Flow diagram of continuous flow system for urinary estrogen analysis as used in Leeds. Abbreviations: TY – Tygon; AF – Acidflex; other letters refer to Technicon catalogue.

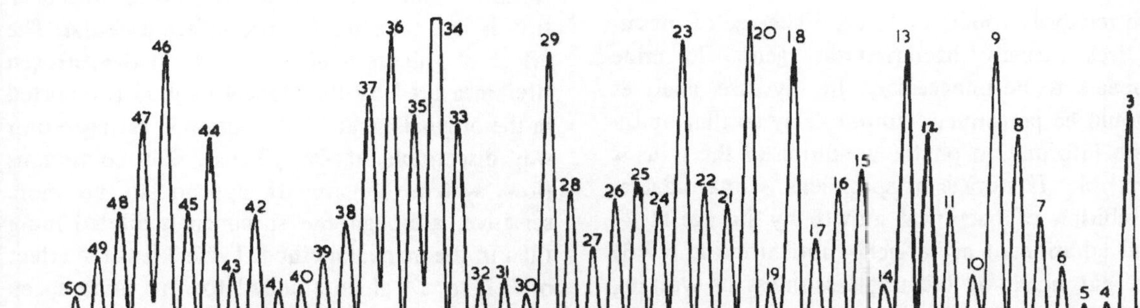

FIGURE 4. Chart recording from urinary estrogen analysis. For key to numbers, see Table 2.

be defined because the exact composition of the mixture of estrogens which are measured is unknown. Urine from males or nonpregnant females, analyzed as described, gave no deflection of the recorder pen.

7. Precision

Repeated analysis of a single 24-hr specimen of urine daily provided the following results: mean 16.0 mg/24 hr, standard deviation 0.7 mg, coefficient of variation 4.5%, number of analyses 36.

8. Specificity

The specificity of the analysis is derived from the specificity of the Kober reaction. Naturally occurring estrogens, their conjugates, and 3-methyl ethers are measured, while other steroids make no contribution. Since separation of individual estrogens is not included in the method, results must be reported as total estrogens rather than as a particular estrogen.

9. Normal Range

Analysis of 108 urines by both this method and that of Oakey et al. (1967) enabled a relationship:

Estrogen measured by AutoAnalyzer®=
1.2 + 1.4 (estrogen measured by Oakey et al.[47])

This relationship was used to calculate a new normal range from that of Oakey et al. (1967). This is shown in Figure 5. Conjugated estrogens are relatively robust molecules. The use of preservatives (usually bacteriastatic agents) in urine appears to be unnecessary. In any case, analyses should be performed without delay so that up-to-date information on the condition of the fetus is available. If storage of specimens is unavoidable, inhibition of bacterical growth by storage at 4° and addition of acetic acid (until urine is pH 3) prevent much of the unpleasantness of working with stale urine.

D. Interfering Substances

Attempts to measure very low levels of estrogen in urine from ovariectomized women indicated that certain materials, including laxatives, can interfere with urinary estrogen determinations.[111] In pregnancy where estimations are made on much smaller portions of a urine specimen, interference from such drugs presents much less of a problem.

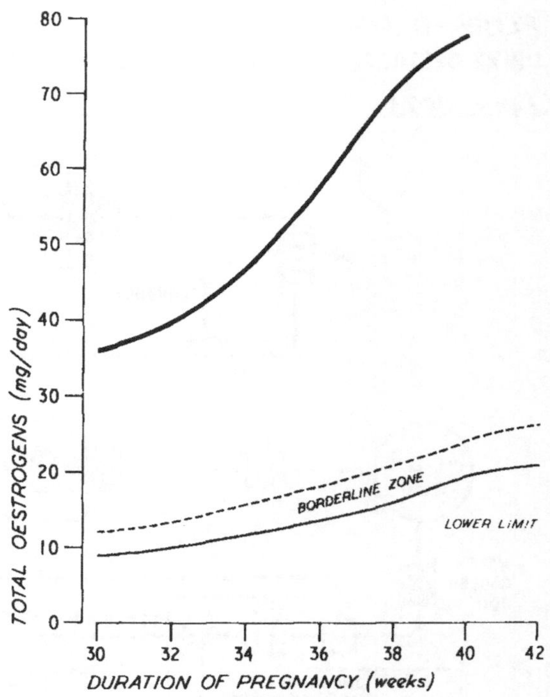

FIGURE 5. Estrogen excretion (mg/24 hr) measured by the continuous flow method in normal pregnancy. Patients seen at Leeds Maternity Hospital. The borderline zone is also shown.

No clear evidence for interference from laxatives has been encountered in this laboratory.

Glucose is the most important compound which gives reason for concern. The effect can be minimized by dilution of the urine if the concentration is more than 1%. If the glucose concentration is less than 0.2%, little effect is found. The effect of glucose added to urine on the estrogen titer measured by the manual method is reported in the original paper.[47] Reduction in estrogen titer was discernible at 4% glucose. The continuous flow system (Figure 3) appears to be more sensitive, although the specimen is diluted more than in the manual method. Table 3 lists the effect of 0.1 to 2% glucose added to four urine specimens, which were assayed by the continuous flow system. The presence of 0.5% glucose clearly reduced the estrogen titer in all four specimens. The manual method is unsuitable for use in cases where patients are being treated with Mandelamine® (methamine mandelate) for urinary tract infection.[47] Similar techniques that involve acid hydrolysis of an undiluted specimen also suffer from this disadvantage.[117-119] Formaldehyde released from this compound complexes with

TABLE 3

Effect of Glucose on Estrogen Concentration in Human Pregnancy*

Glucose concentration (%)	Total estrogens (mg estriol/24 hr) Urine 1	Urine 2	Urine 3	Urine 4
0	23.9	26.2	25.0	31.5
0.2	22.7	24.8	23.0	29.7
0.5	21.0	22.9	21.0	28.5
1.0	16.7	20.1	19.5	27.0
2.0	13.8	17.5	18.6	21.0

*Urine measured by continuous flow technique (Figure 3).

the phenolic estrogens to form tars (phenol + formaldehyde = bakelite!!). The continuous flow system, as used in Leeds (Figure 3), appears to be much less sensitive to this compound, possibly because of the greater dilution of the specimen prior to acid treatment. For example, a patient on treatment with Mandelamine had total estrogens of 2.1 mg/24 hr measured by Oakey et al.,[47] but 29 mg/24 hr measured by the method of Howarth and Robertshaw[92] (Burton, P., personal communication).

E. Compounds Which Suppress Estrogen Excretions

It has been demonstrated that ingestion of corticosteroids by the mother may lead to reduced estrogen excretion in pregnancy.[120-122] This suppression appears to be affected by inhibition of adrenocorticotrophin (ACTH) secretion from the fetal pituitary gland.[32]

Estrogen excretion of pregnant women who are receiving corticosteroids for coexisting asthma, ulcerative colitis, sarcoidosis, or for dermatological conditions may not reflect the condition of the fetus.[122-127] This effect was studied in 17 patients who received between 10 and 160 mg of cortisol (or its equivalent in synthetic steroids) daily during their pregnancies.[127] From comparison of estrogen excretion and the outcome of the pregnancy, it was concluded that estrogen excretion reflected the condition of the fetus provided the dose was less than 75 mg cortisol (or equivalent) daily. At doses greater than this, estrogen excretion was low whether or not the fetus was likely to survive.

Penicillin has also been implicated in reducing estrogen secretion. Willman and Pulkkinen[128] and Pulkkinen and Willman[129] reported reduced urinary excretion and plasma concentration of estriol during treatment with penicillin. No mechanism was suggested for this effect. Interference in the chemical estimation was ruled out, implying an interference with estriol biosynthesis. It is known, however, that penicillin and its derivatives will cross the placenta to the fetal circulation.[130]

F. Renal Function

The methods described in detail both relate to urinary estrogen measurements. Results are expressed in mg/24-hr urine. It is, therefore, pertinent to consider the effect of poor renal function on estrogen excretion.

The importance of renal function in estrogen excretion was demonstrated by the results of Brown et al.[21] discussed earlier. In practical terms, estrogen excretion was found to increase when subjects moved from an upright to a supine position.[131] This increase was associated with an increase in glomerular filtration rate, emphasizing the importance of renal function in estrogen excretion. Improvement in renal function probably accounts for the decrease in blood estrogen concentrations in patients admitted for rest in bed.[19] A more detailed study of the problem was made by Carrington et al.,[132] who measured urinary and plasma estrogens and creatinine clearance in 97 patients. These authors concluded that the low estrogen excretion observed in patients with poor renal function was due to inadequate clearance of estrogens rather than to inadequate production. Plasma estrogen concentrations in these patients were found to be higher than usual.

The effect of diuretics on estrogen excretion is the subject of controversy. Timonen et al.[133] concluded that diuretics increase estrogen excretion. Bird and Reeves[134] were unable to confirm this.

G. Interpretation of Results

The basic question to which the assay has to provide an answer is, Will the fetus survive or not? From an analysis of estrogen excretion measurements in more than 400 patients whose pregnancies had complications associated with a high perinatal mortality, Heys et al. found that 75% of the 16 intrauterine deaths that occurred were preceded by subnormal estrogen excretion.[114] Normal estrogen excretion was associated with subsequent intrauterine death in only one patient. This finding has been the basis of the interpretation of estrogen assays in our service work. It has been found useful also to define a borderline zone, which although within the normal range, draws attention to the approach of the lower limit of the normal range (Figure 5).

Patients attending the prenatal clinic are selected at 30 to 38 weeks of gestation by the obstetricians for estrogen assay if their complications fall into any of the groups defined earlier. If estrogen excretion is within the normal range, specimens are collected and measured at weekly intervals. If the trend of values is upwards, then assurance is gained that the pregnancy is proceeding to a successful end. If the estrogen excretion trend is downwards, then specimens are collected more frequently than once each week. If the downward trend continues, then the patient is brought into hospital where specimens are collected daily. If the excretion of estrogens falls below normal, then serious consideration must be given to premature delivery of the fetus. However, such a decision must be based on more considerations than estrogen excretion alone. Nevertheless, the association between subnormal estrogen estimations and subsequent intrauterine death cannot be lightly disregarded.

Used in this way, estrogen estimations serve two purposes. First, subnormal values provide warning of likely intrauterine death. Second, if the estrogen excretion is within the normal range, despite clinical evidence of complications, this information provides reassurance to the obstetrician that the fetus is unlikely to die. An example of the reassurance is provided by consideration of 42 women whose fetuses were thought to be "small for dates."[1] Of this group of patients, 39 delivered babies who were small for dates. The patients could be divided into two groups on the basis of their estrogen excretion. Seventeen patients had estrogen excretion values which were normal or in the borderline zone. Of these patients, there was only one intrauterine death. Twenty-five patients showed subnormal estrogen excretion; in this group there were six intrauterine deaths. Therefore, although all these patients were placed clinically in a high-risk group, intrauterine deaths were associated in six out of seven instances only with subnormal estrogen excretion.

A more difficult situation occurs in the patient whose trend of estrogen excretion falls into neither of the categories described above, but is chronically subnormal from about the 32nd week of pregnancy. The estrogen excretion points to a high risk of early intrauterine death, but the obstetrician will naturally be reluctant to intervene and hasten delivery since the baby will be immature.

The first response of the laboratory to an estrogen excretion below the normal range for gestation should be to request that another 24-hr specimen be collected by the patient at the earliest opportunity. The need for a complete 24-hr collection should be emphasized at this time, inquiry should be made whether the patient is receiving any drugs, and the expected date of delivery should be confirmed. If subnormal estrogen excretion is again encountered when urine collection is confirmed as complete and the duration of gestation is known, then the following possibilities must be considered for discussion with the obstetrician:

a. Fetal death has already occurred
b. Fetal death is imminent
c. The fetus is anencephalic
d. The patient is taking corticosteroids for a coexisting condition
e. Fetal adrenal hypoplasia
f. Placental sulfatase defect

Low levels of estrogen excretion (less than 3 mg/24 hr) are encountered after fetal death, since the dead fetus cannot provide precursors. When fetal death is imminent, values higher than this may be encountered and may be just below the lower limit of normal range for the period of gestation (Figure 5). Low levels (less than 3 mg/24 hr) are encountered when the fetus is anencephalic, since the small fetal adrenals do not provide precursors to the placenta for estrogen biosynthesis. This condition is usually diagnosed by palpation or by X-ray and is often associated with hydramnios. Estrogen excretion in fetal adrenal hypoplasia is low for reasons similar to those associated with anencephaly.[40] However, fetal adrenal hypoplasia is not amenable to diagnosis in utero by X-ray or manual techniques and, without postnatal corticosteroid therapy, inevitably leads to perinatal death. The reasons for subnormal estrogen excretion in some women treated with corticosteroids have been discussed already. The last and possibly rarest condition, placental sulfatase deficiency, is associated with estrogen excretion of less than 2 mg/24 hr.[38,135-137] This low estrogen production is probably due to the absence of a placental enzyme vital to estrogen biosynthesis (Figure 2) which converts the androgen sulfate to free androgen. The condition was first recorded by France and Liggins.[38] Since then, at least three other patients have been demonstrated to exhibit this defect.[136,137] Before birth, it is difficult to distinguish this condition from fetal adrenal hypoplasia. Comparison of the conversion to estrogen of injected dehydroepiandrosterone sulfate and dehydroepiandrosterone may be of value for this purpose. The infants born to women with placental sulfatase defect have usually required induction of delivery and have all been males. The follow-up and intensive investigation of such cases may throw some light on the mechanism of induction of parturition in the human.

An important aspect of the interpretation of the results of estrogen estimations must be to decide whether a particular change in the quantity of estrogen excreted has significance in relation to the condition of the fetus. In other words, what magnitude of change in the daily excretion rate can be accepted without indicating the need for clinical action. This problem has often been raised, but concrete data are scarce. Frandsen[73] reported that 92% of measured estrogen excretions were within 30% of the value for the preceding day and 81% were within 20% of the preceding day's value. Rather similar variations were noted by Klopper et al.,[138] who measured the daily excretion of 6 women (at 38 weeks of gestation) in a metabolic ward which ensured complete urine specimens were obtained. The coefficient of variation for the 24-hr period was 17.9%. However, these variations must also include a contribution from the analytical techniques used which are likely to have coefficients of variation of 10%. No study of this kind has yet been carried out with the continuous flow method where the precision is much greater. The unknown magnitude of this daily variation is largely responsible for the importance of assessing fetal function on the trend of values established over three or more estimations. A downward trend over three consecutive specimens must give rise to apprehension about the condition of the fetus, even though each successive value is within 20% of the preceding one.

Most laboratories have standardized collection of a 24-hr urine specimen which is essentially a compromise between shorter (e.g., 2, 4, or 6 hr) collection periods and longer (e.g., 48 hr) ones. Klopper et al.[138] showed that variability of estrogen excretion increased as the period of collection was reduced. The longer periods of collection give a more precise index of estrogen excretion, but involve personal and logistic problems for women collecting urine as outpatients, with consequent failures to achieve complete collection of all the urine passed.[139] It has been suggested that estrogen excretion should be expressed relative to creatinine excretion which was believed to be approximately 1.0 g/24 hr and to be independent of diet.[140] Recent evidence from several laboratories[141,142] has challenged the concept that creatinine excretion is constant. Although gross deviations from complete collection can be recognized by measurement of urinary creatinine,[143] simple inspection of the urine volume and comparison with the volumes of other specimens provided by the same patient often indicate failure to make a proper collection. Watney et al.[144] consider that all specimens of less than 1 liter are likely to be inadequate.

It has also been suggested that the estrogen/creatinine ratio measured on a urine sample collected over a short time period can be used to predict the 24-hr estrogen excretion.[145] If this prediction could be made with confidence, then delays could be reduced. These authors found that the coefficient of variation of the estrogen/creatinine ratio was less than the coefficient of variation of the 24-hr estrogen excretion, when both were calculated from excretion during a 2-hr collection period. However, from the data provided, it would appear that there were wide discrepancies between actual estrogen excretion per 24 hr and that calculated from extrapolation of the estrogen/creatinine ratio measured in a 2-hr urine sample. The findings of Shelley et al.[146] also confirm this view. Despite this evidence, claims have been made that estrogen/creatinine ratios measured on samples of urine collected for short periods "correlate well" with the estrogen/creatinine ratio measured on the entire 24-hr collection.[147,148] Before it can be concluded with confidence that the estrogen/creatinine ratio determined on a small urine sample can be validly and accurately used as a fetoplacental function test, direct evidence should be presented in relation to the prediction of intrauterine fetal death.[149] The report by Mackay et al. that the single specimen approach "may have a place" in the assessment of fetoplacental function deserves further investigation.[150]

Interpretation of the quantities of estrogen in maternal plasma in terms of the likelihood of imminent fetal death is fraught with problems, no less difficult than those outlined for urinary assays. Blood can be obtained from the patient without the 24-hr delay associated with collection of urine. The corollary of this is that levels of estrogen in blood reflect conditions at a single moment, whereas estrogen excretion is a value integrated over the period of collection, usually 24 hr. As has been noted earlier, both nonconjugated and conjugated forms of estrone, estradiol-17β, and estriol have been measured. In some cases,[107,151,152] evidence of the presence or absence of significant variations during the day has been recorded. It is certain that much more needs to be done and experience gained before assays of estrogen in blood can be preferred to urinary assays for the purpose of detecting imminent fetal death. On the basis of measurements of nonconjugated estrone, nonconjugated estradiol-17β, nonconjugated estriol, and estrone sulfate in peripheral plasma, Loriaux et al.[26] suggested that estimation of any of these estrogens will serve equally well to assess fetoplacental function. Whether or not this turns out to be an accurate prediction remains to be seen. That a careful approach to this problem is required is illustrated by the report of Townsley et al.[153] These authors found serum estradiol-17β concentration to be misleadingly low in 5 out of 13 pregnant women who had normal urinary estrogen excretion and who subsequently delivered normal infants.

VI. QUALITY CONTROL

"Measurements should not only be done, they must manifestly and undoubtedly be seen to be done well." This modification of Lord Justice Hewart's pronouncement is clearly relevant today.

The time has now passed when it was sufficient for a laboratory merely to complete an assay of a steroid hormone. In medicine, obstetrics, and gynecology, the results of steroid hormone assays can be of crucial significance to management and diagnosis of disease and to saving life. Estrogen estimations are carried out at the request of obstetricians who may decide whether or not to initiate a premature delivery, largely on the basis of the results. Other factors are considered but the results of estrogen estimations must weigh heavily in reaching decisions. It is mandatory, therefore, that maximum effort be made to achieve and maintain a high standard of analysis and that the obstetricians be aware of the dimensions of laboratory error. Accuracy and precision are the two components which must be defined.

As has been discussed above, the accuracy of total estrogen methods cannot be defined rigorously. The different methods available provide different values when applied to measurement of a particular urine specimen. This is shown clearly by the returns of the U.K. National Quality Control Survey of Pregnancy Oestrogen Measurements. The mean values of analyses by groups of laboratories using either (1) the method of Brown et al.,[79] (2) the method of Oakey et al.,[47] or (3) methods based on Brombacher et al.[91] (including both manual and continuous flow techniques) are in the ratio 1.0:1.25:1.55. In this context it is interesting to note that results reported (from a single laboratory) using the method of Brown and Coyle[71] are usually only 50 to 60% of the results reported by laboratories using Brown et al.[79] The reasons for these discrepancies are not known. It seems reasonable to assume, however, that destruction of labile estrogens during acid hydrolysis and the greater manipulative losses in the more complex procedures are involved. Moreover, within a group of laboratories all of which claim to use one method, there are significant differences in the values reported. No particular method shows less interlaboratory variation than the others considered.

These variations are reasons to emphasize that, at present, "normal values" do not represent a population greater than that defined by local patients studied with a particular method in a particular laboratory. Ranges for normal patients must, therefore, be constructed by each laboratory and should *not* be taken from the literature.

Since clinical decisions are largely based on a trend of values, precision of analysis is probably more important than accuracy in this context. The laboratory must be able to define the precision of analysis carried out at a given frequency. For example, laboratories making estrogen measurements daily should assay a particular specimen daily over a period of time. The precision must be high enough for confident clinical action on the basis of laboratory results. We have found that urine adjusted to pH 3 with glacial acetic acid provides material suitable for repeated assays. This acidified urine is stored overnight at 4°C, but during the day is kept in the laboratory. In the author's experience storage of urine at −20°C offers no advantage and in fact, provides results slightly more variable than those from acidified urine, stored at 4°C, analyzed at the same time. This is considered to be due to expansion and leakage during thawing of the nonhomogenous frozen sample.

In our service work two quality control specimens are analyzed twice each day (Figure 4 and Table 2). One specimen has an estrogen titer of 28 mg/2 1.; the concentration of the other specimen is 16 mg/2 l. Therefore, assessment can be made at levels well within the normal range and at the lower limit of the normal range. The results of analysis of these specimens are plotted daily on a large-scale chart displayed in the laboratory. At intervals of 2 weeks, the mean, standard deviation, and coefficient of variation for analyses of both urine specimens are calculated for results obtained during the preceding 4 weeks. Comparison of the daily results of analysis of the quality control urine enables identification of poor performance at the end of the run.

During the period February 1972 to February 1973, the standard deviation of the mean, measured with the continuous flow system described, fluctuated between 0.5 and 0.8 mg. Consequently, a change in estrogen excretion greater than 2.2 mg is unlikely ($p<0.05$) to be due to technical error.

REFERENCES

1. **Heys, R. F., Scott, J. S., Oakey, R. E., and Stitch, S. R.,** Estriol excretion in abnormal pregnancy, *Obstet. Gynecol.*, 33, 390, 1969.
2. **Barnard, W. P. and Logan, R. W.,** The value of urinary oestriol estimation in predicting dysmaturity, *J. Obstet. Gynaecol. Br. Commonw.*, 79, 1091, 1972.
3. **Beischer, N. A., Bhargava, V. L., Brown, J. B., and Smith, M. A.,** The incidence and significance of low oestriol excretion in an obstetric population, *J. Obstet. Gynaecol. Br. Commonw.*, 75, 1024, 1968.
4. **Klopper, A.,** Estriol in liquor amnii, *Am. J. Obstet. Gynecol.*, 112, 459, 1972.
5. **Breuer, H.,** in *Estrogen Assays in Clinical Medicine,* Paulsen, C. A., Ed., University of Washington Press, Seattle, 1965, 251.
6. **Zucconi, G., Lisboa, B. P., Simonitsch, E., Roth, L., Hagen, A. A., and Diczfalusy, E.,** Isolation of 15α-hydroxy-oestriol from pregnancy urine and from the urine of newborn infants, *Acta Endocrinol.*, 56, 413, 1967.
7. **Fishman, J., Schut, H., and Solomon, S.,** Metabolism production and excretion rates of 15α-hydroxy-estriol in late pregnancy, *J. Clin. Endocrinol.*, 35, 339, 1972.
8. **Knuppen, R., Haupt, O., and Breuer, H.,** Isolation and identification of 15α-hydroxyoestrone from the urine of pregnant women, *Biochem. J.*, 96, 33c, 1965.
9. **Knuppen, R., Haupt, O., and Breuer, H.,** Isolation and identification of 15β-hydroxyoestrone and 15β-hydroxy-oestradiol-17β from the urine of pregnant women, *Steroids,* 8, 403, 1966.
10. **Beling, C. G.,** Gel filtration of conjugated urinary oestrogens and its application in clinical assays, *Acta Endocrinol.*, Suppl. 79, 1963.
11. **Smith, E. R. and Kellie, A. E.,** Oestrogen conjugates of human late-pregnancy urine, *Biochem. J.*, 104, 83, 1967.
12. **Hähnel, R.,** The quantitative relationship of oestrogen-3-glucosiduronates and oestrogen-16(or 17)-glucosiduronates in human late pregnancy urine, *J. Endocrinol.*, 38, 417, 1967.
13. **Cohen, S. L. and Oran, E.,** The conjugated estrogens. II. A new method for isolating estriol monoglucosiduronide from normal human pregnancy urine, *Can. J. Biochem.*, 50, 1245, 1972.

14. **Ahmed, J. and Kellie, A. E.**, The excretion of oestrogen conjugates in late pregnancy urine, *J. Steroid Biochem.*, 3, 31, 1972.
15. **Hobkirk, R., Anuman-Rajadhon, Y., Nilsen, M., and Blahey, P. R.**, Contribution of estriol to total urinary estrogens during pregnancy, *Clin. Chem.*, 16, 235, 1970.
16. **Cohen, S. L.**, The excretion of 'labile' oestrogens during human pregnancy, *Acta Endocrinol.*, 67, 677, 1971.
17. **Cohen, S. L.**, A method for the rapid colorimetric assay of total estrins in pregnancy urine, *J. Clin. Endocrinol.*, 26, 994, 1966.
18. **Smith, O. W.**, Free and conjugated estrogens in blood and urine before and during parturition in normal pregnancy, *Acta Endocrinol.*, Suppl. 104, 1966.
19. **Roy, E. J.**, The concentration of oestrogens in maternal and foetal blood obtained at Caesarian section, and the effect of hospitalization on maternal blood oestrogen levels, *J. Obstet. Gynaecol. Br. Commonw.*, 69, 196, 1962.
20. **Rado, A., Crystle, D. C., and Townsley, J. D.**, Concentration of estrogens in maternal peripheral plasma in late pregnancy, during labor and post-partum, *J. Clin. Endocrinol.*, 30, 497, 1970.
21. **Brown, C. H., Saffan, B. D., Howard, C. M., and Preedy, J. R. K.**, The renal clearance of endogenous estrogens in late pregnancy, *J. Clin. Invest.*, 43, 295, 1964.
22. **Svendsen, R. and Sørensen, B.**, The concentration of unconjugated oestrone and oestradiol-17β in plasma during pregnancy, *Acta Endocrinol.*, 47, 237, 1964.
23. **Sybulski, S.**, Determination of free estradiol-17β levels in pregnancy plasma by competitive protein binding method, *Am. J. Obstet. Gynecol.*, 110, 304, 1971.
24. **Tulchinsky, D., Hobel, C. J., Yeager, E., and Marshall, J. R.**, Plasma estrone, estradiol-17β, estriol, progesterone and 17-hydroxyprogesterone in human pregnancy. I. Normal pregnancy, *Am. J. Obstet. Gynecol.*, 112, 1095, 1972.
25. **Giebenhain, M. E., Tagatz, G. E., and Gurpide, E.**, Serum levels of unconjugated estetrol (1,3,5(10)-estratriene-3, 15α,16α,17β-tetrol) during human pregnancy, *J. Steroid Biochem.*, 3, 707, 1972.
26. **Loriaux, D. L., Ruder, H. J., Knab, D. R., and Lipsett, M. B.**, Estrone sulphate, estrone, estradiol and estriol plasma levels in human pregnancy, *J. Clin. Endocrinol.*, 35, 887, 1972.
27. **Schindler, A. E. and Siiteri, P. K.**, Isolation and quantitation of steroids from normal human amniotic fluid, *J. Clin. Endocrinol.*, 28, 1189, 1968.
28. **Michie, E. A. and Livingstone, J. R. B.**, Oestriol concentration in amniotic fluid, *Acta Endocrinol.*, 61, 329, 1969.
29. **Berman, A. M., Kalchman, G. G., Chattoraj, S. C., and Scommegna, A.**, Relationship of amniotic fluid estriol to maternal urinary estriol, *Am. J. Obstet. Gynecol.*, 100, 15, 1968.
30. **Pinkus, G. S. and Pinkus, J. L.**, Fluorometric determination of total estrogens in amniotic fluid of normal and complicated pregnancies, *Obstet. Gynecol.*, 36, 528, 1970.
31. **Young, B. K., Jirku, H., and Levitz, M.**, Estriol conjugates in amniotic fluid at midpregnancy and term, *J. Clin. Endocrinol.*, 35, 208, 1972.
32. **Oakey, R. E.**, The progressive increase in estrogen production in human pregnancy: an appraisal of the factors responsible, *Vitam. Horm.*, 28, 1, 1970.
33. **Liggins, G. C.**, *Endocrinology of the Foeto-maternal Unit in Human Reproductive Physiology*, Shearman, R. P., Ed., Blackwell, Oxford, 1972.
34. **Cassmer, O.**, Hormone production of the isolated human placenta, *Acta Endocrinol.*, Suppl. 45, 1959.
35. **Frandsen, V. A. and Stakemann, G.**, The site of production of oestrogenic hormones in human pregnancy. Hormone excretion in pregnancy with anencephalic foetus, *Acta Endocrinol.*, 38, 383, 1961.
36. **Diczfalusy, E. and Mancuso, S.**, Oestrogen metabolism in pregnancy, in ***Foetus and Placenta***, Klopper, A. and Diczfalusy, E., Eds., Blackwell, Oxford, 1969.
37. **Bolté, E., Mancuso, S., Eriksson, G., Wiqvist, N., and Diczfalusy, E.**, Studies on the aromatisation of neutral steroids in pregnant women. 2. Aromatisation of dehydroepiandrosterone and of its sulphate administered simultaneously into a uterine artery, *Acta Endocrinol.*, 45, 560, 1964.
38. **France, J. T. and Liggins, G. C.**, Placental sulfatase deficiency, *J. Clin. Endocrinol.*, 29, 138, 1969.
39. **Anderson, A. B. M., Laurence, K. M., and Turnbull, A. C.**, The relationship in anencephaly between the size of the adrenal cortex and the length of gestation, *J. Obstet. Gynaecol. Br. Commonw.*, 76, 196, 1969.
40. **Roberts, G. and Cawdery, J. E.**, Congenital adrenal hypoplasia, *J. Obstet. Gynaecol. Br. Commonw.*, 77, 654, 1970.
41. **Marrian, G. F.**, Early work on the chemistry of pregnanediol and the oestrogenic hormones, *J. Endocrinol.*, 35, vi, 1966.
42. **Brown, J. B.**, Some observations on the Kober colour and fluorescence reactions of the natural oestrogens, *J. Endocrinol.*, 8, 196, 1952.
43. **Bauld, W. S.**, Some errors in the colorimetric estimation of oestriol, oestrone and oestradiol by the Kober reaction, *Biochem. J.*, 56, 426, 1954.
44. **Nocke, W.**, A study of the colorimetric estimation of oestradiol-17β, oestradiol-17α, oestrone, oestriol and 16-*epi*oestriol by the Kober reaction, *Biochem. J.*, 78, 593, 1961.
45. **Brown, J. B.**, A chemical method for the determination of oestriol, oestrone and oestradiol in urine, *Biochem. J.*, 60, 185, 1955.
46. **Breuer, H.**, in *Estrogen Assays in Clinical Medicine*, Paulsen, C. A., Ed., University of Washington Press, Seattle, 1965, 88.

47. **Oakey, R. E., Bradshaw, L. R. A., Eccles, S. S., Stitch, S. R., and Heys, R. F.**, The rapid estimation of oestrogens in pregnancy to monitor foetal risk, *Clin. Chim. Acta*, 15, 35, 1967.
48. **Ittrich, G.**, Eine neue Methode zur Chemischen Bestimmung der Oestrogen Hormone im Harn, *Z. Physiol. Chem.*, 312, 1, 1958.
49. **Salokangas, R. A. A. and Bulbrook, R. D.**, The determination of small quantities of urinary oestrone, oestradiol-17β and oestriol using Ittrich's extraction method, *J. Endocrinol.*, 22, 47, 1961.
50. **Bradshaw, L. R. A.**, A supplement to the Kober reaction for urinary oestrogens, *Nature*, 190, 809, 1961.
51. **Bates, R. W. and Cohen, H.**, Experimental basis for selecting the optimal conditions for quantitative fluorimetry of natural estrogens, *Endocrinology*, 47, 166, 1950.
52. **Preedy, J. R. K. and Aitken, E.**, The determination of estrone, estradiol-17β and estriol in urine and plasma with column partition chromatography, *J. Biol. Chem.*, 236, 1300, 1961.
53. **Munson, A. K., Mueller, J. R., and Yannone, M. E.**, Quantification of free 17β-estradiol in pregnancy plasma by gas-liquid chromatography using electron-capture detection, *Biochem. Med.*, 3, 187, 1969.
54. **Wotiz, H. H., Charransol, G., and Smith, I. N.**, Gas chromatographic measurement of plasma estrogens using an electron capture detector, *Steroids*, 10, 127, 1967.
55. **Mead, R. A., Haltmeyer, G. C., and Eik-Nes, K. B.**, A method for the determination of free estradiol in peripheral plasma by gas-phase chromatography, *J. Chromatogr. Sci.*, 7, 554, 1969.
56. **Attal, J. and Eik-Nes, K. B.**, Measurement of free estradiol-17β in blood plasma by gas-liquid chromatography using an electron capture detector, *Anal. Biochem.*, 26, 398, 1968.
57. **Schindler, A. E. and Herrmann, W. L.**, Determination of steroids by gas-liquid chromatography. I. The measurement of estriol in pregnancy urine, *Gynaecologia*, 161, 446, 1966.
58. **Ratanasopa, V., Schindler, A. E., Lee, T. Y., and Herrmann, W. L.**, Measurement of estriol in plasma by gas-liquid chromatography, *Am. J. Obstet. Gynecol.*, 99, 295, 1967.
59. **Aleem, F. A., Neill, D. W., and Pinkerton, J. M. H.**, A method for estriol estimation in amniotic fluid and its use in the study of normal and abnormal pregnancy, *Steroids*, 13, 651, 1969.
60. **Lee, P. C. and Wood, P. J.**, Gas-liquid chromatographic determination of oestriol and oestetrol in late pregnancy urine, *Clin. Chim. Acta*, 30, 221, 1970.
61. **Cooper, W., Coyle, M. G., and Mills, J. A.**, Determination of plasma oestriol in normal pregnancy and labour using gas-liquid chromatography, *J. Endocrinol.*, 51, 447, 1971.
62. **Richardson, S. J.**, The quantitative estimation of oestriol using gas-liquid chromatography, *Clin. Chim. Acta*, 29, 473, 1970.
63. **Smith, P. D. and Stitch, S. R.**, A rapid semi-automated method for the determination of urinary oestriol in late pregnancy by gas-liquid chromatography, *Clin. Chim. Acta*, 36, 439, 1972.
64. **Touchstone, J. C.**, Routine quantitative gas chromatography of urinary estriol, *J. Gas Chromatogr.*, 2, 170, 1964.
65. **Yousem, J. L.**, Simple G.L.C. method for estimation of urinary estriol in pregnant women, *Am. J. Obstet. Gynecol.*, 88, 375, 1964.
66. **Sanghvi, A., Taddeini, L., Nagarajan, R., and Wight, C.**, Estriol determination in pregnancy urine by gas chromatography, *Clin. Chim. Acta*, 31, 255, 1971.
67. **Diczfalusy, E. and Diczfalusy, A., Eds.**, Steroid assay by protein binding, *Acta Endocrinol.*, Suppl. 147, 1970.
68. **Odell, W. M. and Daughaday, W. H., Eds.**, *Principles of Competitive Protein-binding Assays*, J. B. Lippincott, Philadelphia, 1971.
69. **Bauld, W. S.**, A method for the determination of oestriol, oestrone and oestradiol-17β in human urine by partition chromatography and colorimetric estimation, *Biochem. J.*, 63, 488, 1956.
70. **Brown, J. B.**, Urinary excretion of oestrogens during pregnancy, lactation and the re-establishment of menstruation, *Lancet*, i, 704, 1956.
71. **Brown, J. B. and Coyle, M. G.**, Urinary excretion of oestriol in pregnancy. 1. A shortened procedure, *J. Obstet. Gynaecol. Br. Commonw.*, 70, 219, 1963.
72. **Klopper, A. I. and Wilson, G. R.**, A method for the assay of oestriol in pregnancy urine, *J. Obstet. Gynaecol. Br. Commonw.*, 69, 533, 1962.
73. **Frandsen, V. A.**, *The Excretion of Oestriol in Normal Human Pregnancy*, Munksgaard, Copenhagen, 1963.
74. **Eberlein, W. R., Bongiovanni, A. M., and Francis, C. M.**, A simplified method for the routine measurement of urinary estriol, *J. Clin. Endocrinol.*, 18, 1275, 1958.
75. **Greene, J. W. and Touchstone, J. C.**, Urinary estriol as an index of placental function *Am. J. Obstet. Gynecol.*, 85, 1, 1963.
76. **Bachman, C.**, Photometric determination of estrogens. II. A new color reaction for estriol, *J. Biol. Chem.*, 131, 463, 1939.
77. **Montagu, K.**, A short method of estimating urinary excretion of oestrogens for therapeutic use in pregnancy, *J. Obstet. Gynaecol. Br. Commonw.*, 71, 92, 1964.
78. **Crowley, M. F., Garbion, K. J. T., and Rosser, A.**, Oestrogen determination in pregnancy urine, *Clin. Chim. Acta*, 38, 91, 1972.
79. **Brown, J. B., MacLeod, S. C., Macnaughtan, C., Smith, M. A., and Smyth, B.**, A rapid method for estimating oestrogens in urine using a semi-automatic extractor, *J. Endocrinol.*, 42, 5, 1968.

80. **Huang, W. Y.**, A rapid and reliable method of urinary oestriol determination in pregnancy, *Steroids*, 11, 453, 1968.
81. **Osawa, Y. and Slaunwhite, W. R.**, Studies on phenolic steroids in human subjects, XIII. A rapid assay of urinary estrogen conjugates in pregnancy, *Steroids*, 15, 73, 1970.
82. **Lee, L. and Hahnel, R.**, Rapid fluorimetric assay of estrogen conjugates in pregnancy urine, *Clin. Chem.*, 17, 1194, 1971.
83. **Rourke, J. E., Marshall, L. D., and Shelley, T. F.**, A simple rapid assay of estrogens in pregnancy, *Am. J. Obstet. Gynecol.*, 100, 331, 1968.
84. **Strickler, H. S., Holt, S. S., Acevedo, H. F., Saier, E., and Grauer, R. C.**, The determination of urinary estrogens in pregnancy using an automated fluorimetric assay, *Steroids*, 9, 193, 1967.
85. **Thysen, B., Meyer, C. J., and Gatz, M.**, Semiautomated assay for rapid determination of oestrogens in late pregnancy urine, *Obstet. Gynecol.*, 36, 799, 1970.
86. **Ua Conaill, D. and Muir, G. G.**, An automated method for the determination of oestrogens in pregnancy, *Clin. Chem.*, 14, 1010, 1968.
87. **Muir, G. G., Ua Conaill, D., and Ryan, M.**, The use of fluorimetry in the automated assay of Ittrich's total oestrogens in the urine of pregnant women, *Steroids*, 13, 719, 1969.
88. **Muir, G. G. and Ryan, M.**, Automated hydrolysis of total estrogens in urine from pregnant women, *Clin. Chem.*, 17, 1007, 1971.
89. **Barnard, W. P. and Logan, R. W.**, An automated method for urinary oestriol determination, *Clin. Chim. Acta*, 29, 401, 1970.
90. **Ittrich, G.**, Untersuchungen über die extraktion des roten Kober-farbstoffs durch organische Lösungsmittel zur Ostrogenbestimmung im Harn, *Acta Endocrinol.*, 35, 34, 1960.
91. **Brombacher, P. J., Gijzen, A. H. J., and Verheesen, P. E.**, Simple and rapid determination of total oestrogens, *Clin. Chim. Acta*, 20, 360, 1968.
92. **Howarth, A. T. and Robertshaw, D. M.**, Urinary estrogen determination in pregnancy, *Clin. Chem.*, 17, 316, 1971.
93. **Hainsworth, I. R. and Hall, P. E.**, A simple automated method for the measurement of oestrogens in the urine of pregnant women, *Clin. Chim. Acta*, 35, 201, 1971.
94. **Campbell, D. G. and Gardner, G.**, The automated analysis of urinary oestrogens during pregnancy, *Clin. Chim. Acta*, 32, 153, 1971.
95. **Beischer, N. A. and Brown, J. B.**, Current status of estrogen assays in obstetrics and gynecology. Part 2. Estrogen assays in late pregnancy, *Obstet. Gynecol. Survey*, 27, 303, 1972.
96. **Goebelsmann, U., Thorneycroft, I. H., Nakamura, R. M., and Mishell, D. R.**, Estriol in pregnancy. I. A radioimmunoassay for urinary estriol, *Am. J. Obstet. Gynecol.*, 112, 802, 1972.
97. **Wotiz, H. H. and Martin, H. F.**, Studies in steroid metabolism. XI. Gas chromatographic determination of oestrogens in human pregnancy urine, *Anal. Biochem.*, 3, 97, 1962.
98. **Scommegna, A. and Chattoraj, S. C.**, Gas chromatographic estimation of urinary oestriol in pregnancy, *Am. J. Obstet. Gynecol.*, 99, 1087, 1967.
99. **Smith, O. W. and Arai, K.**, Blood estrogens in late pregnancy: an evaluation of methods with improved recovery, *J. Clin. Endocrinol.*, 23, 1141, 1963.
100. **Nachtigall, L., Bassett, M., Hogsander, U., Slagle, S., and Levitz, M.**, A rapid method for the assay of plasma estriol in pregnancy, *J. Clin. Endocrinol.*, 26, 941, 1966.
101. **Mathur, R. S., Leaming, A. B., and Williamson, H. O.**, A simplified method for estimation of estriol in pregnancy plasma, *Am. J. Obstet. Gynecol.*, 113, 1120, 1972.
102. **Touchstone, J. C. and Murawec, T.**, Free and conjugated estrogens in blood plasma during human pregnancy, *Biochemistry*, 4, 1612, 1965.
103. **Kroman, H. S., Bender, S. R., Brest, A. N., Moskovitz, M. L., and King, M. O.**, Estrogens in human pregnancy plasma. I. Studies with gas chromatography, *Clin. Chem.*, 12, 670, 1966.
104. **Corker, C. S. and Naftolin, F.**, A rapid method for the measurement of oestriol in pregnancy plasma by competitive protein binding analysis, *J. Obstet. Gynaecol. Br. Commonw.*, 78, 330, 1971.
105. **Tulchinsky, D. and Abraham, G. E.**, Radioimmunoassay of plasma estriol, *J. Clin. Endocrinol.*, 33, 775, 1971.
106. **Tulchinsky, D., Hobel, C. J., and Korenman, S. G.**, A radioligand assay for plasma unconjugated estriol in normal and abnormal pregnancy, *Am. J. Obstet. Gynecol.*, 111, 311, 1971.
107. **Tulchinsky, D. and Korenman, S. G.**, The plasma estradiol as an index of fetoplacental function, *J. Clin. Invest.*, 50, 1490, 1971.
108. **Tulchinsky, D., Hobel, C. J., Yeager, E., and Marshall, J. R.**, Plasma estradiol, estriol and progesterone in human pregnancy. II. Clinical applications in Rh-isoimmunization disease, *Am. J. Obstet. Gynecol.*, 113, 766, 1972.
109. **Fishman, J. and Guzik, H.**, Radioimmunoassay of 15α-hydroxyestriol in pregnancy plasma, *J. Clin. Endocrinol.*, 35, 892, 1972.
110. **Schindler, A. E. and Herrmann, W. L.**, Estriol in pregnancy urine and amniotic fluid, *Am. J. Obstet. Gynecol.*, 95, 301, 1966.
111. **Brown, J. B., Bulbrook, R. D., and Greenwood, F. C.**, An additional purification step for a method for estimating oestriol, oestrone and oestradiol-17β in human urine, *J. Endocrinol.*, 16, 49, 1957.

112. **Biggs, J. S., Klopper, A., and Wilson, G. R.,** Oestriol in amniotic fluid: a gas chromatographic method, *J. Endocrinol.*, 44, 579, 1969.
113. **Klopper, A., Michie, E., and Aleem, F.,** The estimation of oestriol in amniotic fluid: a comparison of three methods, *J. Obstet. Gynaecol. Br. Commonw.*, 78, 444, 1971.
114. **Heys, R. F., Scott, J. S., Oakey, R. E., and Stitch, S. R.,** Urinary oestrogen in late pregnancy. Oestriol excretion as a guide to impending foetal death before term, *Lancet*, 1, 328, 1968.
115. **Klopper, A.,** The assessment of placental function in clinical practice, in *Foetus and Placenta*, Klopper, A. and Diczfalusy, E., Eds., Blackwell, Oxford, 1969.
116. **Brown, J. B. and Beischer, N. A.,** Current status of estrogen assay in gynecology and obstetrics. Part I. Estrogen assays in gynecology and early pregnancy, *Obstet. Gynecol. Survey*, 27, 205, 1972.
117. **Touchstone, J. C., Stojkewycz, M., and Smith, K.,** The effect of methenamine mandelate (Mandelamine®) on determination of urinary estriol, *Clin. Chem.*, 11, 1019, 1965.
118. **Mackay, E. V., Macafee, C. A. J., and Anderson, C.,** Depression of urinary oestrogen values during methenamine mandelate administration, *Aust. N.Z. J. Obstet. Gynaecol.*, 7, 94, 1967.
119. **Eraz, J. and Hausknecht, R.,** Diminished urinary estriol due to Mandelamine® administration during pregnancy, *Am. J. Obstet. Gynecol.*, 104, 924, 1969.
120. **Warren, J. C. and Cheatum, S. G.,** Maternal urinary oestrogen excretion: effect of adrenal supression, *J. Clin. Endocrinol.*, 27, 433, 1967.
121. **Scommegna, A., Nedoss, B. R., and Chattoraj, S. C.,** Maternal urinary estriol excretion after dehydroepiandrosterone sulfate infusion and adrenal stimulation and suppression, *Obstet. Gynecol.*, 31, 526, 1968.
122. **Brown, J. B., Beischer, N. A., and Smith, M. A.,** Excretion of urinary oestrogens in pregnant patients treated with cortisone and its analogues, *J. Obstet. Gynaecol. Br. Commonw.*, 75, 819, 1968.
123. **Wray, P. M. and Russell, C. S.,** Maternal urinary oestriol levels before and after death of the foetus, *J. Obstet. Gynaecol. Br. Commonw.*, 71, 97, 1964.
124. **Simmer, H. H., Dignam, W. J., Easterling, W. E., Frankland, M. V., and Naftolin, F.,** Neutral C_{19}-steroids and steroid sulphates in pregnancy. III. Dehydroepiandrosterone sulphate, 16α-hydroxydehydroepiandrosterone and 16α-hydroxydehydroepiandrosterone sulphate in cord blood and blood of pregnant women with and without treatment with corticoids, *Steroids*, 8, 179, 1966.
125. **Wallace, S. J. and Michie, E. A.,** A follow-up study of infants born to mothers with low oestriol excretion during pregnancy, *Lancet*, ii, 560, 1966.
126. **Morrison, J. and Kilpatrick, N.,** Low urinary oestriol excretion in pregnancy associated with oral prednisone therapy, *J. Obstet. Gynaecol. Br. Commonw.*, 76, 719, 1969.
127. **Oakey, R. E.,** The interpretation of urinary oestrogen and pregnanediol excretion in pregnant women receiving corticosteroids, *J. Obstet. Gynaecol. Br. Commonw.*, 77, 922, 1970.
128. **Willman, K. and Pulkkinen, M. O.,** Reduced maternal plasma and urinary oestriol during ampicillin treatment, *Am. J. Obstet. Gynecol.*, 109, 893, 1971.
129. **Pulkkinen, M. and Willman, K.,** Maternal oestrogen levels during penicillin treatment, *Br. Med. J.*, 4, 48, 1971.
130. **MacAuley, M. A., Molloy, W. B., and Charles, D.,** Placental transfer of methicillin, *Am. J. Obstet. Gynecol.*, 115, 58, 1973.
131. **Dickey, R. P., Carter, W. T., Besch, P. K., and Ullery, J. C.,** Effect of posture on estrogen excretion during pregnancy, *Am. J. Obstet. Gynecol.*, 96, 127, 1966.
132. **Carrington, E. R., Oesterling, M. J., and Adams, F. E.,** Renal clearance of estriol in complicated pregnancies, *Am. J. Obstet. Gynecol.*, 106, 1131, 1970.
133. **Timonen, S., Hirvonen, E., and Sokkanen, R.,** Urinary volume and excretion of oestrogens in late pregnancy, *Acta Endocrinol.*, 49, 393, 1965.
134. **Bird, C. C. and Reeves, B. D.,** Effect of diuretic administration on urinary estriol levels in late pregnancy, *Am. J. Obstet. Gynecol.*, 105, 552, 1969.
135. **France, J. T., Seddon, R., J., and Liggins, G. C.,** A study of a pregnancy with low estrogen production due to placental sulfatase deficiency, *J. Clin. Endocrinol.*, 36, 1, 1973.
136. **Fliegner, J. R. H., Schindler, I., and Brown, J. B.,** Low urinary oestriol excretion during pregnancy associated with placental sulphatase deficiency or congenital adrenal hyperplasia, *J. Obstet. Gynaecol. Br. Commonw.*, 79, 810, 1972.
137. **Oakey, R. E., Cawood, M. L., and Macdonald, R. R.,** unpublished results, 1972.
138. **Klopper, A., Wilson, G., and Cooke, I.,** Studies on the variability of urinary oestriol and pregnanediol output during pregnancy, *J. Endocrinol.*, 43, 295, 1969.
139. **Osofsky, H. J., Nesbitt, R. E. L., and Hagen, J. H.,** High risk obstetrics. IV. Estrogen/creatinine ratios in routine urine samples as a method of screening a high risk obstetric population, *Am. J. Obstet. Gynecol.*, 106, 692, 1970.
140. **Folin, O.,** Approximately complete analyses of thirty 'normal' urines, *Am. J. Physiol.*, 13, 45, 1903.
141. **Edwards, O. M., Bayliss, R. I. S., and Millen, S.,** Urinary creatinine excretion as an index of the completeness of 24-hr urine collections, *Lancet*, ii. 1165, 1969.
142. **Chattaway, F. W., Hullin, R. P., and Odds, F. C.,** The variability of creatinine excretion in normal subjects, mental patients and pregnant women, *Clin. Chim. Acta*,. 26, 567, 1969.

143. **Cummings, R. V., Rourke, J. E., and Shelley, T. F.,** Serial rapid assay of total urinary estrogens in pregnancy, *Am. J. Obstet. Gynecol.*, 104, 1047, 1969.
144. **Watney, P. J. M., Hallum, J., Ladell, D., and Scott, P.,** The relative usefulness of methods of assessing placental function, *J. Obstet. Gynaecol. Br. Commonw.*, 77, 301, 1970.
145. **Dickey, R. P., Besch, P. K., Vorys, N., and Ullery, J. C.,** Diurnal excretion of estrogen and creatinine during pregnancy, *Am. J. Obstet. Gynecol.*, 94, 591, 1966.
146. **Shelley, T. F., Cummings, R. V., Rourke, J. E., and Marshall, L. D.,** Estrogen-creatinine ratios. Clinical application and significance, *Obstet. Gynecol.*, 35, 184, 1970.
147. **Pariente, C., Goldberg, S., and Lewenthal, H.,** Measurement of oestriol excretion in pregnancy, *Lancet*, i, 79, 1971.
148. **Salvadori, B., Vadora, E., and Coppola, F.,** Diurnal variation in the oestrogen/creatinine ratio in pregnancy, *Lancet*, ii, 659, 1971.
149. **Siddiqui, A. and Watson, D.,** Urine oestriol measurement from a 12-hour specimen, *Lancet*, ii, 659, 1971.
150. **Mackay, E. V., Macafee, C. A. J., and Anderson, C.,** The value of single urinary specimens for oestrogen estimation during pregnancy, *Aust. N.Z. J. Obstet. Gynaecol.*, 8, 17, 1968.
151. **Selinger, M. and Levitz, M.,** Diurnal variation of total plasma estriol levels in late pregnancy, *J. Clin. Endocrinol.*, 29, 995, 1969.
152. **Macourt, D., Corker, C. S., and Naftolin, F.,** Plasma oestriol in pregnancy, *J. Obstet. Gynaecol. Br. Commonw.*, 78, 335, 1971.
153. **Townsley, J. D., Gartman, L. J., and Crystle, C. D.,** Maternal serum 17β-estradiol levels in normal and complicated pregnancies: a comparison with other estrogen indices of fetal health, *Am. J. Obstet. Gynecol.*, 115, 830, 1973.

PROGESTERONE AND METABOLITES

O. J. Lucis

TABLE OF CONTENTS

I. Introduction . . . 45
A. Metabolism of Progesterone . . . 47
B. Metabolism of 17α-Hydroxyprogesterone . . . 47
C. Endocrine Dysfunctional States . . . 48

II. Currently Used Analytical Methods . . . 50
A. Competitive Protein Binding Assays for Progesterone (CPB) . . . 50
1. General Considerations . . . 50
2. Quantitation of Plasma Progesterone Without Chromatography . . . 51
3. Quantitation of Plasma Progesterone After Chromatography . . . 53
4. Quantitation of Plasma Progesterone – Progesterone Displacement System . . . 55
B. Radioimmunoassay for Progesterone . . . 56
1. General Considerations . . . 56
2. Radioimmunoassay for Progesterone . . . 57
C. Urinary Steroid Determinations . . . 59
1. Introduction . . . 59
2. Colorimetric Determination of Urinary Pregnanediol and Pregnanetriol . . . 60
3. Determination of Urinary Pregnanediol, Pregnanetriol, and Pregnanetriolone by Gas-liquid Chromatography . . . 64
4. Simultaneous Determination of Urinary Pregnanolone, Pregnanediol, and Pregnanetriol by Gas-liquid Chromatography . . . 66
5. Qualitative Separation of Urinary Pregnanetriol, Pregnanetriolone, and Pregnanetetrol 70

III. Interpretation of Laboratory Values . . . 72

References . . . 77

I. INTRODUCTION

Progesterone is secreted in varying quantities by the gonadal and adrenal tissue. During childhood the rate of progesterone production is low and it increases at the time of puberty and during the age of reproduction. In nonpregnant women of reproductive age the corpus luteum becomes the major source of progesterone.

In the ovary, during the follicular phase, the cells of the theca interna of the Graafian follicle possess all enzyme systems necessary for biosynthesis of estradiol-17β. Within this phase of follicular development, rather little progesterone is produced and the plasma progesterone levels range from less than 0.15 to 0.42 ng/ml. Along with progesterone, the plasma also contains low levels of 17α-hydroxyprogesterone (0.4 ± 0.2 ng/ml), 20α-dihydroprogesterone (pregn-4-ene-20α-ol-3-one) (approximately 0.2 ng/ml), and Δ5-pregnenolone (approximately 0.9 ng/ml).

The maturation of the Graafian follicle is culminated at midcycle when a sharp rise in plasma estrogens, FSH, and LH occurs; the follicle ruptures and the ovum is released.

In the ovary after the rupture of the follicle,

the theca interna cells regress and the granulosa cells acquire a good blood supply and develop the corpus luteum.

The plasma progesterone levels show signs of increase already on the day of "midcycle total urinary estrogen peak." On the average, in women with regular ovulatory cycles, the plasma progesterone reaches a plateau of 10 to 20 ng/ml some 4 to 6 days after ovulation. In individual cycles the highest plasma progesterone concentration is observed between 6 and 9 days after ovulation. Toward the end of the luteal phase, plasma progesterone levels decrease rapidly to less than 1 ng/ml at the onset of menstruation. During the luteal phase the plasma 20α-dihydroprogesterone may be of the order of 1.5 ng/ml and 17α-hydroxyprogesterone may range from 0.3 to 2.8 ng/ml. The plasma levels of Δ-5 pregnenolone show only some inconsistent increase. It has been observed that plasma 17α-hydroxyprogesterone is consistently increased on the day of midcycle gonadotrophin peak to a level of 1 ng/ml or above.

In plasma, progesterone circulates primarily in bound form associated with various proteins such as transcortin, albumin, and orosomucoid. Of the total plasma progesterone, only some 1 to 2% appears to be in the free form. The biological half-life of progesterone is short and it is rapidly metabolized by the liver and other tissues. Administration of radioactive progesterone to nonpregnant women has shown that 6 to 27% is excreted in urine as pregnanediol, 1.5 to 5% as pregnanolone, and 0.5 to 2% as pregnanediones. Only traces may appear in the urine in unchanged form.

Physiologically, progesterone has a number of extragenital and genital effects. During an ovulatory menstrual cycle, the increased secretion of progesterone in the luteal phase is responsible for elevation of the basal body temperature of about 0.5 to 0.8°C. During the follicular phase, the body temperature is fairly constant until ovulation, when there may be a slight drop followed by a rise which lasts for 9 to 13 days. The optimal fertile period is believed to precede the temperature rise by about 2 days. Prior to menstruation, the basal body temperature declines to the level of the follicular phase. If conception occurs, the basal body temperature remains elevated to about mid-pregnancy and then declines, probably due to the antagonistic action of estrogens. Metabolically, progesterone tends to antagonize mineralocorticoids and promotes sodium and water excretion. At the time of puberty and during pregnancy, progesterone promotes the development of the alveoli of the breasts.

During the menstrual cycle the estrogen-primed *proliferative endometrium, in response* to increased progesterone levels in the luteal phase, is transformed to secretory endometrium, which is essential for nidation of the fertilized ovum and for the maintenance of pregnancy.

The rise in progesterone level after ovulation antagonizes the estrogen effect on the cervical glands and on the cervical mucus. The estrogen-induced glandular secretion is inhibited by progesterone and the cervical mucus looses its spinnbarkeit, increases in viscosity and cellularity, and looses its capability to form a crystallization (ferning) pattern on a smear. In response to progesterone, the vaginal cytology also becomes altered. The uterine muscle (myometrial) tone and activity are reduced by progesterone, which tends to induce smooth muscle relaxation and reduces vascular tone and the motility of the intestinal tract.

Recent studies of plasma hormonal levels during pregnancy have revealed significant relationships.[22] After conception, plasma chorionic gonadotrophin, progesterone, and 17α-hydroxyprogesterone show a continuous increase. By the 3rd to 4th week of gestation, plasma progesterone and 17α-hydroxyprogesterone reach a peak concentration. As the pregnancy progresses, the plasma progesterone levels decline and attain a nadir by the 9th week of gestation. The plasma 17α-hydroxyprogesterone shows a trend to decrease as the pregnancy progresses. The levels of plasma chorionic gonadotrophin increase continuously up to the 12th week of gestation. It appears that the functional life span of corpus luteum during pregnancy is limited and the task of progesterone secretion is taken over by the placental trophoblasts at some time around the 8th week of gestation. According to Johansson[1] the mean plasma progesterone during the 5th week of gestation is in the order of 24.8 ± 7.3 ng/ml; by the 9th week of gestation the level declines to 16.7 ± 7.4 ng/ml. This physiological transition period where progesterone production makes a changeover from corpus luteum to placenta coincides with the time when spontaneous abortions frequently occur.

After the 9th week of pregnancy, plasma progesterone increases continuously, reaching a plateau of about 125 ± 38 ng/ml by the 32nd week. Some sporadic and relatively small increases are noted during the last weeks of pregnancy. Following parturition and expulsion of the placenta, the plasma progesterone concentration rapidly returns to low levels.

Determinations of plasma 20α-dihydroprogesterone during pregnancy have shown that this progesterone derivative remains low during early pregnancy and by the 24th week increases to a first plateau of 5 to 10 ng/ml; a second plateau of 12 to 25 ng/ml is attained by the 32nd week and persists until delivery.[36] The plasma 17α-hydroxyprogesterone after the 20th week of pregnancy may reach a level of 3 ng/ml. In advanced pregnancy the increase in plasma 17α-hydroxyprogesterone is probably related to conversion of maternal progesterone by the fetus.

A. Metabolism of Progesterone

In progesterone metabolism the principal centers of metabolic attack are the ketonic groups at carbons 3 and 20 of the steroid molecule, where, by reduction, 3α and β as well as 20α- and β-oriented hydroxy groups may be formed. Reduction of the double bond between carbon atoms 4 and 5 can lead to formation of pregnane and allopregnane derivatives. In the metabolic sequence the attack on the double bond tends to precede the reduction of the 3-keto group. After saturation of the double bond, one or more of these reductions can take place, thus leading to a wide variety of metabolites shown in Figure I.

The principal metabolite of progesterone is pregnanediol (5β-pregnane-3α,20α-diol) which is excreted in urine conjugated with glucuronic acid.

Quantitation of urinary pregnanediol in nonpregnant women offers a means for determining the duration and functional activity of the corpus luteum. A wide variation is observed in excretion of pregnanediol among individuals. In general, the urinary excretion of this metabolite mirrors the cycle of corpus luteum activity rising shortly after ovulation and declining at the time of corpus luteum involution. This pattern follows well the changes seen in plasma progesterone levels. The excretion of pregnanediol during the follicular phase of a normal ovulatory cycle may vary from 0.2 to 4 mg/24 hr. During the luteal phase, the excretion increases up to 7 mg/24 hr. In women taking combined or sequential oral contraceptives, as well as in postmenopausal women, the excretion of pregnanediol tends to be below 1 mg/24 hr.

The metabolic reduction of the functional groups in the progesterone molecule reduces and finally destroys the biological activity of progesterone. Some progestational activity is still retained for 20α-dihydroprogesterone or 20β-dihydroprogesterone. Further changes in the ring A inactivate the metabolites. Apart from these metabolic pathways, progesterone may undergo hydroxylation at 6β, 16α, and 15α positions. It has been observed that 6β-hydroxylation may occur in the liver and in other tissues, including the ovary. The corpus luteum possesses the capacity to produce 16α-hydroxyprogesterone. During pregnancy the maternal progesterone may be transformed to 15α-hydroxyprogesterone by entering the fetal compartment. The clinical significance of these products in diagnostic assays is still not established.[22]

In pregnant women assays of urinary pregnanediol and pregnanolone (5β-pregnan-3α-ol-20-one) have found a place in monitoring progesterone production. The excretion of these metabolites shows a wide variation among individuals; nevertheless, there is a definite pattern in the increments of excretion as the pregnancy advances. At the 20th week of gestation, the quantity of pregnanolone excreted is in the order of 2 to 5 mg/24 hr, whereas pregnanediol may range from 10 to 20 mg/24 hr. After the 20th week of gestation, the excretion of both these metabolites increases and by the 30th week pregnanolone excretion may be in a range of 5 to 17 mg/24 hr and pregnanediol 20 to 65 mg/24 hr. Normally by the 36th week of gestation, the maximal excretion of these metabolites is attained when the pregnanolone may range from 10 to 27 mg/24 hr and pregnanediol 35 to 100 mg/24 hr. Between the 36th week and parturition, the excretion of these metabolites tends to decrease.

B. Metabolism of 17α-Hydroxyprogesterone

The metabolic transformation of 17α-hydroxyprogesterone follows a pattern similar to that of progesterone. The principal metabolite formed is pregnanetriol (5β-pregnane-3α,17α,20α-triol), which appears in urine as the glucosiduronate. In women small quantities of 17α-hydroxyprogesterone are partly secreted by the adrenal glands and partly by the ovaries. During the follicular phase of the menstrual cycle the excretion of

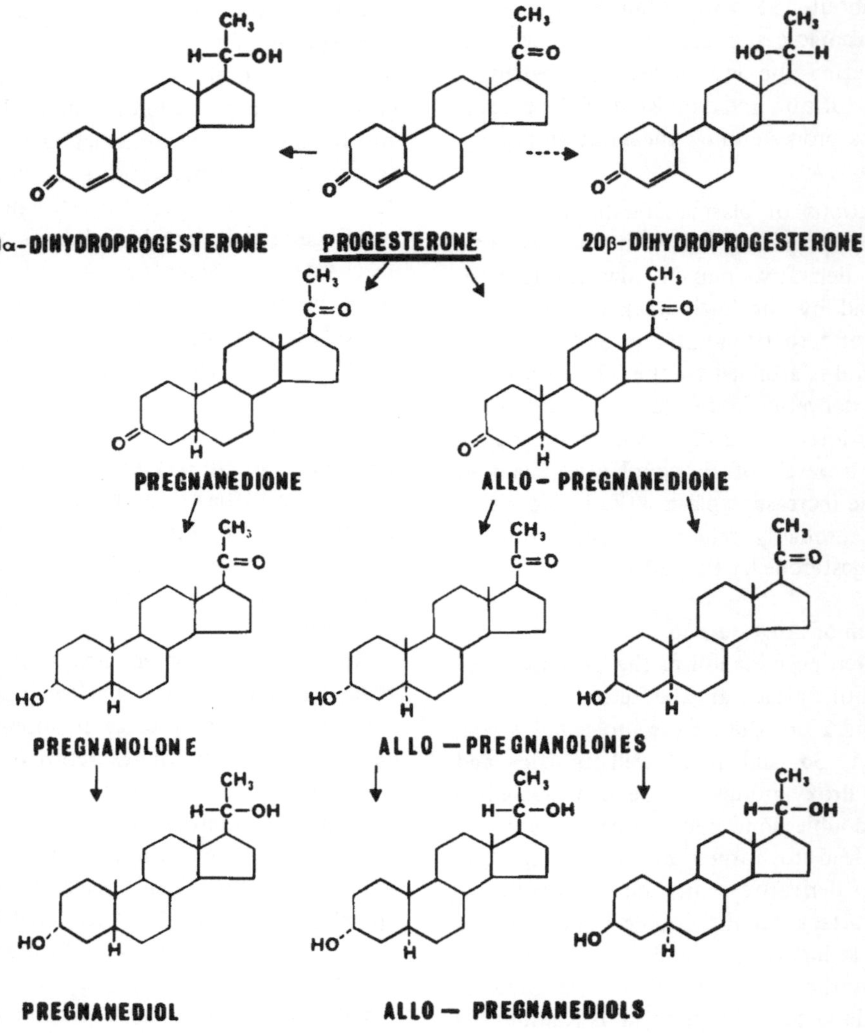

FIGURE 1. The metabolism of progesterone.

pregnanetriol may range from 0.1 to 1 mg/24 hr and in the luteal phase an increase to 2 mg/24 hr is observed. Women taking oral contraceptives of sequential or combination type show a lower excretion of urinary pregnanetriol which tends to be below 1 mg/24 hr.

During early pregnancy up to the 20th week of gestation, urinary pregnanetriol excretion remains low. As the pregnancy advances, a continuous increase in urinary pregnanetriol is observed and near term the excretion may range from 4 to 13 mg/24 hr. The rise in pregnanetriol during normal pregnancy shows some parallelism with urinary total estrogen and estriol excretion, thus reflecting the function of the fetal component of the fetoplacental unit.

C. Endocrine Dysfunctional States

Progesterone in adrenocortical cells acts as one of the intermediates for further formation of androgens and corticosteroids. In cases of congenital adrenal hyperplasia due to steroid 21-hydroxylase deficiency, the normal biosynthetic pathways become distorted, resulting in an excessive production of 17α-hydroxyprogesterone. Accumulation of this intermediate in the adrenocortical cells leads to an excessive conversion to androgens, formation of an abnormal steroid: 21-deoxycortisol (4-pregnen-11β, 17α diol-3,20 dione) and an increased secretion of 17α-hydroxyprogesterone (see Figure 2). Biochemically as well as clinically, the steroid 21-hydroxylase deficiency may vary in the degree of severity from mild to severe. As a

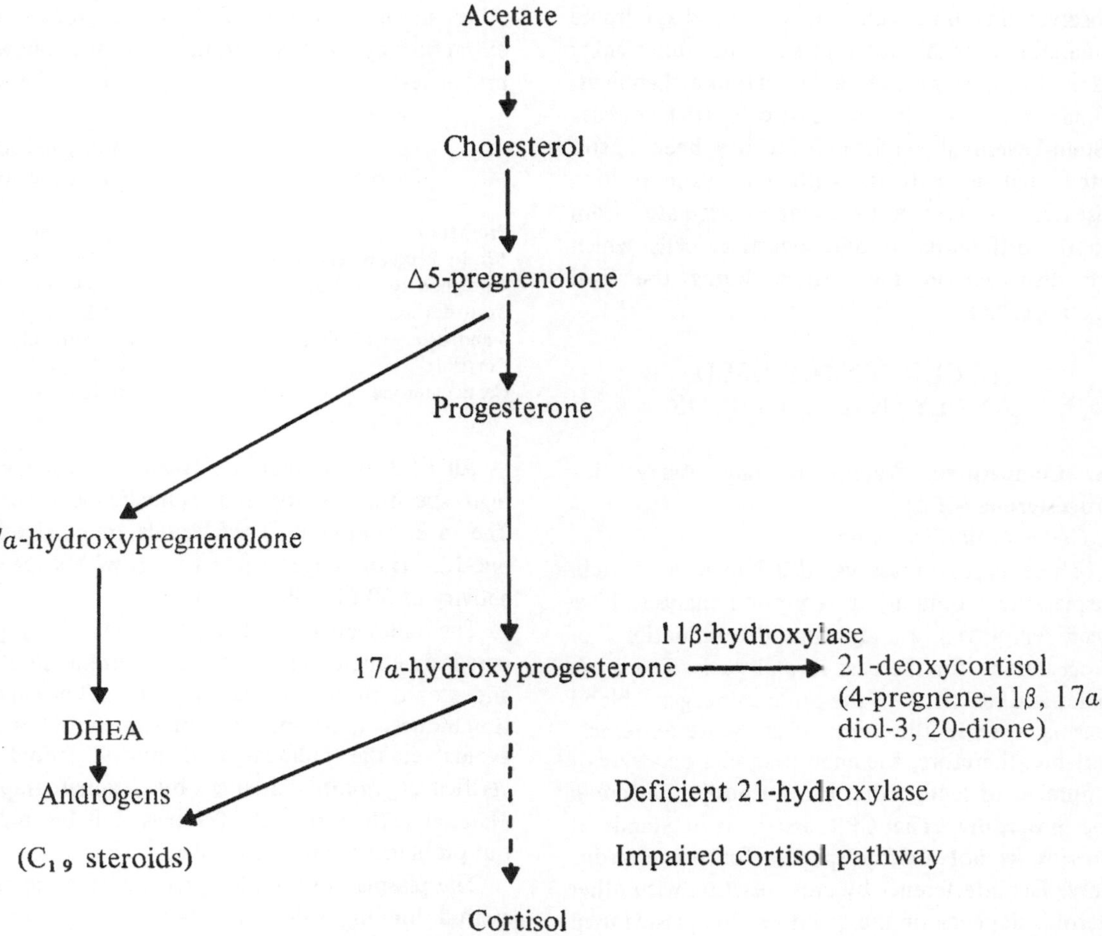

FIGURE 2. Abbreviated diagrammatic scheme of adrenal steroidogenesis in cases of steroid 21-hydroxylase deficiency.

result of this deficiency, there is an impairment in production of cortisol. The inadequacy in cortisol secretion impairs the inhibitory feedback mechanism of pituitary corticotrophin release. Excessive endogenous corticotrophin secretion results in overstimulation of the adrenal cortex, leading to adrenocortical hyperplasia and an excessive conversion of adrenal cholesterol to progesterone. Depending on the severity of 21-hydroxylase deficiency, the diseased adrenal glands may secrete varying ratios of C_{19} steroids, 17α-hydroxyprogesterone, progesterone, and 21-deoxycortisol. As a result of this, increased quantities of urinary 17-ketosteroids, pregnanetriol, pregnanediol, and metabolites of 21-deoxycortisol can be expected.

Metabolically, 21-deoxycortisol is transformed by the liver mainly to pregnanetriolone (5β-pregnan-3α, 17α, 20α-triol-11-one), but pregnanetetrol (5β-pregnan-3α, 11β, 17α, 20α-tetrol) is also formed to some extent. Both of these metabolites appear in urine as glucosiduronates. So far in the urine from normal subjects, these metabolites have not been detected. Excretion of pregnanetriolone in cases of untreated congenital adrenal hyperplasia due to 21-hydroxylase deficiency can be detected within the first few days after birth. As the infants grow older, the urinary excretion of pregnanetriolone increases and measurable quantities of pregnanetriol as well as pregnanetetrol also appear in urine.

Treatment of these patients with cortisol or with cortisol analogs suppresses the adrenocortical activity and reduces urinary excretion of these metabolites.[32]

Formation of 21-deoxycortisol and excretion of small quantities of pregnanetriolone have been observed in some cases of Cushing's syndrome (bilateral adrenal hyperplasia), in functioning adrenal tumors such as carcinoma, in arrhenoblastoma, and also in polycystic ovarian disease (Stein-Leventhal syndrome). It has been postulated that in patients with polycystic ovarian disease, 21-deoxycortisol may originate from poorly differentiated adrenocortical cells which are displaced in the ovaries during the fetal organogenesis.

II. CURRENTLY USED ANALYTICAL METHODS

A. Competitive Protein Binding Assays for Progesterone (CPB)

1. General Considerations

These types of assays, also known as protein displacement binding or saturation analyses, have been employed for routine determinations of progesterone for the past several years. The properties and reactivity of the principal reagent, blood plasma, used in this form of assay are extremely variable; therefore, the analyst should be aware of a number of factors that can seriously interfere in the procedure. The CPB assay, as it stands at present, is not strictly specific for progesterone only. The interference by cross-reaction with other steroids depends on the quality of the plasma used as a source of corticosteroid binding globulin (CBG) and the quantity of various steroids present in the final extract. In all systems so far described, some elevation of the final progesterone result is observed by the presence of

- 17α-hydroxyprogesterone
- 4-pregnen-20α-ol-3-one
- 4-pregnen-20β-ol-3-one
- 5α-pregnan-3,20-dione
- 5β-pregnan-3,20-dione
- Testosterone

It is essential to exclude cortisol and corticosterone from the extract, since traces of these steroids alter the final result significantly. Experience has shown that the most suitable solvent for specimen extraction so far is petroleum ether with a boiling range of 30 to 60°C. The extraction properties of this solvent may vary among different manufacturing batches. Therefore, it is necessary to redistill the petroleum ether and to check its extraction properties routinely. Petroleum ether successfully used for plasma progesterone extraction by various authors shows the following properties:

Steroid	Efficiency of single extraction
Progesterone	85–90%
17α-hydroxyprogesterone	14–40%
4-pregnen-20α-ol-3-one	64–90%
Testosterone	29–45%
4-androsten-3,17 dione	Approximately 87%
Cortisol	Less than 1%
Corticosterone	Less than 1%

All CPB assays require ^3H-labeled ligand with high specific activity and radiochemical purity. The most commonly used ligands are corticosterone-1,2-^3H or progesterone 1,2-^3H with a specific activity of 30 Ci/millimol or more.

The radioactive stock solution is usually prepared in ethanol or in ethanol benzene solution and stored in the freezer at −10°C. Since radiation-induced decomposition of the steroid is not avoidable, the radiochemical purity should be verified at monthly intervals by chromatography. Material with purity of 90% may still be usable, but problems can be anticipated.

The plasma or serum for preparation of corticosteroid binding solution (CBG) may be derived from various species. Human plasma obtained from women treated with estrogens or estrogen-containing oral contraceptives is used most frequently. In these subjects, the levels of corticosteroid binding protein are increased as a result of estrogen therapy, thus providing a plasma suitable for preparation of CBG reagent. For laboratory purposes the plasma stored in aliquots in a freezer at −20°C remains stable for several months. Some investigators prefer to use plasma from dogs treated with estrogens and dexamethasone for preparation of the CBG reagent. If the CBG solution is prepared using progesterone 1,2-^3H, then plasma from pregnant guinea pigs is found to be a suitable one. In guinea pigs, the CBG has a higher affinity for binding progesterone than corticosterone.

In the preparation of CBG reagent, it is advisable to dissolve the dry residue of the radioactive tracer first in water or buffer solution and then add the required quantity of stock plasma. For progesterone assays the CBG reagent is commonly prepared 12 to 24 hr in advance and kept for equilibration at room temperature until used. Experience regarding the stability of the CBG reagent varies. In some instances, this reagent remains stable and usable for as long as 14 days if stored in a refrigerator at 5°C. An improvement of stability has been observed if sodium azide (0.02% w/v) is added to the solution.

A common problem with CBG reagents is the variability in sensitivity. One source of interfering factors has been traced to endogenous steroids bound to CBG. Pichon and Milgrom[2] introduced a purification step by adsorbing the endogenous plasma steroids on dextran-coated charcoal. The diluted supernatant plasma is then used for preparation of the CBG reagent. Attempts to adsorb endogenous steroids by other methods so far have given only variable success.

The optimal quantity of plasma necessary for preparation of the CBG reagent has to be established experimentally. In most methods the concentration of plasma in this reagent ranges from 50 to 200 μl/100 ml of water or buffer. The concentration of the radioactive ligand depends on the specific activity of the radioactive stock solution and in terms of mass, the commonly used range is up to 0.4 ng/ml. If the mass of the ligand is kept constant for the CBG reagent and the quantity of plasma is reduced, a higher sensitivity but a narrower range of reading can be expected.

The nonradioactive progesterone standard solution is usually prepared in ethanol or methanol in convenient concentrations to deliver 0 to 10 ng for assay. Prior to the preparation of standard solutions, the purity of the crystalline progesterone should be verified. Another important factor in these assay procedures is the glassware. It should be free from detergent residues or other contaminants. Most workers prefer to use precleaned disposable glassware. It has been repeatedly observed that plastic interferes with the assay by giving high blank values and erratic results. The interference from plastic may even originate when the venous blood is drawn in plastic syringes. So far the types of plastic that contribute toward errors have not been clearly identified. If problems in assay develop, the analyst should keep this factor in mind.

For proper analyses, the handling of blood samples is of importance. It is known that human erythrocytes contain 20α-hydroxy steroid dehydrogenase and at room temperature this enzyme may convert the plasma progesterone to 20α-dihydroprogesterone (4-pregnen-20α-ol-3-one). The enzymatic activity is minimized if the blood specimen is kept under refrigerated conditions until centrifugation and separation. For determination of plasma progesterone, citrated, oxalated, or heparinized blood has been used. Even serum from clotted blood can be employed. If the assays of progesterone are performed within days, the plasma or serum can be stored in the refrigerator. In case longer storage is required, freezing of the specimen is advisable.

The assays of progesterone should be performed in duplicate or triplicate. Each run should include reagent blanks and one or two sets of quality control sera. The quality control specimens may be prepared by pooling plasma or serum which contains selected levels of progesterone and stored frozen in appropriate aliquots. In addition to standard curves, these quality control specimens should serve as a guide for statistical analysis on the reliability of the results.

In the performance of a CPB assay, the incubation timing and temperature have to be uniform for all specimens. A vulnerable step in the procedure is the time of contact with Florisil®- or dextran-coated charcoal which are used as adsorbants for the free radioactive ligand in the system. Even though CBG has a strong binding to the attached radioactive ligand at +4°C, a disturbance of the equilibrium by the added adsorbant will cause a continuous loss of the ligand from the CBG in favor of Florisil- or dextran-coated charcoal.

For the determination of ^3H radioactivity, any standard liquid scintillation phosphor which accommodates the required quantity of aqueous solution can be used. The type of liquid scintillation spectrometer is of no major importance, provided that the efficiency for counting is statistically adequate.

Since the publication of the general principles of CPB assay by Murphy,[3] a number of modified procedures have been developed. All of these procedures have merits and it is extremely difficult to recommend a particular method. For use as guidelines in assays of plasma progesterone, three CPB procedures are included.

2. Quantitation of Plasma Progesterone Without Chromatography[1,4-6]

a. Principle

Plasma samples are extracted with petroleum

ether and progesterone present in the extract is determined by protein displacement binding assay. The protein binding solution is prepared by using human plasma obtained from women treated with ethinylestradiol. Tritiated corticosterone is employed as a tracer in this system. Unbound corticosterone is adsorbed by Florisil.

b. Reagents and Materials

1. Corticosterone-1, 2-^{3}H (specific activity 57.2 Ci/millimol).
2. Progesterone reference standard, tested for purity by thin-layer chromatography.
3. Petroleum ether, Mallinckrodt® 4980 lot RYC and RVR with boiling range 30 to 60°C.
4. Human plasma for protein binding solution was obtained from a young woman treated with 0.1 mg ethinylestradiol daily for 2 weeks. The plasma was stored at −15°C in 1-ml aliquots.
5. Corticosterone binding globulin solution. Corticosterone-1, 2-^{3}H, 15 ng in ethanol is evaporated under nitrogen and dissolved in 100 ml of glass distilled water. To this water solution 50 μl of the estrogen-treated patient's plasma is added and the solution is mixed. Before use, the prepared solution is kept at room temperature for 1 hr or overnight at 6°C.
6. Florisil 60-100 mesh (AB Kistner, Sweden) is washed four times with glass distilled water, the fine particles are decanted, and the remaining Florisil is dried at 120°C overnight. Prepared Florisil can be stored in a desiccator.
7. Liquid scintillation phosphor used contained 100 g naphthalene, 7.0 g 2,5-diphenyloxazole, and 0.3 g 1,4-bis-2 (5-phenyl-oxazolyl)-benzene per 1,000 ml dioxane.

c. Procedure

1. Plasma sample 0.25 ml is pipetted into 8 ml extraction tube with a well-fitted glass stopper. Prepare in duplicate.
2. Petroleum ether 2.5 ml is added to each tube, which is stoppered and shaken for 1 min. When the phases are separated on standing or after centrifugation, the petroleum ether layer is transferred to a culture tube using a Pasteur pipette. It is important that the extract remain free from plasma droplets. A transfer of plasma interferes with the assay. The extract contained in the culture tubes is evaporated.
3. Prepare duplicate standard tubes containing 0.0 to 5.0 ng of progesterone and 2.5 ml of fresh petroleum ether, which is subsequently evaporated.
4. To each tube add 1 ml corticosterone binding globulin solution. Mix gently.
5. Incubate for 5 min in a water bath at 40°C.
6. Transfer the tubes to an ice bath for 10 min and shake the tubes frequently.
7. To each tube add 80 mg of Florisil and mix for 30 sec over a vortex mixer and replace the tube in the ice bath.
8. When all samples are mixed, remove 0.5 ml of supernatant for liquid scintillation counting. The timing of contact with Florisil should be exactly the same for each tube.
9. Construct the calibration curve and calculate progesterone in the specimens.

d. Comments

Sensitivity – The sensitivity of the method is reported as 0.1 ng. Plasma progesterone values of 0.5 ng/ml or less are regarded as being below the limit of sensitivity. Therefore, this method is limited for plasma progesterone assays during the luteal phase of the menstrual cycle and during pregnancy.

Specificity of the method depends on the extraction properties of petroleum ether. Since corticosterone and cortisol interfere with the assay system, these should be excluded from the extract. Testosterone and 20α-dihydroprogesterone have similar displacing properties and their presence in the extract at concentrations above 0.2 ng interferes with the assay. Experimentally it was shown that corticosterone and 17α-hydroxyprogesterone are more effective than progesterone in competing for the binding sites in the system. Problems of interference were not encountered with pregnanediol, pregnenolone, dehydroepiandrosterone, estradiol-17β, estrone, estriol, and androstenedione.

Recoveries of various steroids in a range of 0.2 to 20 ng added to female plasma and extracted once with 10 vol of petroleum ether (Mallinckrodt 4980 lot RVR) are shown in Table 1.

An interference of other steroids in overestimation of progesterone by this method may be anticipated to be in the order of 5 to 10%.

The use of this method is not recommended for assays of progesterone in male plasma. A concentration of testosterone over 1 ng/ml significantly interferes with the method, leading to an over-

TABLE 1

Recoveries of Steroids Added to Plasma

Steroid	Recovery % – S.D.
Progesterone	85.5 ± 3.4
17α-hydroxyprogesterone	14.0 ± 1.5
Pregn-4-ene-20α-ol-3-one	64.0 ± 2.0
Testosterone	28.5 ± 2.0
Estradiol-17β	5.1 ± 0.6
Corticosterone	Less than 0.25
Cortisol	Less than 0.25
11-deoxycortisol	Less than 0.25

estimation. The plasma levels of testosterone in healthy adult males are reported to be in a range from 3 to 11 ng/ml.

Progesterone values obtained by this method in healthy male subjects ranged from 0.6 to 1.4 ng/ml, with a mean of 0.89 ng/ml. The accepted level of plasma progesterone in males is approximately 0.3 ng/ml. This indicated that about 10% of the testosterone adds to the overestimate of progesterone.

Accuracy – Addition of crystalline progesterone to human plasma resulted in reasonable recoveries.

Progesterone added	Progesterone measured
ng	ng ± S.D.
0.50	0.53 ± 0.12
1.00	0.96 ± 0.14
2.00	1.97 ± 0.25
5.00	4.81 ± 0.55

Similar values were obtained when progesterone was recovered from water.

Precision – The coefficient of variation was the highest for the range of 0.1 to 0.5 ng/ml (25%). Assays of plasma progesterone in ranges of 0.5 to 2.0 ng/ml, 2 to 5 ng/ml, and 5 to 10 ng/ml showed the respective coefficients of variation in the order of 8.5%, 10.3%, and 14.5%.

3. Quantitation of Plasma Progesterone After Chromatography[3,7,8]

a. Principle

To blood plasma or serum a tracer quantity of progesterone-7α-^{3}H is added and the specimen is extracted with petroleum ether. The residue from the extract is purified by thin-layer chromatography on silica gel using an ether:benzene system. The area containing progesterone is extracted with diethyl ether which is evaporated and the extract is redissolved in ethanol. Aliquots from the ethanol solution are taken for determination of recovery and for competitive protein binding assay. The assay system uses human plasma for CBG and labeled corticosterone as an indicator. The free corticosterone is adsorbed by Florisil.

b. Reagents and Materials

1. Progesterone- 7α-^{3}H (specific activity 14.3 Ci/millimol or better). The purity should be verified using paper chromatography; ligroin: methanol:water system (5:4:1). For use in the procedure, prepare a dilute solution of 1,800 dpm/10 μl in ethanol.
2. Corticosterone-1,2-^{3}H (specific activity 48 Ci/millimol or better). The purity of the material should be verified by crystallization with a carrier or chromatography. Material with 96% purity can be used in the procedure. A stock solution of 1.0 μCi/100 μl is freshly prepared every second month.
3. Precoated thin-layer sheets, silica gel with fluorescent indicator Eastman Kodak® #6060, are washed prior to use twice with methanol by ascending chromatography.
4. Florisil 60–100 mesh, A. B. Kistner, Stockholm, is washed four times with distilled water, the fine particles are decanted. The washed Florisil is dried at 100°C.
5. Vortex mixer.
6. Micrometer syringes.
7. Automatic micropipettes – 200, 500, and 1,000 μl capacity.
8. Petroleum ether redistilled and the fraction distilling between 40 to 50°C is collected for use.
9. Liquid scintillation phosphor suitable for water-soluble radioactive material such as BBOT-toluene-Triton X-100®, Instagel (Packard) or Bray's solution.

c. Corticosterone Binding Globulin Solution (CBG)

For preparation of CBG solution plasma samples obtained from ovariectomized women treated with equine-conjugated estrogens are aliquoted in 0.3-ml portions and stored at −20°C.

In a 100-ml volumetric flask approximately 35 ng of corticosterone-1,2-^{3}H is evaporated to dryness and redissolved in 90 ml of distilled water; 0.2 ml of CBG plasma is added and the flask filled to volume with distilled water. The solution is thoroughly mixed by inversion, kept at room temperature overnight, and used the next day.

d. Procedure

Extraction of Plasma Specimens

1. To each of the marked glass round-bottom tubes 10 μl (18,000 dpm) of progesterone-^{3}H is delivered by micrometer syringe to the bottom of the tube and the solvent is evaporated under nitrogen in a water bath at 40°C.
2. To each tube add 1.0 ml plasma specimen and heat the tubes in a water bath at 40°C for 15 min.
3. Add 8.0 ml petroleum ether and mix the contents on a vortex mixer for 2 min.
4. Centrifuge the tubes for 10 min to separate the layers.
5. Transfer the petroleum ether phase using a disposable Pasteur pipette into a marked conical glass tube 11 x 1.5 cm.
6. Repeat the extraction procedure on the plasma sample with an additional 8.0 ml petroleum ether and vortex mixing for 1 min.
7. Combine the petroleum ether extracts and evaporate under stream of nitrogen by keeping the conical tubes in a constant temperature bath at 30°C.
8. Wash the sides of the tubes with 0.75 ml ethanol, evaporate the solvent, and repeat the washing with 0.2 ml ethanol.
9. The crude extract obtained may be kept at room temperature for up to 3 days until TLC purification.

e. TLC Purification

1. An activated TLC plate is divided into six lanes with three sample lanes 3.5 cm wide and three standard lanes approximately 3 cm wide.
2. The sample extracts are dried and applied on the chromatogram using four successive applications of 2.5, 7.5, 5.0, and 2.5 μl of ethanol.
3. To each of the standard lanes approximately 0.25 μg progesterone is applied.
4. Prior to chromatography, the visibility of the progesterone standard is verified under a shortwave ultraviolet light source.
5. The chromatogram is developed in ether: benzene system (2:1 v/v).
6. After drying, the progesterone areas are visualized under ultraviolet light and a horizontal line is drawn 1.1 cm above and 1.1 cm below the centers of the standard spots.
7. Each marked progesterone area from the sample lanes is cut out and divided into pieces approximately 0.7 x 1.2 cm which are placed in marked 10 x 2 cm round-bottom tubes.
8. Elution of the samples is achieved by the addition of 1.0 ml of distilled water followed by 3.0 ml of water-saturated ether. The contents are mixed for 1 min and the ether phase is transferred with a Pasteur pipette into a marked 7.5 x 1.2 cm glass tube. A second volume of 3.0 ml of ether is added to each tube, the extraction is repeated, the ether phases are combined and evaporated under nitrogen. The tubes are washed down with 0.75 ml of dry ether and the solvent evaporated.
9. Determination of procedural losses. To each tube add 1 ml of ethanol. Out of this, 0.2 ml are placed in a liquid scintillation counting vial for determination of recovery. The remainder of the sample is kept in a cold room until used for assay within 72 hr.

f. Protein Displacement Binding Assay

1. Preparation of the blank. Blank TLC sheets are chromatographed the same way as the sample sheets. Ten areas 2.2 x 3.5 cm are eluted and pooled, 80% of the eluate is distributed between ten standard curve tubes. The solvent is evaporated.
2. To the prepared standard tubes 0.0, 0.25, 0.50, 0.75, and 1.0 ng of progesterone are added using a micrometer syringe.
3. To each tube 1,800 dpm of progesterone-7α-^{3}H is added.
4. Evaporate the solvent.
5. Add to each tube 1.0 ml of CBG solution.
6. Heat the tubes at 40°C for 5 min.
7. Cool for 10 min in an ice bath.
8. Add 40 mg Florisil to each tube and mix the contents for 30 sec on the vortex mixer. Replace the tubes in the ice bath.
9. Withdraw 0.5 ml of the supernatant and transfer it into liquid scintillation tubes containing

TABLE 2

Steroid R_F Values on TLC

Steroid	$R_F \pm$ (S.D.)
Progesterone	0.46 ± 0.02
4,16-pregnadien-3,20-dione	0.42 ± 0.03
4-androsten-3,17-dione	0.38 ± 0.02
5-pregnen-3β-ol-20-one	0.35 ± 0.02
4-pregnen-20β-ol-3-one	0.31 ± 0.02
4-pregnen-17α-ol-3,20-dione	0.30 ± 0.02
4-pregnen-20α-ol-3-one	0.29 ± 0.02
4-androsten-17β-ol-3-one	0.28 ± 0.02
4-pregnen-6β-ol-3,20-dione	0.21 ± 0.02
4-pregnen-16α-ol-3,20-dione	0.10 ± 0.02
Cortisol	Less than 0.2
Corticosterone	Less than 0.2
Deoxycorticosterone	Less than 0.2

(From Reeves, B. D. et al., *Acta Endocrinol.*, 63, 225, 1970. With permission.)

scintillation phosphor.

10. Determine the radioactivity and convert into disintegrations per minute.

11. Construct the calibration curve and determine the quantity of progesterone taking in account the procedural losses. The formula for calculating the total amount of progesterone in each sample is

$$\frac{\text{ng from calibration curve} \times 1.25 \times 100}{\text{percentage recovery}}$$

g. Variables Affecting the Method

Extraction

By hand shaking the plasma, samples with petroleum ether yielded 88.0 ± 5.2% recovery. The variation was reduced by mixing the specimen on a vortex mixer; this revealed a recovery of 86.5 ± 1.8% (S.D.).

TLC

The silica gel is known to contribute toward interferences in protein displacement binding. The precoated TLC plates previously washed by ascending chromatography were found to be the most suitable. Problems with uneven runs were encountered when the chromatogram was developed by the "sandwich technique." This was minimized by clipping the TLC sheet to a glass plate and developing it in a Desaga chamber using a fresh ether-benzene mixture each time. In 148 experiments the mean $R_F \pm$ S.D. for progesterone was 0.46 ± 0.02.

In the elution step it was found that the use of ether-water partition minimized the method interfering factors and optimized the recovery of progesterone. The authors emphasized that for an exact compensation toward interfering factors the same batch of TLC plates should be used.

The R_f values for several steroids in the above TLC system are shown in Table 2.

The internal standard (progesterone-7α-^{3}H) was found to be necessary for improvement of accuracy. In 141 experiments the overall recovery was reported as 73.7 ± 7.6% (S.D.).

h. Precision

Using this method, the authors observed a standard deviation of replicate samples between 0.1 and 0.2 ng and for the water blank 0.06 ng.

i. Accuracy

Experiments with plasma samples revealed recoveries of 94 and 112%. Statistical analysis of variance indicated significant differences among assays at various concentrations as well as among assays run at different times.

Sensitivity of the method approached 0.2 ng.

j. Specificity

So far no natural steroid has been encountered which would remain associated with progesterone after purification. Using pooled male and female plasma samples that had been extracted with petroleum ether 5 to 6 times, it was observed that some variable and low quantities of "progesterone" were detected by this method. Prior to extraction, the male plasma revealed a progesterone concentration of 0.33 ± 0.14 ng/ml (1 S.D.) and the progesterone "free" male plasma 0.27 ± 0.10 ng/ml. The female follicular phase plasma contained progesterone 0.51 ± 0.17 ng/ml, whereas the progesterone "free" plasma showed values of 0.32 ± 0.20 ng/ml. This suggests that assays conducted at low levels of progesterone are of questionable value.

4. Quantitation of Plasma Progesterone – Progesterone Displacement System[2]

a. Principle

Blood plasma or serum is extracted with petroleum ether and the extract is assayed by protein displacement binding using a pregnant guinea pig plasma and a tritiated progesterone system. The free progesterone is adsorbed by dextran-coated charcoal.

b. Reagents and Materials

1. Progesterone-1,2-^{3}H (specific activity 33.5 Ci/millimol or better), prepurified by alumina column chromatography using benzene-ethanol gradient. The purity of the labeled progesterone is periodically verified by TLC silica gel chromatography, benzene:ethyl acetate (3:2) system.
2. Crystalline progesterone, twice recrystallized for use as a reference standard.
3. Petroleum ether, Merck; glass distilled and the fraction distilling between 40 and 60°C is collected. Distilled fresh each week.
4. Norite A (Prolabo).
5. Dextran T70 (Pharmacia).
6. Buffer solution, pH 7.4, 0.1 *M* phosphate buffer.
7. Progesterone binding solution. Plasma is collected from pregnant guinea pigs (40 to 60 days of gestation) and is aliquoted in 0.5-ml samples and kept frozen until use. Before use, the aliquot is thawed and incubated for 1 hr at 37°C with 10 vol of dextran-coated charcoal, followed by centrifugation. The supernatant plasma is further diluted 160-fold with buffer solution. Progesterone-^{3}H is added to the solution to yield a concentration of 22,000 dpm per ml. This is followed by incubation for 15 min at 37°C and for 30 min in an ice bath.
8. Dextran-coated charcoal suspension for removal of progesterone from guinea pig plasma; 0.17% dextran and 1.7% Norite-A® is prepared in phosphate buffer solution (0.05 *M*, pH 7.2). A fresh solution should be prepared every 2 weeks and stored in the refrigerator.
9. Dextran-coated charcoal suspension for use in protein displacement assay: 0.5% dextran and 0.5% Norite-A is suspended in phosphate buffer (0.05 *M*, pH 7.2). A fresh solution should be prepared every 2 weeks and stored in a refrigerator.

c. Procedure

Extraction

Extract 1 ml of the plasma specimen with 10 ml of petroleum ether by mixing the tube with a vortex mixer. The phases are separated by freezing the sample at −20°C. An aliquot of the petroleum ether phase is used for the protein displacement binding assay. In case the plama progesterone level is high, an aliquot of less than 5 ml of the petroleum ether should be used. The volume of the petroleum ether is then adjusted to 5 ml with fresh solvent to compensate for the nonspecific interference of the solvent in the assay system.

Protein Displacement Binding Assay

1. Prepare duplicate standard tubes containing 0.0 and 0.5 to 10 ng of unlabeled progesterone. Add fresh petroleum ether, 5 ml per tube. Evaporate solvent from standard tubes and from sample tubes under nitrogen.
2. Add 1 ml of progesterone binding solution to each tube.
3. Incubate for 25 min at 37°C followed by incubation for 30 min at 4°C.
4. Add to each tube 100 μl of charcoal suspension.
5. Mix the tubes over a Vortex mixer for 15 sec.
6. Place the tubes in an ice bath for 60 min without agitation.
7. Centrifuge at 0°C and 700 g for 5 min.
8. Withdraw 0.5 ml of supernatant for liquid scintillation counting.
9. Construct the standard curve and calculate the progesterone concentration in the specimens.

d. Comments

Extraction of progesterone varied with each batch of petroleum ether. The recovery varied between 76.2 and 95.5%.

Sensitivity was reported as 0.1 ng and the coefficient of variation for standards between 0.1 and 5 ng was smaller than 5%.

Specificity. In this system interference can be expected if plasma specimens contain 4-pregnen-20α-ol-3-one or testosterone. Experience has shown that 4-pregnen-20α-ol-3-one can introduce an overestimation of progesterone by about 16%.

Precision. Duplicate determinations in the same assay had a coefficient of variation of 8.3% and between assays of 18.5%.

Adsorption of unbound steroid. Dextran-coated charcoal showed a rapid uptake of tritiated progesterone during 45 to 60 min of contact; thereafter, the uptake was slow.

B. Radioimmunoassay for Progesterone

1. General Considerations

In recent years, considerable advances have been made in production of antisera for radioimmunoassays. Using 11-deoxycortisol-21-hemisuccinate, human serum albumin conjugate for immunization of adult ewes, Abraham et al.[9] obtained antiserum that reacted with progesterone (100%), 17α-hydroxyprogesterone (90%), 11-deoxycortisol (90%), and 11-deoxycorticosterone (35%). More specific antisera have been prepared by immunization of rabbits with progesterone-3-oxime-bovine serum albumin complex[10] and with progesterone-11-succinate bovine serum albumin conjugate.[11,12] Commercially available antisera prepared by these techniques are marketed by Endocrine Sciences, Tarzana, California. Experience with antisera in radioimmunoassays of serum progesterone has shown a sensitivity which is about 20 times greater than that of the competitive protein binding method. On the other hand, the range of the assay is considerably narrower.

Of all haptens for progesterone specific antibody production, so far the promising one appears to be 11α-hydroxyprogesterone-hemisuccinate. This seems to be due to preservation of progesterone structural features in both the ring A and the ring D of the hapten. It has been shown by Kutas and her co-workers[11] and Spieler and his co-workers[12] that antisera produced in their laboratories could be employed for assays of serum progesterone on crude ether extracts without further chromatographic separation.

In preparations of diluted antiserum for assay, borate buffer 0.05 *M*, pH 8 or sodium phosphate buffer 0.05 *M*, pH 7.2 has been used. For assay purposes, a dilution of antiserum that binds 50 to 60% of the labeled progesterone is expected to give satisfactory results.

Like competitive protein binding methods, the radioimmunoassay is subject to unexpected interferences. The cleanliness of the glassware and the purity of the reagents are of the utmost importance. It is advisable to use deionized glass distilled water for preparation of the reagents. Some workers prefer to use disposable glassware; some prefer to use disposable polystyrene tubes for the antigen-antibody reaction. For extraction of plasma, progesterone, ether or hexane has been employed.

The optimal incubation time and temperature of the sample with antiserum should be established by the analyst. The reaction may be complete by 2 hr at room temperature, 1 hr at 3 to 5°C, or overnight at 4°C.

Quantitation of progesterone in the samples may be done by estimating the bound radioactivity to the antibody or the unbound radioactivity and by comparing the results with standards. In the case where ammonium sulfate is employed for precipitation of progesterone-antibody complex, the supernatant that contains unbound progesterone is used for quantitation of radioactivity. In their assay system, Spieler et al.[12] separated the antibody-bound progesterone from unbound progesterone by dextran-coated charcoal suspension (0.25% Norite A and 0.05% dextran in sodium phosphate buffer 0.05 *M*, pH 7.2).

The dextran-coated charcoal adsorbs the unbound progesterone and after centrifugation, the supernatant contains the antibody-bound progesterone. For quantitation of antibody-bound radioactivity aliquots of the supernatant are used.

In setting up a radioimmunoassay, consideration should be given to all the relevant precautions and sources of error which are discussed under competitive protein binding assays.

2. Radioimmunoassay for Progesterone[10]

a. Principle

Plasma samples are mixed with a known quantity of progesterone-1,2-^{3}H and extracted with hexane. The extract is purified by microcolumn chromatography on alumina. The assay is performed with antiserum obtained from rabbits previously immunized with progesterone conjugated to bovine serum albumin. Separation of the free and bound progesterone in the system is achieved by ammonium sulfate precipitation. The ammonium sulfate precipitates the antibody-bound label and the supernatant contains the unbound label. Calculation of the final result is done from a standard graph, which is constructed by plotting the percent of bound progesterone against the quantity of progesterone in the standard tubes. A formula is provided for correction of procedural losses and for compensation of the chemical weight of tritiated progesterone indicator.

b. Reagents and Materials

1. The authors used antiserum prepared in their laboratory; however, specific progesterone antiserum of rabbit origin is obtainable commercially from Endocrine Sciences, Tarzana, Calif.

2. Progesterone-1,2-^{3}H (specific activity 29.6 Ci/millimol or better), New England Nuclear Corp., Boston, Mass. If purification of progesterone-1,2-^{3}H is necessary, this can be achieved by

paper chromatography on Whatman No. 1 chromatographic paper using hexane:methanol:water (100:70:30) system. The rechromatographed progesterone is eluted with absolute ethanol.

3. Progesterone-1,2-^{3}H stock solution prepared in ethanol (approximately 200,000 dpm/ml).

4. Progesterone reference standard (Sigma Chemicals Co.).

5. Progesterone reference standard solutions in methanol: 0.25 ng/ml, 0.50 ng/ml, 1.00 ng/ml, 2.00 ng/ml.

6. Methanol, reagent grade.

7. Absolute ethanol, reagent grade.

8. Hexane, reagent grade.

9. Dichloromethane, reagent grade.

10. Alumina Al_2O_3 for chromatography (Savory and Moore, Brockmann grade 2-3, mesh 100-150).

11. Elution solvents: (a) 0.28% absolute ethanol in hexane; (b) 0.45% absolute ethanol in hexane.

12. Preparation of glassware. All glassware has to be thoroughly washed, soaked in dilute HCl overnight, rinsed with distilled water, and dried. Before use, all glassware is rinsed with methanol:dichloromethane (1:1) and dried.

13. Preparation of alumina columns. Disposable glass pipettes are used to make 11-cm long and 0.5-cm internal diameter columns by shortening the constricted portion. A piece of glass wool is inserted and the column is packed with 3 cm of Al_2O_3. Prior to use, the columns are washed twice with ethanol, four times with methanol, and three times with methanol:dichloromethane (1:1). The flow rate of the columns should be 1.6 ml/5 min. Columns can be used repeatedly for several months. Verification of the column performance should be done periodically using progesterone-1,2-^{3}H.

14. Rabbit antiserum stock solution. Antiserum diluted 1:10 with 0.05 *M* borate buffer, pH 8 and stored at −10°C.

15. Antiserum working solution is prepared just before use by further dilution 1:450 with 0.05 *M* borate buffer containing 0.16% bovine serum albumin and 0.065% gamma globulin.

Note: The dilution of antiserum will depend on its titer which has to be established experimentally.

c. Procedure

Extraction and Purification

1. Plasma sample size for assay

Male	0.5 ml
Female, follicular phase	0.5 ml
Female, luteal phase	0.05 ml

2. In conical 5-ml tubes fitted with glass stoppers, pipette 20,000 dpm of progesterone-1,2-^{3}H and evaporate to dryness. An equal quantity is added to a counting vial for determination of radioactivity.

3. In the prepared tubes pipette the appropriate plasma volume.

4. Add 2 ml of hexane, stopper the tubes, and mix over a vortex mixer for 2½ min.

5. Centrifuge the tubes to separate layers, at 1,000 x g for 5 min.

6. Transfer the hexane phase directly to alumina columns and wash the column with 1.7 ml of 0.28% absolute ethanol in hexane.

7. For elution of progesterone add 3.4 ml of 0.45% absolute ethanol in hexane and collect the eluate.

8. Mix the eluate and transfer duplicate aliquots of 1 ml containing approximately 4,000 dpm into 2 ml conical glass tubes. An aliquot (0.5 ml) is transferred into a counting vial for determination of radioactivity.

9. Water for blanks is preextracted twice with 5 vol of dichloromethane before extraction with hexane.

Radioimmunoassay

1. Prepare standards in duplicate by pipetting into 2 ml conical glass tubes 0.2 ml vol of methanol containing 0, 25, 50, 100, and 200 pg of progesterone. To each standard tube add progesterone-1,2-^{3}H (4,000 dpm). For determination of the total radioactivity in the standard tube, the same quantity of progesterone-1,2-^{3}H is pipetted into a counting vial.

2. Evaporate the solvent from all tubes using a vacuum assembly.

3. To each tube add 0.25 ml of diluted antiserum and cover the tubes with parafilm.

4. Mix the contents over a vortex mixer and keep the tubes at room temperature for 2 hr.

5. To separate the free from the bound

progesterone, to each tube add 0.25 ml saturated $(NH_4)_2SO_4$ solution and mix immediately on a vortex. Keep the tubes for 10 min at room temperature and then centrifuge at 2,000 x g for 10 min.

6. Transfer 0.2 ml of the supernatant to counting vials without disturbing the precipitate.

7. To the counting vials containing the dried column eluates or dried standards 0.2 ml of half-saturated $(NH_4)_2SO_4$ is added.

8. To all the counting vials add 10 ml of liquid scintillation phosphor suitable for water-soluble radioactive material such as BBOT-toluene-Triton-X-100, Instagel (Packard) or Bray's solution.

d. Calculation

The standard curve is constructed by plotting the percent of bound progesterone against the quantity of unlabeled progesterone standards.

For correction of losses during the processing and for the compensation of the mass of progesterone-1,2-^{3}H in standards and in the samples the following equation is used.

$$\text{Picograms of progesterone in sample} = \left\{(\text{quantity from standard graph}) - \frac{A-S}{C}\right\} \times \frac{T}{A}$$

where

A = dpm in the aliquots of eluates used for analysis.

S = total dpm in the standards.

C = specific activity of progesterone-1,2-^{3}H expressed in dpm/pg.

T = total dpm added to original plasma samples prior to extraction.

The corrected quantity of progesterone in the sample is converted to picograms of progesterone per ml of plasma.

e. Comments

The sensitivity of the method appears to be in the order of 25 pg per sample, which may be adequate for estimation of progesterone in 0.5 ml plasma samples if the concentration of progesterone is above 50 pg/ml. The high sensitivity allows assays of plasma progesterone in male plasma and in plasma from women in the follicular phase.

Using commercial antisera, the sensitivity of the method may approximate 50 pg per sample and the range for an adequate assay may extend to 200 pg per sample. For assay purposes, commercial antiserum from rabbits immunized with progesterone-3-oxime-bovine serum albumin conjugate in dilution 1/5,000 produced acceptable standard curves. Antiserum obtained from rabbits after immunization with progesterone-11-succinate-bovine serum albumin conjugate and marketed by Endocrine Sciences shows considerably higher sensitivity and improved specificity. At present, all antisera for radioimmunoassays of progesterone show some degree of cross-reactivity with other steroids. The principal natural steroid metabolites that can interfere with the procedure are 5β-pregnan-3,20-dione and 5α-pregnan-3,20-dione. A considerably smaller cross-reaction is observed with 17α-hydroxyprogesterone, 11-deoxycorticosterone, and 20α-dihydroprogesterone.

In performing radioimmunoassays, it should be considered that the end result may be altered by factors such as nonsteroidal interfering substances, steroids with high affinity for the antibody, and by steroids with low affinity but high plasma concentration. The methodology in radioimmunoassays for progesterone is still in a stage of development. Recently prepared antisera using 11α-hydroxyprogesterone hemisuccinate have sufficient specificity and sensitivity for use in plasma progesterone assays without chromatographic purification of plasma extract.[11,12]

Commercial antisera for progesterone assays (Endocrine Sciences) are prepared from stock serum diluted with 0.05 *M* borate buffer, pH 8.0 and freeze-dried before shipping. Before use, the antiserum is reconstituted by the addition of water. The reconstituted serum appears to be quite stable if stored at −10°C.

C. Urinary Steroid Determinations

1. Introduction

Pregnanediol glucuronide monosodium salt was isolated by Venning and Browne in 1936 from butanol extract of late pregnancy urine. In later years it was shown that urinary pregnanediol determinations served as an index of production of progesterone in the organism. Up until the 1960's the most common procedure for quantitation of pregnanediol was a sulfuric acid color reaction. Because of the nonspecificity of the sulfuric acid reaction, interferences with various other urinary chromogens were frequently encountered. Numerous attempts have been made in the past to improve the reliability of urinary pregnanediol assays. The methodological approach of Klopper et al.,[14] using alumina column chromatography, proved to be the most widely accepted one. Their procedure was modified by a number of investigators. Employing benzene and benzene-ethanol mixtures a system was devised which separated the 17-oxosteroids and pregnanediol as the first components removed from the alumina column. By increasing the ethanol concentration in benzene, the more polar metabolites such as pregnanetriol and other trihydroxy or tetrahydroxy compounds were eluted. It was observed that the pregnanediol fraction after alumina column chromatography was usually relatively pure for the sulfuric acid color reaction; however, in some instances this fraction contained nonspecific substances which interfered with the color development. To obtain a further purification, the pregnanediol fraction was acetylated and rechromatographed on a silica gel or alumina column. Repeated chromatography of the extracts made the pregnanediol procedure cumbersome and time-consuming.

During the past decade with the refinement of gas-liquid chromatography, a number of GLC procedures have been described for quantitation of urinary pregnanediol, pregnanetriol, and other metabolites.[20] Technically, the GLC method is regarded as considerably superior to colorimetric procedures since it provides a means of separation of the metabolites as well as quantitation. The advantages of gas-liquid chromatography in the determination of progesterone metabolites have been reviewed by Van der Molen[16] as well as by Klopper.[15] For the determination of pregnanediol and other pregnane derivatives in urine from children, men, and nonpregnant women, a prepurification of the urine extract by alumina has been found advantageous in removing a number of interfering substances.[17-19] Steroid fractions thus obtained can be further resolved and quantitated by GLC as free steroids or by formation of derivatives such as trimethyl-silylethers or acetates. In cases of pregnancy, GLC of urinary extracts offers advantages in the simultaneous separation and quantitation of pregnanediol, pregnanolone, and pregnanetriol. Results of this procedure have been applied to a fetal health survey.[24-26]

For the diagnostic confirmation of congenital adrenal hyperplasia, the detection of pregnanetriolone and pregnanetetrol in urine is of importance. These abnormal steroids can be isolated qualitatively by rapid chromatographic screening and quantitated, if necessary, by gas-liquid chromatography.

2. Colorimetric Determination of Urinary Pregnanediol and Pregnanetriol[14,15,21,23]

a. Principle

The urinary glucosiduronic acid esters of steroids are hydrolyzed with β-glucuronidase. After hydrolysis, the steroids are extracted with dichloromethane and the solvent is evaporated. The residue obtained is chromatographed using an alumina column and separated into pregnanediol and pregnanetriol fractions. The pregnanediol fraction is acetylated and rechromatographed on a silica gel column. Eluted pregnanediol diacetate is reacted with sulfite-sulfuric acid reagent and the color intensity is determined spectrophotometrically. The pregnanetriol fraction obtained after alumina column chromatography is reacted with concentrated sulfuric acid and the color intensity is determined spectrophotometrically.

b. Apparatus and Glassware

1. Spectrophotometer with cuvettes.
2. Rotary flash evaporator.
3. Glass columns 15-cm tall and 1-cm inside diameter with sintered glass disc, and equipped with a reservoir.
4. Pyrex® glass wool.
5. Bunsen burner.
6. Soft glass tubing (8 mm).
7. Soft glass rod (3 mm).
8. Round-bottom flasks with standard taper necks.
9. Separatory funnels, graduated cylinders, pipettes, and Erlenmeyer flasks.
10. Glass culture tubes (15 x 125 mm).
11. Centrifuge tubes 10-ml capacity.

c. Reagents

1. Glacial acetic acid, reagent grade.
2. Acetate buffer pH 5.0, 1.0 *M*.
3. β-glucuronidase (Ketodase, Warner Chilcott Co.).
4. Dichloromethane, reagent grade, freshly distilled.
5. Sodium hydroxide, 0.1 *N*.
6. Alumina (Brockmann grade I, deactivat-

TABLE 3

Fractionation of Urine Extract by Column Chromatography

1. Preparation for chromatography

Glass column (1 cm dia.)	3 g Alumina (Brockmann I + 3% H_2O)
	10 ml benzene

2. Separation of the extract

Elution solvent	Eluate
(1) Dry residue + benzene 10 ml	Discard
(2) 0.8% ethanol in benzene 25 ml	Discard
(3) 3.0% ethanol in benzene 15 ml	Pregnanediol fraction (to be used for further purification)
(4) 20.0% ethanol in benzene 20 ml	Pregnanetriol fraction (to be used for colorimetry)

ed with 3 ml of distilled water per 100 g of alumina).

7. Benzene, reagent grade, glass distilled.

8. Ethanol, reagent grade, ketone free.

9. Ethyl acetate, reagent grade, glass distilled.

10. Acetic anhydride, reagent grade.

11. Pyridine, reagent grade.

12. Silica gel, Davison Chemical Co., grade 923, 100-200 mesh. Fill two small (about ½ pint) containers with silica gel and heat for 2 hr at 110°C with the lid off. Replace the lid and place immediately in a desiccator having about ½ in. of concentrated sulfuric acid in the bottom. Allow to cool. Always keep the silica gel in the desiccator when not in actual use. Each batch of silica gel should be standardized for chromatographic efficiency.

13. Sulfur dioxide-sulfuric acid reagent. Bubble SO_2 gas from a Matheson lecture bottle into 500 ml of concentrated sulfuric acid at the rate of 2 to 3 bubbles/sec for ½ hr under the fume hood. The reagent is stable in a stoppered container for at least 3 weeks at room temperature.

14. Steroid standard solutions.

Pregnanediol 100 μg/ml in ethanol.

Pregnanetriol 100 μg/ml in ethanol.

Store in a refrigerator.

d. Procedure

Hydrolysis and Extraction

From a freshly collected 24-hr urine specimen (without preservative) an aliquot is filtered through Whatman No. 2 filter paper. A 50-ml specimen of the filtrate is adjusted to pH 4.7 with glacial acetic acid and the pH is stabilized by the addition of 2.5 ml of 1 *M* acetate buffer (pH 5.0). Steroid glucuronides in the specimens are hydrolyzed enzymatically using 30,000 Fishman units of β-glucuronidase ("Ketodase," Warner Chilcott Co.). The specimens are incubated at 37°C for 24 hr. After incubation the hydrolyzed steroids are extracted three times with 30-ml portions of freshly distilled dichloromethane. The extracts are pooled and washed with 25 ml of 0.1 *N* sodium hydroxide followed by washings with 10-ml portions of distilled water until neutrality. This procedure removes phenolic compounds and urinary pigments. The dichloromethane is evaporated under reduced pressure in a rotating evaporator at a temperature below 50°C. With each set of specimens the quality control urine pool is carried through the procedure.

Column Chromatography Using Alumina

Glass columns, 15-cm tall and 1-cm internal diameter, are loosely packed with a small piece of glass wool. Using 3 g of previously standardized alumina (Brockmann grade 1 deactivated with 3 ml of distilled water per 100 g), a slurry is made in 10 ml of benzene, which is transferred to each of the columns and allowed to settle. The excess solvent is drained and a small piece of glass wool is placed on top of the settled alumina. The urine extracts and the standards are quantitatively transferred on the columns (see Table 3). The column chromatography is done as described below. Desired fractions are collected in 100-ml round-bottom flasks and the solvent is evaporated under reduced pressure. The dry residue is quantitatively transferred to centrifuge tubes using ethanol as solvent, which is subsequently evaporated to dryness under a stream of air by heating the tubes at 50°C. Pregnanediol and pregnanetriol reference standards in quantities of 20, 40, and 60 μg and the reagent blanks are carried through the chromatographic procedure to compensate for losses.

Standardization of Alumina

Prepare a column as described above. Add 200 μg of pregnanediol in 1 ml of benzene and allow to run until the liquid disappears into the alumina. Elute the column with 9 ml of benzene and discard the eluate. For second elution use 25 ml of 0.8% ethanol in benzene, collect the eluate, and evaporate the solvent. To the dry residue add 2 ml of concentrated sulfuric acid; the acid should not turn yellow. For the third elution, use 15 ml 3.0% ethanol in benzene and collect 5-ml aliquots, evaporate to dryness, and add 2 ml of concentrated sulfuric acid. The most intense yellow color produced by pregnanediol should appear in the first and second fraction.

If pregnanediol appears in the eluate with 0.8% ethanol in benzene, the alumina needs to be reactivated by heating at 120°C for 2 hr and restandardized.

It is more likely that the alumina will be found to be too active and the elution of pregnanediol will not be complete with 3.0% ethanol in benzene. In this case, add more water to deactivate alumina. The addition of 1 ml of water to 100 g of alumina should shift the pregnanediol elution by about 5 ml of 3.0% ethanol in benzene.

After the addition of water to the alumina, the sample should be shaken vigorously in a closed container until the alumina is completely homogeneous and no lumps are visible.

It is advisable to prepare a sufficiently large batch of alumina and to store the prepared alumina in a closed container to avoid changes in water content.

Once alumina is standardized for pregnanediol, verify the elution optimum for pregnanetriol.

Colorimetric Determination of Pregnanetriol

1. Add 3.0 ml concentrated sulfuric acid to the tubes containing the dry residues of reagent blank, pregnanetriol standards, and urinary pregnanetriol fractions.

2. The tubes are left to stand at room temperature for 2 hr.

3. The optical density of the yellow color is determined at 400, 435, and 470 nm.

Calculation

Using the Allen correction, determine the corrected optical density:

$$\text{O.D.}_{435} - \frac{(\text{O.D.}_{400} - \text{O.D.}_{470})}{2} = \text{O.D.}_{\text{corr.}}$$

Construct the calibration curve and determine the quantity of pregnanetriol in the specimens. Calculate the urinary pregnanetriol in mg/24-hr urine collection.

e. Acetylation and Column Chromatography of the Pregnanediol Fraction

To the tubes containing the dry residues of pregnanediol standards and fractions, add 0.2 ml acetic anhydride and 0.2 ml of pyridine. Stopper the tubes and swirl to dissolve the residue. Keep the tubes at room temperature overnight. Evaporate the reagents under a stream of air in a fume hood by keeping the tubes in a water bath at 50°C.

f. Silica Gel Column Chromatography

1. Soft glass tubing, 8 mm in diameter, is cut into pieces 30 in. long. A mark is made at 15 in. with a wax pencil. The glass is heated in a Bunsen burner flame until soft, then removed and carefully pulled out to a fine capillary. The capillary end is broken and sealed in the flame. The finished microcolumn should be about 14 in. long with an approximately 10-in. tapered capillary tip. Twelve columns can be conveniently run at one time.

2. The columns are set in place on the support and filled to within 3 in. of the top with benzene. The column is then tapped gently to dislodge all air from the tip. A small wisp of glass wool is gently pushed into the column with a 3-mm glass rod, and tamped into the bottom to form a plug about ¼ in. long. The column is then tapped again with a glass rod to dislodge any remaining air.

3. Weigh out 1.0 g of silica gel and transfer to a 15 x 125 mm culture tube for each column being run. Immediately add 5 ml of benzene to

prevent absorption of water by the silica gel. The mixture is slurried by sucking it up with a Pasteur pipette several times. This procedure also removes all air from the silica gel. Using the eyedropper, a small portion is added to the column. The silica gel will sift slowly through the benzene and settle on the glass wool. When the layer has built up to about 1/4 to 3/8 in., the tip of the column is broken off. Add slurried silica gel and benzene as fast as the benzene runs off the column. When all the slurry has been transferred, 2 ml of benzene is added to the tube, and any remaining silica gel is transferred to the column. The column is then tapped gently to settle the silica gel. After the benzene has run off, the walls of the column are washed down with another 1.0 ml of benzene.

4. While the benzene is dripping out of the column, three tests tubes (15 x 125 mm) per column are prepared as follows:

a. Into the first, place 5 ml of benzene.

b. Into the second, place 2 ml of 5% ethyl acetate in benzene.

c. Into the third, place 5 ml of 5% ethyl acetate in benzene.

5. When the column has stopped dripping, quantitatively transfer the residue from the pregnanediol fraction to the column, using the 5 ml of benzene in the first tube. Discard the benzene when it has run through the column. The second fraction (b) is also discarded.

6. The third fraction (c) containing pregnanediol diacetate is collected in a test tube and evaporated to dryness.

7. After use all columns are discarded.

g. Color Development and Quantitation

1. Add 4.0 ml of SO_2-H_2SO_4 reagent to each sample tube, the standard tube, and to clean tube (blank). Roll the acid around the side (inside) wall of the tube to completely dissolve all residue.

2. Keep the tubes at room temperature overnight. Transfer 3 ml of the solution to cuvettes and read the optical densities in the spectrophotometer at 395, 430, and 465 nm using the blank tube to zero the instrument at each setting. The color developed is stable for several hours.

3. Using the Allen correction determine the corrected optical density.

$$O.D._{430} - \frac{(O.D._{395} - O.D._{465})}{2} = O.D._{corr.}$$

Construct the calibration curve and determine the quantity of pregnanediol in the specimen. Calculate the urinary pregnanediol in mg/24-hr urine collection.

h. Normal Values

Pregnanediol

Nonpregnant women	mg/24 hr
Proliferative phase	0.5–1.5
Luteal phase	2–5
Postmenopausal	0.2–1.0
Pregnant women	mg/24 hr
Weeks of gestation	
10–12	5–15
12–18	5–25
18–24	13–33
24–28	20–42
28–32	27–47

Pregnanetriol normal range is up to 2 mg/24 hr. In cases of congenital adrenal hyperplasia, pregnanetriol may be as high as 15 mg/24 hr.

i. Comments

Occasionally the urine may contain chromogenic substances that are carried through the chromatographic procedures and cause a false elevation of the color in sulfuric acid.

The nature of the interfering chromogenic substances is not known. In some cases, these chromogens may be metabolites of drugs. If an interference is observed, the urine should be collected from the patient several days after discontinuation of the medication.

The analyst should be aware that the aliquots of urine or the aliquot of the chromatographically separated steroid fraction should be chosen so that the quantity of steroid in the sample falls in the range of the standard curve. This applies particularly for urine specimens from pregnant subjects.

Hot acid hydrolysis of urinary steroid conjugates causes destruction of the compounds and losses of pregnanetriol may be as high as 50%.

3. *Determination of Urinary Pregnanediol, Pregnanetriol, and Pregnanetriolone by Gas-liquid Chromatography*[17-19,35]

a. Principle

The urinary glucosiduronic acid esters of steroids are hydrolyzed with β-glucuronidase. After hydrolysis, the steroids are extracted with dichloromethane and the solvent is evaporated. Residue obtained is fractionated using an alumina column into pregnanediol and pregnanetriol-pregnanetriolone fractions. These fractions are further separated and quantitated by gas-liquid chromatography.

b. Apparatus

1. Constant temperature oven or water bath at 37°C.
2. Glass columns with sintered glass discs, 1 cm I.D., 15 cm tall equipped with a reservoir and a stopcock.
3. F and M gas chromatograph Model 402.
   Hydrogen flame detector.
   Column 4 foot "U" shaped glass, packed with Diatoport-S containing 3.8% SE-30.
   Hamilton syringes 5 or 10 μl capacity.

c. Reagents

1. Glacial acetic acid, reagent grade.
2. Acetate buffer, pH 5.0; 1.0 *M.*
3. β-glucuronidase (Ketodase, Warner Chilcott Co.).
   Dichloromethane, reagent grade, freshly distilled.
4. Sodium hydroxide, 0.1 *N.*
5. Alumina (Brockmann grade I, deactivated with 3 ml of distilled water per 100 g of alumina).
6. Benzene, reagent grade, glass distilled.
7. Ethanol, reagent grade, ketone-free.

d. Gases

1. Helium, moisture-free.
2. Hydrogen, prepurified, moisture-free.
3. Compressed air, prepurified, moisture-free.
4. Nitrogen, prepurified, moisture-free.

e. Standards

1. Internal standard: 5α Cholestane (Aldrich Co.) ethanol solution 100 μg/ml.
2. GLC standards: 5β-pregnan-3α,20α-diol (Pregnanediol-P_2) ethanol solution 100 μg/ml.
   5β-pregnan-3α,17α,20α-triol (Pregnanetriol-P_3) ethanol solution 100 μg/ml.
   5β-pregnan-3α,17α,20α-triol-11-one (Pregnanetriolone) ethanol solution 100 μg/ml.

f. Quality Control Materials

Example: Human urine pool containing 2.5 mg pregnanediol per liter and 1.34 mg pregnanetriol per liter, aliquoted in 100-ml portions and stored frozen at −20°C.

g. Procedure

Hydrolysis and Extraction

From a freshly collected 24-hr urine specimen (without preservative) an aliquot is filtered through Whatman No. 2 filter paper. Of the filtrate, a 50-ml specimen is adjusted to pH 4.7 with glacial acetic acid and the pH is stabilized by the addition of 2.5 ml of 1 *M* acetate buffer (pH 5.0). Steroid glucuronides in the specimens are hydrolyzed enzymatically using 30,000 Fishman units of β-glucuronidase (Ketodase®, Warner Chilcott Co.). The specimens are incubated at 37°C for 24 hr. After incubation, the hydrolyzed steroids are extracted three times with 30-ml portions of freshly distilled dichloromethane. The extracts are pooled and washed with 25 ml of 0.1 *N* sodium hydroxide followed by washings with 10-ml portions of distilled water until neutrality. This procedure removes phenolic material and urinary pigments. The dichloromethane is evaporated under reduced pressure in a rotating evaporator at a temperature below 50°C. With each set of specimens the quality control urine pool is carried through the procedure.

Column Chromatography

Glass columns, 15 cm tall and 1 cm internal diameter, are loosely packed with a small piece of glass wool. Using 3 g of alumina (Brockmann grade I deactivated with 3 ml of distilled water per 100 g), a slurry is made in 10 ml of benzene, which is transferred to each of the columns and allowed to settle. The excess solvent is drained and a small piece of glass wool is placed on top of the settled alumina. The urine extracts and the standards are quantitatively transferred on the columns. The column chromatography is done as shown in Table 4. Desired fractions are collected in 100-ml round-

TABLE 4

Fractionation of Urine Extract Column Chromatography

1. Preparation for chromatography	
GLASS COLUMN (1 cm dia.)	3 g Alumina (Brockmann I – 3% H_2O)
	10 ml Benzene
2. Separation of the extract	
Elution solvent	Eluate
(1) Dry residue – benzene 10 ml	Discard
(2) 0.8% ethanol in benzene 25 ml	Discard
(3) 3.0% ethanol in benzene 15 ml	Pregnanediol fraction
(4) 20.0% ethanol in benzene 40 ml	Pregnanetriol, pregnanetriolone fraction

bottom flasks and the solvent is evaporated under reduced pressure. The dry residue is quantitatively transferred to centrifuge tubes using ethanol as solvent. Pregnanediol, pregnanetriol, and pregnanetriolone reference standards in quantities of 20, 40, and 60 μg are carried through the chromatographic procedure to compensate for losses. To each specimen and standard tube for pregnanediol assay add 10 μg of 5α cholestane as internal standard. To tubes containing pregnanetriol and pregnanetriolone add 10 μg of pregnanediol as internal GLC standard. Evaporate the solvent under a stream of nitrogen by heating the tubes at 50°C.

Gas-liquid Chromatography

The separation and quantitation of pregnanediol, pregnanetriol, and pregnanetriolone are achieved by means of an "F and M" gas chromatograph Model 402 equipped with a hydrogen flame ionization detector. Standards and urine extracts are dissolved in 50 μl of ethanol containing the relevant internal standard, and by means of a Hamilton syringe 1 to 2 μl portions are injected on the gas chromatography column. The separation of the steroid metabolites is done on a 4-ft "U" shaped glass column packed with Diatoport-S containing 3.8% SE-30. The column is maintained at 235 ± 5°C and helium is employed as a carrier gas. As an internal standard for the separation of pregnanediol, cholestane is used. For pregnanetriol and pregnanetriolone, separation pure pregnanediol serves as an internal standard.

Relative retention times

5α cholestane	1.00 (internal standard)
Pregnanediol	0.64
Pregnanediol	1.00 (internal standard)
Pregnanetriol	1.57
Pregnanetriolone	2.07

Quantitation

The peak heights of the GLC standards and of the unknown extracts are measured in millimeters. A graph is constructed on a linear graph paper by plotting standard peak height vs. the quantity of standard. The quantity of steroid in the unknown sample is estimated from the standard graph and it is further expressed as mg/24-hr urine collection.

h. Precautions

To avoid errors and losses, the alumina used for column chromatography should be prepared in reasonable quantity and standardized prior to use. It should also be checked periodically for deterioration in quality. The standard maintenance procedures should be applied for the gas chromatograph and the Hamilton syringes.

i. Standardization of Alumina

Prepare a column as described. Add 200 μg of pregnanediol in 1 ml of benzene and allow to run until the liquid disappears into the alumina. Elute the column with 9 ml of benzene and discard the eluate. For the second elution, use 25 ml of 0.8% ethanol in benzene, collect the eluate, and evaporate the solvent. To the dry residue, add 2 ml of concentrated sulfuric acid; the acid should not turn yellow. For the third elution, use 15 ml 3.0% ethanol in benzene and collect 5-ml aliquots, evaporate to dryness, and add 2 ml of concentrated sulfuric acid. The most intense yellow color produced by pregnanediol should appear in the first and second fraction.

If pregnanediol appears in the eluate with 0.8% ethanol in benzene, the alumina needs to be reactivated by heating at 120°C for 2 hr and restandardized.

It is more likely that the alumina will be found to be too active and the elution of pregnanediol will not be complete with 3.0% ethanol in benzene. In this case add more water to deactivate alumina. The addition of 1 ml of water to 100 g of alumina should shift the pregnanediol elution by about 5 ml of 3.0% ethanol in benzene.

After addition of water to alumina, the sample should be vigorously shaken in a closed container until the alumina is completely homogeneous and no lumps are visible.

It is advisable to prepare a sufficiently large batch of alumina and to store the prepared alumina in a closed container to avoid changes in water content.

Once alumina is standardized for pregnanediol, verify the elution optimum for pregnanetriol and pregnanetriolone.

j. Comments

This procedure has been applied routinely in a clinical chemistry service laboratory for a number of years. The quality control urine pool as an example revealed pregnanediol with a mean of 2.5 mg/l ± 0.4 mg/l (2 S.D.) and for pregnanetriol a mean of 1.34 mg/l ± 0.4 mg/l (2 S.D.).

In clinically and biochemically normal women during the reproductive age the urinary pregnanediol excretion ranged as follows:

Follicular phase	Luteal phase
0.2–4.1 mg/24 hr	0.8–7.0 mg/24 hr

Women taking combined or sequential oral contraceptives excreted urinary pregnanediol in a range from 0.1 to 0.5 mg/24 hr. The urinary pregnanetriol excretion in women during reproductive age ranged from 0.1 to 2.0 mg/24 hr, whereas in women on oral contraceptives the excretion ranged from 0.1 to 1.0 mg/24 hr.

Pregnanetriolone was detected in urine only from patients with congenital adrenal hyperplasia due to steroid 21-hydroxylase deficiency.

4. Simultaneous Determination of Urinary Pregnanolone, Pregnanediol, and Pregnanetriol by Gas-liquid Chromatography[24-26]

a. Principle

The urinary glucosiduronic acid esters of progesterone and 17α-hydroxyprogesterone metabolites are hydrolyzed with β-glucuronidase. The hydrolyzed steroids as well as any "free" steroids present before hydrolysis are then extracted with chloroform. Interfering materials such as estrogens, bile acids, and urinary pigments are removed by washing with alkali and water. The chloroform extract is subsequently acetylated and 5α-cholestane is added as an internal standard. After dissolving the mixture in CS_2, an aliquot is injected into a gas chromatograph for qualitative and quantitative analysis, the latter of which is performed using the peak-height ratio technique.

b. Special Apparatus

1. Size 2 International Centrifuge with a head (No. 240) suitable for the 100-ml centrifuge tubes (Kontes Glass Co., Cat. No. K-41105).
2. Mechanical shaker, such as an Eberbach Preciprocating Shaker (Burrell Corp., Pittsburgh, Pa.), is useful for the extractions.
3. Perkin-Elmer® Model 801 Gas Chromatograph equipped with a hydrogen flame ionization detector and 6 or 12 ft by 0.075 in. I.D. glass coil columns with built-in sample injection ports. The columns are prepared and packed with 1.5% UC-98 on 100-120 mesh Gas-Chrom-Q. The support is coated with the evaporation technique.

c. Reagents

1. Acetic acid, Glacial (J. T. Baker Laboratory Chemicals, Cat. No. 9507).
2. Acetic anhydride, A. R. (Mallinckrodt Chem. Co., Cat. No. 4532).
3. Anhydrous sodium sulfate, A. R. (Mallinckrodt Chem. Co., Cat. No. 8024).
4. Acetyl chloride, A. R. (Mallinckrodt Chem. Co., Cat. No. 2456).

5. Boric Acid, Tablets (USP).

6. Benzene, A.R. (J. T. Baker Laboratory Chemicals, Cat. No. 9154), distilled once and stored over sodium sulfate.

7. Carbon disulfide, A.R. (Mallinckrodt Chem. Co., Cat. No. 4351).

8. Chloroform (J. T. Baker Laboratory Chemicals, Cat. No. 9180), distilled once over potassium carbonate and stored in brown glass bottles.

9. Gases

a. Air-zero Gas (J. T. Baker Specialty Gases, Phillipsburg, N. J.).

b. Hydrogen, prepurified (J. T. Baker Specialty Gases, Phillipsburg, N. J.).

c. Nitrogen, prepurified (J. T. Baker Specialty Gases, Phillipsburg, N. J.).

d. Nitrogen, high purity dry (Linde), for evaporation of extracts.

10. Gas-Chrom-Q; 100-120 mesh (Applied Science Laboratories, State College, Pa.).

11. Ketodase, Warner-Chilcott brand of β-glucuronidase. Filter through Whatman No. 1 filter paper and store in refrigerator.

12. pH paper: range 3.0 to 5.5

13. Potassium carbonate, anhydrous granular (J. T. Baker Laboratory Chemicals, Cat. No. 3012).

14. Sodium acetate buffer, 1 *M* pH 4.5: 58.0 g sodium acetate (J. T. Baker Laboratory Chemicals, Cat. No. 3460); 32.5 g glacial acetic acid, q.s. add 1,000 ml with distilled water.

15. Sodium hydroxide (J. T. Baker Laboratory Chemicals, Cat. No. 3722).

16. Sodium hydroxide, 0.1 *N*.

Sodium hydroxide, 10% w/v.

17. Sulfuric acid (Fisher Scientific Co., Cat. No. A-300).

18. Sulfuric acid, 5 *N*.

19. UC-W98 (Applied Science Laboratories, State College, Pa.).

d. Standards

Internal Standard

5α-Cholestane (Steraloids, Inc., Cat. No. C-700). A solution containing 20 μg/ml in absolute alcohol.

GLC Standard

1. *Stock GLC Std.*

3α-hydroxy-5β-pregnan-20-one (Pregnanolone – P_1O) Mann Research Lab., Inc., Cat. No. 522.

5β-pregnan-3α,20α-diol (Pregnanediol – P_2) Mann Research Lab., Inc., Cat. No. 3575.

5β-pregnan-3α,17α-triol (Pregnanetriol – P_3) Steraloids, Inc., Cat. No. 2493.

Weigh out separately pregnanolone, pregnanediol and pregnanetriol, and acetylate, as will be described subsequently. The acetylated dry residue is dissolved in absolute ethanol to a concentration of 0.8 mg/ml. Prepare a 5α-cholestane solution of 0.8 mg/ml in absolute ethanol. Store these stock solutions in refrigerator at 4 to 5°C.

2. *Working GLC Standards*

In a Teflon®-lined screw cap glass tube prepare a mixture of all four steroid standards. From each stock GLC standard (2.4 mg) 3 ml are transferred to the tube and the solvent is evaporated to dryness under a stream of nitrogen in a water bath at 45 to 50°C. The dry residue is dissolved in exactly 6 ml of CS_2 and a Teflon-lined cap is used to stopper the tube tightly. The concentration of each compound is 0.4 μg/μl. This solution is stable for as long as 4 months at room temperature, but must be protected from light by wrapping the screw cap tube with black photographic tape. Decomposition or hydrolysis of the acetates will be noted when more than four peaks appear in the gas chromatogram of this standard.

e. Procedure

A 24-hr urine specimen is collected with boric acid tablets as preservative. In general, 5-ml aliquots are taken and diluted to 20 ml with distilled water if the urine sample is from a second or third trimester pregnancy. Otherwise, a 10- or 20-ml aliquot is used, depending on whether the urine sample is from a first trimester pregnancy, nonpregnant women, or men.

Hydrolysis – Enzymatic hydrolysis is carried out with β-glucuronidase. Two procedures can be used, an 18- to 24-hr hydrolysis (Method 1) or a 1-hr hydrolysis (Method 2).

Method 1 – The urine aliquot is pipetted into a 100-ml centrifuge tube and diluted to 20 ml with distilled water if the aliquot is less than 20 ml. The pH is adjusted to approximately 4.5 with 5 *N* H_2SO_4 or 10% NaOH. Then 2 ml of 1 *M* acetate buffer followed by 6,000 units (1.2 ml) of Ketodase is added. The centrifuge tube is stoppered and, after mixing, placed in an incubator at 37.5 to 38°C for 18 to 24 hr.

Method 2 – Only 5- or 10-ml urine aliquots are used in this procedure. After placing the urine aliquot into the 100-ml centrifuge tube, Ketodase, in a concentration of 4,000 units/ml of urine, is added. In this way, 4 ml of Ketodase is used for 5 ml of urine and 8 ml of Ketodase is used for 10 ml of urine. The solution is then diluted to 20 ml with distilled water. The pH is adjusted to 4.5 as in Method 1, and 2 ml acetate buffer is added. The centrifuge tube is stoppered and incubated for 60 min at 60°C. After the incubation time is completed, the tube is removed from the incubator and immediately cooled under running tap water for about 4 min.

Extraction – First, 20 ml of chloroform is added to the hydrolyzed urine. The centrifuge tube is tightly stoppered and then shaken gently for 10 min. After shaking, the phases are separated by centrifuging for 5 min at 1,500 to 2,000 rpm. The chloroform is then transferred to a 50-ml centrifuge tube (Kimble Products, Cat. No. 45190) by means of a 9-in. disposable pipette with a rubber bulb. The aqueous phase is extracted again as above. After separation of the phases, the aqueous layer is removed and discarded by suction with a glass capillary siphon connected to vacuum. The chloroform extract in the 50-ml centrifuge tube is transferred to the 100-ml centrifuge tube. The 50-ml centrifuge tube is rinsed with 15 ml 0.1 *N* NaOH. The NaOH is then added to the combined chloroform extract. The 50-ml tube is rinsed twice with 3 to 4 ml of chloroform which is also added to the chloroform extract. The 100-ml centrifuge tube containing the chloroform-alkali mixture is tightly stoppered and shaken vigorously some 12 times, then centrifuged for 5 min at 1,500 to 2,000 rpm. The aqueous phase is removed as above. The washing is repeated once with 15 ml of distilled water. After centrifuging and discarding the aqueous phase, approximately 7 g of granular anhydrous sodium sulfate is added to the washed chloroform extract. A convenient dispenser for this amount (7 g) of sodium sulfate may be fashioned by fusing a 200 x 6 mm glass rod to the lip of a 10-ml Pyrex® beaker. The tube is stoppered and agitated for 1 to 2 min. The chloroform extract is filtered into a 125-ml Erlenmeyer flask through a plug of Pyrex glass wool (Corning® Glass Works, Cat. No. 3950) at the bottom of a 60-mm glass funnel. The centrifuge tube, stopper, sodium sulfate, and the funnel are rinsed at least three times with chloroform from a glass wash bottle.

The chloroform extract is then evaporated to dryness under a stream of nitrogen in a 45 to 50°C water bath. The residue is quantitatively transferred with chloroform, using a disposable Pasteur pipette, into a 13-ml glass-stoppered conical bottom centrifuge tube containing exactly 1 ml (20 μg) of the 5α-cholestane internal standard. The chloroform extract containing the internal standard is evaporated to dryness in the water bath at 45 to 50°C under a stream of nitrogen or air. The walls of the tube are rinsed three times with small volumes of chloroform, evaporating to dryness each time.

Acetylation – The dry extract can be acetylated using 1 ml of acetic anhydride at 100°C for 1 hr, or acetyl chloride/benzene (v/v) at 60°C for 15 min. When the latter is used, 0.5 ml of benzene is added to dissolve the dry extract, rinsing down the walls of the tube, followed by 0.5 ml of acetyl chloride. In both procedures, the tube is tightly stoppered to prohibit any moisture from entering. After the reaction is completed by either method, the excess reagents are evaporated under a stream of nitrogen in the water bath at 45 to 50°C until dry. The walls of the tube are rinsed down once with 0.5 ml of chloroform and then dried as above. The dry acetylated residue is then dissolved with 0.2 ml of CS_2 for GLC analysis.

GLC analysis – The general operating parameters for the Perkin-Elmer Model 801 Gas Chromatograph are as follows: Column temperature, 250°C (6-ft column) or 270°C (12-ft column). Nitrogen carrier gas flow rate is adjusted to obtain a retention time of about 4 min for 5α-cholestane. Air and hydrogen pressures as well as flow rates for the flame ionization detector must be optimized prior to analysis, as sensitivity is related to these variables. To optimize these parameters and establish peak-height ratios for quantitation, the GLC standard containing equal amounts of 5α-cholestane, pregnanolone acetate, pregnanediol diacetate, and pregnanetriol diacetate is injected into the gas chromatograph using a 10-μl Hamilton syringe. Once the operating conditions are optimized and the standard peak-height ratio established, the unknown extracts are analyzed. In general, 1- to 5-μl volumes of the unknown extracts are injected. In order to monitor the overall chromatographic system, it is recommended to repeat the injection of the GLC standard in the middle and after completing the analysis of a group of extracts.

Quantitation – The peak-heights of the 5α-cholestane and of the three acetates of the GLC standard and the unknown extracts are measured in mm and applied to the following equation:

$$R_{std} \times R_{unk} \times \frac{IS}{A} \times TV \times K = \text{mg/24 hr of steroid}$$

(free alcohol) in which

$$R_{std} = \frac{\text{Peak height of } 5\alpha\text{-cholestane in GLC standard}}{\text{Peak height of steroid acetate in GLC standard}}$$

$$R_{unk} = \frac{\text{Peak height of steroid acetate in unknown sample}}{\text{Peak height of } 5\alpha\text{-cholestane in unknown sample}}$$

IS = Mass (in μg) of the internal standard (5α-cholestane) added to the unknown sample

A = Urine aliquot used (in ml)

TV = Total urine volume (in liters)

$$K = \frac{\text{Molecular weight of free steroid}}{\text{Molecular weight of steroid acetate}}$$

which is

0.883 for pregnanolone
0.792 for pregnanediol
0.800 for pregnanetriol

f. Comments

The described method was developed to eliminate many of the usual sources of error encountered in similar methodologies. The utilization of a 100-ml centrifuge tube in place of a separatory funnel affords a better than 95% extraction efficiency.

Acetate derivatives were selected because of the simplicity of the acetylation reaction, the stability of the derivatives, and the enhanced resolution, therefore, specificity, of the compounds when subjected to GLC analysis. The acetylation also eliminates a considerable amount of irreversible adsorption on the column. However, the tertiary hydroxyl group of pregnanetriol at C-17 does not acetylate under the acetylation conditions employed. For this reason, it may adsorb when the column has not been correctly prepared or when the column has reached the end of its useful life. An indication of this is given when the 5α-cholestane/pregnanetriol diacetate ratio of the GLC standard is greater than 4.

The utilization of the internal standard technique avoids the tedious and time-consuming chore of analyzing and plotting calibration curves, as well as the errors in injection accuracy and concentration changes due to evaporation of the solvent used to dissolve the extracts.

The use of glass wash bottles for the chloroform and glass wool for the filtration eliminated serious contamination encountered when polyethylene and filter paper were tried. It was found that chloroform extracted from polyethylene a substance that has an identical retention time as that of the internal standard on the column used. A similar contamination was also observed in various types of the commonly used filter paper. It is advisable to frequently monitor the method for the possible presence of any contamination from the material or reagents used.

This GLC procedure offers reliability and rapidity. When using the 1-hr hydrolysis method, four specimens can be analyzed in a period of 3 hr.

Recoveries of steroids added to male urine are found to be better than 95%. In serial dilution experiments using urine specimens which were carried through the procedure, the found values agreed with those which were calculated. The mean differences between duplicate determinations were better than 5%.

Comparison of enzymatic hydrolysis with hot mineral acid hydrolysis (15% HCl (v/v) for 10 min with simultaneous toluene extraction) revealed a destruction of steroids. The loss in steroid after hot acid hydrolysis for pregnanediol was 4 to 8%, pregnanolone 24 to 27%, and for pregnanetriol 56%.

In order to obtain accurate results with pregnancy urine specimens, it is advisable to use enzymatic hydrolysis. Interferences in GLC separation of pregnanolone acetate may be expected in the presence of 11β-hydroxyandrosterone and 11β-hydroxyetiocholanolone. These metabolites usually constitute not more than 6% of the total 17-oxosteroids. In assays of pregnancy urines the interference is regarded as minimal. Interference by the isomer of pregnanediol, 5β-pregnan-3α,20β-diol, is unlikely since it is rarely found in pregnancy urine. The 5α-isomers may be present in small quantities.

g. Range of Values

In the clinically and biochemically normal adult individual the range of values in mg/24 hr is as follows:

	Males	Females
Pregnanolone	–	0.0–0.9
Pregnanediol	0.2–0.8	0.2–7.0
Pregnanetriol	0.3–1.8	0.3–1.9

In women during the follicular phase, excretion of pregnanolone is low and frequently not detectable. Measurable quantities of pregnanolone appear during the luteal phase. In the follicular phase the excretion of pregnanetriol is usually less than 0.8 mg/24 hr.

During normal pregnancy the excretion of these metabolites varies with the age of gestation (see Table 5).

During normal pregnancy, the pattern of pregnanolone excretion follows closely that of pregnanediol. The highest excretion of these two steroids is frequently observed during the 36th and 37th weeks of gestation. Urinary pregnanetriol rises in advancing pregnancy and tends to parallel that of the total urinary estrogen excretion. Simultaneous assays of urinary pregnanediol, pregnanolone, and pregnanetriol in conjunction with total urinary estrogens are found to be useful in fetal health surveys.

5. Qualitative Separation of Urinary Pregnanetriol, Pregnanetriolone, and Pregnanetetrol[17-19,27,28]

a. Principle

The urinary steroids are enzymatically hydrolyzed with β-glucuronidase. After hydrolysis the steroids are extracted with dichloromethane and the solvent is evaporated. The residue obtained is fractionated using an alumina column. The pregnanetriol fraction is further separated by thin-layer chromatography on silica gel G. Pregnanetriol, pregnanetriolone, and pregnanetetrol are detected on the chromatograms with phosphoric acid reagent. The characteristic blue-violet fluorescence of pregnanetriolone under UV light can be used for semiquantitative estimation of this steroid.

TABLE 5

Approximate Ranges of Steroid Excretion During Normal Pregnancy

Age of gestation weeks	Pregnanediol mg/24 hr	Pregnanolone mg/24 hr	Pregnanetriol mg/24 hr
10–20	5–20	1–3	1–2.5
20–30	20–65	5–17	Up to 6.5
35	30–80	7–27	Up to 11.0
37	35–95	8–32	Up to 13.0
40	20–55	6–20	Up to 12.0

b. Apparatus

1. Thin-layer chromatography equipment for use with glass plates.
2. Thin-layer glass plates coated with silica gel G.
3. Ultraviolet light source (365 mμ).
4. Rotating flash evaporator.
5. Water bath, thermostatically controlled.
6. Glass columns, 1 cm inside diameter, 15 cm tall with sintered glass disc, and equipped with a reservoir.
7. Heating oven, thermostatically controlled.

c. Reagents

1. Glacial acetic acid, reagent grade.
2. Acetate buffer pH 5.0, 1.0 *M*.
3. β-Glucuronidase (Ketodase, Warner Chilcott Co.).
4. Sodium hydroxide, 0.1 *N*.
5. Alumina (Brockmann grade I, deactivated with 3 ml of distilled water per 100 g of alumina).
6. Benzene, reagent grade, glass distilled.

7. Ethanol, reagent grade, ketone free.

8. Phosphoric acid 85%, reagent grade.

9. Phosphoric acid reagent. To 20 ml of phosphoric acid (85%) add 60 cm^2 of Whatman No. 1 filter paper cut into small pieces. The mixture is slowly heated until the paper is dissolved and the acid becomes yellow. After cooling, add 4 ml of distilled water. The reagent is stable for 1 week at room temperature.

10. Steroid standards (Ikapharm, Ramat Gan, Israel). Pregnanetriol, pregnanetriolone, and pregnanetetrol solutions in ethanol (100 μg/ml).

d. Procedure

Hydrolysis and Extraction

From a freshly collected 24-hr urine specimen (without preservative) an aliquot is filtered through Whatman No. 2 filter paper. Of the filtrate, a 50-ml specimen is adjusted to pH 4.7 with glacial acetic acid and the pH is stabilized by the addition of 2.5 ml of 1 *M* acetate buffer (pH 5.0). Steroid glucuronides in the specimens are hydrolyzed enzymatically using 30,000 Fishman units of β-glucuronidase (Ketodase, Warner Chilcott Co.). The specimens are incubated at 37°C for 24 hr. After incubation, the hydrolyzed steroids are extracted three times with 30-ml portions of freshly distilled dichloromethane. The extracts are pooled and washed with 25 ml of 0.1 *N* sodium hydroxide followed by washings with 10-ml portions of distilled water until neutrality. The dichloromethane is evaporated under reduced pressure in a rotating evaporator at a temperature below 50°C.

Column Chromatography

Glass columns, 15 cm tall and 1 cm inside diameter, are loosely packed with a small piece of glass wool. Using 3 g of alumina (Brockmann grade I deactivated with 3 ml of distilled water per 100 g), a slurry is made in 10 ml of benzene which is transferred to each of the columns and allowed to settle. The excess solvent is drained and a small piece of glass wool is placed on top of the settled alumina. The urine extracts and the standards are quantitatively transferred on the columns. The column chromatography is done as described below. Desired fractions are collected in 100-ml round-bottom flasks and the solvent is evaporated under reduced pressure. The dry residue is quantitatively transferred to centrifuge tubes using dichloromethane as solvent, which is subsequently evaporated to dryness under a stream of nitrogen by heating the tubes at 50°C. Pregnanetriol, pregnanetriolone, and pregnanetetrol reference standards in quantities of 20, 40, and 60 μg are carried through chromatographic procedure.

Separation of the Extract

Elution solvent	Eluate
1. Dry residue – benzene, 10 ml	Discard
2. 0.8% ethanol in benzene, 25 ml	Discard
3. 3.0% ethanol in benzene, 15 ml	Pregnanediol fraction
4. 20.0% ethanol in benzene, 40 ml	Pregnanetriol, pregnanetriolone, and pregnanetetrol fraction

Standardization of Alumina

Prepare a column as described. Add 20 μg of pregnanediol in 1 ml of benzene and allow to run until the liquid disappears into the alumina. Elute the column with 9 ml of benzene and discard the eluate. For the second elution use 25 ml of 0.8% ethanol in benzene, collect the eluate, and evaporate the solvent. To the dry residue add 2 ml of concentrated sulfuric acid; the acid should not turn yellow. For the third elution, use 15 ml 3.0% ethanol in benzene and collect 5-ml aliquots, evaporate to dryness, and add 2 ml of concentrated sulfuric acid. The most intense yellow color produced by pregnanediol should appear in the first and second fraction.

If pregnanediol appears in the eluate with 0.8% ethanol in benzene, the alumina needs to be reactivated by heating at 120°C for 2 hr and restandardized.

It is more likely that the alumina will be found to be too active and the elution of pregnanediol will not be complete with 3.0% ethanol in benzene. In this case add more water to deactivate alumina. Addition of 1 ml of water to 100 g of alumina should shift the pregnanediol elution by about 5 ml of 3.0% ethanol in benzene.

TABLE 6

Chromatographic Mobility

	Pregnanetetrol R_F	Pregnanetriolone R_F	Pregnanetriol R_F
Chloroform (saturated with water/formamide:methanol (3:7)	0.27	0.53	0.67
Chloroform:ethanol (4:1)	0.33	0.50	0.60
Chloroform:methanol:water (90:10:1)	0.45	0.60	0.70

TABLE 7

Colors and Fluorescence of Steroids

	Color in daylight	Sensitivity ($\mu g/cm^2$)	Fluorescence 365 mμ	Sensitivity ($\mu g/cm^2$)
Pregnanetetrol	Violet-purple	1	Pink-red	0.5
Pregnanetriolone	Violet	2	Blue-violet	0.5
Pregnanetriol	Violet	2	Pink-red	1

After addition of water to the alumina, the sample should be vigorously shaken in a closed container until the alumina is completely homogeneous and no lumps are visible.

It is advisable to prepare a sufficiently large batch of alumina and to store the prepared alumina in a closed container to avoid changes in water content.

Once the alumina is standardized for pregnanediol, verify the elution optimum for pregnanetriol, pregnanetriolone, and pregnanetetrol.

Thin-layer Chromatography

Aliquots of the pregnanetriol fraction are chromatographed on TLC, silica gel G-covered glass plates to separate pregnanetriol from pregnanetriolone and pregnanetetrol. Authentic steroid standards are applied as pilots.

See Table 6 for TLC chromatographic systems.

Detection of Steroids on TLC

Using a fine capillary pipette, apply the phosphoric acid reagent on the chromatogram. Heat the plate for 5 to 10 min at 95 to 100°C. After cooling, place the plate over a tray with ice cubes and visualize the steroids under ultraviolet light source (approximately 365 nm) (see Table 7).

e. Comments

In cases of congenital adrenal hyperplasia where the excretion of pregnanetriolone and pregnanetetrol may be expected, thin-layer separation of these abnormal steroids is useful for rapid screening and for confirmation of the gas-liquid chromatography results.

III. INTERPRETATION OF LABORATORY VALUES

Adaptation of competetive protein binding assays and radioimmunoassays for routine determinations of plasma progesterone have introduced a new tool for hormonal investigations of problems associated with ovarian function. Measurements of plasma progesterone levels during the follicular phase by competitive protein binding procedures are close to the sensitivity limit of the method. As a result of this, difficulties have arisen in obtaining accurate results. The radioimmunoassay shows considerably greater sensitivity and the levels of progesterone during the follicular

TABLE 8

Plasma Progesterone in Women with Regular Menstrual Cycles

Method or Authors	Follicular phase: Days of cycle	Follicular phase: ng/ml ± S.D. (range)	Days of cycle	Luteal phase ng/ml ± S.D. (range)
Double isotope dilution[34]		1.13 ± 0.49		10.40 ± 3.20
Gas chromatography[38]	1–9	(<0.15–1.10)	16–24	(7.90–20.60)
	7–16	(<0.15–4.20)	22–30	(1.00–17.90)
Competitive protein binding[33]	1–6	1.90 ± 0.11		8.20 ± 7.40
	7–13	1.00 ± 0.20		
Johansson[1]		0.32 ± 0.25	4 to 6 days after ovulation	(10–20)
Pichon and Milgrom[2]	1–13	0.81 ± 0.38	16–22	12.5 ± 2.96
Radioimmunoassay[10]	1–9	0.26 ± 0.19	16–29	8.28 ± 6.34
		(0.12–0.79)		(1.41–17.47)
Abraham et al.[9]		0.55 ± 0.10	Mid-phase	8.56 ± 4.66
		(0.41–0.65)		(1.48–13.50)
Abraham et al.[29]		(0.3–0.8)	Day of LH peak	(1–2)
			Plateau	(10–20)

phase fall well within the limits of assay. As shown in Table 8, the values of plasma progesterone detected during the follicular phase by various methods reveal very low levels with a considerably wide standard deviation. This may be a result of technical interferences as well as variations between different individuals and day-to-day variations of progesterone level in the same individual.

Using radioimmunoassay, Abraham et al.[29] have shown that some 10 days prior to midcycle, LH peak plasma progesterone tends to be in the order of 0.8 ng/ml and reaches a low level of about 0.3 ng/ml some 5 days prior to the LH surge. On the day of LH peak, plasma progesterone rises above 1 ng/ml and continues to increase as the corpus luteum develops. Radioimmunoassays as well as competitive protein binding assays have confirmed that in the normal luteal phase, progesterone reaches its high plateau level up to 20 ng/ml by about the 6th day after ovulation.

Plasma progesterone levels in adult men tend to be in a range of those seen in women during the follicular phase (see Table 9). Published data suggest that on the average progesterone determined by radioimmunoassay appears to be lower than values obtained by other methods. The competitive protein binding methods in assays of progesterone in male plasma may be interfered with by the presence of testosterone which, by cross-reactivity in the system, tends to elevate the final progesterone result. So far determinations of progesterone in male plasma for clinical investigations are limited to studies of endocrine physiology and to some pathological conditions such as congenital adrenal hyperplasia and cases of steroid-producing tumors. A firm diagnostic value of these determinations is not yet established.

In women with menstrual irregularities and

TABLE 9

Plasma Progesterone in Men

Method and authors	Mean ng/ml ± S.D.
Double isotope dilution[34]	0.28 ± 0.13
Gas chromatography[38]	0.28
Competitive protein binding	
Martin et al.[33]	0.25 ± 0.16
Pichon and Milgrom[2]	0.46 ± 0.14
Reeves[7]	0.43 ± 0.11
Yoshimi and Lipsett[39]	0.33 ± 0.17
Sasaki et al.[37]	0.35 ± 0.14
Radioimmunoassay	
Abraham et al.[9]	0.23 ± 0.06
Furuyama[10]	0.22 ± 0.10

with problems of fertility, assays of plasma progesterone and/or urinary pregnanediol provide information on the activity of corpus luteum. It has been suggested that plasma progesterone determinations might replace urinary pregnanediol assays in the evaluation of the luteal phase. Comparisons of these parameters have shown an overall parallelism; however, the values of urinary pregnanediol do not exactly coincide with those of plasma progesterone. The plasma progesterone assay provides an instantaneous value at a given time of the day, while urinary pregnanediol excretion represents the final metabolic product elimination over a 24-hr period. The plasma level of progesterone is influenced by the rate of secretion, the rate of peripheral metabolism, as well as by plasma protein binding. So far a diurnal variation in plasma progesterone for nonpregnant women has not been established; however, day-to-day variations are well appreciated.

In clinical investigations of the ovarian function, single determinations of plasma progesterone or urinary pregnanediol are considered to be of little value. A functional profile of the endocrine activity by serial determinations is frequently essential. If urinary steroid determinations are included in the investigations, the completeness of a 24-hr urine collection should be ascertained. A common problem encountered in laboratories is the incomplete urine collection which is submitted for assay. If such a collection represents a fraction of the 24-hr urine output, erroneous results are obtained. To overcome this source of error to some extent, an additional determination of urinary creatinine is used. This additional parameter allows some correction of the deficiency and an expression of urinary steroid value relative to gram of urinary creatinine.

For diagnostic purposes, determinations of progesterone or its metabolites provide only a part of the endocrine profile which should be supplemented by urinary or plasma estrogen levels, plasma gonadotrophin levels, basal body temperature chart, cervical mucus properties, vaginal cytology, and the histology of the endometrial biopsies.

The normal length of the menstrual cycle may vary from 25 to 38 days. During the years of reproduction irregular but ovulatory menstrual cycles are found in some 13.5%. The incidence of ovulatory polymenorrhea tends to increase with age, whereas in adolescents ovulatory oligomenorrhea is observed more frequently.

In women with ovulatory dysfunctional uterine bleeding, the ovarian hormone production shows some alterations. Midcycle spotting in otherwise normal menstrual cycles appears to be due to accentuated fluctuations in the estrogen titer. The decrease is estrogen after the midcycle peak appears to be responsible for withdrawal spotting. Endometrial biopsies at the time of spotting show proliferative or early secretory endometrium.

In most cases elevation of body temperature roughly coincides with spotting and the pregnanediol excretion pattern is normal.

Regular ovulatory polymenorrhea with cycles as short as 18 to 21 days is observed in adolescents and in premenopausal women. This pattern may be due to a shortened proliferative phase and a premature corpus luteum involution. The length of corpus luteum activity can be determined by plasma progesterone levels or urinary pregnanediol excretion.

Ovulatory oligomenorrhea may be associated with a prolonged proliferative phase and with a luteal phase of normal length. In cases of corpus luteum insufficiency the menstrual pattern is altered and it may manifest itself as premenstrual spotting, menorrhagia, or polymenorrhea. These conditions appear to be related to inadequate progesterone secretion. In some women prolonged corpus luteum activity is observed where progesterone secretion is extended. These women may present with menorrhagia or oligomenorrhea as a result of continued progesterone stimulation. The basal body temperature shows poor and late decline and urinary pregnanediol levels remain elevated.

Women with anovulatory dysfunctional uterine bleeding may present with oligomenorrhea, hypermenorrhea, and menometrorrhagia. In these cases, the corpus luteum fails to develop and the progesterone secretion remains low; consequently, the urinary pregnanediol excretion is also low.

Determinations of plasma progesterone along with urinary pregnanediol find a place in the follow-up of the patient's response to therapy for anovulation. If the failure of ovulation is caused by hypothalamic-pituitary dysfunction in gonadotrophin release, ovulation and corpus luteum formation have been induced by administration of gonadotrophin or clomiphene. Low plasma progesterone and low urinary pregnanediol are found in women who are taking combined or sequential oral contraceptives. Continuous low-dose progesterone preparations used as oral contraceptives tend to suppress midcycle LH and FSH peaks, serum progesterone, and urinary pregnanediol during the luteal phase. Some women, however, show an elevation of serum progesterone and of urinary pregnanediol during the luteal phase. The rise in progesterone secretion appears to be lower than in a normal ovulatory cycle.

During pregnancy large quantities of progesterone are produced by the placenta. Calculations of progesterone secretion have revealed variable estimates. In early pregnancy up to 92 mg of progesterone/day may be produced. In midpregnancy the secretion rate appears to be in a range of 76 to 263 mg/day and in late pregnancy some 190 to 322 mg/day are produced. These calculated quantities may not be accurate because of problems presented by the multiple origins of progesterone from other anatomical compartments apart from the placenta. The transformation of progesterone to pregnanediol seems to depend on the stage of gestation and on the changes in pathological pregnancy. During the early stages of normal gestation (8 to 14 weeks) some 9.2% of labeled progesterone is recovered as urinary pregnanediol and at least 6% of the dose appears in the 6-hydroxylated steroid fraction. In late pregnancy (30 to 36 weeks) the metabolic conversion of progesterone to pregnanediol is calculated to be in order of 20%. After the 36th week the conversion tends to decrease. Interpretation of progesterone transformation to urinary pregnanediol is complicated by daily variations in urinary excretion. In the same subject the variation may be in order of 24%. The fecal excretion of pregnanediol appears to be another variable and it may account for as high as 30%. Small quantities of pregnanediol are also excreted through the skin.

The fetus near term significantly contributes toward the metabolism of progesterone. Near term up to 80 mg of progesterone per day may be transformed to other steroids by the fetal compartment.

Progesterone production appears to be related to the size of the placenta, but fails to show a correlation with the fetal weight. In cases of multiple pregnancy with twins, triplets, and quadruplets, blood progesterone and urinary pregnanediol excretion have been observed above normal range. Pregnant diabetic subjects tend to excrete urinary pregnanediol in the upper range of the normal nondiabetic pattern. This may be related to a greater placental weight in the diabetic group. In pregnant diabetics, impending fetal death is signaled by a fall in urinary estriol excretion, but pregnanediol levels tend to remain unaltered. It has been repeatedly observed that in cases of fetal death in utero, urinary pregnanediol levels and plasma progesterone remain unaltered until after delivery of the placenta. It should be emphasized that normal plasma progesterone and urinary pregnanediol excretion do not exclude intrauterine fetal death due to nonplacental causes. If the fetal death is the final result of placental insufficiency, low pregnanediol and low plasma progesterone can be expected.[40]

Follow-ups of premature labor have shown a tendency to a lower excretion of urinary pregnanolone and pregnanediol in the presence of a normal excretion of pregnanetriol and total estrogens.

The relationship of plasma progesterone and urinary pregnanediol excretion to the onset of labor is still not clear. Most investigators have failed to observe a fall in progesterone levels. In cases of women with twins, triplets, and quadruplets, blood progesterone levels continuously increase and appear higher than in normal pregnancy, yet the gestation period tends to be shorter.

The prognostic value of urinary pregnanediol determinations in cases of threatened abortion is still disputed. Acevedo et al.[26] have observed a good prognosis in cases that present clinical symptoms of threatened abortion but who have normal excretion of pregnanediol and pregnanolone. The findings of low pregnanediol and pregnanolone levels in threatened abortion may be linked with insufficient function of the corpus luteum of pregnancy. These may also be associated with trophoblastic changes and genetic anomalies of the conceptus.

In pregnant subjects with problems of Rh isoimmunization, urinary pregnanediol excretion tends to be within the normal range.[40] In some cases the hypertrophic placenta may produce larger quantities of progesterone and the plasma levels of this hormone are found to be higher than in normal pregnancy. So far determinations of pregnanediol or plasma progesterone in isosensitized patients have not been proven to be of a prognostic value. In toxemia of pregnancy, a large

TABLE 10

Urinary Steroid Excretion in Untreated Congenital Adrenal Hyperplasia (21-Hydroxylase Deficiency)

Age	(mg/24 hr) Pregnanetriol	(mg/24 hr) Pregnanetriolone	(mg/24 hr) Pregnanetetrol
2–10 days	Up to 0.02	Up to 0.2	–
2–7 days	Up to 1.0	Up to 2.0	Up to 0.05
2–12 months	Up to 2.0	Up to 2.0	Up to 0.2
1–6 years	Up to 6.0	Up to 5.0	Up to 1.0
7–15 years	4–60	1–20	Up to 1.5
Adults	5–60	2–20	0.5–1.5

day-to-day variation of pregnanediol excretion is observed. Severe cases of toxemia on the average show about 50% reduction of pregnanediol, as compared to normal gestation of the same duration. In mild toxemia, the progesterone secretion rate may be higher than in normal pregnancy. The wide variations in pregnanediol values in toxemia of pregnancy limit the practical usefulness of this determination.

After termination of abdominal pregnancy with the placenta left *in situ,* elevated levels of urinary pregnanediol and of plasma progesterone may persist for several weeks. With the decrease in metabolic function of the placental tissue, continuously declining progesterone production may last for some 9 weeks.

Increased excretion of pregnanediol and pregnanetriol has been observed in cases of hydatidiform mole. After evacuation of the mole, most patients show a rapid decrease in pregnanediol excretion. If remnants of the mole are retained in the uterus, increased pregnanediol and chorionic gonadotrophin may persist for up to 7 weeks.

The increased excretion of urinary pregnanetriol in cases of hydatidiform mole, apparently, is related to augmented 17α-hydroxyprogesterone secretion by the chorionic gonadotrophin-stimulated ovaries.

There is also a possibility that molar trophoblasts *in situ* may also contribute toward the formation of 17α-hydroxyprogesterone. In cases of choriocarcinoma along with high levels of urinary chorionic gonadotrophin, increased quantities of pregnanediol are found.

Determination of 17α-hydroxyprogesterone and 21-deoxycortisol metabolites in urine has been found of diagnostic value in cases of virilizing congenital adrenal hyperplasia of 21-hydroxylase deficiency type. Infants born with this enzymatic defect show detectable quantities of pregnanetriolone in urine during the first few days of life (see Table 10). As the infant grows older, there is a trend toward an increase in the excretion of pregnanetriol, pregnanetriolone, and an appearance of pregnanetetrol. Along with these steroid metabolites, excretion of 17-oxosteroids is also notably increased. Clinically, these infants present with pseudohermaphroditism in females and with macrogenitosomia in males. Glucocorticoid replacement therapy or dexamethasone suppression markedly reduces pregnanetriol and 17-oxosteroid secretion. Pregnanetriolone and pregnanetetrol excretion are suppressed to a considerably lesser extent. Most intensively suppressed adult patients may still excrete some 0.1 to 0.5 mg of pregnanetriolone per day. In some patients treated with glucocorticoids, the decrease in pregnanetriolone correlates well with the decrease in 17-oxosteroids and pregnanetriol. In others, the replacement therapy may be sufficient to reduce 17-oxosteroids and pregnanetriol without appreciably changing pregnanetriolone excretion. It has been suggested by Finkelstein[30] that in congenital adrenal hyperplasia of 21-hydroxylase deficiency type, the adrenal cells may contain an abnormal 11β-hydroxylase which converts 17α-hydroxyprogesterone to 21-deoxycortisol. Kinoshita et al.[18] have observed that administration of metyrapone in patients with congenital adrenal hyperplasia increases pregnanetriol excretion, with a striking reduction in the levels of pregnanetriolone and pregnanetetrol. This suppressive effect of the drug is attributable to the 11β-hydroxylase inhibiting activity. The presence of pregnanetriolone in the urine of virilized infants is regarded as highly suggestive of congenital adrenal hyperplasia.

In virilized adult patients, excretion of milligram quantities of pregnanetriolone along with pregnanetriol is characteristic of congenital adrenal hyperplasia of steroid 21-hydroxylase deficiency type.

Infants and children with congenital adrenal hyperplasia of steroid 11β-hydroxylase deficiency type fail to excrete pregnanetriolone; however, urinary 17-oxosteroids, THS, and pregnanetriol are increased.

Pregnanetriolone has been occasionally detected in urine from patients who present with post-pubertal virilization. In these cases, the levels of pregnanetriolone tend to be below 1 mg/24 hr and urinary pregnanetriol may be elevated or in the upper range of normal. Some patients show a marked increase of urinary pregnanetriolone and pregnanetriol excretion in response to ACTH stimulation. It is believed that certain patients described as "post-pubertal virilizing syndrome" are affected by a mild form of congenital adrenal hyperplasia which becomes clinically more apparent after puberty. This group may also include some patients with polycystic ovarian disease of the Stein-Leventhal type. Certain patients with polycystic ovarian disease do excrete measurable quantities of pregnanetriolone which are usually below 0.4 mg/day. Excretion of pregnanetriol in these patients is frequently within the normal range. In a few of these patients, stimulation with ACTH increased and administration of dexamethasone decreased the excretion of pregnanetriolone and pregnanetriol.

Certain patients with Cushing's syndrome due to bilateral adrenal hyperplasia or adrenal tumors are found to excrete pregnanetriolone in a range from traces to 0.5 mg/24 hr. The significance of this finding is still not clear.

REFERENCES

1. **Johansson, E. D. B.,** *Acta Endocrinol.*, 61, 592, 1969.
2. **Pichon, M. F. and Milgrom, E.,** *Steroids,* 21, 335, 1973.
3. **Murphy, B. E. P.,** *J. Clin. Endocrinol.*, 27, 793, 1967.
4. **Johansson, E. B. D.,** *Acta Endocrinol. Suppl. 147,* 64, 188, 1970.
5. **Neill, J. D., Johansson, E. D. B., Dalta, J. K., and Knobil, E.,** *J. Clin. Endocrinol.*, 27, 1167, 1967.
6. **Johansson, E. J. B., Neill, J. D., and Knobil, E.,** *Endocrinology,* 82, 143, 1968.
7. **Reeves, B. D., de Souza, M. L. A., Thompson, I. E., and Diczfalusy, E.,** *Acta Endocrinol.*, 63, 225, 1970.
8. **Murphy, B. E. P.,** *Acta Endocrinol. Suppl. 147,* 64, 37, 1970.
9. **Abraham, G. E., Swerdloff, R., Tulchinsky, D., and Odell, W. D.,** *J. Clin. Endocrinol.*, 32, 619, 1971.
10. **Furuyama, S. and Nugent, C. A.,** *Steroids,* 17, 663, 1971.
11. **Kutas, M., Chung, A., Bartos, D., and Castro, A.,** *Steroids,* 20, 697, 1972.
12. **Spieler, J. M., Webb, R. L., Saldarini, R. J., and Coppola, J. A.,** *Steroids,* 19, 751, 1972.
13. **Mayes, D. and Nugent, C. A.,** *J. Clin. Endocrinol.*, 28, 1169, 1968.
14. **Klopper, A., Michie, E. A., and Brown, J. B.,** *J. Endocrinol.*, 12, 209, 1955.
15. **Klopper, A.,** *Methods in Hormone Research,* Dorfman, R. I., Ed., 2nd ed., Academic Press, New York, 1968, 229.
16. **Van der Molen, H. J.,** Gas phase chromatography of progesterone and related steroids, in *Gas Phase Chromatography of Steroids,* Eik-Nes, K. B. and Horning, E. C., Eds., Springer-Verlag, Berlin, 1968, 190.
17. **Raman, P. B., Avramov, R., McNiven, N. L., and Dorfman, R. I.,** *Steroids,* 6, 177, 1965.
18. **Kinoshita, K., Isurugi, K., Kumamoto, Y., and Takayasu, H.,** *J. Clin. Endocrinol.*, 26, 1219, 1966.
19. **Lucis, O. J., Cummings, G. T., Tsay, Y. H., and Minihan, B.,** *Clin. Biochem.*, 2, 59, 1968.
20. **Eik-Nes, K. B. and Horning, E. C.,** *Gas Phase Chromatography of Steroids,* Springer-Verlag, Berlin, 1968.
21. **Goldzieher, J. W. and Nakamura, Y.,** *Acta Endocrinol.*, 41, 371, 1962.
22. **Fuchs, F. and Klopper, A., Eds.,** *Endocrinology of Pregnancy,* Harper and Row, New York, 1971.
23. **Stern, M. I.,** *J. Endocrinol.*, 16, 180, 1957.
24. **Vela, B. A., Acevedo, H. F., and Campbell, E. A.,** *Am. J. Obstet. Gynecol.*, 103, 179, 1969.
25. **Acevedo, H. F., Strickler, H. S., Gilmore, J., Vela, B. A., Campbell, E. A., and Arras, B. J.,** *Am. J. Obstet. Gynecol.*, 102, 867, 1968.
26. **Acevedo, H. F., Vela, B. A., Campbell, E. A., Strickler, H. S., Gilmore, J., Moraca, J. I., and Arras, B. J.,** *Am. J. Obstet. Gynecol.*, 104, 964, 1969.
27. **Finkelstein, M.,** *Methods in Hormone Research,* Vol. 1, Dorfman, R. I., Ed., Academic Press, New York, 1968, 451.
28. **Cox, R. I.,** *Acta Endocrinol.*, 33, 477, 1960.
29. **Abraham, G. E., Odell, W. D., Swerdloff, R. S., and Hopper, K.,** *J. Clin. Endocrinol.*, 34, 312, 1972.
30. **Finkelstein, M.,** *Hormonal Steroids, Biochemistry, Pharmacology and Therapeutics,* Martini, L. and Pecile, A., Eds., Academic Press, New York, 1965, 625.

31. **Johansson, E. D. B.,** *Acta Endocrinol.*, 61, 607, 1969.
32. **Lucis, O. J., Hollenberg, C. H., MacDonald, S. A., and Blahey, P.,** *Can. Med. Assoc. J.*, 94, 1, 1966.
33. **Martin, B. T., Cooke, B. A., and Black, W. P.,** *J. Endocrinol.*, 46, 369, 1970.
34. **Riondel, A., Tait, J. F., Tait, S. A. S., Gut, M., and Little, B.,** *J. Clin. Endocrinol.*, 25, 229, 1965.
35. **Lucis, O. J. and Lucis, R.,** *Bull. W.H.O.*, 46, 443, 1972.
36. **Rubin, B. L., Maralit, M., and Kinard, J. H.,** *J. Clin. Endocrinol.*, 31, 511, 1970.
37. **Sasaki, C., Nowaczynski, W., Kuchel, O., Chavez, C., Ledoux, F., Gauthier, S., and Genest, J.,** *J. Clin. Endocrinol.*, 34, 650, 1972.
38. **Van der Molen, H. J. and Groen, D.,** *J. Clin. Endocrinol.*, 25, 1625, 1965.
39. **Yoshimi, T. and Lipsett, M. B.,** *Steroids*, 11, 527, 1968.
40. **Van Leusden, H.,** *Vit. Horm.*, 30, 281, 1972.

GONADOTROPINS AND PROLACTIN

C. Robyn and M. Vekemans

TABLE OF CONTENTS

I. Introduction . . . 79

II. Biological Properties . . . 81
- A. Bioassays for LH . . . 82
- B. Bioassays for FSH . . . 88
- C. Bioassays for Prolactin . . . 88
- D. Design of Bioassays . . . 91
- E. Extraction Procedures . . . 93

III. Immunological Properties . . . 94
- A. Principle of the Radioimmunological Method . . . 94

IV. Choice of a Reference Preparation . . . 105
- A. Standards for Bioassays . . . 106
- B. Standards for Radioimmunoassays . . . 106

V. Reliability Criteria of Bio- and Immunoassays . . . 107
- A. Accuracy . . . 107
- B. Precision . . . 108
- C. Sensitivity . . . 110
- D. Specificity . . . 110
- E. Practicability . . . 111

VI. Interpretation of Results . . . 112
- A. During Fetal Life . . . 112
- B. During Childhood and at Puberty . . . 112
- C. Between Puberty and Menopause . . . 113
- D. At and After the Menopause . . . 117
- E. During Pregnancy and in the Postpartum . . . 117

Acknowledgment . . . 118

References . . . 121

I. INTRODUCTION

Gonadotrophins and prolactin are protein hormones of the anterior pituitary. By immunohistochemical staining[1] both gonadotrophins, luteinizing hormone (LH) and follicle stimulating hormone (FSH) were found in the same cell type, a part of Romeis' δ cells[2] corresponding to the S1 mucoids of Pearse.[3] Prolactin has been localized in the acidophilic ε cell type by combining Herlant's tetrachrome staining and immunofluorescence.[4]

Both gonadotrophins are glycoproteins of very similar molecular weight, 33,000 for HLH and 35,000 for HFSH. They can be dissociated in two chemically dissimilar subunits, α and β, bound by noncovalent linkages and containing some 200 amino acid residues. The α subunits of FSH, LH, thyrotrophin (TSH), and chorionic gonadotrophin (HCG) are immunologically and physicochemically homologous (Figure 1). The β subunits of these four glycoprotein hormones, on the contrary, are hormone and antigen specific. The subunits α and β of LH are themselves virtually inactive; however, when reassociated under proper conditions, the biological activity is restored.[5-7]

Human prolactin is of another physicochemical character than gonadotrophins; it is a single polypeptide chain of some 200 amino acids without carbohydrate attachment (Figure 2). Its primary structure is very different from that of human growth hormone (HGH) but exhibits some 80% homology with ovine prolactin.[8,9] There is little if any immunological cross reaction between HPro and HGH, but extensive, although not complete, cross reactivity between ovine and human prolactin.[10]

The classic definition of a hormone is implicit in its bioassay.[11] This is even more true for gonadotrophins and prolactin. At the origin the definition concerned a biological activity rather than a hormone per se since the biologically active substances were not isolated. It is even surprising to see how these endocrine concepts of decisive importance in reproductive physiology were based on relatively unpure extracts tested in a few hypophysectomized rats. Some 30 years ago it was found that these protein hormones also had antigenic properties: antihormone sera could be raised in heterologous species. This was at the origin of the development of highly sensitive immunoassay methods for the measurement of gonadotrophins and prolactin in tissues and in biological fluids.

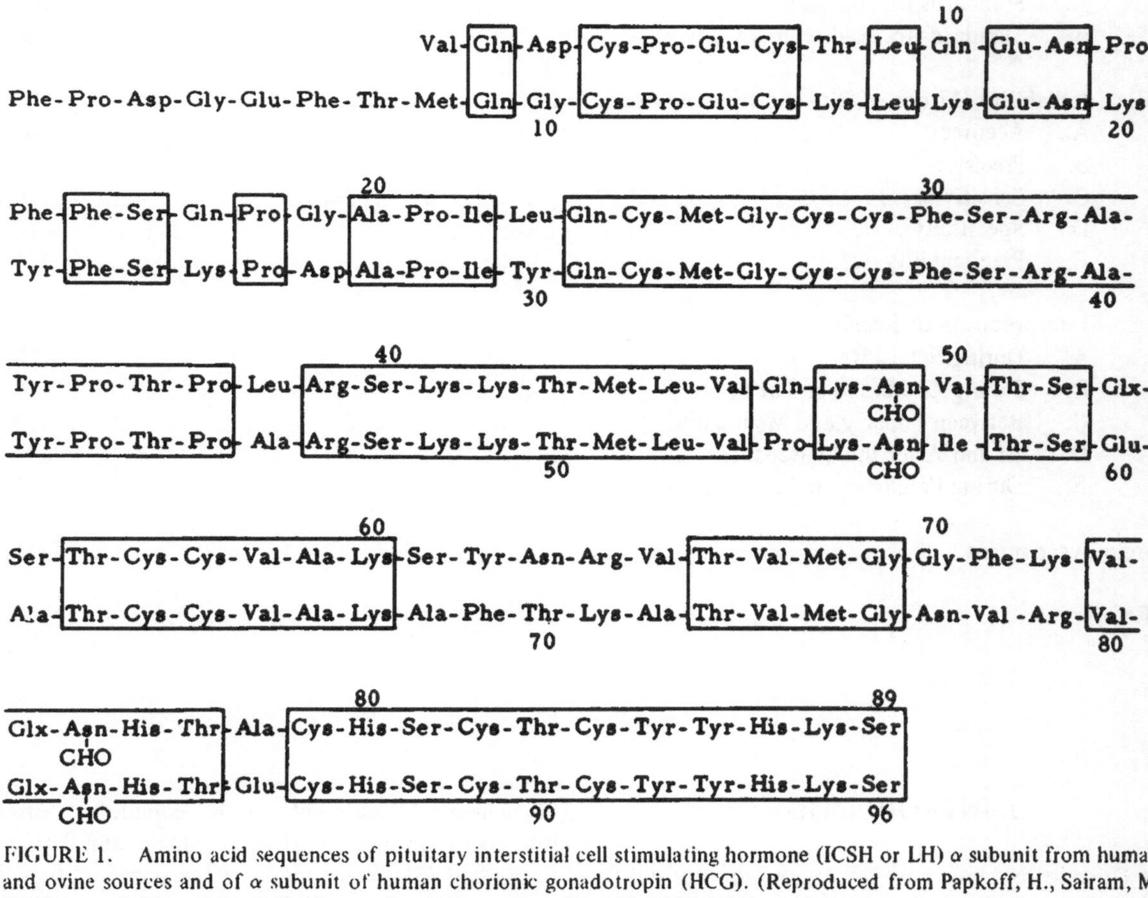

FIGURE 1. Amino acid sequences of pituitary interstitial cell stimulating hormone (ICSH or LH) α subunit from human and ovine sources and of α subunit of human chorionic gonadotropin (HCG). (Reproduced from Papkoff, H., Sairam, M. R., Farmer, S. W., and Li, C. H., *Recent Progr. Horm. Res.*, 29, 563, 1973. With permission.)

16 21

OP: Thr-Pro-Val-Cys-Pro-Asn-Gly-Pro-Gly-Asp - Cys-Gln-Val-Ser -Leu-Arg Asp-Leu-Phe-Asp-Arg
PP: Leu-Pro-Ile -Cys-Pro-Val -Gly-Pro-Ala -Asx - Cys-Glx-Val-Ser -Ile -Arg Asx-Leu-Phe-Asx-Arg
HP: Leu-Pro-Ile -Cys-Pro-( )-Gly-Ala-Ala -Arg Cys-Glx-Val-Thr-Leu-Arg Asp-Leu-Phe-Asp-Arg

43

Ala-Val-Met-Val-Ser-His-Tyr-Ile-His -Asn-Leu-Ser-Ser-Glu-Met-Phe-Asn-Glu-Phe-Asp-Lys-Arg
Ala-Val-Leu-Ile -Ser-His-Tyr-Ile-Asx-Asx-Leu-Ser-Ser-Glx-Met-Phe-Asx-Glx-Phe-Asx-Lys-Arg
( )

48

Tyr-Ala-Gln-Gly-Lys Gly-Phe-Ile-Thr-Met - Ala-Leu-Asn-Ser-Cys-His-Thr-Ser-Ser-Leu-Pro-
Tyr-Ala-Glx-Gly-Arg Gly-Phe-Ile-Thr-Lys Ala-Ile -Asx-Ser-Cys-His-Thr-Ser-Ser-Leu-Ser-
Tyr-Thr-His-Gly-Arg Gly-Phe-Ile-Thr-Lys Ala-Ile -Asx-Ser-Cys-His-Thr-Ser- Leu-Pro-

69 88

Thr-Pro-Glu-Asp-Lys Glu-Gln-Ala-Gln-Gln-Thr-His-His-Glu-Val-Leu-Met-Ser-Leu-Ile-Leu-Gly-Leu-Arg
Thr-Pro-Glx-Asx-Lys Glx-Glx-Ala-Glx-Glx-Asx-His-His-Asx-Val-Leu-Ile -Ser-Leu-Ile-Leu-Ser -Leu-Arg
-Pro-Glx-Asx-Lys ( )

102

Ser-Trp-Asn-Asp-Pro-Leu-Tyr-His-Leu-Val-Thr-Glu-Val-Arg
Ser- + -Asx-Asx-Pro-Leu-Tyr-His-Leu-Val-Thr-Glx-Val-Arg
Ser- + -Glx-Glx-Pro-Leu- His-Leu-Val-Thr-Glx-Val-Arg

105 114 124

Gly-Met-Lys Gly-Val-Pro-Asp-Ala-Ile-Leu-Ser-Arg Ala-Ile-Glu-Ile -Glu-Glu-Glu-Asn-Lys-Arg
Gly-Met-Glx - Glx-Ala-Pro-Asx-Ala-Ile-Leu-Ser-Arg Ala-Ile-Glx-Ile -Glx-Glx-Glx-Asx-Lys-Arg
Gly-Met-Glx - Glx-Ala-Pro-Glx -Ala-Ile-Leu-Ser-Arg Ala-Ile-Glx-Val-Glx-Glx- Thr-Lys-Arg

141

Leu-Leu-Glu-Gly-Met-Glu-Met - Ile-Phe-Gly-Gln-Val-Ile-Pro-Gly-Ala-Lys
Leu-Leu-Glx-Gly-Met-Glx-Lys Ile-Val-Gly-Glx-Val-Ile-Pro-Gly-His-Lys
( )

158 163

Glu-Thr-Glu-Pro-Tyr-Pro-Val-Trp-Ser-Gly-Leu-Pro-Ser-Leu-Gln-Thr-Lys Asp-Glu-Asp-Ala-Arg
Glu-Thr-Val-Pro-Tyr-Ser-Val- + -Ser-Gly-Leu-Glx-Ser-Leu-Glx-Asx-Met - Asx-Glx-Asx-Ala-Arg
Glx- Val-Pro-Tyr-Pro-Ile - + -Ser-Gly-Leu-Glx-Ser- Glx-Asx-Met - Asx-Glx- Ala-Arg

175 180 186

His-Ser-Ala-Phe-Tyr-Asn-Leu-Leu-His-Cys-Leu-Arg Arg Asp-Ser-Ser-Lys Ile-Asp-Thr-Tyr-Leu-Lys
( )-Ser-Ala-Phe-Tyr-Asx-Leu-Leu-His-Cys-Leu-Arg Arg Asx-Ser-His-Lys Ile-Asx-Asx-Tyr-Leu-Lys
( )-Ser-Ala-Leu-Tyr-Asx-Leu-Leu-His-Cys-Leu-Arg Arg Asx-Ser-His-Lys Ile-Asx-Asx-Tyr-Leu-Lys

198

Leu-Leu-Asn - Cys-Arg Ile-Ile-Tyr-Asn-Asn-Asn-Cys
Leu-Leu-Lys Cys-Arg Ile- Tyr-Ser -Asx-Asx-Cys
Leu-Lys-Asx - Cys-Arg Ile- His -Asx-Asx-Asx-Cys

FIGURE 2. Comparison of amino acid sequences of pituitary prolactin from human (HP), ovine (OP), and porcine (PP) sources. (Reproduced from Lewis, U. J., Discussion, in *Human Prolactin*, Pasteels, J. L. and Robyn, C., Eds., Excerpta Medica, Amsterdam, 1973, 38. With permission.)

II. BIOLOGICAL PROPERTIES [12]

The luteinizing hormone is the substance which, in the female, acts upon a preformed follicle, causing it to ovulate and be replaced by a corpus luteum, and causing the corpus luteum to secrete progesterone. In hypophysectomized immature female rats, LH promotes the repair of the interstitial cells but without any follicular growth. The follicle stimulating hormone stimulates growth of ovarian follicles; alone it cannot cause the Graafian follicles to ovulate. In hypophysectomized immature female rats, FSH induces growth and maturation of the follicles by

mitotic proliferation of granulosa cells, transformation of the surrounding stroma into a layer of theca cells, and secretion of follicular fluid, without repair of the interstitial cells and without any sign of luteinization. It has been extensively debated whether FSH alone can promote ovarian steroidogenesis and estrogen secretion. Petrušz et al.[13] have shown that human urinary FSH inducing no growth of the ventral prostate and seminal vesicles in hypophysectomized male rats is capable of promoting an increase in uterine weight and vaginal cornification, which are signs of an enhanced ovarian steroidogenesis. However, traces of LH increase markedly such FSH effect on steroidogenesis. It is likely that in physiological conditions the balance between LH and FSH secretions is responsible for the control of steroidogenesis in the growing follicle (Figure 3).

Prolactin should be regarded as a hormone of metabolic as well as of reproductive significance. Prolactin did not become specialized early in vertebrate phylogeny for the regulation of a single physiological process as LH and FSH.[14] Nicoll and Bern reported 82 different actions.[14] The most obvious and also the best known prolactin actions are related to reproduction, including parental care. This hormone promotes mammary development and lactation. It stimulates progesterone secretion from ovaries of certain species of rodents and acts on male gonads and sex accessory organs. Among birds, pigeon crop sac development and crop milk production are the most widely known actions of prolactin on reproductive processes. It is analogous to the action of prolactin on the mammary gland in mammals and to the stimulation of cutaneous "milk" secretion in certain teleosts. Survival of teleosts in fresh water and effects on skin mucus gland secretion and blood electrolyte changes accompanying water drive in amphibians are also under the control of prolactin. It seems that in other species such as reptiles, birds, and mammals, this hormone acts on the plasma levels of sodium chloride. It has been recently shown that in humans intramuscular injection of prolactin reduced urinary excretion of sodium, potassium, and water. It is also established that injections of hypotonic solution reduce and that injections of hypertonic solution enhance prolactin secretion in humans. The biological effects of prolactin include stimulation of body growth in amphibians (larval growth), reptiles, birds, and mammals and growth promotion on several target cells and tissues. In addition, this hormone exhibits several metabolic actions similar to those of growth hormone.

Finally, prolactin acts on ectodermal or integumentary structures in all classes of vertebrates: mammary gland, broad patch of birds, epidermal sloughing in reptiles, skin changes associated with water drive in amphibians, and mucus production in teleosts.

So far no compelling arguments exist for selecting any of these categories of biological effects as being of primary significance in "vertebrate function and organization."[15] The role of prolactin in ion and water movements and on membrane permeability may be fundamental actions of prolactin at the biochemical level. They may, at least partially, account for the wide variety of responses seen.

A. Bioassays for LH

1. In Vivo Methods

Some of the bioassays for LH are now mainly of historical interest; these include the repair of interstitial tissue in ovaries of hypophysectomized immature female rats,[16] ovarian hyperemia in rats,[17,18] enlargement of seminal vesicles in hypophysectomized immature male rats,[19] and the weaver finch test in the Paradise Wydah.[20] Three methods are still in current use: ventral prostate weight increase, ovarian ascorbic acid depletion, and ovarian cholesterol depletion in rats.

a. Ventral Prostate Weight (VPW) in Rats

The ventral prostate of the hypophysectomized male rat appears to be the best organ for the measurement of LH and is closely related to the biological activity on which the definition of this hormone has been based.

The assay was originally described by Greep et al.[21,22] using hypophysectomized immature male rats. Hypophysectomy is performed between day 21 and day 23 by the parapharyngeal approach or by transauricular aspiration.[23] Alternatively, hypophysectomized immature rats of the Sprague-Dawley strain can be purchased from Hormone Assay Laboratories Inc., Chicago, Illinois, arriving within 48 hr after hypophysectomy. The hypophysectomized animals should be on a special diet including maize (Indian corn), orange, tea infusion, and food in pellets for rats, and kept in cages containing shavings of

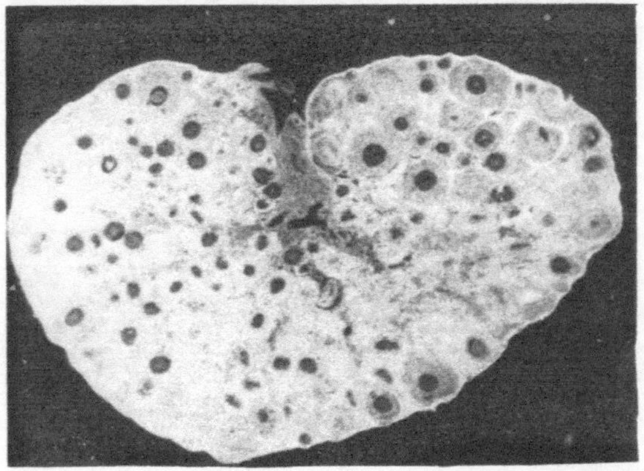

FIGURE 3A.

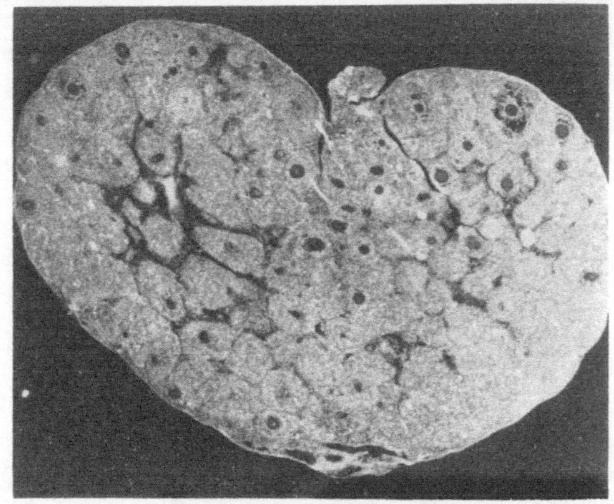

FIGURE 3B.

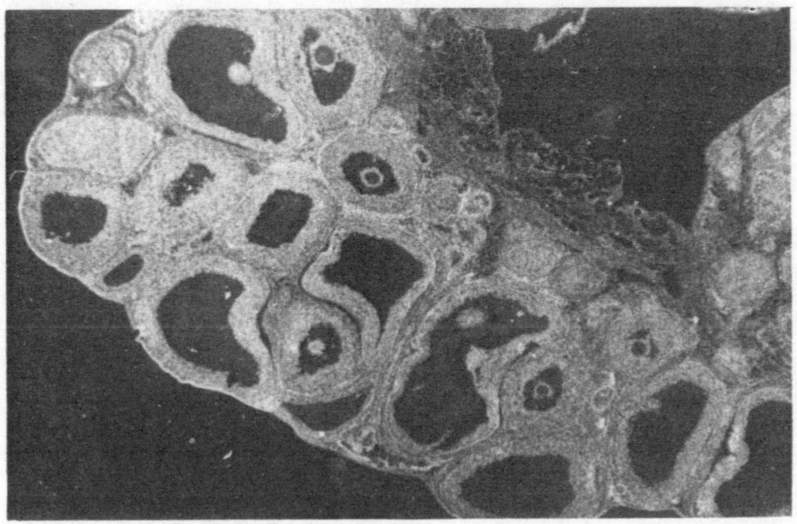

FIGURE 3C.

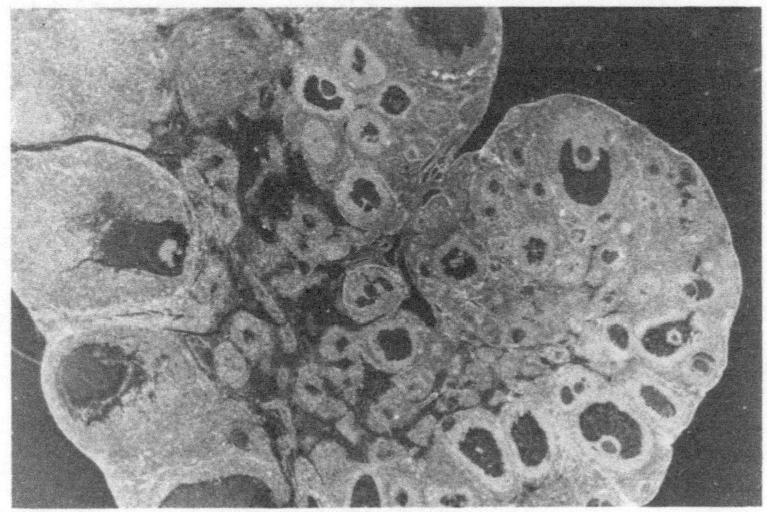

FIGURE 3D.

FIGURE 3. Representative ovarian histological picture from hypophysectomized immature rats: control (a and e), interstitial cell stimulation (LH) activity free of follicle stimulating (FSH) activity (b and f), FSH activity free of LH activity (c), combined FSH and LH activities (d).

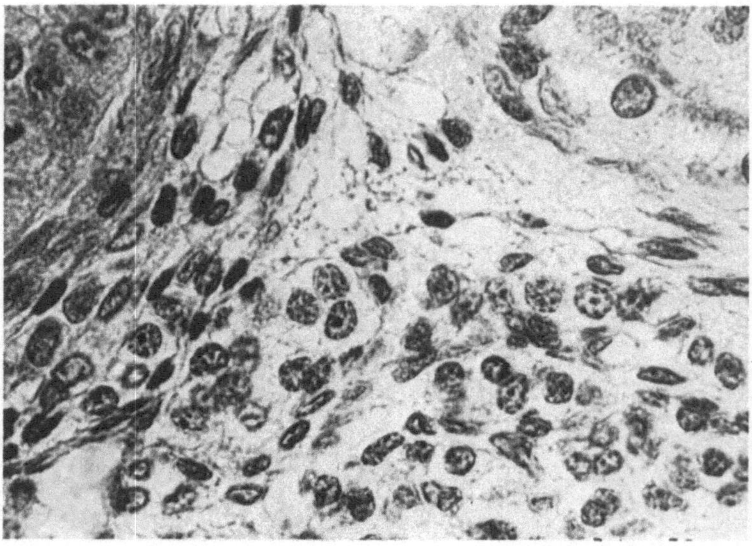

FIGURE 3E.

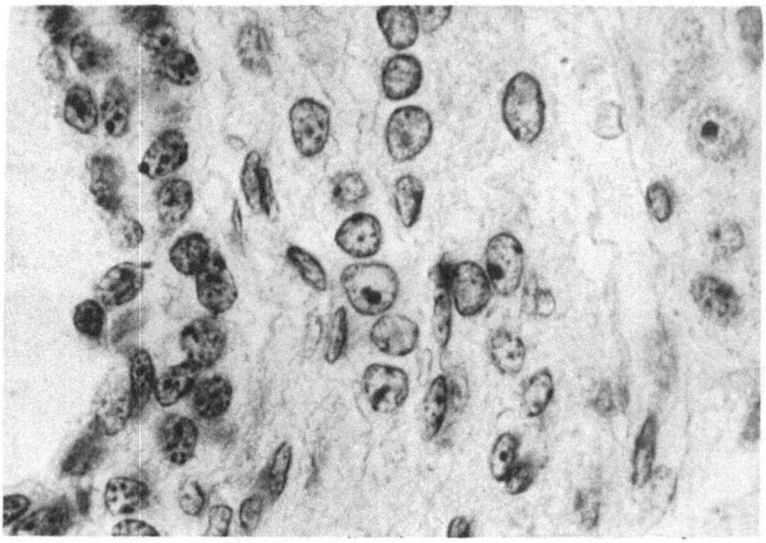

FIGURE 3F.

wood. The animals are injected subcutaneously with 1.0 ml of the test solution on each of 4 consecutive days and sacrificed with chloroform on the 5th day. Use a 5-ml syringe with a 21-gauge 1-in. needle; start the injection with the lowest doses and rinse the syringe carefully when proceeding from a standard solution to an unknown solution or from an unknown solution to another. All animals receiving the same dose of the standard or unknown are kept in the same cage. Do not put more than 5 to 10 animals together in the same cage. The animals are killed cage by cage in the same order as they were injected. The total accessory reproductive organs are cut off and put into Bouin solution for 1 to 2 days. Then the seminal vesicles and dorsal and ventral lobes of the prostate are dissected free from connective tissue and from the urethra under a magnifying glass. The ventral prostate is blotted on filter paper and weighed on a torsion balance at the nearest 0.2 mg.

Diczfalusy[24] has described a method for the assay of human chorionic gonadotrophin in intact immature male rats. It can be used for the measurement of human pituitary LH, the results being very similar if not identical to those obtained by the ventral prostate weight test in hypophysectomized male rats. However, there is a theoretical objection to such a method since in nonhypophysectomized animals the endogenous gonadotrophins may interfere with the gonadotrophin preparations to be tested. The method proposed by Diczfalusy[24] and modified by Robyn and Diczfalusy[25] is based on the increase in weight of the total accessory reproductive organs. Immature Sprague-Dawley rats of 21 to 23 days of age are injected subcutaneously with 1.0 ml of the test solutions on each of 3 consecutive days and are sacrificed on the 4th day. The total accessory reproductive organs are cut off and treated as indicated above. Here, however, after careful dissection, seminal vesicles and ventral and dorsal prostates are weighed all together after blotting on filter paper (Figure 4). The influence of variables such as species and strain differences, age of animals, body weight, route of administration, schedule of dosage, time of sacrifice, index of response, seasonal variations, and fixation and drying of accessory sex organs are checked very carefully.[24] The experimental error is significantly lower when the differences between litters are balanced by assigning one animal from each litter to each of the treatment or dosage groups. In this case dissection of the fixed organs has to be conducted litter by litter.

b. Ovarian Ascorbic Acid Depletion (OAAD) in Rats

Parlow[26,27] described a bioassay method for LH based on the depletion of ascorbic acid in ovaries of intact immature rats pretreated with pregnant mare serum gonadotrophin (PMS) and HCG. Female Sprague-Dawley (or Holtzman but not Witsar) rats weighing 40 to 50 g at 21 days of age are pretreated with a single subcutaneous injection of 25 I.U. of PMS followed 56 to 65 hr later by a single subcutaneous injection of 25 I.U. of HCG. The ovaries of such animals are heavily luteinized, each weighing more than 100 mg. The bioassay itself is conducted from 6 to 8 days after the injection of HCG: 1 ml of the test solution per 100 g body weight is injected intravenously via a tail vein of the lightly anesthetized rat. A single ovary is removed for the ascorbic acid determination 3 hr later (± 10 min). The ovary is dissected free from connective tissue, weighed on a torsion balance to the nearest 0.2 mg, and ground in a mortar with a trace of neutral sand and 10 ml of 2.5% metaphosphoric acid solution. After centrifugation, the ascorbic acid concentration in the supernatant is measured by a modification of the method of Mindlin and Butler.[28] This method is based on the reduction of 2:6 dichlorophenol indophenol to a colorless form by ascorbic acid. Metaphosphoric acid solution at 5% is stored at 4°C no longer than 2 weeks. A stock solution of ascorbic acid standard (0.1 mg/ml) is prepared in 2.5% metaphosphoric acid and stored at −20°C for several months in aliquots of ± 20 ml. Just before the determination prepare 3 to 4 dilutions of the ascorbic acid standard in 2.5% metaphosphoric acid: 4.0, 6.0, 8.0, and 10.0 μg/ml. Dissolve 20 mg

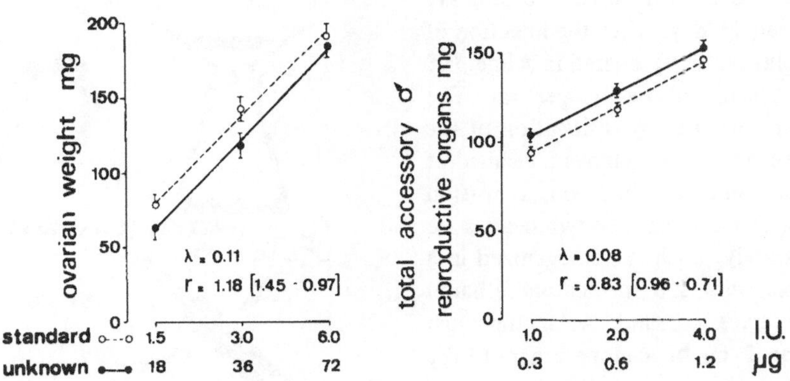

FIGURE 4. 3 + 3 point parallel line assays for FSH and LH using as endpoints the weight of the ovaries of intact immature female rats treated simultaneously with HCG, and the weight of total accessory reproductive organs of intact immature male rats, respectively. The vertical bars represent one standard deviation from each side of the mean. The doses of the standard are expressed in international units (I.U.) and those of the unknowns in micrograms. The relative potency (r) with the 95% fiducial limits in parentheses and the index of precision (λ) are indicated for the LH and the FSH assay.

of 2:6 dichlorophenol indophenol ($C_{12}H_6O_2NCl_2Na$) in 100 ml distilled water (at 85 to 95°C) and dilute to 500 ml with distilled water. Prepare a 0.33 *M* sodium acetate solution in distilled water and adjust pH to 7.0 with 0.5 *N* acetic acid. Store both solutions at 4°C no longer than 2 weeks. Mix equal volumes of both solutions just before use and dispense 4 ml of this indophenol-acetate solution to a series of test tubes of 10-ml capacity. Add 1 ml of test sample prepared in 2.5% metaphosphoric acid solution and mix with a vortex mixer. Read at 520 nm after rinsing the cuvette with the solution to be measured. Record the optical density 45 sec after addition of the test sample. The reading of metaphosphoric acid alone gives the "0" value. Run first the standard curve with the known ascorbic acid concentrations. The ascorbic acid concentration of the unknown is calculated by interpolating the values of optical density in the unknown in the standard curve expressed in micrograms per milliliter and per ovary (when multiplied by 10).

c. Ovarian Cholesterol Depletion (OCD) in Rats

An assay method for LH based on ovarian cholesterol depletion in intact immature rats pretreated with PMS and HCG has been described by Bell et al.[29] Intact immature female Wistar rats of 21 to 24 days of age and weighing 30 to 50 g receive a single subcutaneous injection of 50 I.U. of PMS followed 72 hr later by a single subcutaneous injection of 25 I.U. of HCG. The bioassay proper is performed 11 days after the injection of HCG. The test solutions are prepared in saline: 0.5 ml are injected intraperitoneally per rat. The animals are killed 5 hr later by dislocation of the cervical spine. The ovaries are removed, cleaned of connective tissue, and weighed on a torsion balance to the nearest 0.2 mg. The two ovaries are then treated separately. Each is homogenized in a mortar and pestle with 2.0 ml acetone:ethanol (1:1 v/v) and a trace of sand. After thorough grinding a further 2 ml of acetone:ethanol (1:1 v/v) is added and the homogenate filtered through a Whatman® No. 2 filter paper. The total cholesterol content of the filtrate is determined by the method described by Searcy and Berquist.[30] The results are expressed in micrograms of cholesterol per 100 mg ovarian weight.

The procedure appears to be dependent on the strain of the animals and the conditions under which they are kept.[31] The dose-response curve is of nonmonotonic nature. Great care should be given to the standardization of this method using an individual colony of rats; establishment of optimal pretreatment conditions is essential.

2. In Vitro Methods

The in vitro bioassay methods for LH are based on the use of target tissue receptors in the rat testis or in the rat ovary, or on the use of testosterone production by incubated mice or rat testis under stimulation with gonadotrophins.

a. Radioligand-receptor Assay of LH

Catt et al.[32,33] and Reichert et al.[7] described an assay for LH based on the binding of labeled LH to the cellular receptor site of rat testis for this hormone (Figure 5).

Testicular receptor fraction – Sprague-Dawley rats of 250 to 350 g are sacrificed by cervical dislocation. The tunica albuginea is removed and the remaining testicular tissue weighed on a torsion balance to the nearest 0.2 mg. Testicular homogenate is prepared using a McShan-type hand homogenizer (MacAllister Co., Cambridge, Mass.): one testis is dissociated in 2 to 4 ml of medium 199 solution with $NaHCO_3$ (Grand Island

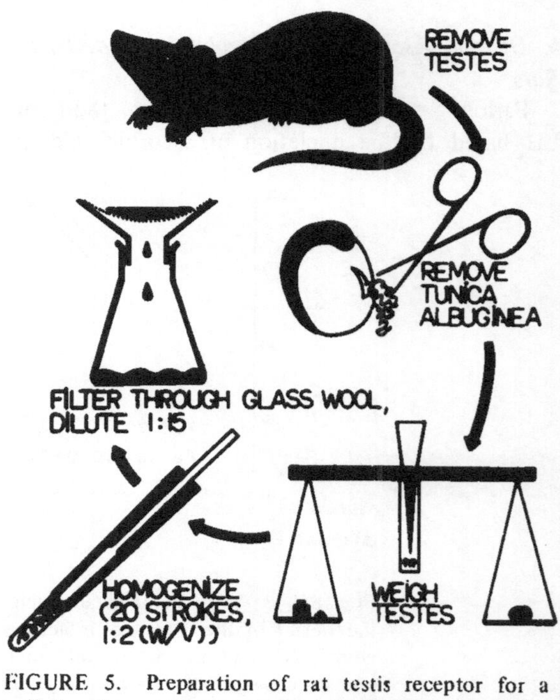

FIGURE 5. Preparation of rat testis receptor for a radioligand receptor assay of luteinizing hormone (LH). (Reproduced from Reichert, L. E., Leidenberger, F., and Trowbridge, C. G., *Recent Progr. Horm. Res.*, 29, 497, 1973. With permission.)

Biological Co., New York) per gram of tissue with 20 to 30 strokes of the pestle (loosely fitting). After homogenization the suspension is filtered through a thin layer of glass wool to remove the tubules and cellular debris and diluted to 1/15 with tris HCl buffer (0.1 *M* at pH 7.4, made 5 m*M* with $MgCl_2$ and 0.1 *M* with sucrose). Keep the diluted suspension at 4°C or in an ice bath until the assay; it provides testicular receptor fraction for 40 to 60 tubes.

Radioiodinated tracer – Iodination of a highly purified HLH or HCG preparation is conducted by the chloramine T method according to Greenwood et al.[34] as modified by Catt et al.[33] or Reichert et al.[7]

Binding assays – The assays are conducted in disposable flint glass tubes (12.5 X 75 mm). Add the standard or the unknown sample (100μl) diluted in tris HCl buffer containing 1% of egg albumin (crystallized two times; Sigma Co., St. Louis, Mo.) and 1.0 ml of the testicular receptor fraction. This mixture is incubated at 37°C for 30 min and then chilled prior to addition of 50μl of the radioiodinated tracer (100 to 300,000 cpm); Reichert et al.[7] emphasize the importance of this step. Incubation with the tracer is conducted at 4°C during 15 min at 1,500 to 3,000 g. The supernatant is discarded, the homogenate is washed with 1 to 4 ml cold tris HCl buffer, and centrifugation is repeated. The supernatant is aspirated again and the radioactivity is counted for 1 min in an automatic gamma counter. Some 10% of the tracer is bound to the testicular homogenate.

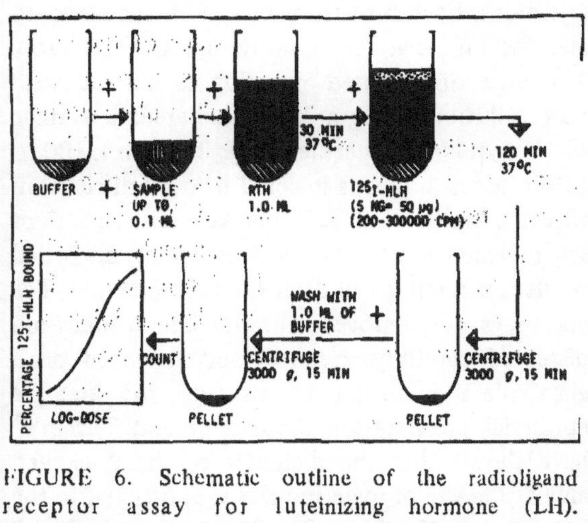

FIGURE 6. Schematic outline of the radioligand receptor assay for luteinizing hormone (LH). (Reproduced from Reichert, L. E., Leidenberger, F., and Trowbridge, C. G., *Recent Progr. Horm. Res.*, 29, 497, 1973. With permission.)

The unlabeled LH of the standard or unknown sample will compete for the testicular receptor with the radioiodinated LH: the dose-response curves obtained are similar to those described for radioimmunoassays (Figure 6). Since this procedure of assays has been very recently introduced the methodology as described above[7] has several variants proposed by the same authors or by others. Catt et al.[33] use phosphate buffered saline instead of the tris HCl buffer and other incubation conditions. It is advisable to set up locally the optimal conditions for such types of assays.

b. In Vitro Production of Testosterone by the Mice Testis.

Dufau et al.[35] described a bioassay method for LH based on the measurement of testosterone production by the rat testis in vitro. Van Damme et al.[36] described a similar method using mice testis (Figure 7).

Incubation conditions – Testes from adult mice (3 to 4 months old) sacrificed by cervical dislocation are removed and decapsulated very

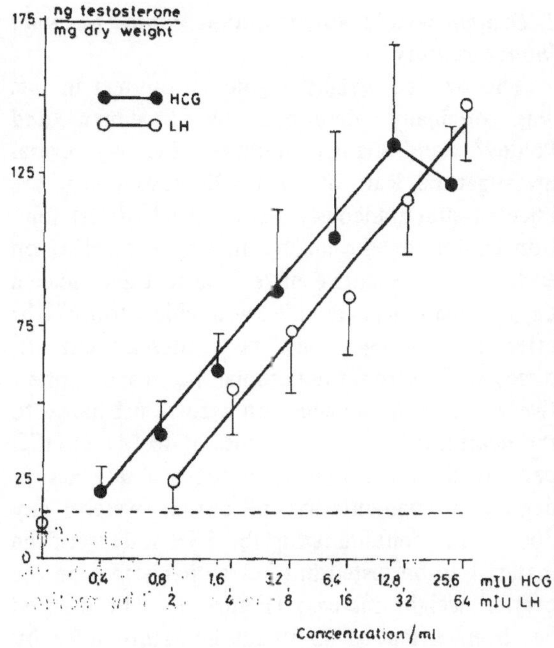

FIGURE 7. Dose response-curves with mouse testes for human chorionic gonadotrophin (HCG) and luteinizing hormone (LH) from a human menopausal gonadotrophin (HMG) preparation. The mean of six incubations with standard deviation is represented. (Reproduced from Van Damme, M.-P., Robertson, D. M., Romani, P., and Diczfalusy, E., *Acta Endocrinol.* (Kbh.), 74, 642, 1973. With permission.)

carefully. The decapsulated testes are incubated at 34°C with the test solutions prepared with Krebs Ringer bicarbonate in vials gassed with cyanogen and placed in a water-bath moving at 50 to 100 cycles per minute. After 3 hr incubation, the testes are removed and the media diluted (1:500) with distilled water.

Radioimmunoassay of testosterone – The testosterone of the incubation medium is measured by a radioimmunoassay[35-37] or by a competitive protein-binding method.[38]

B. Bioassays for FSH

Assay methods such as the determination of the minimum effective dose (MED) in reestablishing follicular growth in the ovaries of hypophysectomized rats[16,39] and the technique based on the increase in testicular weight in hypophysectomized immature rats treated with an excess of HCG are now of historical interest. They are of much less practicability than the ovarian augmentation test in rats or mice, although the specificity is equivalent.

1. Ovarian Weight Augmentation Method in Intact Immature Rats

The ovarian weight augmentation test in rats was originally described by Steelman and Pohley[40] and slightly modified since by several investigators. Rats of 21 to 23 days of age are injected subcutaneously with 1.0 ml of test solution containing the augmenting dose of HCG on each of 3 consecutive days (one to three times a day). The animals are killed with chloroform 72 hr after the first injection. The ovaries are cut off, dissected free from the fallopian tubes and connective tissue, and weighed on a torsion balance to the nearest 0.2 mg. A total dose of 40 I.U. of HCG per animal was found to produce a satisfactory degree of augmentation of ovarian weight; any dose of LH contaminating the FSH activity of the material to be tested has no further effect on the ovarian weight increase (Figure 4). The method has been extended to intact immature mice by Brown.[41] However, the success of the latter method appears to depend on the strain of animals employed[42] or on environmental factors.[43]

When differences between litters are balanced by assigning one animal from each litter to each of the dosage groups, the variance is significantly reduced.

C. Bioassays for Prolactin

As shown in Table 1, bioassay methods have been proposed for the measurement of prolactin but none of them are entirely satisfactory. All in vivo methods are insufficiently sensitive and thus none of them are applicable to plasma without preliminary extraction. The in vitro bioassays are tedious and of very low practicability. Recognition of prolactin in humans as separate from growth hormone has been delayed until the last 2 to 3 years, in part because of the lack of a suitable, sensitive, specific bioassay.

1. Pigeon Crop Gland Assay

Until recently, most of the knowledge on prolactin has been based on the pigeon crop gland assay.[57] The crop sacs are thin, transparent epithelial membranes. During the latter half of the incubation period of a clutch of eggs, they increase in weight both in males and females, become thickened, and secrete a "crop milk" which the birds feed to their young. This process is under the control of pituitary prolactin.

Several procedures for assay have been proposed. Systematic subcutaneous injections of the test solutions followed by weighing of the crop sac[44,58] were abandoned due to the very low sensitivity of this procedure (5 I.U. of prolactin). The most sensitive crop sac assay is the micro method using local intracutaneous injections in contact with the crop sac.[59] Pigeons of a uniform strain are divided into groups in order to build a 2 + 2 or 3 + 3 point assay design. The feathers are plucked on both sides of the crop at the sites of injections. Each bird is injected with the standard and the unknown on opposite sides of the crop. The area of repeated injection is marked with eosin solution. Using a 250-μl tuberculine syringe with a hypodermic needle (gauge 27, ½ in.), 100 μl of the test solution is injected into a follicle from which a feather has been plucked. The injections are repeated every day and the animals are killed by decapitation 96 hr after the first injection. The crop sacs are removed and the lateral lobes are dissected carefully, cleaned, and stretched on a glass plate in order to visualize the area of epithelial proliferation. Grosvenor and Turner[60] have shown that the diameter of the thickened area at the site of injection was proportional to the logarithm of the dose of prolactin injected (Figure 8). The measurement of the diameter is done under diffuse light. It is advisable to have a

TABLE 1

Bioassay Methods for Prolactin

Authors	Animal	End organ	Site of injection	Variable	Sensitivity I.U.
In vivo					
Riddle et al. (1933)[44]	Pigeon	Crop sac	Systemic (subcutaneous)	Crop sac weight	5.0
Reece and Turner (1937)[45]	Pigeon	Crop sac	Local (intradermal)	Proliferative area	5.0×10^{-2}
von Berswordt-Wallrabe et al. (1965)[46]	Pigeon	Crop sac	Local (intradermal + corticoids)	Proliferative area	
Ben David (1967)[47]	Pigeon	Crop sac	Local (intradermal)	Incorporation of ^3H methyl-thymidine	1.0×10^{-3}
Lyon (1942)[48]	Rabbit	Mammary gland	Intraductal	Histologic grading	1.0
Mishkinsky et al. (1966)[49]	Rabbit	Mammary gland	Intracerebral	Milk production	
Turner (1939)[50]	Rat	Mammary gland	Subcutaneous	Histologic grading	2.0
In vitro					
Prop (1961)[51]	Virgin mouse	Mammary gland		Histologic grading	1.0×10^{-2}
Mishkinsy et al. (1967)[52]	Virgin rabbit	Mammary gland		Histologic grading	1.0×10^{-3}
Franz and Kleinberg (1970)[53]	Pregnant mouse	Mammary gland		Histologic grading	4.0×10^{-4} (15 ng)
Forsyth and Myres (1971)[54]	Pseudo-pregnant virgin rabbit	Mammary gland		Histologic grading	1.2×10^{-3} (50 ng)
Loewenstein et al. (1971)[55]	Pregnant mouse	Mammary gland		*N*-acetyl lactosamine synthetase	1.0×10^{-4} (4 ng)
Turkington (1971)[56]	Mid-pregnant mouse	Mammary gland		Incorporation of P^{32} into casein	1.0×10^{-4} (5 ng)

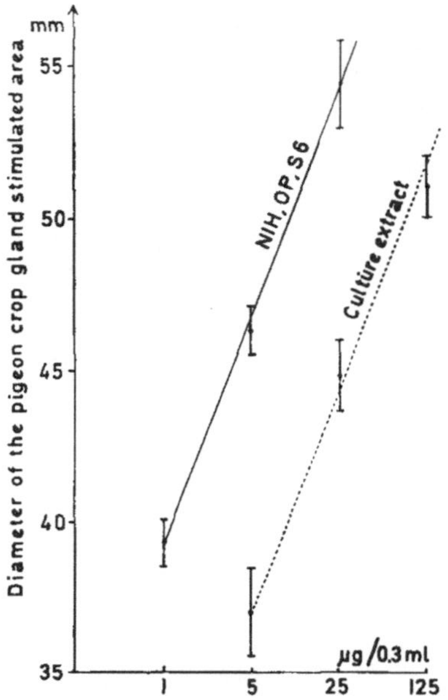

FIGURE 8. Pigeon crop gland assay of prolactin activity from an extract prepared from human fetal pituitary culture medium. The standard is an ovine pituitary prolactin preparation (OP-S-6) from the National Institute of Health (NIH). Vertical bars represent standard error of the means. (Reproduced from Pasteels, J. L., Tissue culture of human hypophyses: evidence of a specific prolactin in man, in *Lactogenic Hormones,* Wolstenholme, G. E. W. and Knight, J., Eds., Churchill Livingstone, Edinburgh, 1972, 269. With permission.)

complete randomization of doses among sites and animals in order to avoid systematic influence of subjective factors. In some instances, an intense inflammatory reaction masks the specific epithelial proliferation and disturbs the measurement of its dimension. Von Berswordt-Wallrabe et al.[46] succeeded in preventing this inflammatory reaction by intradermal administration of prednisone. A similar effect may be obtained by systemic injection of the corticosteroid.[61,62] This procedure is still very useful to estimate the biological activity of pituitary extracts or for the establishment of standard preparations of human prolactin.[63]

2. Radioligand Receptor Assay of Prolactin

Very recently Friesen et al.[206] and Parke and Forsyth[207] described radioreceptor assays for prolactin using radioiodinated human or ovine prolactin. However, complete description of a standardized assay procedure has not yet been achieved.

3. In Vitro Bioassay Using Midpregnant Mouse Breast Tissue

This procedure has been developed by Kleinberg and Frantz[64] in order to estimate prolactin concentration in human plasma without preliminary extraction using a mammalian end organ.[64]

Swiss albino mice, aged greater than 10 weeks, nulliparous, between 25 and 35 g are weighed every day at the same time and killed by an overdose of ether vapor at day 8 or 9 of gestation. It has been found that 7 days after mating a 10% increase in body weight denotes pregnancy in 80% of the mice.

The skin of the abdominal wall is reflected from a midline incision and the thoracic and inguinal mammary glands are removed and put in a Petri dish containing Hank's balanced salt solution (GIBCO, Grand Island, New York) containing penicillin (250 U/ml) and streptomycin (250 µg/ml). Under a dissecting microscope, the mammary glands are dissected free from excess connective tissue and transferred to a filter paper soaked with Hank's solution. Blocs of tissue (2 X 2 X 1 mm) are cut with uncoated stainless steel razor blades. Each assay is conducted with the mammary gland of only one mouse because of the considerable variation of the response to prolactin found between glands of different animals even when they are of the same strain. Thoracic and inguinal glands of one animal yield about 120 fragments. Dissection is conducted as rapidly as possible. Four fragments, each of a different gland, are put on a square of lens paper on top of a stainless steel grid suspended in the central compartment of a plastic organ culture dish (Falcon Plastics, Oxnard, California) containing 1 ml of medium and surrounded by some 5 ml of water in order to avoid extensive evaporation of the medium. The lens paper and grids are washed with water, 95% ethanol, and ether before heat sterilization. The cultures are kept for 4 days at 36°C in 95% oxygen-5% carbon dioxide. Then the blocs of mammary tissue are fixed in Bouin solution and cut. The histologic sections are stained with hematoxylin-phloxine-safranine.

The culture medium is Medium 199 distributed by GIBCO with pure beef insulin (10 µg/ml) and 30% human plasma added. Ovine pituitary prolactin (National Institute of Health, Bethesda, Md.) is used as reference preparation. The scale of concentrations in the incubation medium ranges from 0 to 50 ng/ml in these experimental conditions.

Unknown human plasma samples are tested at 10 and 30% final dilution in the incubation medium. A pool of human male plasma with less than 1 ng/ml of radioimmunoassayable growth hormone and no detectable prolactin activity is used to adjust the plasma concentration to 30% in all dishes containing the standard solutions or diluted plasma.

A histological grading is established on an arbitrary scale from 0 to 4:

0: in absence of prolactin, there is complete absence of red stained material in the lumina and a low and immature-appearing cytoplasm

1: a secretory material is present in small amounts in some 25% of the lumina

2 and 3: intermediate stages between 1 and 4

4: there is an abundant secretory material, stained darkly red in almost all lumina, and a cytoplasm with mature secretory appearance

It is recommended to take as score for a given dose of standard or unknown sample the mean of the scores of two independent observers for each of the four fragments incubated in the same conditions and to test each sample in at least two different assays. A mean standard curve obtained from 44 assays is represented in Figure 9. Kleinberg and Frantz[64] reported that human growth hormone in blood exhibited in this assay system an intrinsic lactogenic activity like that of growth hormone extracted from pituitaries.

When the immunoassayable growth hormone is elevated in a plasma sample it is advisable to test it with and without preincubation (2 hr or more at room temperature) with an antigrowth hormone serum; 0.1 ml of antiserum or nonimmunized rabbit serum is mixed with 0.9 ml of the human plasma to be tested.

D. Design of Bioassays

The most satisfactory assays are those in which the potency of an unknown is compared to that of a standard preparation at the same time. Gaddum[65] has reviewed the various experimental designs intended for such types of bioassays.

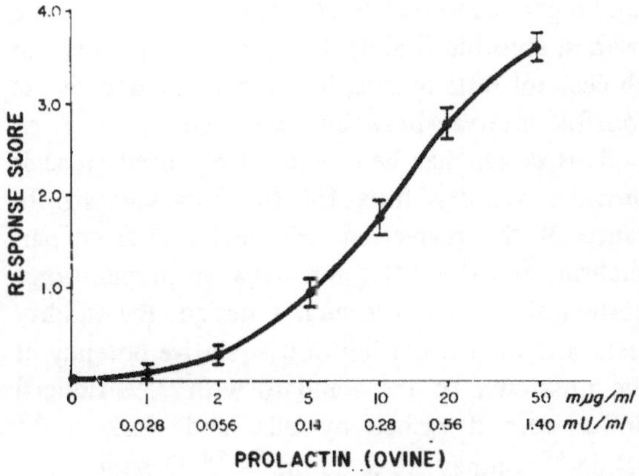

FIGURE 9. Standard curve of in vitro assay of prolactin using midpregnant mouse breast tissue based on combined experience of 44 assays. Vertical bars represent standard error of the means. (Reproduced from Kleinberg, D. L. and Frantz, A. G., 1971 (64). *J. Clin. Invest.*, 50, 1557, 1971. With permission.)

1. The Three Point Assay (2 + 1)

This is the simplest model, acceptable only for routine clinical studies. Three groups of animals are used: two receive the standard preparation and one the unknown specimen at a dose expected to give a response in the animals intermediate between the two doses of standard. The animals should be allocated to groups by an unbiased randomization procedure. However, such design provides no validity test of any kind. The calculation of the potency and approximate limits of error proceeds similarly to the calculation used in a four point assay.[24] A slope ratio assay can also be conducted according to a three point design;[58] the regression lines are obtained by joining the points representing the mean response to the standard and that to the unknown to the point for the blanks. Relative potencies and fiducial limits are calculated as described by Finney.[58]

2. The Four Point Assay (2 + 2)

The simplest symmetrical design for parallel line assays was first described by Gaddum.[66] Two doses of standard and two doses of unknown are included in an assay. The high and low doses of the two preparations have the same difference on the logarithmic scale, and the total number of animals is equally divided between doses. Use twofold intervals between doses for the ventral prostate weight increase, the ovarian weight augmentation method in intact immature rats, the in vitro bioassay methods for LH, the pigeon crop sac test, and the in vitro bioassay of prolactin using midpregnant mouse breast tissue. However, for the ovarian ascorbic acid depletion and for the ovarian cholesterol tests in rats, it is recommended to use fourfold intervals between dose levels.

This design has been extensively used since it provides validity tests for the bioassay: significance of the regression, the deviation from parallelism, and the difference between preparations. Statistical analysis for such a design, the validity tests, and the calculation of the relative potency of the unknown to the standard with 95% fiducial limits were described by Bliss and Marks,[67,68] Schild,[69] Finney,[58] Gaddum,[66,70] Diczfalusy,[24] and Borth et al.[71] The design can also be conducted using a litter mate control: one animal from each litter is assigned to each of the four dosage groups. In such a randomized block design, missing entries due to the loss of one animal or to incomplete litters may be calculated according to Snedecor[72] and Finney.[58] Thus, in litter mate controlled tests, it is always possible to preserve the symmetry of the assays; this facilitates the variance analysis of the data. The variances within groups should be examined to discover whether there is any evidence of heteroscedasticity. Assessment of the significance for the validity tests and of the precision of the relative potency of the bioassay, in theory, requires that the variance should be independent of the level of response to any dose. Evidence for heteroscedasticity may be established by Bartlett's test.[72] However, experience shows that even quite large departure from homoscedasticity would not seriously affect conclusions drawn from a good assay.[58] A significant difference between preparations at a level of probability of 0.05 would not per se invalidate the assay. It is, however, a danger signal and the precision of the potency estimation will be reduced.

Analytical dilution assays are based upon the similarity between standard and unknown; this implies the parallelism of the regression curves when log dose is used as a metamer. Any significant ($p<0.05$) deviation from parallelism indicates fundamental invalidity of the assay: qualitative biological differences exist between unknown and standard. Such results should be rejected.

3. The Six Point Assay (3 + 3)

In this symmetrical design three doses of standard and three doses of unknown are included: the three doses of each preparation have the same difference on the logarithmic scale and the total number of animals is equally divided between doses. In addition to tests on the difference between preparations and on deviation from parallelism, the validity of the 3 + 3 point assay is also based on tests on the difference in curvature and opposite curvature for the two regression lines. Complete statistical analysis, the validity tests, and the calculation of the relative potency of the unknown to the standard with 95% fiducial limits were described by Finney[58] and Borth et al.[71] Although providing more detailed information on the validity of the assay than the two previous designs, the six point assay requires a large number of animals, at least five or six per dose. Therefore, they are generally too tedious and cumbersome for clinical studies.

4. Multiple Assays

If several unknowns are to be assayed against the same standard preparation, all preparations may be included in one multiple assay. Naturally, such a multiple assay will be a little more confusing to perform. However, it permits a more economical use of the animals. If three unknowns are tested against the same standard in a 3 + 3 point multiple assay, 12 groups of animals will be required instead of 18 groups when single 3 + 3 point assays are conducted: three groups for the standard and three groups for each of the three unknowns. A further advantage is that the regression coefficient will be estimated from the combined evidence of all animals and thus more precisely determined. Statistical analysis, validity tests, and calculation of the relative potencies of multiple assays have been described by Finney[58] and Borth.[73] If a multiple assay requires too much labor, too much space, or more animals than are available at any one time, it can be divided into a series of small multiple assays conducted separately on all preparations but each including only a fraction of the total amount of animals to be involved; the results may eventually be combined at the end.[58]

5. Adjustment for Body Weight

Sometimes a positive correlation exists between the body weight and the response of the target organ. A common practice is to express the weight of the organ as a proportion of the body weight and to use this proportion as a response metameter instead of the absolute organ weight. Such an adjustment may often be very effective in improving the precision of an assay.[58] Initial body weight should rather be considered since the body weight may be affected by the substance under assay.

By this method only a single concomitant variant, i.e., initial body weight, is taken into consideration. When simultaneous adjustment for two or more concomitants is desirable a method based on covariance analysis should be employed.[58,67] Application of such a procedure to a bioassay for prolactin has been described by Finney.[58]

E. Extraction Procedures

1. Extraction of Gonadotrophins from Urine

Such extraction methods were required for the measurements of LH and FSH in urine samples by bioassay methods which are of low sensitivity. Most of these concentration methods are laborious and a considerable loss of gonadotrophic activity may occur during their performance. In addition, such assays on urine specimens require instructing the patient how to conduct a 24-hr urine collection, which has to be started at a specific time of the day with an empty bladder. Then all urine passed up to and including the same time the next day is collected into a single refrigerated container. It is recommended to perform a creatinine determination on 24-hr urine to serve as a further check of the completeness of the collection.

a. Kaolin Adsorption with Acetone Precipitation[74,75]

The extract is prepared as follows:

1. Adjust the urine specimen to pH 4.5 with glacial acetic acid.
2. Add 20 g of dry kaolin and stir briskly.
3. Filter the mixture and discard the filtrate; filtration may be hastened by addition of the filter aid Hyflo Supercel® to the kaolin.[76,77]
4. Wash the residual cake with water (2 l) containing 0.5% glacial acetic acid. Discard the washing water.
5. The gonadotrophin is eluted with 100 ml of 2 *N* NH_4OH passed through the kaolin cake twice and followed by 50 ml of distilled water. Adjust the eluate to pH 5.5 with glacial acetic acid.
6. Add 2 volumes of acetone, stir, and incubate in the refrigerator for 30 min.
7. Centrifuge and discard the supernatant.
8. Take the precipitate directly in distilled water for assay (or dry it overnight).

Further purifications of crude kaolin extracts have been described by Albert et al.[78-81] using ammonium acetate and ethanol, by Loraine and Brown[82] using tricalcium phosphate, and by Hipkin[83] using Sephadex®. These procedures reduce the toxicity of the extracts but aggravate the loss in gonadotrophic activity. However, this step seems to be essential to conduct valid assays with the ovarian ascorbic acid depletion test.[84] The mean percentage recovery achieved with such methods has been established at 76% by Loraine and Brown.[82]

b. Tannic Acid Precipitation[85]

The principle is to elute the activity from a tannic acid precipitate by differential elution.

1. Adjust the urine specimen to pH 4.0 by acetic acid.
2. Add 10 g of NaCl per 24-hr sample.
3. Add tannic acid (20 ml of a 10% tannic acid solution in distilled water) 1 min after Hyflo Supercel (10 to 12 g) and stir vigorously.
4. Filter and discard the filtrate. Pass diluted tannic acid (225 ml of a 0.1% solution containing 2 g Hyflo Supercel) through the cake in order to remove the urine.
5. Pass 96% alcohol (225 ml) through the cake in order to remove the excess tannic acid.
6. A solution containing 10% ammonium acetate in ammonium hydroxide at pH 11 in 80% alcohol is applied to remove the protein-bound tannic acid.
7. The active constituents are removed by lowering the percentage of alcohol from 80 to 40% in the same solution.
8. Cool the eluate, adjust the pH to about 7.0 with acetic acid, and precipitate the active constituents by increasing the alcohol concentration to 85%.
9. Collect the precipitate and wash it with alcohol and ether.

According to Johnsen,[85] 1/5, 1/10, 1/20, 1/40, 1/60, 1/80, and 1/320 of the 24-hr sample is injected into two mice, when nothing is known about the activity of the specimen. These test solutions are prepared in 0.1 *M* borate buffer.[85] The extracts prepared by tannic acid precipitation are less toxic to rats than those obtained by the majority of the other extraction methods.[85] The toxicity, however, is greater than that of the purified extracts obtained by the kaolin-acetone method.[80] Furthermore, according to Albert et al.[78,79,87] the LH potency assayed by the ovarian ascorbic acid depletion method is not affected by treatment by tannic acid. Recovery experiments conducted by Herbst et al.[88] indicated a recovery of 100% for the LH activity and 50% for the FSH activity from a pool of normal male urine added with the second international reference preparation of human menopausal gonadotrophins.

2. Extraction of Prolactin

Most of the bioassay methods for prolactin require an extraction of prolactin from urine or serum in order to be able to measure the concentration of the hormone in these body fluids.

a. Urinary Extract

As proposed by Coppedge and Segaloff,[89] add 1% by weight of NaCl and acidify with glacial acetic acid (pH 4.0) the total 24-hr urine. Then add 4 volumes of 95% alcohol, stir briskly, and store overnight at 4°C. The supernatant is siphoned off and the precipitate is recovered by centrifugation. Wash the precipitate (2X) with 95% alcohol and with ether (2X) and then allow to dry overnight at room temperature. Extract three times with 15 ml of distilled water; the three extracts recovered by centrifugation are dialyzed at 4°C against 6-hourly changes of 0.5% NaCl. Then add 100 mg NaCl and 4 volumes of 95% alcohol, stir vigorously, and store overnight at 4°C. The supernatant is decanted and the precipitate washed and dried as before. Just before the assay extract the dry precipitate with 40 ml of distilled water for 2 hr. The supernatant is decanted after centrifugation and used in the appropriate bioassay.

b. Serum or Plasma Extract

Simkin and Goodart[90] described an extraction procedure for serum prolactin as follows:

1. Add 20 ml of acetone and 0.1 ml of glacial acetic acid to 10 ml freshly taken whole blood or to 10 ml serum.
2. Add to the filtrate 60 ml acetone and 10 ml ether; store overnight at 4°C.
3. The next day decant the supernatant, wash the precipitate with ether, and allow it to dry.
4. Dissolve the dry precipitate in 1.6 ml of 1/20 *N* sodium hydroxide and adjust the pH to 7.0.
5. The volume of test solutions injected for a local pigeon crop sac assay is 0.5 ml.

III. IMMUNOLOGICAL PROPERTIES

When injected into individuals of heterologous species, gonadotrophins induce the production of antibodies. Since 1930, it has been known that sera of animals injected repeatedly with gonadotrophic extracts antagonized the biological activities of such crude preparations.[91,92] It took some 30 years before the practical possibility of using the immune response for the development of highly sensitive assays arose. First, a method based on inhibition of passive agglutination of sheep red blood cells coated with HCG and using anti-HCG sera has been proposed for detection and quantitative evaluation of HCG in urine of pregnant women.[93] Then radioimmunoassay methods derived from those described for insulin by Berson and Yalow[94] were developed for the measurement of pituitary gonadotrophins in serum or urine[95-108] and more recently for prolactin.[109-119]

A. Principle of the Radioimmunological Method

The method depends on the reaction between a hormone labeled with ^{131}I or ^{125}I and an antibody specific for the hormone. Unlabeled hormone in unknown samples or in standard preparations competes against the labeled antigen for binding to the antibody and thereby diminishes the binding of the labeled antigen (Figure 10). The degree of inhibition observed with the unknown is compared to that obtained with the standard tested at different doses. Thus, the ratio of labeled hormone bound to antibody (B) to the free labeled hormone (F) falls progressively in proportion to the increase in dose of the unlabeled hormone (Figure 11). Another representation consists of plotting, instead of B/F, the ratio between the amount of labeled hormone bound to the antibody in the presence of unlabeled hormone (B) and the amount of labeled hormone bound in the absence of unlabeled hormone (Bo) against the dose or the log of the dose (Figure 11).

Two modalities exist for the addition of labeled hormone (Figure 10):

1. Labeled hormone and unlabeled hormone are mixed and incubated together with the antihormone antibody = equilibrium conditions.

2. Unlabeled hormone and antihormone antibody are incubated first for some hours or days, then labeled hormone is added for a second incubation period = nonequilibrium conditions.

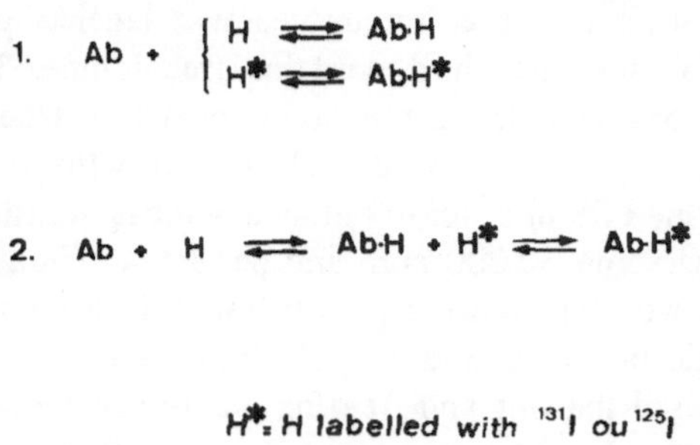

FIGURE 10. Schematic representation of radioimmunoassay of hormone: Ab is the antihormone antibody, H is the unlabeled hormone of the standard or unknown, and H* is the labeled hormone. At equilibrium, Ab H* (B) is separated from H* (F) and counted in a gamma spectrometer. In equilibrium conditions, Ab is incubated with both H and H*; in nonequilibrium conditions, Ab and H are incubated prior to addition of H*.

Nonequilibrium conditions or delayed addition of the labeled hormone increases the sensitivity of the assay.

The following aspects of the methodology will be analyzed: iodination of the hormones, production of antihormone sera, methods of separating antibody bound from free hormone, and description of a double antibody radioimmunoassay method.

1. Iodination of Hormones

The chloramine T method was developed some 11 years ago by Hunter and Greenwood[120] and is now almost universally used for the preparation of labeled proteins and peptides of high specific activity to be used in radioimmunoassays.[34] The isotopic abundance of ^{125}I preparation is always superior to that of ^{131}I. Freedlender[121] has shown that atom for atom, as used in radioimmunoassays, ^{125}I gives rather more than twice the number of counts given by ^{131}I. "At the present time, the general approach is first to try ^{125}I, and reports that a given hormone does not iodinate well with this isotope have not been generally substantiated."[122]

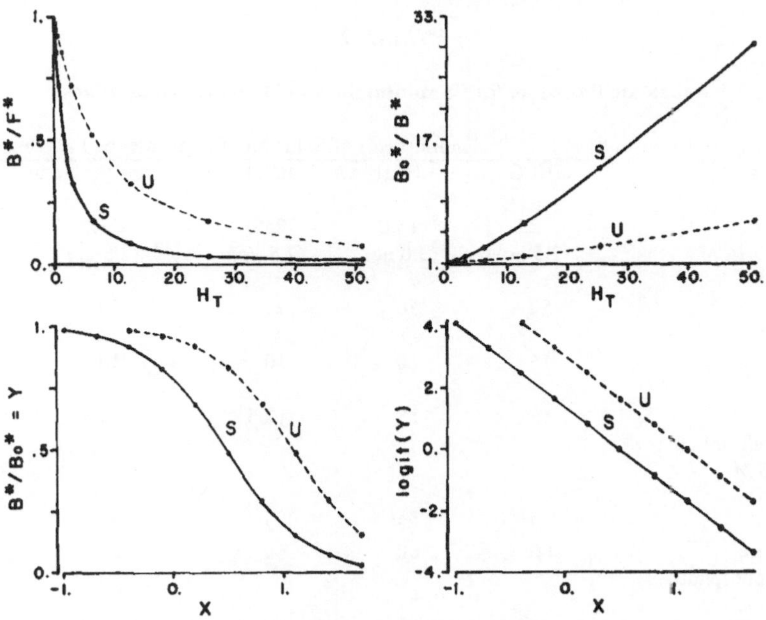

FIGURE 11. Four possible graphical representations of radioimmunoassay data: theoretical dose response curves of a standard (S) and an unknown (U). The doses are separated by twofold intervals. (Reproduced from Midgley, A. R. Jr., Niswender, G. D., and Rebar, R. W., *Acta Endocrinol.* [Suppl.] (Kbh.), 142, 163, 1969. With permission.)

Several procedures have been recommended for the iodination. The amounts and the sequence of combination of the reagents used in this laboratory are indicated in Table 2. The hormones are dissolved in phosphate buffered saline or distilled water and distributed in aliquots needed for one labeling and stored at −25°C in 1.0 ml multidose vials with a rubber stopper (Arthur H. Thomas, Philadelphia, Pa., Cat. No. 1780-A11).

The iodine 125 (IMS 30) may be ordered from the Radiochemical Center, Amersham, Great Britain. The chloramine T solution is prepared immediately before use. All solutions are injected in the multidose vial using a Hamilton syringe of appropriate capacity and a disposable needle (gauge 25, 1½ in.; Becton, Dickinson and Co., Rutherford, N.J.). The transfer of the solution of iodinated hormone and free iodide on top of a Sephadex column is carried out by a disposable tuberculin syringe using a disposable needle (gauge 25, 1½ in.).

For iodinated HCG, HLH, and HFSH, the chromatography for removal of unreacted radioiodide and reactants is conducted on a column of Sephadex G 50 (ϕ 1 cm; 16 cm) equilibrated with phosphate buffered saline (pH 7.5 and 0.05 *M*) and saturated with bovine serum albumin (2.0 ml of a 5% solution in phosphate buffered saline). One-milliliter aliquots are collected in tubes containing 1 ml of a 0.2% gelatine solution in 0.01 *M* sodium phosphate and 0.14 *M* sodium chloride at pH 7.0 (PBS). Two peaks of radioactive material are eluted: the first is the ^{125}I-hormone and the second contains the unreacted radioiodide. For ovine and human prolactin, the chromatography is conducted on a Sephadex G 100 column (ϕ 1 cm; 31 cm length); three peaks are eluted: the first corresponds to damaged hormone or to aggregates, the second is the immunoreactive labeled prolactin, and the third contains free iodine. The fractions containing the immunoreactive labeled hormone are combined and diluted with 0.2% gelatine-PBS in order to obtain a working solution, giving some 50,000 cpm and per 100-μl aliquots. This working solution is distributed in disposable plastic bottles stored at −25°C; in order to avoid repeated thawing and freezing one bottle contains enough tracer for one assay. The specific activity of the iodinated hormones lies between 50 and 300 μCi/μg. The immunoreactivity of the tracer as assessed according to Hunter[122] is maintained for 2 to 3 weeks.

2. Antihormone Sera

The two main advantages of radioimmunoassay procedures are their sensitivity and specificity,

TABLE 2

Iodination Procedure for Gonadotrophins and Prolactin Preparations

Reagents	Amounts (μl) of Solution of Reagents for Labeling of				
	HCG	HLH	HFSH	Ovine prolactin	Human prolactin
1. Hormone 0.01 *M* phosphate 0.14 *M* saline buffer pH 7.5	2.5 (2.5 μg)	15.0 (3.0 μg)	25.0 (2.5 μg)	2.5 (2.5 μg)	10.0 (1.0 μg)
2. Phosphate buffer pH 7.5, 0.5 *M*	50	50	25	50	50
3. Iodine 125 1 mCi/10 μl	15	10	10	10	10
4. Chloramine T 20 mg in 10 ml of phosphate buffer pH 7.5, 0.05 *M*	15	15	12.5	12.5	10*
Time of reaction (mix by flicking)	2′	2′	2′	45″	45″
5. Sodium metabisulfite 24 mg in 10 ml of phosphate buffer pH 7.5, 0.05 *M*	60	60	60	60	60
6. Transfer solution	200	200	200	200	200
100 mg KI in 10 ml of a 16% sucrose solution in distilled water	100	100	100	100	100

* For human prolactin: 35 mg of chloramine T in 10 ml phosphate buffer (pH 7.5, 0.05 *M*)

both of which depend predominantly on the use of antibodies that possess high avidity and selectivity for the material to be assayed.[123] Successful production of the ideal antiserum is considered to be the result of good fortune;[124] it is an art rather than a science (Table 3). Gonadotrophins usually available in adequate amounts for immunization[124] are good immunogens, with the exception of human menopausal gonadotrophins preparations.

Human prolactin has been isolated from pituitary glands by Hwang et al.[211] and by Lewis et al.[8] It thus became possible to develop human homologous radioimmunoassays for this hormone.[110,112] However, since highly purified human pituitary prolactin was available only in very limited amounts, several groups of investigators have used prolactin of various species, human prolactin from in vitro cultures of pituitary, or human prolactin from amniotic fluid for labeling and obtaining antisera. These different combinations of hormone and antisera are indicated in Table 4. At present, samples of highly purified human pituitary prolactin (VLS '1) and of anti-human prolactin serum are distributed by the National Pituitary Agency (Institute of Arthritis, Metabolism and Digestive Diseases, Bethesda, Md.) for radioimmunoassay purposes.

Rabbits are the preferred species to raise antigonadotrophic and antiprolactin sera: they are easier to handle and 40 to 80 ml of blood can be collected from an ear vein at 2- to 6- month

TABLE 3

A "Standard" Immunization Schedule for Rabbits.* (The Corresponding Information for Guinea Pigs is Given in Brackets.)

1. Take six or more young adult animals who have had at least 1 week to adjust to their environment.
2. Dissolve or, if insoluble, make a fine suspension of the immunogen in isotonic saline or distilled water at a concentration of 2 mg/ml (8 mg/ml). Take 1 ml (0.125 ml) of this aqueous solution per animal and emulsify with 3 ml (0.375 ml) of Freund's complete adjuvant as described in the text.
3. Primary injection. Inject 1-ml amounts of the emulsion deeply into the muscles of all four limbs (0.5 ml subcutaneously as a single injection, close to the midabdominal line).
4. "Booster" injections. Emulsions are prepared exactly as above and half the amounts used for the primary injection are given at 4-weekly intervals. The first booster injection consists of 1 ml intramuscularly into each hind limb, the second into each fore limb, and the third into the two hind limbs again (0.25 ml subcutaneously on each of three monthly occasions to guinea pigs).
5. Ten and again 14 days after the third booster injection, 25 to 50 ml of blood are taken from an ear vein and the sera from the two bleeds pooled (bleed 4 to 10 ml by cardiac puncture after 14 days).
6. Assess each individual antiserum and dispose of any animals that have failed to produce reasonable levels of antibody.
7. Three to 6 months later reinject survivors with the lower dose of immunogen spread over all 4 limbs. Bleed after 10 and 14 days as in item 5, and assess each antiserum.
8. Continue with injections and bleeds at 1- to 6-month intervals till high avidity antibodies have been obtained. Animals failing to produce satisfactory antisera should be disposed of after 18 months, since there is little prospect of further improvement.

* It is appreciated that many other schedules have been employed and have produced satisfactory results. In the authors' experience, however, this relatively simple scheme has resulted in the production of high avidity antibodies against a large number of peptide hormones including bovine parathyroid hormone, ACTH, vasopressin, oxytocin, angiotensin II, and digoxin-bovine albumin conjugate.

From Hurn, B. A. L. and Landon, J., Antisera for radioimmunoassay, in *Radioimmunoassay Methods,* Kirkham, K. E. and Hunter, W. M., Eds., Churchill Livingstone, Edinburgh, 1971, 121. With permission.

TABLE 4

Radioimmunoassay Methods for Human Prolactin : Hormone Preparations for Labeling and Antihormone Sera

Authors	Iodinated prolactin	Antiprolactin serum	
		Immunogen	Animal
1. Homologous Assay			
Bryant et al. (1971)[109]	Human prolactin from fetal pituitary cultures (M.C.)	Human prolactin from fetal pituitary cultures (M.C.)	Rabbit
Friesen et al. (1972)[110]	Human pituitary prolactin	Human pituitary prolactin	Rabbit
Sassin et al. (1972)[111]	Human pituitary prolactin	Human pituitary prolactin	Rabbit
Sinha et al. (1973)[112]	Human pituitary prolactin	Human pituitary prolactin	Rabbit
Cole and Boyns (1973)[113]	Human pituitary prolactin	Human prolactin from amniotic fluid	Rabbit
Badawi et al. (1973)[10]	Human pituitary prolactin	Human prolactin from fetal pituitary cultures (M.C.)	Rabbit
2. Mixed Homologous assay			
Hwang et al. (1971)[114]	Monkey pituitary prolactin	Monkey pituitary prolactin	Rabbit
L'Hermite et al. (1972)[115]	Ovine pituitary prolactin	Ovine pituitary prolactin	Rabbit
3. Heterologous assay			
Rodbard (1973)[116]	Human pituitary prolactin	Ovine pituitary prolactin	Rabbit
4. Mixed Heterologous assay			
Jacobs et al. (1971)[117]	Porcine pituitary prolactin	Ovine pituitary prolactin	Rabbit
Guyda et al. (1971)[118]	Ovine pituitary prolactin	Monkey pituitary prolactin	Rabbit
Greenwood et al. (1972)[119]	Porcine pituitary prolactin	Ovine pituitary prolactin	Rabbit

intervals. Three to ten animals should be immunized at once. Animals that have completely failed to respond after 3 months of immunization are better killed: it is likely that they will never respond. The most avid antisera are obtained by injections of a stable emulsion of an aqueous solution of the immunogen in complete Freund's adjuvant (Difco Laboratories, Detroit, Michigan) in equal volumes. The emulsion is better obtained by use of an MSE Homogenizer (Measuring and Scientific Equipment, London, S.W.1, England). Intramuscular injections at one or several sites on the back or in the four limbs are usually efficient. The dose of immunogen varies widely from one author to the other; it ranges from micrograms to milligrams.[124] Ross et al.[125] raised potent antisera against HCG subunits using 20 to 50 μg of immunogen per rabbit and for one series of injections. The emulsion was injected intracutaneously at 30 to 50 sites on the back, lateral surfaces of fore and hind limbs, and into two toe pads on each hind foot (Table 5).

The animals are bled from an ear vein 10 to 14 days after two or three booster injections. The serum is separated and merthiolate is added to a final dilution of 1:10,000. Aliquots are stored at −25°C. The serum aliquots are heated at 56°C for 20 min before they are to be used.

The major difficulty is that HCG, LH, and FSH share antigenic determinants: usually the antibodies raised against one of these hormones react with all three. This is equally the case for antibodies binding iodinated hormones and for antibodies neutralizing the biological activity.[126] Antiprolactin sera are obtained in similar conditions as those described for gonadotrophins. However, human prolactin is still scarcely available for immunization purposes. Detailed assessment of titer, specificity, and affinity of every batch of antiserum to be used in a radioimmunoassay is required.

a. Titer

The titer is the final dilution of the antiserum used in the assay tubes and not of the antiserum dilution added to these tubes. The titer is

TABLE 5

Procedures that May Be of Value in the Event of Failure to Produce Satisfactory Antisera by a Standard Immunization Schedule

1. Immunize larger numbers of a different breed of rabbit or change to guinea pigs, or eventually, chickens.
2. Change to a different batch of immunogen (especially if synthetic material is being used) or consider chemical modification of the immunogen.
3. Conjugate the immunogen to a substrate such as rabbit serum albumin. If this fails another substrate or another conjugation reaction should be tried.
4. Modify the adjuvant system employed. Always continue to emulsify with Freund's complete adjuvant, but first adsorb the immunogen to, for example, microparticles of carbon or aluminum hydroxide.
5. Change the sites of injection, for example, to the intraperitoneal, intrasplenic, and intranodal route.
6. Find out whether other workers have succeeded. If they have, there may be sufficient antiserum already available. Ask yourself whether it is still worth trying to produce your own.

From Hurn, B. A. L. and Landon, J., Antisera for radioimmunoassay, in *Radioimmunoassay Methods,* Kirkham, K. E. and Hunter, W. M., Eds., Churchill Livingstone, Edinburgh, 1971, 121. With permission.

evaluated by incubating serial dilutions of the antiserum with the amount of iodinated hormone to be used in a regular assay. The dilution selected is that which binds some 20 to 50% of the total amount of tracer. For antigonadotrophic and antiprolactin sera, this final dilution ranged from 1:100,000 to 1:1,000,000.

b. Specificity

In pratical terms, specificity of an antiserum for radioimmunoassay purposes is defined as the degree to which substances other than the hormone used as an immunogen or to be measured in the assay compete for binding.[127] "It is important to note that an antiserum capable of being used in a highly specific radioimmunoassay for one hormone still may bind a different radiolabeled hormone."[128]

Very extensive cross reactions exist between HCG and HLH. Therefore, radioimmunoassays using anti-HCG sera and labeled HCG are of similar specificity for the measurement of serum HLH than radioimmunoassays using anti-HLH sera and labeled LH. In such assays the interference of HFSH is usually not significant. The competition of HLH in radioimmunoassays for FSH is more striking and almost systematic. Very frequently, anti-FSH sera have to be absorbed with HCG before a satisfactory degree of specificity is attained.[127,128] Most anti-human prolactin sera exhibit little or no immunological cross reaction with growth hormone or human chorionic somatomammotrophin; the interference of these lactogenic substances in radioimmunoassays of human prolactin is usually negligible.

c. Affinity

Affinity is defined as the avidity with which an antiserum binds its corresponding antigen. It is frequently but not systematically correlated with the titer. It has been assumed that the binding of a hormone to an antihormone antibody is reversible and obeys the law of mass action.[127] The reversibility of hormone/antihormone reactions is much less apparent when tested by bioassay methods.[129,130] The populations of antibody molecules and antigenic sites involved in the neutralization of a biological activity and in the binding of a labeled hormone in a radioimmunoassay are very likely different.

As proposed by Odell et al.[127,128] the association constant may be estimated as a Michaelis constant for enzymes. Increasing amounts of hormone are added to a fixed amount of antiserum; the concentration of free hormone (moles/liter) at 50% saturation may be used to estimate an average association constant. For antigonadotrophic sera the association constants ranged from 1 to 3×10^{-9}M. In a radioimmunoassay, the slope of the dose-response curve is related to affinity of the antiserum: the higher the association constant, the steeper the dose-response curve, the greater the precision.

3. Methods of Separating Antibody-bound from Free Hormone

It is recommended that laboratory workers select separation methods which fit their resources and needs and use the same system of separation for as many different assays as possible.[131,209]

The separation methods may be considered under six categories (Table 6):

1. Electrophoretic and chromatoelectrophoretic methods: separation is achieved by differential migration in an electrical field on substrates with or without affinity for the labeled antigen.

2. Gel filtration: free hormones are much smaller than hormone-antihormone complexes. It is possible to separate them by gel filtration on cross-linked dextrans (Sephadex).

3. Nonspecific precipitation of antigen-antibody complexes: selective precipitation of antigen-antibody complexes may be obtained by various substances (sodium sulfite, ammonium sulfate, ethanol with natrium chloride, dioxane, etc.).

4. Immunoprecipitation of antigen-antibody complexes (double antibody method): one of the most successful separation techniques is based on the use of a second antibody raised in an heterologous species against the antihormone antibody. At the dilutions of first antibody in classical radioimmunoassays, the precipitates formed with the second antibody are too slight for accurate handling. Thus, normal rabbit or guinea pig serum in final dilutions ranging from 1/50 to 1/3,000 is added to the antihormone serum dilution.[131] In our group we have the greatest experience with the double antibody method.

5. Techniques based on adsorption of the labeled antigen: separation is achieved after conclusion of the primary immunologic reaction by selective adsorption of the free hormone into an insoluble substrate such as a dextran-treated charcoal, talc, Florisil®, microgranules of silica (QUSO®), or resins.

6. Solid phase antibodies: the separation is obtained by the use of insoluble antihormone antibodies which can be removed very simply by centrifugation. The antibodies may be insolubilized by polymerization with ethynl chloroformate or glutaraldehyde. Antihormone antibodies may also be coupled to polytetrafluoroethylene discs, to plastic (polypropylene or polystyrene) tubes, to cross-linked dextrans (Sephadex), or to beaded agarose (Sepharose®).

Advantages and inconveniences of these six categories of separation methods are listed in Table 6 as previously discussed.[132]

4. Performance of a Radioimmunoassay Using the Second Antibody Method

The following description is valid for radioimmunoassays of LH, FSH, and prolactin. Only the concentrations of the antisera will vary from one laboratory to another. All assay conditions described are not necessarily optimal; they are illustrative of the procedure of assay used in our laboratory.

a. Handling of Antihormone Sera

The antiserum is diluted to 1/1,000 in PBS with 0.05 *M* ethylenediaminetetraacetic (EDTA). This stock solution is stored at −25°C in small amounts necessary to prepare working solutions for 1 to 2 months of assays. The subsequent dilution, at least 1/400 of the stock solution, is prepared in a 1/600 solution of nonimmunized rabbit serum (NRS) in PBS with 0.05 *M* EDTA. This working solution is stored at −25°C for 1 to 2 months in aliquots necessary to run an assay of some 275 tubes. The dilution of NRS is established on the basis of its proportion with the second antibody dilution giving optimal antigen-antibody precipitation.

At the time of the assay, an aliquot of the working solution is thawed and 2 volumes are mixed with 3 volumes of cold 0.2% gelatine-PBS. This solution is kept in water and ice while dispensing 500 μl per assay tube using a Hamilton syringe with a disposable needle (Becton, Dickinson and Co., gauge 21, 1½ in.).

b. Handling of the Second Antibody

Usually antigonadotrophin and antiprolactin sera are raised in rabbits. The second antibody is, therefore, obtained by immunization of sheep against rabbit immunoglobulins isolated by ammonium sulfate precipitation according to Deutsch.[133] The immunoglobulins equivalent to 5 ml of complete rabbit serum are emulsified in 5 ml of complete Freund's adjuvant and injected intramuscularly in the posterior thighs of one

TABLE 6

Methods Separating Antibody-bound from Free Hormone

Type of method	Substrate	Advantages	Disadvantages
1. Electrophoresis or chromatoelectrophoresis	Cellulose acetate	Hormone damaged by iodination or by incubation may be recognized; it is not bound to the paper strip in absence of antihormone serum	Large space in cold room
	Chromatography paper (Whatman® 3 MM or 3 MC; Toyo® ≠ 514; DEAE paper)		Tedious technician work; strip counters (poor efficiency for 125 I); blind cutting of strips; hormones of very high specific activity are required.
2. Gel filtration	Sephadex®		Space requiring; time consuming; tedious technician work
3. Nonspecific precipitation of antigen-antibody complexes	Sodium sulfite, ammonium sulfate, ethanol, dioxane	Constant substances; readily available and cheap; rapid and simple	High blank values due to precipitation of large amounts of proteins other than antibodies; lower degree of precision; certain conditions are critical (ionic strength, type of diluent).
4. Immunoprecipitation or "double antibody method"	Ovine or goat antirabbit immunoglobulins or rabbit antiguinea pig immunoglobulins	Clean separation; little occlusion of free hormone in the precipitate (low blank value); great practicability	Skill and care for decantation of supernatant; possible impairment by precipitation of human immunoglobulins; variability in batches of second antibody; adjustment of dilution for each batch; expensive when the second antibody is not "home made."

TABLE 6 (continued)

Type of method	Substrate	Advantages	Disadvantages
5. Techniques based on adsorption of the labeled hormone	Dextran treated charcoal; talc; microgranules of silica (QUSO®), Florisil®, resins	Adequate for peptide hormones and steroids; cheap and rapid	Less clean separation than No. 4; non specific effects induced by difference in protein concentrations; affinity of the substrates for gonadotrophins-antigonadotrophins complexes
6. Solid phase antibodies	Discs in polytetra fluoroethylene; plastic tubes (poly-propylene or -styrene); polymerized antibodies; Cross-linked dextran; agarose	High practicability; very little technician work (automatization)	Storage of large number of tubes; lower precision; more antiserum is used

sheep, 5 ml on each side. Booster injections prepared as the primary injection are given every 4 weeks. The first bleeding takes place 10 to 14 days after the third injection; 500 ml of blood may be collected in a siliconated glass bottle (Baxter, S.A. Travenol Laboratories, Brussels) under vacuum. Serum is isolated (200 to 250 ml), heated in a water bath at 56°C for 30 min, and tested at 1/50, 1/100, 1/200, and 1/400 final dilutions for optimal precipitation with the final dilution of NRS (1/3,000). After addition of merthiolate at a final concentration of 1:10,000 it is stored at -25°C in enough aliquots for preparing 2.5 to 5 l of a working solution. When necessary an aliquot of sheep antirabbit immunoglobulin (anti-RGG) serum is diluted to the required working concentration in PBS containing 0.05 *M* EDTA and merthiolate at a final concentration of 1:10,000. Dispense amounts of the anti-RGG working solution in bottles necessary for one assay of some 275 tubes and store at -25°C.

At the time of distribution, thaw rapidly a bottle of anti-RGG working solution, keep it in a container surrounded by ice in water, and dispense 200 μl in the assay tubes using a Hamilton syringe with a disposable needle (Becton, Dickinson and Co., gauge 21, 1½ in.) and a Chaney Adaption. The syringe is used with a B-D Cornwall Continuous Pipetting Outfit (Becton, Dickinson and Co.).

c. Handling of Labeled Hormone

Thaw rapidly a bottle of labeled hormone working solution, shake gently, keep in a container surrounded by water with ice, and dispense 100 μl in the assay tubes using a Hamilton syringe with a disposable needle (Becton, Dickinson, Gauge 21 × 1 in.) and Chaney Adaption.

d. Handling of the Standard Preparation

The laboratory standard should be identical to the unknown; at least the population of hormone molecules detected in the radioimmunoassay should be identical or very similar both in the laboratory standard and in the unknown. This condition is the same as that required for standardization of bioassays. However, detecting differences in antigenic composition between gonadotrophin or between prolactin preparations by radioimmunoassays is much more frequent than detecting differences in their biological properties. Such qualitative differences are revealed by deviation from parallelism between the dose-response curves obtained in bioassays or in radioimmunoassays with the preparations tested. However, if the curves do not deviate significantly from parallelism, this is not an indication that the preparations are identical.

For LH – The laboratory standard is a semipurified preparation (HHG-B[134]) obtained from human pituitaries as described by Bettendorf et al.[135] A stock solution at 30 mI.U./μl is prepared from the lyophilized material in 0.2% gelatine-PBS with merthiolate at a final concentration of 1/10,000. Eight dilutions ranging from 0.3 to 38.4 mI.U./200 μl are prepared from the same aliquot of stock solution avoiding serial dilutions; these working solutions are distributed in aliquots necessary for three standard curves (±2.5 ml) and stored at -25°C for 4 to 6 months.

For FSH – The laboratory standard is a semipurified gonadotrophin (LER 907) preparation distributed by the National Pituitary Agency (National Institute of Arthritis, Metabolism and Digestive Diseases). A stock solution at 100 μg/ml is prepared as described for the laboratory standard of LH from a vial containing 1 mg of LER 907. The eight dilutions for the standard curves ranging from 10 ng to 1,280 ng/200 μl are obtained from the same aliquot of stock solution.

For human prolactin – The laboratory standard is a large pool of sera collected from women within the first 5 days postpartum. One milliunit (mU) is the amount of immunoreactive prolactin contained in one microliter of this pool. Seven dilutions are prepared for the standard curve; their immunoactivity ranges from 3 to 192 mU/200 μl.

For use, thaw the seven or eight tubes of a group in order to recombine a standard curve, shake gently, and dispense 200 μl per assay tube using a Hamilton syringe with a disposable needle (Becton-Dickinson, Gauge 21, 1½ in.) and Chaney Adaption.

e. Procedure for Setting up a Routine Assay

All manipulations are repeated in triplicate. Number all assay tubes (ϕ 12 mm × 75 mm) starting from 1 and place them in plastic-coated racks holding 30 tubes in 10 rows of 3. Each row of three tubes corresponds to a triplicate.

Total counts: tubes No. 1 to 3 i.e., the first

row of three tubes, each receives only 100 μl of labeled hormone.

Background: in tubes No. 4 to 6 dispense 200 μl of 0.2% gelatine-PBS and 500 μl of the 1/3,000 dilution of NRS. In tubes No. 7 to 9 the buffer solution is replaced by a pool of human serum.

Excess of antibody: tubes No. 10 to 12 each receives 200 μl of 0.2% gelatine-PBS and 500 μl of a 1/3,000 to 1/10,000 dilution of the antihormone serum; several dilutions are to be tested repeatedly and the one binding the maximum amount (75 to 95%) of labeled hormone is selected.

Buffer control: tubes No. 13 to 18 each receives 200 μl of 0.2% gelatine-PBS and 500 μl of the antihormone serum at the working dilution to be used in the assay.

Standard curve: in tubes No. 19 to 42 dispense 200 μl of the eight dilutions of the laboratory standard: three tubes for each dilution. Add 500 μl of the antihormone serum at the dilution to be used in the assay.

Unknown samples: from tube 43 dispense 200 μl of the unknown samples: three tubes for each sample. Add 500 μl of the antihormone serum at the dilution to be used in the assay. Do not exceed 40 to 50 samples per standard curve.

Serum controls: the last six tubes receive 200 μl of a pool of high LH serum concentration (three tubes) or 200 μl of a pool of low LH serum concentration (three tubes). Add 500 μl of the antihormone serum at the dilution to be used in the assay.

Repeat this series of tubes in order to involve all samples to be assayed. It is recommended that one dispenses first the 0.2% gelatine-PBS, the unknown samples, and the control sera to all series. Then the diluted antiserum is added rack by rack. Mix all tubes with a Vortex-Genie® mixer (Scientific Industries Inc., Springfield, Mass.); as soon as a rack is completed store immediately at 4°C. Never keep more than one rack at room temperature. After 24 to 48 hr add 100 μl of labeled hormone. As soon as a rack is completed mix all tubes with a Vortex-Genie mixer and store immediately at 4°C. After another incubation of 24 hr, dispense 200 μl of the second antibody rack by rack, mix with a Vortex-Genie mixer, and store again at 4°C. Do not add the second antibody to the "total counts." When large numbers of samples have to be handled at the same time, the distribution of the standard solutions, the unknowns, the antihormone serum, the labeled hormone, and the second antibody justify the use of an automated pipetting unit (Micromedic Systems, Inc., Philadelphia). After at least 48 hr, 4.0 ml of cold PBS is dispensed with a B-D Cornwall continuous pipetting outfit to all tubes except the "total counts." All tubes are transferred to a prerefrigerated (4°C) centrifuge of large capacity and centrifuged at 2,500 to 3,000 rpm for 30 min. To our knowledge the refrigerated centrifuges of the largest capacity are distributed by Hettich (200 tubes) and Lrist-Hereaus (270 tubes). Decant the tubes one by one in a glass funnel and collect the radioactive solution in a glass container. Allow the tubes to drain on absorbing paper. However, this is usually responsible for an important contamination of surfaces, which may be avoided by the use of a small sponge in plastic foam (ϕ 13.5 mm) connected to a vacuum flask. Keep the tube upside down, thrust the piece of plastic foam inside the tube, and let the last drop of radioactive solution be sucked out. Avoid shaking the tubes thoroughly during this operation.

All tubes are finally transferred to an Automatic Gamma Spectrometer of large capacity (300 to 500 tubes) such as that distributed by Wallac-LKB (LKB Producter AB, Stockholm, Sweden). All tubes are counted for 60 sec. The results of counting and the sequence number of each tube counted are printed and punched by a teletype (Teletype Inc., Skokie, Illinois) on a paper tape using the ASCII Code after all information required for the identification of the paper tape, of the assay, and of the unknown samples is obtained.

If one wishes to compare one or several preparations to the same standard, it is recommended to set up the assay as described above until tube No. 42. Then 8 to 10 dilutions of each of the preparations to be tested are dispensed (200 μl) in triplicate one after another. Do not exceed five or six preparations or repeat the standard curve.

5. Calculation of Results and Statistical Analysis of Radioimmunoassay Data

The dose-response curves obtained in radioimmunoassays are not straight lines, even when considering the log of the dose. Furthermore, the useful range of such curves is usually much greater than that of dose-response curves obtained in bioassays; the number of doses of standard tested exceeds only occasionally three in bioassays, but in radioimmunoassays it easily reaches eight or ten, although in both types of assays twofold intervals between doses are commonly used. Rodbard et al.[136] have shown that the relationship between the logit of B/Bo and the log of the dose is linear (Figure 10) and may be expressed as follows:

$$\text{Logit } B/Bo = \ln \frac{B/Bo}{1 - B/Bo} = b \ln \text{dose} + a \qquad (1)$$

In Equation 1, b is the slope of the regression line and a is the Y intercept. The hormone concentration (dose) in an unknown sample is calculated by introduction of the corresponding response (B/Bo) in the equation of the standard curve and considering the amount of serum or plasma dispensed in each assay tube. This relationship has been found not only valid for assays of LH and FSH but also for assays of prolactin. Although it is mathematically exact at or near the region of antibody saturation[137] where the conditions for a radioimmunoassay are optimal,[138] it is necessary to test in every assay whether there is a significant departure from linearity by inspection of the graph of logit B/Bo vs. ln dose or by an F test[139] when using a computer program for calculation of the results. Even with the help of a computer it is always advisable to introduce into the program a graphical representation of the dose-response curve after logit transformation.

The variance in B/Bo shows striking heteroscedasticity; it is a rather direct, essentially linear function of B/Bo.[140,141] This heterogeneity is aggravated by the logit transformation. Therefore, even if in theory the curve of logit B/Bo vs. ln dose is linear, it cannot be treated by least squares regression analysis, unless weighted values of logit B/Bo are used: the simplest weight is the reciprocal of the variance.[140] Such methods are tedious. However, computer programs now available calculate the slope of the regression line (b) and the Y intercept (a), predict the variance of the response variable, and calculate the parameters for quality control of the assay including 95% confidence limits for the estimated activity of unknowns.[141,142] However, classical variance analysis on unweighted values[139] usually gives a reasonable evaluation of the variance around the linear regression, of the departure from linearity (F test), of the significance of the regression (t test), and of the index of precision (λ). Calculation of the slope of the regression line (b), of the Y intercept (a), and of the hormone concentration in the unknown samples can be achieved easily by the use of an electronic desk calculator of limited capacity. Precision on the estimates of unknown sample may be simply expressed by the standard deviation calculated on each of the corresponding triplicates. When dose-response curves obtained with several hormonal preparations are to be compared, relative potencies and their 95% fiducial limits, validity tests on regression, linearity and parallelism, and the index of precision (λ) may be calculated according to classical techniques as detailed in the Geigy tables[139] or according to the multiple assay model described by Finney[58] and Borth[73] for bioassays.

IV. CHOICE OF A REFERENCE PREPARATION

In bioassays and radioimmunoassays of gonadotrophins and prolactin, reference preparations are required to determine comparatively bio- or immunopotencies of unknown samples since these hormones have not been isolated yet as physicochemically pure substances.

Laboratory standards are selected in a given laboratory for internal use in bioassays, in radioimmunoassays, or in both. The basic principle is that the biological or immunological activity in the standard should be identical to or, at least, very similar in qualitative terms to that in the unknown; dose-response curves should be parallel in the specific assay employed.

In order to compare bioassay or immunoassay results in physiological or clinical studies under widely varying conditions, it is necessary to express such results everywhere in the world with reference to a preparation accessible to every investigator. An *international standard* is a preparation to which an international unit has been assigned on the basis of an extensive international collaborative study. An *international reference preparation* is established to serve the same function but without a full international study, or where a collaborative study has shown that the preparation is not entirely suitable to serve as an international standard. However, an international unit may be assigned to an international reference preparation, especially when it is desirable to prevent the emergence of a multiplicity of systems for designating a potency.[143] *One international unit* is defined as the specific activity contained in a defined weight of a current international standard. International standards and international reference preparations are provided free of charge by the World Health Organization (W.H.O.) and are intended for the calibration by comparative assays of laboratory or national standards.

Working standards and research standards are not provided by the W.H.O.; they are made available by the National Institute for Biological Standards and Control (Medical Research Council, Holly Hill, London). They serve the same function as the international standards when, for one or the other reason, such standards are not available.

The international standards, the international reference preparations, the working standards, and the research standards are distributed in sealed vials containing a defined amount of units (maximum interampule variation = ± 1%). No attempt should be made to weigh out portions of the freeze-dried material. The entire content of the vial should be dissolved in water or other suitable buffer solution.

Valid calibration of a laboratory standard against the international standard (or international reference preparation) lies in the assumption that the active substances to be compared are identical in both preparations or so similar that the laboratory standard behaves in the system in the same way as a dilution of the international standard. It should be kept in mind that "with the exception of a few highly purified actions existing in small amounts in specialized laboratories, most preparations of glycoprotein hormones of human origin, crude as well as purified, are likely to contain more than one biologically active or immunoreactive substance. The components are sometimes called 'forms of the hormone' or 'molecular species,' they may share certain structural details, and, therefore, have some similar properties, but the fact remains that they are different chemical substances, which may include fragments, precursors, metabolites, and extraction artifacts in addition to the native hormone."[144] There is some evidence that a similar heterogeneity also exists in preparations of human prolactin and for gonadotrophins and prolactin in biological fluids such as serum and urine. Significant qualitative differences in biological properties, and/or in antigenic composition of hormones in biological fluids or in extracts of various purity are sometimes detected by bioassays and more frequently by radioimmunoassays. If international standards (or international reference preparations) are of such heterogeneity, there is a high probability of finding such qualitative differences in the forms of hormone present when testing them in radioimmunoassays against laboratory standards or biological fluids (serum, plasma, urine). In such conditions, the use of these international preparations for comparing the results of LH, FSH, and prolactin assays obtained in different laboratories is open to criticism. However, it has the advantage of avoiding the proliferation of local standards of similar heterogeneity but different from each other in composition, which has been the source of increasing confusion in the field of gonadotrophins during the last 10 years.

A. Standards for Bioassays

Impure gonadotrophin preparations have been set up as international standards for bioassays.

The second international standard for HCG was established in 1963 and defined on the basis of the same I.U. as the first international standard.[145] One I.U. of HCG is the amount of biological activity contained in 0.001279 mg of the second international standard. Each ampule contains 5.300 I.U.

The second international reference preparation of HMG was established in 1964.[146] One I.U. of FSH and LH is the FSH and LH activities contained in 0.2295 mg of the second international reference preparation of HMG. Each ampule contains 40 I.U. of FSH and 40 I.U. of LH. Converting factors have been reported to relate the FSH and LH activities of the second reference preparation to the first one.[147]

The second international standard of ovine prolactin was established in 1962.[148] Each ampule contains 220 I.U. One I.U. is defined as the biological activity contained in 0.04545 mg of the second international standard.

B. Standards for Radioimmunoassays

All previously described standards were not primarily designed to be used in radioimmunoassays. In most instances they are not even suitable reference preparations for radioimmunoassays of serum or plasma LH, FSH, and prolactin. The species specificity is much greater for the immunoreactivity of these three hormones than for their biological properties; hormones of different species showing the same biological properties may produce little or no reaction in the immunoassay designed for another species. Although human prolactin may be measured in a radioimmunoassay for ovine prolactin, the dose-response curve obtained with the human hormone is much flatter that that obtained with the ovine hormone. Also in radioimmunoassays for the human hormone, dose-response curves obtained with human and ovine prolactin deviate from parallelism. Thus, the second international standard of ovine prolactin cannot be used for the measurement of the human hormone.

It has been established that basic qualitative differences exist in the antigenic composition between pituitary and urinary gonadotrophin preparations, even when highly purified.[149] It has also been shown that serum or plasma gonadotrophins are immunologically closer to the pituitary hormones than the urinary hormones. In several immunoassays differences between serum and pituitary gonadotrophins are not even detectable. Since the second international reference preparation of HMG is of urinary origin, it is not suitable for radioimmunoassay of pituitary or serum LH and FSH; deviations from parallelism of the dose-response curves are found in radioimmunoassays.

So far no international standards are available for radioimmunoassays of human LH, FSH, and prolactin. At present, a great majority of radioimmunoassays of these hormones discriminate poorly; thus, more than one molecular form of the hormone will influence such systems. In order to avoid discontinuity in the estimates obtained in different laboratories with different assay systems and different heterogenous reference preparations, immunoassay standards should be set up consisting of essentially pure hormone peptides in their biologically intact native form.[144] On this basis immunoassay research standards for pituitary LH and FSH will be established for distribution to qualified investigators (National Institute for Biological Standards and Control). The use of the second international reference preparation of HMG for immunoassays should be discontinued as soon as such immunoassay standards can be widely distributed.

The MRC preparation coded 69/104 may be useful for immunoassays of LH and FSH as long as the pure materials are not available. According to the results of a collaborative study based on immunoassays, one ampule of MRC 69/104 should be regarded as containing immunoreactive FSH and LH equivalent to those found in 0.5 mg of the preparation LER-907 distributed by the National Pituitary Agency (National Institute of Arthritis, Metabolic and Digestive Diseases). This conversion factor is recommended for retrospective calculation.[144] A research standard (MRC 71/222) for human prolactin is also distributed by the National Institute for Biological Standards and Control. Each ampule contains about 1 μg of a partially purified human pituitary prolactin. By definition one ampule contains 10 mU. By the pigeon crop gland assay it was found that the preparation distributed in vials exhibited a biological potency of 10.68 I.U./mg.[150]

The W.H.O. Expert Committee on Biological Standardization has recommended that workers distinguish very clearly in every publication between "international units (bioassay)" and "international units (immunoassay)."

V. RELIABILITY CRITERIA OF BIO- AND IMMUNOASSAYS

According to Borth,[151] there are four criteria of reliability: accuracy, precision, specificity, and sensitivity. All four criteria are of equal importance in assessing the overall value of an assay. Practicability will also have to be considered.

A. Accuracy

The accuracy of a measurement is the closeness with which it approaches the true value. Accuracy is usually determined by the percentage recovery of pure compounds added to a sample before analysis. In addition, the linear regression analysis of the amount predicted vs. the amount measured in terms of the standard should give a slope not significantly different from 1.0 and a Y intercept not significantly different from zero. Recovery experiments have been conducted for bioassays of gonadotrophins in biological fluids (urine) involving extraction procedures; the percentage of recovered material ranged from 50 to 100%.[88,152] Similar experiments were also conducted for the radioimmunoassays of FSH in unextracted serum; 100% of the added FSH was recovered.[153] The assessment of accuracy requires a complete, clear, and unequivocal definition of the substance being measured. In bioassays and radioimmunoassays, the activity of an unknown sample is measured comparatively to the activity of a standard preparation. Thus, the definition of what is being measured is implicit in the definition of the standard, which is usually a crude or a semipurified preparation. Such a definition is therefore not often possible. The activity of an unknown preparation tested against the same standard is frequently different from one bioassay or one immunoassay to the other and also from immunoassays to bioassays. Indeed, the biological fluids, serum and urine, do not contain one type of protein hormone molecule but several; complete molecules, breakdown metabolic products, derivatives, subunits, aggregates, and partially synthesized molecules may be present. All these substances may exhibit different amounts of biological and immunological activities considering also that the biologically active sites are of a different nature than the immunologically active ones. All these substances may also have different half-lives, differently influencing bioassays according to the time intervals between the time of injection of the test solution and the time of killing of the animals. Antisera of different specificity may detect differently the various forms of hormone molecules according to their antigenic composition. Accuracy is also based on the assumption that the bioassay or the immunoassay is not affected by differences in composition of incubation media. High serum concentrations affect the ventral prostate weight test in rats;[25,154] in undiluted serum the activity is some two times higher than in saline or in serum diluted 1:15 (v/v). Serum proteins do not affect the ovarian weight augmentation test; when the same preparation was assayed comparatively in undiluted serum and in saline, the relative potency was not significantly different from 1.0.[149]

The relative accuracy of radioimmunoassays with regard to biological activity is fortuitous; it is basically due to the similarity existing between the heterogenous compositon of the preparations tested and that of the standard. For example, biological estimates and immunological estimates are in good agreement when urinary extracts are compared to urinary standards or when partially purified pituitary hormones even differing widely in their LH, FSH, TSH, GH, and prolactin content are compared to partially purified pituitary preparations.[140] However, major discrepancies are usually found between bioassay and immunoassay estimates when urinary preparations are tested against pituitary standards or vice versa.

B. Precision

The term "precision" is concerned merely with the repeatability of the measurements as, for instance, in duplicate determinations; precision is the extent to which a given set of measurements of the same sample agrees with the mean. Precision can be estimated by calculating the standard deviation of the mean of repeated determinations of the same sample. An alternative method is to assay a series of at least 20 samples in duplicate and then to determine the standard deviation of the mean difference between duplicates.[72] In biological assays the work involved in conducting replicate determinations is considerable, and thus evaluation of precision in this way has not been attempted so far. In such conditions it is rather the precision of individual assays that has been estimated, by calculation of the index of precision (λ) and of the 95% fiducial limits of error on the mean relative potency of the bioassay. The term "index of precision (λ)" was introduced by Gaddum.[65] It is an estimate of the standard deviation of the logarithms of the individual effective dose. In assays depending on graded effects the index of precision is calculated by dividing the standard deviation of the responses by the slope of the line connecting the response with the logarithm of the dose. According to Gaddum,[65,66] λ is independent of such factors as assay design and numbers of animals employed. It is therefore a very convenient criterion to compare assay precision from one method to another. The index of precision reported for the different types of bioassays and immunoassays described above is indicated in Table 7. According to Loraine and Bell,[86] assays in which λ is equal to 0.2 or less are very precise and suitable for clinical studies. Assays in which λ is between 0.2 and 0.3 can be used with reasonable confidence. When λ is greater than 0.3, the assays are unsuitable for quantitative work.

All assays listed in Table 7 are very precise, λ being below 0.2, except for the ovarian ascorbic acid and cholesterol depletion tests. The latter is not even supposed to be used in quantitative determination, λ being usually greater than 0.3. However, it has been included in this review for its

TABLE 7

Sensitivity as Reported for Several Bioassay and Immunoassay Methods for Human Gonadotrophins

Hormone	Type of assay	Assay method	Sensitivity (I.U.)	Index of precision (λ)
Luteinizing hormone (LH)	In vivo bioassay	Ventral prostate weight in hypophysectomized immature rat	0.75	0.20
		Total accessory reproductive organs in intact immature rats	0.75	0.12
		Ovarian ascorbic acid depletion test	0.5	0.16–0.20
		Ovarian cholesterol depletion test	0.1×10^{-6}	0.36
	In vitro bioassay	Radioligand–receptor assay	0.010	*
		Production of testosterone by mice testis	0.002	0.22
	Radioimmunoassay	Second antibody;	0.0001–0.002	0.05
		coated charcoal;	0.0008	
		disc solid phase;	0.002	
		chromatoelectrophoresis;	0.0001	
		dioxan precipitation;	0.002†	
Follicle stimulating hormone (FSH)	In vivo bioassay	Ovarian weight augmentation method in intact immature rats	1.00	0.05–0.20
		Ovarian weight augmentation method in intact immature mice	0.20	0.20
	Radioimmunoassay	Second antibody;	0.0001–0.004	0.05
		chromatoelectrophoresis	0.0005	–

* Between-assay variation (coefficient of variation) = 15%
† mIU of second international standard of HCG

very high sensitivity, which is 5,000,000 times that of the ovarian ascorbic depletion test.[29] In radioimmunoassays the variance of the response is not homogenous throughout the range of doses. The logit transformation of the responses as proposed by Rodbard et al.[136] aggravates even this heteroscedasticity. Thus, on a theoretical point of view, the curve relating the logit of the response to the log of the dose cannot be treated by the least squares regression analysis as described for bioassays unless the logit parameters are weighted according to the reciprocal of their variance. In such conditions, precision of radioimmunoassay methods has been only occasionally estimated by the index of precision calculated from individual assays. When calculated, the λ values indicate a greater precision for radioimmunoassays of LH and FSH than for bioassays of these hormones (Table 7). However, for radioimmunoassays of prolactin, the precision of the homologous ovine assay is much less and of the same order of magnitude than that of most bioassays. For most radioimmunoassays, the within-assay variation has been evaluated on the basis of coefficients of variation calculated from replicates of the same sample. The intra-assay variation is less than 10%. However, it is very difficult to establish on this basis a definite opinion as to what degree of precision is acceptable for a given assay method.

Fiducial limits of the mean relative potency[58,66,70,71] are defined for a given probability, commonly P=0.05. The actual potency of a preparation is expected to lie within the fiducial limits of an individual assay on 95 occasions if the assay is repeated 100 times. Most authorities agree that fiducial limits give more information on the error of an individual assay than a λ figure alone. As for the index of precision, the calculation of fiducial limits has been widely used in bioassays but only occasionally in radioimmunoassays. Reproducibility should also be included among the criteria of reliability of assays; it is the extent to which several measurements of the same samples agree with each other when obtained in different assays or in different laboratories. The British Pharmacopoeia (1958) explicitly counsels that, when possible, precision should also be assessed in terms of a simple error mean square calculated from potency estimates from independent assays. Index of precision, fiducial limits of error, and intraassay variations are calculated from the internal evidence of a single assay. "Potency estimation ought then to be independent of assay technique and the assessment of sampling variation expressed by the fiducial limits ought to have universal validity."[58] In bioassays, few examples exist establishing whether repeated assays will agree within the limits indicated by intraassay variances.

For bioassays, the work involved in conducting replicate determinations is very considerable. However, for radioimmunoassays, evaluation of between-assay variation is easily obtainable. It usually ranges between 10 and 15% and is somewhat greater than the intraassay variation.

C. Sensitivity

As defined by Borth,[155] sensitivity is "the smallest single result which, with some assurance, can be distinguished from zero, or, in statistical terms, as the smallest single result where fiducial limits, for say P=0.05, do not include zero." Data on sensitivity for bioassay and immunoassay methods are reported in Table 7. The data of Table 7 indicate that radioimmunoassay methods are 500 to 20,000 times more sensitive than the in vivo bioassays. It is also apparent that the sensitivity of the in vitro bioassays for LH is of the same order of magnitude as most radioimmunoassays. The most sensitive assay for LH is the ovarian cholesterol depletion test, but, as already mentioned, it is also the least precise of all assays reported in this review.

D. Specificity

Specificity is defined as the extent of interference in an assay by substances other than the one to be measured. It is based on "the list of names and potencies of compounds which, if present in the test sample, are known to contribute to the assay result."[144]

Specificity in bioassay resides in the structure of the active parts of the hormone and in their interactions with the specific cell receptor present in the target tissue of the animal. In immunoassay the specificity resides in the structure of the immunoreactive compounds and in their interactions with corresponding antibodies. For both bioassays and radioimmunoassays of each of the gonadotrophins and of prolactin the interference of the other pituitary hormones has been thoroughly investigated.

The specificity of the above described bioassay methods for LH has been confirmed by several investigators; the influence of FSH preparations exists only to the extent of their contamination with LH. The specificity of the bioassay methods for FSH has been established in similar conditions.[33,36,86]

In radioimmunoassay methods, specificity is predominantly influenced by immunologic cross reactions existing between glycoprotein hormones. It has been established that α subunits of HCG, LH, FSH, and TSH are antigenically very similar if not identical. Antisera raised against HCG, LH, FSH, and TSH are usually binding to each of these four iodinated hormones, but with wide differences in titer; these antisera are not monospecific. The specificity of radioimmunoassays may be achieved with such nonmonospecific antisera since it requires only the specific displacement or competition with unlabeled hormone.[127] Most anti-HCG or anti-HLH sera do not exhibit significant displacement with highly purified FSH preparations, which exceeds what can be expected from the LH contaminant.[156-161]

Most anti-FSH antisera exhibit a significant displacement of labeled FSH with unlabeled HCG, LH, or TSH and require specific absorbtion with HCG prior to reaching a sufficient degree of specificity in radioimmunoassays for FSH.[127,128] A specific radioimmunoassay also assumes the presence of only one molecular species of labeled hormone and thus the immunochemical purity of the HCG, LH, FSH, and prolactin preparations used for labeling. The presence of a major component iodinated simultaneously with the hormone implies the existence in the assay of two immunoreactive systems. By running inhibition curves with the purified hormone and other substances which might contaminate the tracer, biphasic curves should appear.[140]

Specificity implies also that in bioassays and immunoassays the dose-response curves obtained with purified preparations of the hormone and the unknown should be parallel. If significant deviation from parallelism leads to the conclusion that two preparations are not identical, the absence of any significant deviation from parallelism does not establish their identity. "Thus the presence of heterogeneous forms of hormone in plasma or tissues is not excluded by the apparent identity of their immunochemical reactivities with that of highly purified or synthetic homones."[162]

Immunoassays are commonly used to investigate several aspects of the physiology of a hormone. Quite erroneous conclusions may be drawn if such methods have no correlation whatsoever with the specific bioassay on which the definition of the hormone is based. Therefore, results of immunoassays on extracts from pituitary glands or on body fluids should follow the same rank order as bioassay results obtained on the same samples. However, one should remember that bioassay methods are not looking to the same structures of the hormones as immunoassay methods; a one-to-one correlation exists only exceptionally between the two assay methods. Discrepancies are related to the reference preparation employed or to the fact that the antiserum is directed towards portions of the molecule without activity in the specific bioassay.

Finally, fluctuations in hormone concentration in body fluids should parallel fluctuations in levels expected from differences in the physiological or pathological state of the individual (serum LH and FSH are lower after hypophysectomy than after castration; there is a midcycle peak of LH; serum LH and FSH decrease in menopausal women after administration of estrogen; serum prolactin increases after administration of psychotropic drugs).

E. Practicability

Practicability includes factors such as speed, cost, and the skill required in the performance of a method. Methods satisfying the most severe criteria of reliability may be quite unsuitable for routine use in clinical investigations conducted on a very large scale. This is in fact the case for all reliable bioassay methods described in this chapter, and even more so when using a stringent assay design; only a very limited number of specimens can be tested per week, the work involved is considerable, and the cost per estimation in terms of animals is very high. Thus, bioassay determinations should be restricted to research projects, to standardization of new gonadotrophin preparations, to the comparison between biological and immunological activities of the sample, or to estimation of the gonadotrophin neutralizing potencies of antisera tested in specific bioassays for LH, FSH, and prolactin. However, it should be kept in mind that hormones are by definition biologically active substances before they are to be considered as antigens.

For the measurement of hormones, radioimmunoassays are of much greater practicability than all available bioassays. Such radioimmunoassays are extensively used for routine purposes in clinical investigations; several hundreds of samples can be handled per week by the same technician. Obviously, the equipment for gamma counting is expensive, but the output is excellent.

VI. INTERPRETATION OF RESULTS

A comprehensive appraisal of normal values for serum or urinary LH, FSH, and prolactin is impaired by the diversity of reference preparations used in bioassays as well as in radioimmunoassays. Tabulation of "conversion factors" to interrelate standards used in different laboratories leads usually to dangerous speculations and to greater confusion. Such "conversion factors" may sometimes be useful but it should be kept in mind that they may differ from one laboratory to another and that even in the same laboratory they may change according to the strain of animals, to the assay method, or to the antiserum. Although general agreement exists in the changes in serum or urinary concentrations occurring in physiopathological or in experimental conditions, sometimes important discrepancies from one laboratory to another may be encountered in the absolute values. In the future, a greater uniformity will be achieved if all investigators are prepared to apply more systematically the recommendations of the Expert Committee for Biological Standardization of the W.H.O.

A. During Fetal Life

Already at 3 months of pregnancy, gonadotrophic and prolactin cells are detectable in the fetal pituitary by immunohistochemical staining using specific antisera. Kaplan et al.[163] have detected by radioimmunological methods the presence of immunoreactive FSH and LH not only in the pituitary but also in the sera of human fetuses as early as the end of the first trimester. As indicated in Table 8, the cord blood contains LH, FSH, and prolactin; the serum concentration of LH is some two times higher than the mean basal levels found during the follicular phase of a cycle. It has been shown that part of this immunological activity results from cross reactions with the small amounts of HCG crossing the placenta. At term, the mean level of HCG in the serum of the mother is at 55 I.U./ml, in contradistinction with a mean serum HCG/LH concentration of 30 mI.U./ml in the fetus. Serum FSH is significantly higher in the mother (8.4 mI.U./ml) than in the fetus (4.6 mI.U./ml). At birth, the serum levels of prolactin in cord blood are very high, actually as high as in the serum of the mother; there is no significant difference between mother and fetus.[164]

In addition, an excellent correlation ($p<0.001$) exists in serum prolactin concentration between the mother and in the fetus. Friesen et al.[110] found that even 24 hr after delivery prolactin serum concentration is still elevated in the child, ranging from 40 to 400 ng/ml; since the half-life of serum prolactin is of some 15 min, this would indicate that at birth prolactin secretion is enhanced in the fetus as compared to the situation during childhood. So far no sex differences have been reported for LH, FSH, and prolactin serum concentrations in the child at the time of delivery. There are also no significant differences in serum LH, FSH, and prolactin between the umbilical vein and the umbilical arteries.[164]

B. During Childhood and at Puberty

Using biological methods, both FSH and LH activities have been demonstrated in the urine of

TABLE 8

Serum HLH, HCG, HFSH, and Prolactin (HPRO) in the Fetus and in the Mother at the Time of Delivery as Measured by Radioimmunoassays. Mean ± s.e.m. Calculated from 20 Cases.

Hormones	Fetal cord blood	Mother (antecubital vein)
HCG/HLH[a]	30.3	55,400
(mIU/ml)	(25.0–36.7)	(45,600–67,000)
HFSH[b]	4.6	8.4
(mIU/ml)	(4,0–5,3)	(7,7–9,2)
HPRO[c]	1,050	1,190
(mU/ml)	(950–1,150)	(1,090–1,290)

[a] HCG/HLH is expressed in mIU/ml with reference to the second international standard of HCG. It has been found that in the cord blood 25% of this immunological activity could be attributed to HCG and 75% to HLH.

[b] HFSH is expressed in mIU/ml with reference to the second international reference preparation of human menopausal ganodotrophin.

[c] HPRO is expressed with reference to a laboratory standard, i.e., a pool of serum collected during the early postpartum: 1 mU is the amount of immunoreactive prolactin contained in 1.0 μl of this pool.

children prior to the age of puberty with considerable individual variations.[165] As indicated in Figure 12, more detailed studies were reported and serum gonadotrophins measured by radioimmunoassays in boys and girls from 1 year until adult age. Immunoreactive FSH and LH begin to rise in girls at a mean age of 9.8 and 10.9, respectively; for boys the corresponding ages are 11.0 and 12.0. Thus, it seems that the rise towards adult levels of both hormones begins some 1 year earlier in girls. As indicated in Figure 11, FSH and LH rise progressively over a period of 4 years. There is a twofold increase for FSH levels and a three- to fourfold increase for LH.[166]

Before serum LH and FSH begin to increase there is already in several cases a significant but variable degree of sexual maturation. However, there is a greater correlation between gonadotrophin concentration and stage of sexual development than between gonadotrophin concentration and chronological age.[167] Since maturation starts earlier in girls than in boys, this is likely to be related to the slightly delayed rise in FSH and LH occurring in boys as compared to girls. The development of urinary excretion of gonadotrophins has been found very similar to that of the serum levels of these hormones.[168]

So far no comprehensive study relating the development of prolactin serum concentration to chronological age or to the stages of sexual maturation has been published. According to Friesen et al.,[110] the mean level between 1 and 15 years is slightly higher (10.8 ng/ml) than between 16 and 84 years (5.0 ng/ml in men and 8.1 in women). Daughaday and Jacobs[169] reported no significant differences between normal children and adults.

C. Between Puberty and Menopause

The development with time of serum and

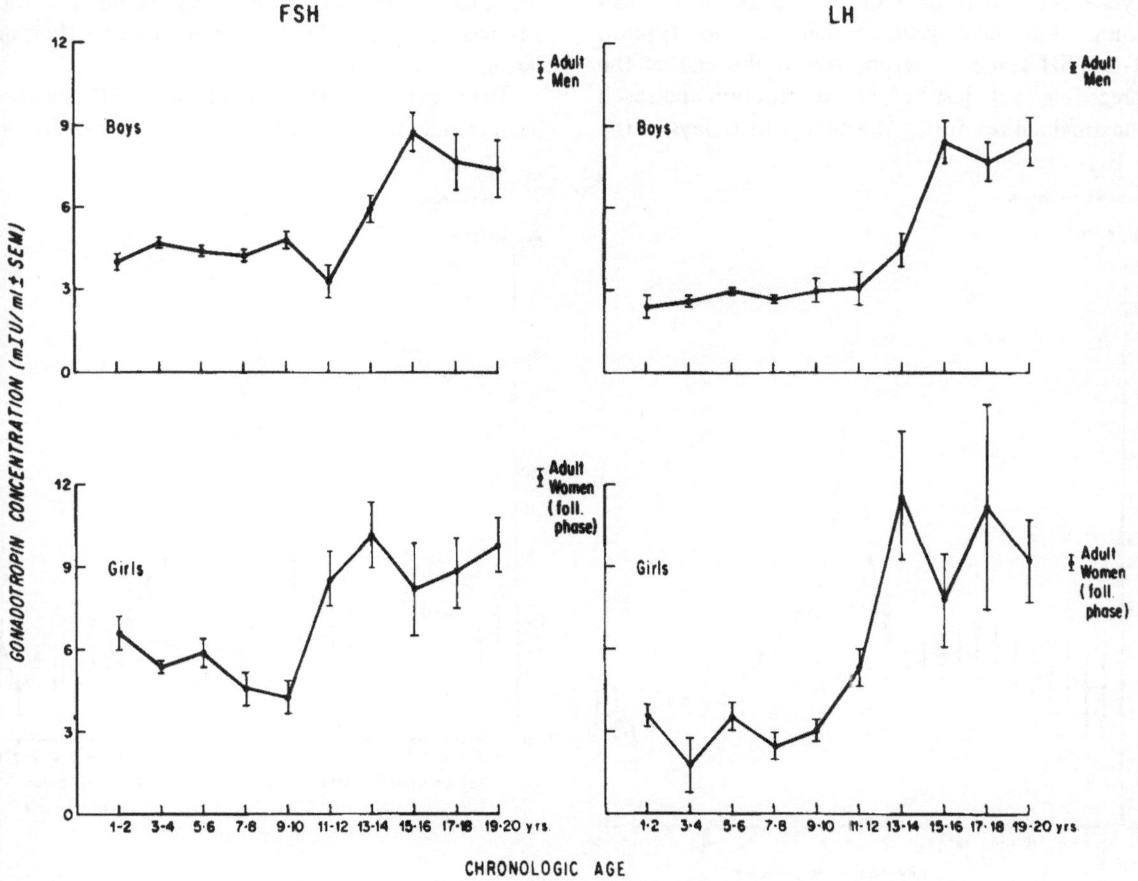

FIGURE 12. Serum LH and FSH concentrations as determined by radioimmunoassays in 198 boys and 111 girls aged 1 through 20 years analyzed at 2-year intervals. Vertical bars represent standard error of the means. (Reproduced from McArthur, J. W., Gonadotrophins in relation to sexual maturity, in *Gonadotrophins*, Saxena, B. B., Beling, C. G., and Gandy, H. M., Eds., Wiley Interscience, New York, 1972, 487. Copyright © 1972 John Wiley and Sons, Inc. Reprinted by permission.)

urinary gonadotrophins is basically different in women and in men. In women, cyclic discharges of LH and to a lesser extent of FSH occur every 28 days (Figure 13). Such rhythmic activity of the pituitary, under control of the hypothalamus, does not exist in men (Figure 14). The mean basal levels of serum LH and the mean values of the LH peaks at midcycle vary widely from author to author; they range from 6 to 30 mI.U./ml and 30 to 120 mI.U./ml, respectively.[170-175] The LH elevations at midcycle persist during 2 to 4 days with 1 of these days at four to ten times the baseline level. Serum concentration of LH increases slightly but progressively during the follicular phase[173] and the rise becomes much steeper 1 or 2 days before the peak value at midcycle. In most instances, the LH values are slightly higher during the follicular phase than during the luteal phase. Frequently, after the peak at midcycle, significant fluctuations (but of smaller amplitude than this peak) are found in serum levels of LH. Finally, occasional cycles with two or more LH peaks have been found, but such cycles are certainly not typical. The FSH levels in serum rise at the end of the preceding cycle just before menstruation and reach maximal values during the first 3 to 6 days of the actual menstrual cycle. There then follows a progressive decrease in serum FSH concentration; the lowest values precede by 1 or 2 days a second rise occurring at midcycle. This second peak of FSH is usually of slightly greater amplitude than the first one occurring in the early follicular phase but of much less amplitude than the midcycle peak of LH. In most cycles, the maximal values of LH and FSH coincide at midcycle. The mean absolute values for the two FSH peaks in serum vary widely from author to author, ranging from 6 to 16 mI.U./ml and 12 to 26 mI.U./ml, respectively. During the luteal phase the FSH levels in serum decrease progressively, reaching minimal values some 10 to 12 days after the midcycle peak.

Comparison between FSH and LH patterns in serum and urine during the menstrual cycle reveals that relative changes are of similar amplitude.[174] However, the FSH levels during the luteal phase remain slightly higher in comparison to those of the follicular phase in urine than in serum, and the LH peak at midcycle is broader in urine than in serum (Figure 13).

The urinary excretion of LH and FSH has also been studied by bioassays. The LH pattern is

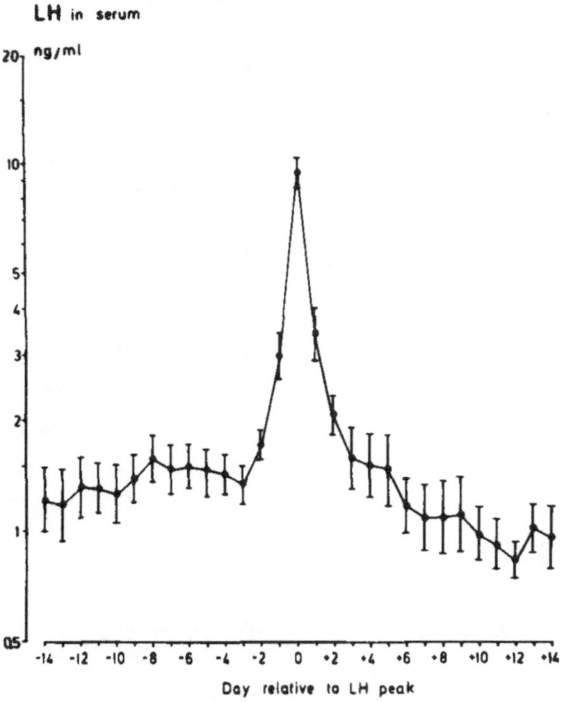

FIGURE 13A.

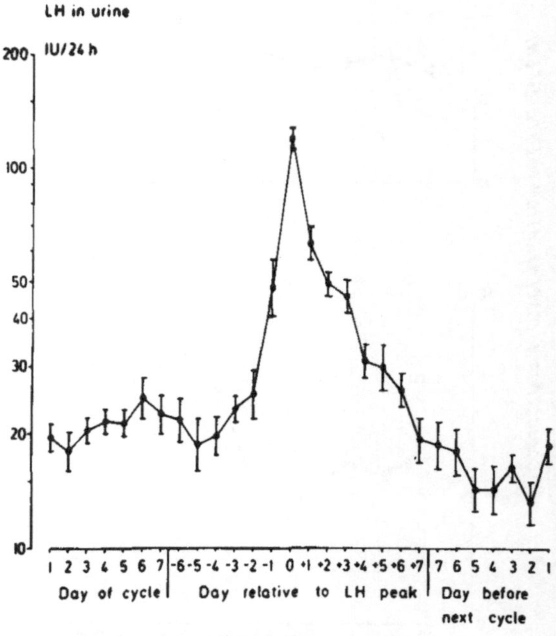

FIGURE 13B.

FIGURE 13. Development of serum and urinary LH and FSH concentrations as determined by radioimmunoassays during 11 and 19 menstrual cycles, respectively. Mean ± standard error of the mean. (Reproduced from Wide, L., Nillius, S. J., Gemzell, C., and Roos, P., *Acta Endocrinol.* [Suppl.] (Kbh.), 174, 1973. With permission.)

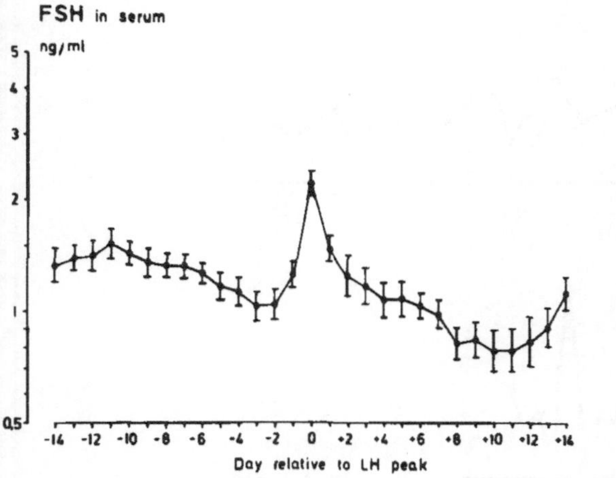

FIGURE 13C.

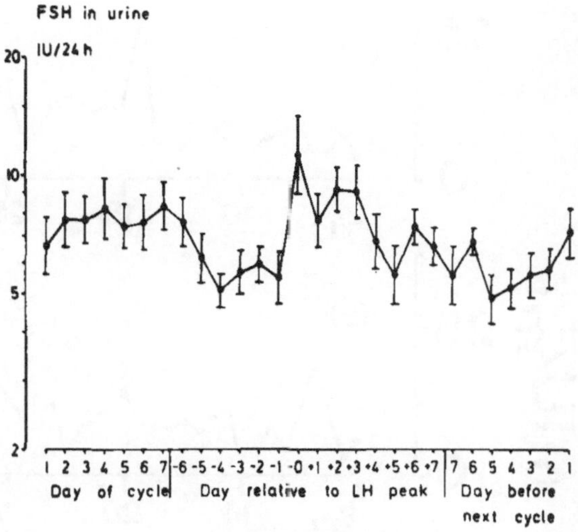

FIGURE 13D.

essentially the same as that obtained by radioimmunoassays.[176-179] However, the amplitude of the peak at midcycle is smaller; there is only a two- to threefold increase, since urinary LH rises from 2.0 I.U./24 hr during the follicular phase to 6 I.U./24 hr at midcycle (Figure 15).

Some authors also found by bioassay a higher level of LH excretion during the luteal phase of the cycle than during the follicular phase;[176,177] these observations are in contradiction with radioimmunoassay data. Urinary FSH as measured by bioassays is high during menses and the early follicular phase. Then the levels drop to low values at midcycle. A second rise is seen at variable times ranging from midcycle to midway of the luteal phase.[179,180-183] In some studies the midcycle peaks of LH and FSH evaluated by bioassays coincide.[177,178] From data obtained during 64 normal cycles, Stevens[179] found a peak of urinary FSH at 4.7 I.U./24 hr early in the follicular phase and low levels at midcycle (3 I.U./24 hr) followed immediately after the LH peak by a second rise (3.8 I.U./24 hr). The FSH levels decrease during the luteal phase and rise again at the end of the cycle (Figure 15). The development of prolactin serum concentration during the menstrual cycle is still being debated. Hwang et al.,[114] Friesen et al.,[110] McNeilly et al.,[184] and Ehara et al.[185] did not find any significant changes in serum levels of prolactin during the cycle. However, Vekemans et al.,[186] Delvoye et al.,[187] and Lequin[188] reported a significant increase at midcycle. When considering the temporal relationship between serum concentration of prolactin and that of pituitary gonadotrophins and ovarian steroids it appears that the rise in prolactin at midcycle is preceded by a rise in estrogens when the mean maximal value in prolactin occurred the day before the LH peak (Figure 16). It also may be seen from Figure 16 that during the luteal phase prolactin drops first to rise again and to delineate a biphasic peak between day +4 and day +8, closely parallel to the biphasic patterns of serum concentrations in estrogens and progesterone. The levels of prolactin fall very late in the luteal phase or early in the follicular phase of the next cycle to reach the minimal levels during the first 4 days of this cycle. The absolute values are still difficult to compare from laboratory to laboratory since very different reference preparations were used (Table 9).

Periodic fluctuations in plasma levels of LH suggesting a pulsatile discharge of the hormone from the pituitary were found in ovariectomized but not in regularly menstruating female monkeys.[189] In women, multiple discharges of LH occur throughout the day during all phases of the menstrual cycle but varying in frequency and magnitude between these phases. Such fluctuations are smaller for FSH than for LH.[174,190,191] They are also found in men.[192-196] Circadian rhythms of plasma FSH and LH with higher values in the morning than in the evening have been reported by Saxena et al.[161] and Faiman and Winter.[197] These data were not supported by several other investigators, who did not find such diurnal rhythms.[198-201] However, it has been established that a definite circadian periodicity in serum concentration exists for prolactin both in women and in men.[111,175,202] As indicated in Figure 17 the levels remain low during the day and rise during the night to

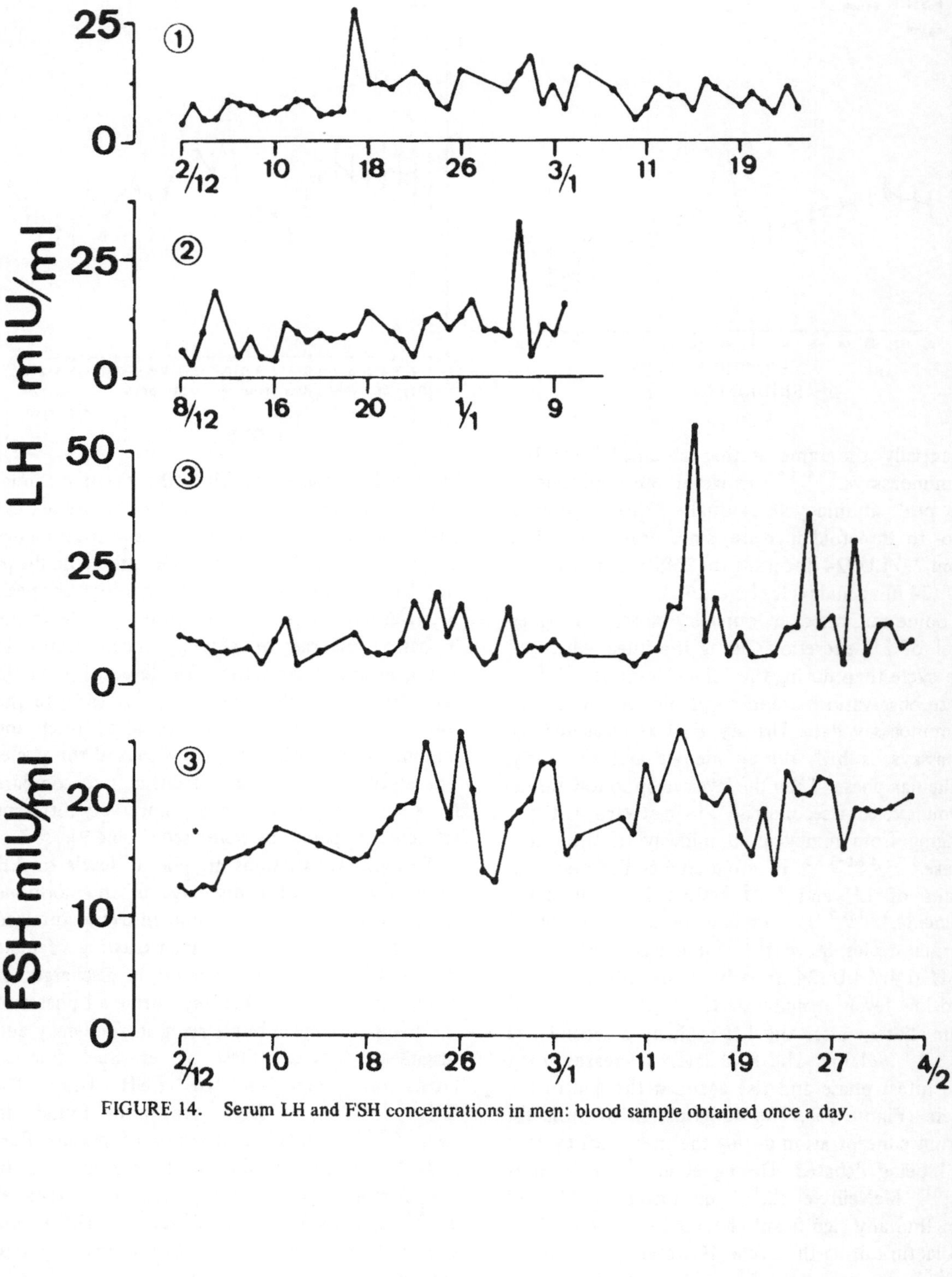

FIGURE 14. Serum LH and FSH concentrations in men: blood sample obtained once a day.

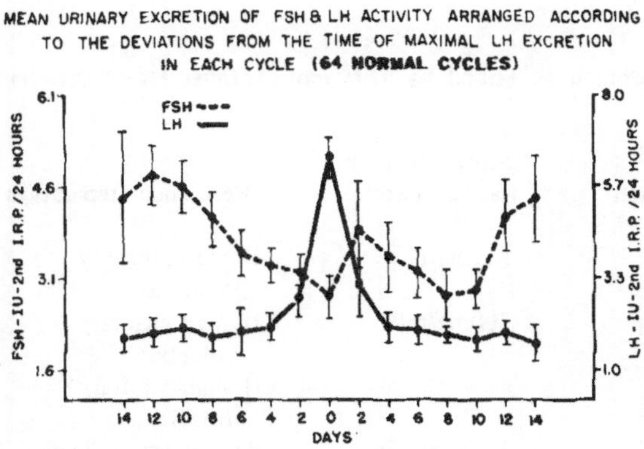

FIGURE 15. Composite pattern of FSH and LH from 64 normal cycles using the LH peak as reference point. Vertical bars represent standard deviation of the mean. (Reproduced from Stevens, V. C., *Acta Endocrinol.* [Suppl.] (Kbh.), 142, 338, 1969. With permission.)

reach peak values between 1 and 5 a.m. The values in men are significantly lower than those in women. Sometimes, prolactin serum concentration is still significantly higher at 8 a.m. than late in the morning or during the afternoon. When blood samples are collected every 30 min for a 24-hr period, irregular variations in serum prolactin concentration, sometimes of large amplitude, are found in addition to the release occurring with circadian periodicity (Figure 18).

"Clearly, a single blood sample taken at a time that is convenient for the investigator may contain levels of hormone which could lead to erroneous conclusions about its physiology or pathology since physiologically (or) pathologically significant release (or inhibition) of the hormone may be occurring at a different time of the day."[203]

D. At and After the Menopause

When considering the relationship existing between serum gonadotrophins and age it appears that no significant changes occur in women between 17 and 45 years.[174] However, in women between 45 and 50, serum LH and FSH begin to rise, reach a maximum between 50 and 60, and decrease slightly between 60 and 75. The relative increase in FSH is greater than that in LH (Figure 19).

In men, Wide et al.[174] also showed a slight but significant age-dependent increase in FSH serum concentration (Figure 20) as well as in urinary excretion of FSH. Such changes were not found for LH.

So far, no significant age-dependent changes in

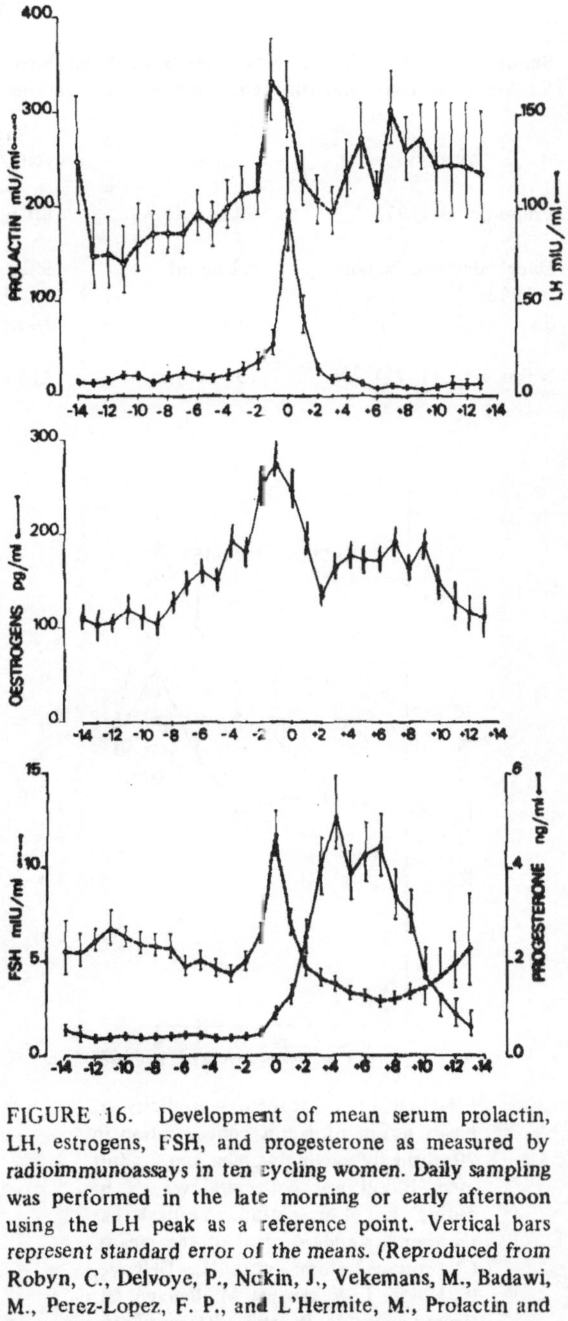

FIGURE 16. Development of mean serum prolactin, LH, estrogens, FSH, and progesterone as measured by radioimmunoassays in ten cycling women. Daily sampling was performed in the late morning or early afternoon using the LH peak as a reference point. Vertical bars represent standard error of the means. (Reproduced from Robyn, C., Delvoye, P., Nokin, J., Vekemans, M., Badawi, M., Perez-Lopez, F. P., and L'Hermite, M., Prolactin and human reproduction, in *Human Prolactin*, Pasteels, J. L. and Robyn, C., Eds., Excerpta Medica, Amsterdam, 1973, 167. With permission.)

serum prolactin levels were reported at the time of the menopause or in older women and men.

E. During Pregnancy and in the Postpartum

There is no report on the development of serum or urinary LH during pregnancy. The extensive cross reactions existing between LH and HCG are

TABLE 9

Serum Prolactin : Normal Mean Levels in Adult Men and Women as Found by Different Authors Using Different Radioimmunoassays and Different Reference Preparations

Authors	Men	Normally cycling women	Pregnant women (last trimester)	Reference preparation
Freisen et al. (1972)[110]	50 ng/ml	8.1 ng/ml	137 ng/nl	Human pituitary prolactin
Daughaday and Jacobs (1972)[169]	6.2 ng/ml	9.0 ng/ml	200 ng/ml	Human pituitary prolactin
Sinha et al. (1973)[112]	13 ng/ml	14 ng/ml	–	Human pituitary prolactin
Nokin et al. (1972)[202]	196 mU/ml	215 mU/ml	2,300 mU/ml	Pool of sera collected during early postpartum

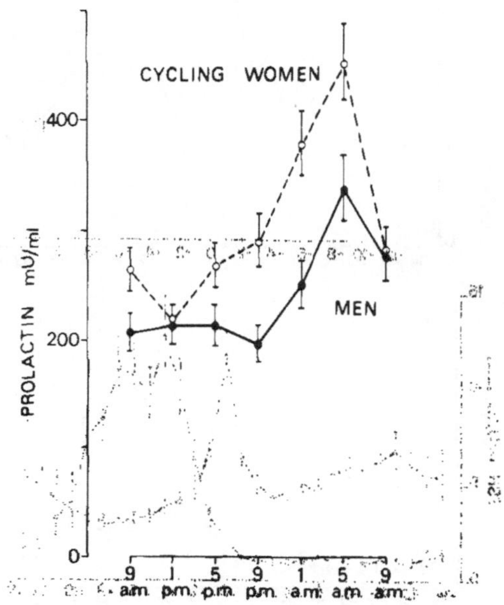

FIGURE 17. Circadian periodicity of mean serum prolactin concentration in 20 normally cycling women and 7 adult men. Blood was collected every 4 hr during a 24-hr period. Vertical bar represents standard error of the means. (Reproduced from Robyn, C., Delvoye, P., Nokin, J., Vekemans, M., Badawi, M., Perez-Lopez, F. P., and L'Hermite, M., Prolactin and human reproduction, in *Human Prolactin*, Pasteels, J. L. and Robyn, C., Eds., Excerpta Medica, Amsterdam, 1973, 167. With permission.)

responsible for this gap in the knowledge of the endocrine balance of pregnancy. The serum levels of FSH are very low during pregnancy and do not exhibit any significant changes until term and during the puerperium.

It can be seen from Figure 21 that serum prolactin concentration increases progressively during pregnancy, reaching the highest levels from the 26th week on until term.[204] As indicated in Table 9, a rise of greater magnitude, 10- to 20-fold instead of 5- to 7-fold, was reported by Hwang et al.[114] and by Daughaday and Jacobs.[169] It has been reported that the circadian periodicity in serum prolactin concentration disappears during pregnancy, at least during the last trimester.[175,205] Serum prolactin decreases within the range of values found in normal adults 1 to 2 weeks after delivery

Acknowledgment

We express our gratitude to Mrs. Preszow-Luftman for preparing the manuscript.

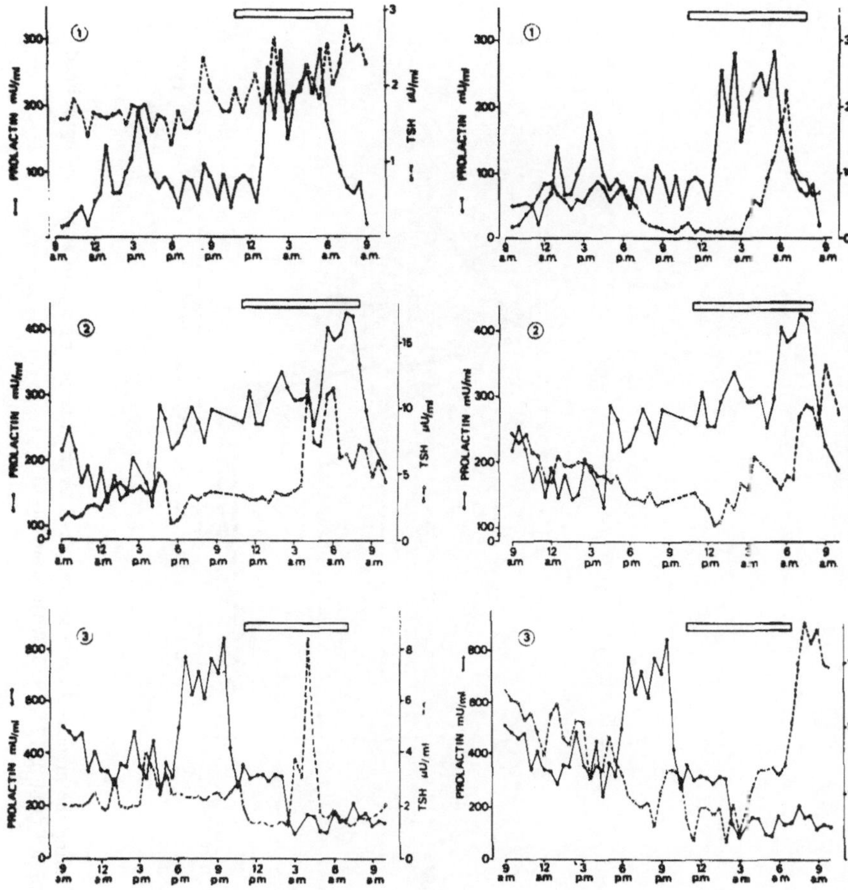

FIGURE 18. Comparison of circadian periodicities of prolactin (●—●) and TSH ((left ○--○) or cortisol (right ○ ○), as measured in serum samples collected through an indwelling catheter every 30 min during a 24-hr period from 2 women (1 and 2) and one man (3). Open rectangles represent the periods of sleep. (Reproduced from Robyn, C., Delvoye, P., Nokin, J., Vekemans, M., Badawi, M., Perez-Lopez, F. P., and L'Hermite, M., Prolactin and human reproduction, in *Human Prolactin*, Pasteels, J. L. and Robyn, C., Eds., Excerpta Medica, Amsterdam, 1973, 167. With permission.)

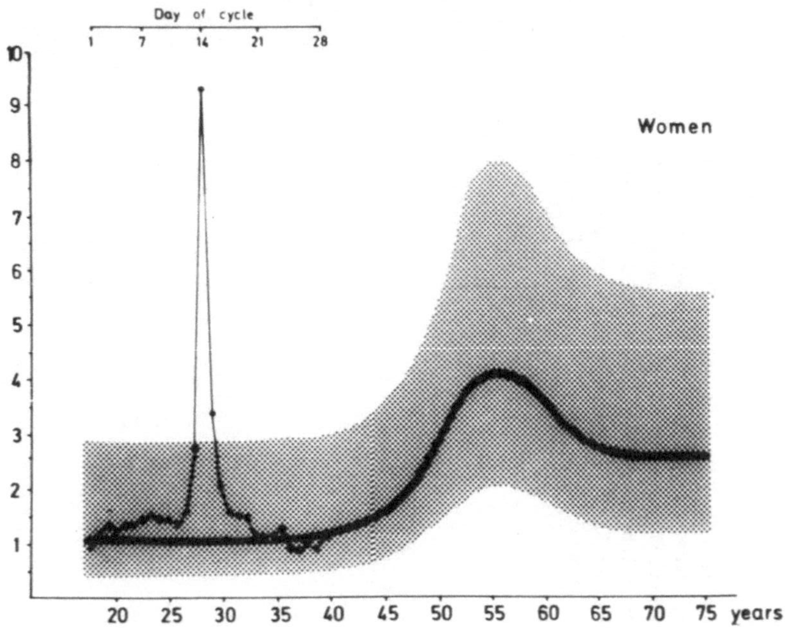

FIGURE 19A.

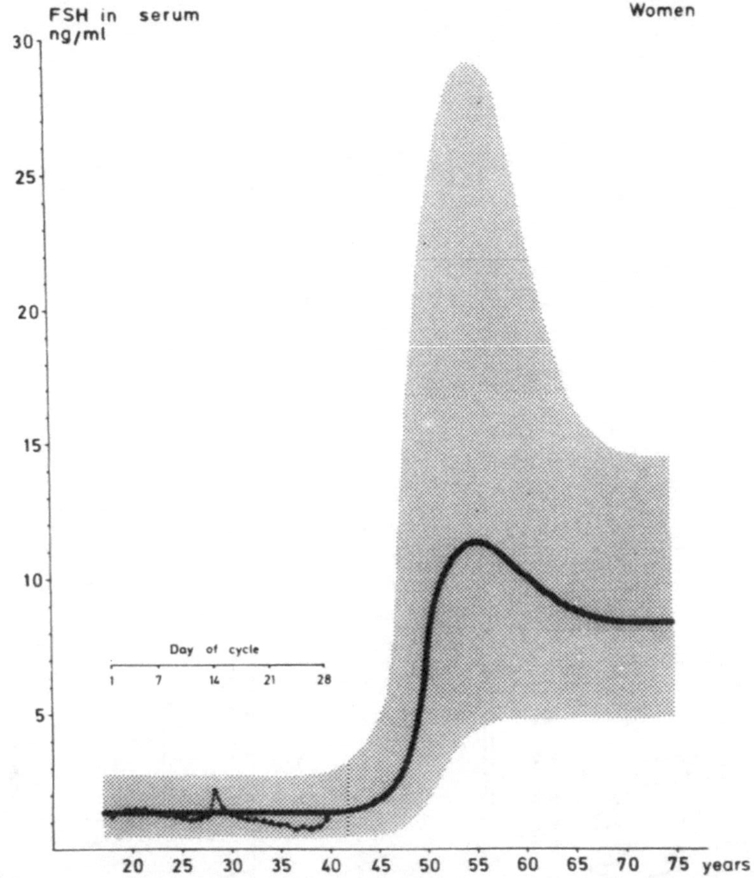

FIGURE 19B.

FIGURE 19. The changes in serum (A) LH and (B) FSH concentrations during the menstrual cycle as compared to the changes by age in women. (Reproduced from Wide, L., Nillius, S. J., Gemzell, C., and Roos, P., *Acta Endocrinol.* [Suppl.] (Kbh.), 174, 1973. With permission.)

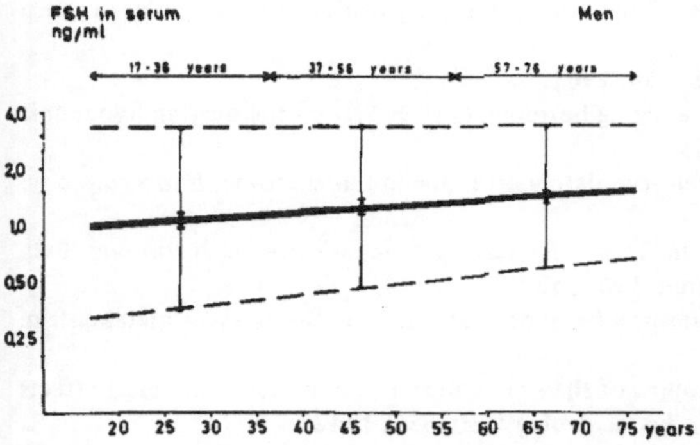

FIGURE 20. The FSH serum concentration in men aged 17 to 76 years. Vertical bars represent standard error of the means and dotted lines represent the 95% confidence limits. (Reproduced from Wide, L., Nillius, S. J., Gemzell, C., and Roos, P., *Acta Endocrinol.* [Suppl.] (Kbh.), 174, 1973. With permission.)

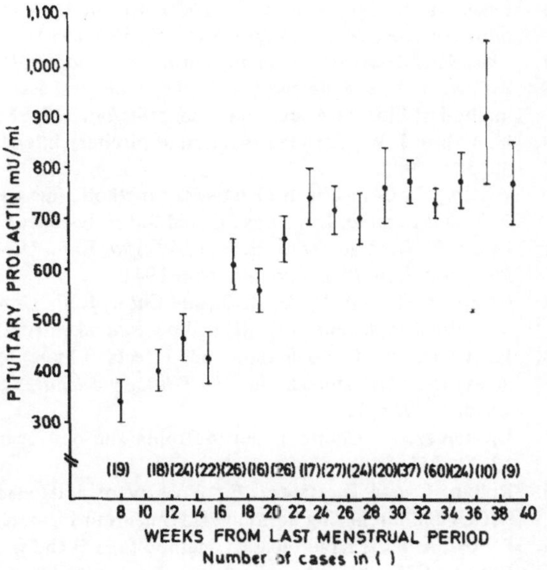

FIGURE 21. Development of prolactin serum concentration during pregnancy. Vertical bars represent standard error of the means. (Modified from Reference 204.)

REFERENCES

1. **Robyn, C., Leleux, P., Vanhaelst, L., Golstein, J., Herlant, M., and Pasteels, J. L.**, Immunohistochemical study of the human pituitary with anti-luteinizing hormone, anti-follicle stimulating hormone and anti-thyrotrophin sera, *Acta Endocrinol.* (Kbh.), 72, 625, 1973.
2. **Romeis, B.**, Die Hypophyse, in *Handbuch der Mikroskopischen Anatomie des Menschen,* Vol. 6(3), von Möllendorf, W., Ed., Springer-Verlag, Berlin, 1940, 1.
3. **Pearse, A. G. E. and van Noorden, S.**, The functional cytology of the human adenohypophysis, *Can. Med. Assoc. J.*, 88, 462, 1963.
4. **Pasteels, J. L., Gausset, P., Danguy, A., Ectors, F., Nicoll, C. S., and Varavudhi, P.**, Morphology of the lactotropes and somatotropes of man and rhesus monkey, *J. Clin. Endocrinol. Metab.*, 34, 959, 1972.
5. **Papkoff, H., Sairam, M. R., Farmer, S. W., and Li, C. H.**, Studies on the structure and function of interstitial cell stimulating hormone, *Recent Progr. Horm. Res.*, 29, 563, 1973.
6. **Ward, D. N., Reichert, L. E., Liu, W. K., Nahm, H. S., Hsia, J., Lamkin, W. M., and Jones, N. S.**, Chemical studies of luteinizing hormone from human and ovine pituitaries, *Recent Progr. Horm. Res.*, 29, 533, 1973.
7. **Reichert, L. E., Leidenberger, F., and Trowbridge, C. G.**, Studies on luteinizing hormone and its subunits: development and application of a radioligand receptor assay and properties of the hormone-receptor interaction, *Recent Progr. Horm. Res.*, 29, 497, 1973.
8. **Lewis, U. J., Singh, R. N. P., and Seavey, B. K.**, Problems in the purification of human prolactin, in *Prolactin and Carcinogenesis,* Boyns, A. R. and Griffiths, K., Eds., Alpha Omega Alpha Publishing, Cardiff, Wales, 1972, 4.
9. **Lewis, U. J.**, Discussion, in *Human Prolactin,* Int. Congr. Ser. No. 308, Pasteels, J. L. and Robyn, C., Eds., Excerpta Medica, Amsterdam, 1973, 38.
10. **Badawi, M., Bila, S., L'Hermite, M., Perez-Lopez, F. R., and Robyn, C.**, Comparative evaluation of radioimmunoassay methods for human prolactin using anti-ovine and anti-human prolactin sera, in *Radioimmunoassay and Related Procedures in Clinical Medicine and Research,* Istanbul, September 1973, in press.
11. **Albert, A.**, Bioassay and radioimmunoassay of human gonadotrophins, *J. Clin. Endocrinol. Metabl.*, 28, 1683, 1968.
12. **Gray, C. H. and Bacharach, A. L.**, *Hormones in Blood,* Academic Press, New York, 1961.
13. **Petrusz, P., Robyn, C., and Diczfalusy, E.**, Biological effects of human urinary follicle stimulating hormone, *Acta Endocrinol.* (Kbh.), 63, 454, 1970.
14. **Nicoll, C. A. and Bern, H. A.**, On the actions of prolactin among the vertebrates: is there a common denominator? in *Lactogenic Hormones,* Wolstenholme, G. E. W. and Knight, J., Eds., Churchill Livingstone, Edinburgh, 1972, 299.
15. **Riddle, O.**, Prolactin in vertebrate function and organization, *J. Natl. Cancer Inst.*, 31, 1039, 1963.

16. **Evans, H. M., Simpson, M. E., Tolksdorf, S., and Jensen, H.,** Biological studies of gonadotropic principles in sheep pituitary substance, *Endocrinology,* 25, 529, 1939.
17. **Ellis, S.,** Bioassay of luteinizing hormone, *Endocrinology,* 68, 334, 1961.
18. **Parlow, A. F. and Reichert, L. E., Jr.,** Biological assay of luteinizing hormone (LH, ICSH) by the ovarian hyperemia method of Ellis: an evaluation, *Endocrinology,* 72, 955, 1963.
19. **McArthur, J. W.,** The identification of pituitary interstitial cell stimulating hormone in human urine, *Endocrinology,* 50, 304, 1952.
20. **Witschi, E.,** Comparison of bio-assay methods (discussion), in *Recent Research on Gonadotrophic Hormones,* Bell, E. T. and Loraine, J. A., Eds., E. and S. Livingstone, Edinburgh, 1967, 52.
21. **Greep, R. O., Van Dyke, H. B., and Chow, B. F.,** Use of anterior lobe of prostate gland in the assay of metakentrin, *Proc. Soc. Exp. Biol. Med.,* 46, 644, 1941.
22. **Greep, R. O., Van Dyke, H. B., and Chow, B. F.,** Gonadotropins of the swine pituitary. I. Various biological effects of purified thylakentrin (FSH) and pure metakentrin (ICSH), *Endocrinology,* 30, 635, 1942.
23. **Jacobsohn, D.,** The techniques and effects of hypophysectomy, pituitary stalk section and pituitary transplantation in experimental animals, in *The Pituitary Gland,* Vol. 2, Harris, G. W. and Donovan, B. T., Eds., Butterworths, London, 1966, 1.
24. **Diczfalusy, E.,** Chorionic gonadotropin and oestrogens in the human placenta, *Acta Endocrinol.* [Suppl.] (Kbh.), 12, 1, 1953.
25. **Robyn, C. and Diczfalusy, E.,** Bioassay of antigonadotrophic sera. Assay of the human chorionic gonadotrophin (HCG) and luteinizing hormone (LH) neutralising potencies, *Acta Endrocrinol.* (Kbh.), 59, 261, 1968.
26. **Parlow, A. F.,** A rapid bioassay method for LH and factors stimulating LH secretion, *Fed. Proc.,* 17, 402, 1958.
27. **Parlow, A. F.,** Bioassay of pituitary luteinizing hormone by depletion of ovarian ascorbic acid, in *Human Pituitary Gonadotropins,* Albert, A., Ed., Charles C Thomas, Springfield, Ill., 1961, 300.
28. **Mindlin, R. L. and Butler, A. M.,** The determination of ascorbic acid in plasma, a macromethod and a micromethod, *J. Biol. Chem.,* 122, 673, 1938.
29. **Bell, E. T., Mukerji, S., and Loraine, J. A.,** A new bioassay method for luteinizing hormone depending on the depletion of rat ovarian cholesterol, *J. Endocrinol.,* 28, 321, 1964.
30. **Searcy, R. L. and Berquist, L. M.,** A new color reaction for the quantitation of serum cholesterol, *Clin. Chim. Acta,* 5, 192, 1960.
31. **Bell, E. T.,** Bioassay of gonadotrophin (discussion), in *Recent Research on Gonadotrophic Hormones,* Bell, E. T. and Loraine, J. A., Eds., E. and S. Livingstone, Edinburgh, 1967, 9.
32. **Catt, K. J., Dufau, M. L., and Tsuruhara, T.,** Studies on a radioligand-receptor assay system for luteinizing hormone and chorionic gonadotropin, *J. Clin. Endocrinol. Metab.,* 32, 860, 1971.
33. **Catt, K. J., Dufau, M. L., and Tsuruhara, T.,** Radioligand-receptor assay of luteinizing hormone and chorionic gonadotropin, *J. Clin. Endrocrinol. Metab.,* 34, 123, 1972.
34. **Greenwood, F. C., Hunter, W., and Glover, J. S.,** The preparation of I-131 labeled human growth hormone of high specific radioactivity, *Biochem. J.,* 89, 114, 1963.
35. **Dufau, M. L., Catt, K. J., and Tsuruhara, T.,** Gonadotrophin stimulation of testosterone production by the rat testis in vitro, *Biochim. Biophys. Acta,* 252, 574, 1971.
36. **Van Damme, M.-P., Robertson, D. M., Romani, P., and Diczfalusy, E.,** A sensitive in vitro bioassay method for luteinizing hormone (LH) activity, *Acta Endocrinol.* (Kbh.), 74, 642, 1973.
37. **Abraham, G. E. and Grover, P. K.,** Covalent linkage of hormonal haptens to protein carriers for use in radioimmunoassay, in *Principles of Competitive Protein-binding Assays,* Odell, W. D. and Daughaday, W. H., Eds., J. B. Lippincott, Philadelphia, 1971, 134.
38. **Murphy, B. E. P.,** Hormone assay using binding proteins in blood, in *Principles of Competitive Protein-binding Assays,* Odell, W. D. and Daughaday, W. H., Eds., J. B. Lippincott, Philadelphia, 1971, 108.
39. **Simpson, M. E.,** Assay of pituitary follicle-stimulating hormone in hypophysectomized female rats, in *Human Pituitary Hormones,* Albert, A., Ed., Charles C Thomas, Springfield, Ill., 1961, 276.
40. **Steelman, S. L. and Pohley, F. M.,** Assay of the follicle stimulating hormone based on the augmentation with human chorionic gonadotropin, *Endocrinology,* 53, 604, 1953.
41. **Brown, P. S.,** The assay of gonadotrophin from urine of non-pregnant human subjects, *J. Endocrinol.,* 13, 59, 1955.
42. **Bell, E. T.,** Bioassay of gonadotrophins (discussion), in *Recent Research on Gonadotrophic Hormones,* Bell, E. T. and Loraine, J. A., Eds., E. and S. Livingstone, Edinburgh, 1967, 33.
43. **Brown, P. S.,** Bioassay of gonadotrophins (discussion), in *Recent Research on Gonadotrophic Hormones,* Bell, E. T. and Loraine, J. A., Eds., E. and S. Livingstone, Edinburgh, 1967, 13.
44. **Riddle, O., Bates, R. W., and Dykshorn, S. W.,** Preparation, identification and assay of prolactin hormone of anterior pituitary, *Am. J. Physiol.,* 105, 191, 1933.
45. **Reece, R. P. and Turner, C. W.,** Galactin content of the rat pituitary, *Proc. Soc. Exp. Biol. Med.,* 36, 60, 1937.
46. **von Berswordt-Wallrabe, I., Flaskamp, D., and Jantzen, K.,** Demonstration of lactotrophic activity of human urinary extracts in rabbits, rats and pigeons, *Acta Endrocrinol.* [Suppl.] (Kbh.), 119, 126, 1967.
47. **Ben David, M.,** A sensitive bioassay for prolactin based on ^{3}H-methyl-thymidine uptake by the pigeon-crop mucous epithelium, *Proc. Soc. Exp. Biol. Med.,* 125, 705, 1967.

48. **Lyon, R.**, Lactogenic hormone prolongs the time during which deciduomata may be induced in lactating rats, *Proc. Soc. Exp. Biol. Med.*, 51, 156, 1942.
49. **Mishkinsky, J., Lajtos, Z. K., and Sulman, F. G.**, Initiation of lactation by hypothalamic implantation of perphenazine, *Endrocrinology*, 78, 919, 1966.
50. **Turner, C. W.**, *The Mammary Gland, Sex and Internal Secretions; a Survey of Recent Research*, 2nd ed., Allen, E., Ed., Williams and Wilkins, Baltimore, 1939, 740.
51. **Prop, F. J. A.**, Sensitivity to prolactin of mouse mammary glands in vitro, *Exp. Cell Res.*, 24, 629, 1961.
52. **Mishkinsky, J., Dikstein, S., Ben David, M., Azeroual, J., and Sulman, F. G.**, A sensitive in vitro method for prolactin determination, *Proc. Soc. Exp. Biol. Med.*, 125, 360, 1967.
53. **Frantz, A. G. and Kleinberg, D. L.**, Prolactin: evidence that it is separate from growth hormone in human blood, *Science*, 170, 745, 1970.
54. **Forsyth, I. A. and Myres, R. P.**, Human prolactin. Evidence obtained by the bioassay of human plasma, *J. Endocrinol.*, 51, 157, 1971.
55. **Loewenstein, J. E., Mariz, I. K., Peake, G. T., and Daughaday, W. H.**, Prolactin bioassay by induction of *N*-acetyllactosamine synthetase in mouse mammary gland explants, *J. Clin. Endocrinol. Metab.*, 33, 217, 1971.
56. **Turkington, R. W.**, Measurement of prolactin activity in human serum by the induction of specific milk proteins in mammary gland in vitro, *J. Clin. Endocrinol. Metab.*, 33, 210, 1971.
57. **Cowie, A. T. and Folley, S. J.**, Physiology of the gonadotrophins and the lactogenic hormone. III. The anterior-pituitary lactogenic hormone (prolactin), in *The Hormones*, Vol. 3, Pincus, G. and Thimarin, K. V., Eds., Academic Press, New York, 1955, 345.
58. **Finney, D. J.**, *Statistical Methods in Biological Assays*, 2nd ed., Griffin, London, 1964.
59. **Segaloff, A.**, Prolactin, in *Methods in Hormone Research*, Vol. 2, Dorfman, R. I., Ed., Academic Press, New York, 1962, 609.
60. **Grosvenor, C. E. and Turner C. W.**, Pituitary lactogenic hormone concentration and milk secretion in lactating rats, *Endocrinology*, 63, 530, 1958.
61. **Apostolakis, M., Theile, L., and Schöttle, H.**, Clinical applications (discussion), in *Recent Research on Gonadotrophic Hormones*, Bell, E. T. and Loraine, J. A., Ed., E. and S. Livingstone, Edinburgh, 1967, 251.
62. **Herlyn, U., Jantzen, K., Flaskamp, D., Hoffmann, H., and von Berswordt-Wallrabe, I.**, A modification of the pigeon crop sac assay for lactotrophic hormone determinations by means of the addition of prednisolone, *Acta Endocrinol.* (Kbh.), 60, 555, 1969.
63. W.H.O. Memorandia, 1972 (distributed at the time by the National Institute for Biological Standards, London).
64. **Kleinberg, D. L. and Frantz, A. G.**, Human prolactin: measurement in plasma by in vitro bioassay, *J. Clin. Invest.*, 50, 1557, 1971.
65. **Gaddum, J. H.**, Reports on biological standards. III. Methods of biological assay depending on a quantal response, *Med. Res. Counc. Spec. Rep. Ser.* (Lond.), No. 183, 1933.
66. **Gaddum, J. H.**, Bioassays and mathematics, *Pharmacol. Rev.*, 5, 87, 1953.
67. **Bliss, C. I. and Marks, H. P.**, The biological assay of insulin. Some general considerations directed to increasing the precision of the curve relating dosage and graded response, *Q. J. Pharm. Pharmacol.*, 12, 82, 1939.
68. **Bliss, C. I. and Marks, H. P.**, The biological assay of insulin. The estimation of drug potency from a graded response, *Q. J. Pharm. Pharmacol.*, 12, 182, 1939.
69. **Schild, H. O.**, A method of conducting a biological assay on a preparation giving repeated graded responses illustrated by the estimation of histamine, *J. Physiol.* (Lond.), 101, 115, 1942.
70. **Gaddum, J. H.**, Simplified mathematics for bioassays, *J. Pharm. Pharmacol.*, 5, 345, 1953.
71. **Borth, R., Diczfalusy, E., and Heinrichs, H. D.**, Grundlagen der statistichen Auswertung biologischer Bestimmungen, *Arch. Gynäkol.*, 188, 497, 1957.
72. **Snedecor, G. W. and Cochran, W. G.**, *Statistical Methods*, 6th ed., Iowa State University Press, Ames, Iowa, 1962.
73. **Borth, R.**, Simplified mathematics for multiple bioassays, *Acta Endocrinol.* (Kbh.), 35, 454, 1960.
74. **Albert, A.**, Human urinary gonadotropin, *Recent Progr. Horm. Res.*, 12, 227, 1956.
75. **Albert, A.**, The kaolin-acetone method for processing urine for the routine clinical assay of human pituitary gonadotrophin, *Acta Endocrinol.* [Suppl.] (Kbh.), 106, 5, 1966.
76. **Borth, R., Lunenfeld, B., and Menzi, A.**, Comparison of kaolin-acetone methods, in *Human Pituitary Gonadotropins*, Albert, A., Ed., Charles C Thomas, Springfield, Ill., 1961, 13.
77. **Stevens, V. C.**, Diurnal variation in ascorbic acid content of the ovary of pseudopregnant rat, *Endocrinology*, 74, 493, 1964.
78. **Albert, A., Derner, I., Stellmacher, V., Leiferman, J., and Barnum, J.**, Purification of pituitary gonadotropin from urine of postmenopausal women, *J. Clin. Endocrinol. Metab.*, 21, 1260, 1961.
79. **Albert, A., Kobi, J., Leiferman, J., and Derner, I.**, Purification of pituitary gonadotropin from urine of normal men, *J. Clin. Endocrinol. Metab.*, 21, 1, 1961.
80. **Albert, A., Stellmacher, V., and Leiferman, J.**, Method of purification of human pituitary gonadotropin for clinical assay, *J. Clin. Endocrinol. Metab.*, 21, 856, 1961.

81. **Albert, A., Derner, I., Stellmacher, V., Leiferman, J., and Barnum, J.**, Purification of pituitary gonadotropin from urine of eunuchs, *J. Clin. Endocrinol. Metab.*, 21, 99, 1962.
82. **Loraine, J. A. and Brown, J. B.**, A method for the quantitative determination of gonadotrophins in the urine of non-pregnant human subjects, *J. Endocrinol.*, 18, 77, 1959.
83. **Hipkin, L. J.**, Gonadotrophin-inhibiting properties of kaolin-acetone extracts of human urine, *J. Endocrinol.*, 38, 39, 1967.
84. **Rosemberg, E.**, Bioassay of gonadotrophins (discussion), in *Recent Research on Gonadotrophic Hormones*, Bell, E. T. and Loraine, J. A., Eds., E. and S. Livingstone, Edinburgh, 1967, 43.
85. **Johnsen, S. G.**, A clinical routine method for the quantitative determination of gonadotrophins in 24 hours urine samples, *Acta Endocrinol.* (Kbh.), 28, 69, 1958.
86. **Loraine, J. A. and Bell, E. T.**, *Hormone Assays and Their Clinical Application*, 3rd ed., E. and S. Livingstone, Edinburgh, 1971, 14.
87. **Albert, A., Derner, I., Leiferman, J., Stellmacher, V., and Barnum, J.**, Studies on the biologic characterization of human gonadotropins. VII. Urinary gonadotropins of men, postmenopausal women and eunuchs, *J. Clin. Endocrinol. Metab.*, 21, 839, 1961.
88. **Herbst, A. L., Bell, E. T., and Loraine, J. A.**, Recovery of added gonadotrophins from human male urine following extraction by tannic acid, *Endocrinology*, 80, 378, 1967.
89. **Coppedge, R. L. and Segaloff, A.**, Urinary prolactin extraction in man, *J. Clin. Endocrinol. Metab.*, 11, 465, 1951.
90. **Simkin, B. and Goodart, B. A.**, Preliminary observations on prolactin activity in human blood, *J. Clin. Endocrinol. Metab.*, 20, 1095, 1960.
91. **Collip, J. B.**, Placental hormones, *Internat. Clin.*, 4, 51, 1932.
92. **Collip, J. B., Seleye, H., and Thomson, D. L.**, The antihormones, *Biol. Rev.*, 15, 1, 1940.
93. **Wide, L.**, An immunological method for the assay of human chorionic gonadotrophin, *Acta Endocrinol.* [Suppl.] (Kbh.), 70, 1962.
94. **Berson, S. A., Yalow, R. S., Bauman, A., Rothschild, M. A., and Newerly, K.**, Insulin-I^{131} metabolism in human subjects: demonstration of insulin binding globulin in the circulation of insulin treated subjects, *J. Clin. Invest.*, 35, 170, 1956.
95. **Franchimont, P.**, *Dosage des Hormones Hypophysaires Somatotropes et Gonadotropes*, Arscia, Brussels, 1966.
96. **Franchimont, P.**, Dosage radio-immunologique des gonadotrophines folliculostimulante et lutéinisante, in Proceedings of the Conference on Problems Connected with the Preparation and Use of Labelled Proteins in Tracer Studies, *J. Labelled Compd.*, 2, 303, 1966.
97. **Midgley, A. R., Jr.**, Radioimmunoassay: a method for human chorionic gonadotropin and human luteinizing hormone, *Endocrinology*, 79, 10, 1966.
98. **Midgley, A. R., Jr.**, Radioimmunoassay for human follicle-stimulating hormone, *J. Clin. Endocrinol. Metab.*, 27, 295, 1967.
99. **Odell, W. D., Ross, G. T., and Rayford, P. L.**, Radioimmunoassay for human luteinizing hormone, *Metabolism*, 15, 287, 1966.
100. **Neil, J. D., Johansson, E. D. B., Datta, J. K., and Knobil, E.**, Relationship between the plasma levels of luteinizing hormone and progesterone during the normal menstrual cycle, *J. Clin. Endocrinol. Metab.*, 27, 1167, 1967.
101. **Faiman, C. and Ryan, R. J.**, Serum follicle stimulating hormone and luteinizing hormone concentrations during the menstrual cycle as determined by radioimmunoassays, *J. Clin. Endocrinol. Metab.*, 27, 1711, 1967.
102. **Aono, T., Goldstein, D. P., Taymor, M. L., and Dolch, K.**, A radioimmunoassay method for human pituitary luteinizing hormone (LH) and human chorionic gonadotropin (HCG) using ^{125}I-labelled LH, *Am. J. Obstet. Gynecol.*, 98, 996, 1967.
103. **Catt, K. J., Niall, H. D., Tregear, G. W., and Burger, H. G.**, Disc solid-phase radioimmunoassay of human luteinizing hormone, *J. Clin. Endocrinol. Metab.*, 28, 121, 1968.
104. **Donini, P.**, Radioimmunoassay employing polymerized antisera, *Acta Endocrinol.* [Suppl.] (Kbh.), 142, 257, 1969.
105. **Wide, L. and Porath, J.**, Radioimmunoassay of proteins with the use of Sephadex-coupled antibodies, *Biochim. Biophys. Acta*, 130, 257, 1966.
106. **Wide, L.**, Radioimmunoassays employing immunosorbents, *Acta Endocrinol.* [Suppl.] (Kbh.), 142, 207, 1969.
107. **Dolais, J., Grappin, A.-M., Freychet, P., and Rosselin, G.**, Dosage plasmatique de l'hormone lutéinisante humaine (HLH) par methode radioimmunologique. Taux de la LH dans le plasma au cours du cycle menstruel de la femme normale, pendant et après l'arrêt de contraceptif oral, *C. R. Acad. Sci. [D]* (Paris), 267, 1162, 1968.
108. **Rosselin, G. and Dolais, J.**, Dosage de l'hormone folliculo-stimulante (HFSH) par la methode radio-immunologique, *Presse Méd.*, 75, 2027, 1968.
109. **Bryant, G. D., Siler, T. M., Greenwood, F. C., Pasteels, J. L., Robyn, C., and Hubinont, P. O.**, Radioimmunoassay of a human pituitary prolactin in plasma, *Hormones* (Basel), 2, 139, 1971.
110. **Friesen, H., Hwang, P., Guyda, H., Tolis, G., Tyson, J., and Myers, R.**, A radioimmunoassay for human prolactin, in *Prolactin and Carcinogenesis*, Boyns, A. R. and Griffiths, K., Eds., Alpha Omega Alpha Publishing, Cardiff, Wales, 1972, 64.
111. **Sassin, J. F., Frantz, A. G., Weitzman, E. D., and Kapen, S.**, Human prolactin: 24-hour pattern with increased release during sleep, *Science*, 177, 1205, 1972.

112. **Sinha, Y. N., Selby, F. W., Lewis, U. J., and Vanderlaan, W. P.**, A homologous radioimmunoassay for human prolactin, *J. Clin. Endocrinol. Metab.*, 36, 509, 1973.
113. **Cole, E. N. and Boyns, A. R.**, Radioimmunoassay for human pituitary prolactin using antiserum against an extract of human amniotic fluid, *J. Endocrinol.*, 58, 24, 1973.
114. **Hwang, P., Guyda, H., and Friesen, H.**, A radioimmunoassay for human prolactin, *Proc. Natl. Acad. Sci. U.S.A.*, 68, 1902, 1971.
115. **L'Hermite, M., Delvoye, P., Nokin, J., Vekemans, M., and Robyn, C.**, Human prolactin secretion as studied by radioimmunoassay: some aspects of its regulation, in *Prolactin and Carcinogenesis*, Boyns, A. R. and Griffiths, K., Eds., Alpha Omega Alpha Publishing, Cardiff, Wales, 1972, 81.
116. **Rodbard, D.**, personal communication, 1973.
117. **Jacobs, L. S., Snyder, P. J., Wilber, J. F., Utiger, R. D., and Daughaday, W. H.**, Increased serum prolactin after administration of synthetic thyrotropin releasing hormone (TRH) in man, *J. Clin. Endocrinol. Metab.*, 33, 996, 1971.
118. **Guyda, H., Hwang, P., and Friesen, H.**, Immunologic evidence for monkey and human prolactin (MPr and HPr), *J. Clin. Endocrinol. Metab.*, 32, 120, 1971.
119. **Greenwood, F. C., Siler, T. M., Bryant, G. D., and Morgenstern, L. L.**, Radioimmunoassay of human plasma prolactin: applications and problems, in *Prolactin and Carcinogenesis*, Boyns, A. R. and Griffiths, K., Eds., Alpha Omega Alpha Publishing, Cardiff, Wales, 1972, 98.
120. **Hunter, W. M. and Greenwood, F. C.**, Preparation of iodine^{131} labelled growth hormone of high specific activity, *Nature*, 194, 495, 1962.
121. **Freedlender, A. E.**, Practical and theoretical advantages for the use of I^{125} in radioimmunoassay, in *Protein and Polypeptide Hormones*, Margoulies, M., Ed., Int. Congr. Ser. No. 161, Excerpta Medica, Amsterdam, 1969, 351.
122. **Hunter, W. M.**, The preparation and assessment of iodinated antigens, in *Radioimmunoassay Methods*, Kirkham, K. E. and Hunter, W. M., Eds., Churchill Livingstone, Edinburgh, 1971, 3.
123. **Berson, S. A. and Yalow, R. S.**, Immunoassay of protein hormones, in *The Hormones*, Vol. 4, Pincus, G., Thimann, K. V., and Astwood, E. B., Eds., Academic Press, New York, 1964, 557.
124. **Hurn, B. A. L. and Landon, J.**, Antisera for radioimmunoassay, in *Radioimmunoassay Methods*, Kirkham, K. E. and Hunter, W. M., Eds., Churchill Livingstone, Edinburgh, 1971, 121.
125. **Ross, G. T., Vaitukaitis, J. L., and Robbins, J. B.**, Preparation and properties of antisera made to α- and β-subunits of human chorionic gonadotropin, in *Structure-activity Relationships of Protein and Polypeptide Hormones (Part 1)*, Margoulies, M. and Greenwood, F. C., Eds., Int. Congr. Ser. No. 241, Excerpta Medica, Amsterdam, 1971, 153.
126. **Robyn, C.**, Biological and immunological characterization of antigonadotrophic profiles, *Acta Endocrinol.* [Suppl.] (Kbh.), 142, 31, 1969.
127. **Odell, W. D., Abraham, G., Rand, H. R., Swerdloff, R. S., and Fisher, D. A.**, Influence of immunization procedures on the titer, affinity and specificity of antisera to glycopolypeptides, *Acta Endocrinol.* [Suppl.] (Kbh.), 142, 54, 1969.
128. **Odell, W. D., Abraham, G. A., Skowsky, W. R., Hescox, M. A., and Fisher, D. A.**, Production of antisera for radioimmunoassays, in *Principles of Competitive Protein-binding Assays*, Odell, W. D. and Daughaday, W. H., Eds., J. B. Lippincott, Philadelphia, 1971, 89.
129. **Petrusz, P., Robyn, C., Diczfalusy, E., and Finney, D. J.**, Experimental verification of the principle of additivity, *Acta Endocrinol.* (Kbh.), 63, 150, 1970.
130. **Romani, P., Robyn, C., Petrusz, P., and Diczfalusy, E.**, Bioassay of antigonadotrophic sera 5. Further studies on the reliability of the bioassay method for the estimation of human chorionic gonadotrophin neutralizing potency, *Acta Endocrinol.* (Kbh.), in press, 1974.
131. **Daughaday, W. H. and Jacobs, L. S.**, Methods of separating antibody-bound from free antigen, in *Principles of Competitive Protein-binding Assays*, Odell, W. D. and Daughaday, W. H., Eds., J. B. Lippincott, Philadelphia, 1971, 303.
132. **Diczfalusy, E.**, Immunoassays of gonadotrophins, *Acta Endocrinol.* [Suppl.] (Kbh.), 142, 1969.
133. **Deutsch, H. F.**, Preparation of immunoglobulin concentrates, in *Methods in Immunology and Immunochemistry*, Vol. 1, Williams, C. A. and Chase, M. W., Eds., Academic Press, New York, 1967, 319.
134. **Robyn, C., L'Hermite, M., Petrusz, P., and Diczfalusy, E.**, Potency estimates of human gonadotrophins by a bioassay and three immunoassay methods, *Acta Endocrinol.* (Kbh.), 67, 417, 1971.
135. **Bettendorf, G., Apostolakis, M., and Voigt, K. D.**, Darstellung von Gonadotropin aus menschlichen Hypophysen, *Acta Endocrinol.* (Kbh.), 41, 1, 1962.
136. **Rodbard, D., Rayford, P. L., Cooper, J. A., and Ross, G. T.**, Statistical quality control of radioimmunoassays, *J. Clin. Endocrinol. Metab.*, 28, 1412, 1968.
137. **Rodbard, D., Bridson, W., and Rayford, P. L.**, Rapid calculation of radioimmunoassay results, *J. Lab. Clin. Med.*, 74, 770, 1969.
138. **Ekins, R. P., Newman, B. G., and O'Riordan, J. L. H.**, Theoretical aspects of saturation analysis, *Acta Endocrinol.* [Suppl.] (Kbh.), 147, 11, 1970.
139. *Scientific Tables*, Ciba-Geigy, S. A., Ed., Basel, Switzerland, 1972, 176.

140. **Midgley, A. R., Jr., Niswender, G. D., and Rebar, R. W.,** Principles for the assessment of the reliability of radioimmunoassay methods (precision, accuracy, sensitivity, specificity), *Acta Endocrinol.* [Suppl.] (Kbh.), 142, 163, 1969.
141. **Rodbard, D.,** Statistical aspects of radioimmunoassays, in *Principles of Competitive Protein-binding Assays,* Odell, W. D. and Daughaday, W. H., Eds., J. B. Lippincott, Philadelphia, 1971, 204.
142. **Rodbard, D. and Lewald, J. E.,** Computer analysis of radioligand assay and radioimmunoassay data, *Acta Endocrinol.* (Kbh.), 147, 79, 1970.
143. W.H.O., *Notes on Biological Standards,* Division of Biological Standards, World Health Organization, Geneva, 1968.
144. W.H.O. (Human Reproduction Unit), Assay of protein hormones related to human reproduction: problems of specificity of assay methods and reference standards, *Acta Endocrinol.* (Kbh.), 71, 625, 1972.
145. **Bangham, D. R. and Grab, B.,** The second international standard for chorionic gonadotropin, *Bull. W.H.O.,* 31, 11, 1964.
146. **Robyn, C.,** Unités biologiques et immunologiques des gonadotrophines, in *L'ovulation,* Colloque de la société nationale pour l'étude de la stérilité et de la fécondité, Masson and Cie, Paris, 1969, 281.
147. W.H.O., *W.H.O. Tech. Rep. Ser.,* 274, 11, 1964.
148. W.H.O., *W.H.O. Tech. Rep. Ser.,* 259, 12, 1963.
149. **Robyn, C. and Diczfalusy, E.,** Bioassay of antigonadotrophic sera 3. Assay of the human follicle stimulating hormone (FSH) neutralizing potency, *Acta Endocrinol.* (Kbh.), 59, 277, 1968.
150. National Institute for Biological Standards and Control, Medical Research Council, Notes on Research Standard A for Human Prolactin (71/222), August 1972.
151. **Borth, R.,** The chromatographic method for the determination of urinary pregnanediol, in *Ciba Foundation Colloquia on Endocrinology,* Vol. 2, Wolstenholme, G. E. W. and Cameron, M. P., Eds., J. and A. Churchill, London, 1952, 45.
152. **Loraine, J. A. and Brown, J. B.,** A method for the quantitative determination of gonadotrophins in urine of non-pregnant human subjects, *J. Endocrinol.,* 18, 77, 1959.
153. **Faiman, C. and Ryan, R. J.,** Radioimmunoassay for human follicle stimulating hormone, *J. Clin. Endocrinol. Metab.,* 27, 444, 1967.
154. **Parlow, A. F.,** Influence of serum on the prostate assay of luteinizing hormone (LH, ICSH), *Endocrinology,* 73, 456, 1963.
155. **Borth, R.,** Steroids in human blood, *Vitam. Horm.,* 15, 259, 1957.
156. **Odell, W. D., Ross, G. T., and Rayford, P. L.,** Radioimmunoassay for luteinizing hormone in human plasma or serum: physiological studies, *J. Clin. Invest.,* 46, 248, 1967.
157. **Franchimont, P.,** Le dosage radio-immunologique des gonadotrophines, *Ann. Endocrinol.* (Paris), 29, 403, 1968.
158. **Thomas, K. and Ferin, J.,** A new rapid radioimmunoassay for HCG (LH, ICSH) in plasma using dioxan, *J. Clin. Endocrinol. Metab.,* 28, 1667, 1968.
159. **Schlach, D. S., Parlow, A. F., Boon, R. C., and Reichlin, S.,** Measurement of human luteinizing hormone in plasma by radioimmunoassay, *J. Clin. Invest.,* 47, 665, 1968.
160. **Saxena, B. B., Demura, H., Gandy, H. M., and Peterson, R. E.,** Radioimmunoassay of human follicle stimulating and luteinizing hormones in plasma, *J. Clin. Endocrinol. Metab.,* 28, 519, 1968.
161. **Saxena, B. B., Leyendecker, G., Chen, W., Gandy, H. M., and Peterson, R. E.,** Radioimmunoassay of follicle-stimulating (FSH) and luteinizing (LH) hormones by chromatoelectrophoresis, *Acta Endocrinol.* [Suppl] (Kbh.), 142, 185, 1969.
162. **Berson, S. A. and Yalow, R. S.,** Immunologic specificity of peptide hormones, in *Structure-activity Relationships of Protein and Polypeptide Hormones,* Int. Congr. Ser. No. 241, Margoulies, M. and Greenwood, F. C., Eds., Excerpta Medica, Amsterdam, 1971, 38.
163. **Kaplan, S. L., Grumbach, M. M., and Shepard, T. H.,** Gonadotropins in serum and pituitary of human fetuses and infants, *Pediatr. Res.,* 3, 512, 1969.
164. **Vekemans, M. and Robyn, C.,** Serum prolactin and gonadotrophins in the mother and in the child at birth, manuscript in preparation.
165. **Rifkind, A. B., Kulin, H. E., and Ross, G. T.,** Follicle-stimulating hormone (FSH) and luteinizing hormone (LH) in the urine of prepubertal children, *J. Clin. Invest.,* 46, 1925, 1967.
166. **McArthur, J. W.,** Gonadotropins in relation to sexual maturity, in *Gonadotropins,* Saxena, B. B., Beling, C. G., and Gandy, H. M., Eds., Wiley Interscience, New York, 1972, 487.
167. **Penny, R., Guyda, H. J., Baghdassarian, A., Johanson, A., and Blizzard, R. M.,** Correlation of serum follicular stimulating hormone (FSH) and luteinizing hormone (LH) as measured by radioimmunoassay in disorders of sexual development, *J. Clin. Invest.,* 49, 1847, 1970.
168. **Blizzard, R. M.,** Penny, R., Foley, T. P., Jr., Baghdassarian, A., Johanson, A., and Yen, S. S. C., Pituitary-gonadal interrelationships in relation to puberty, in *Gonadotropins,* Saxena, B. B., Beling, C. G., and Gandy, H. M., Eds., Wiley Interscience, New York, 1972, 502.
169. **Daughaday, W. H.** and Jacobs, L. S., Human prolactin, in *Ergebnisse der Physiologie,* Vol. 67, Springer-Verlag, Berlin, 1972, 169.

170. **Odell, W. D., Ross, G. T., and Rayford, P. L.**, Radioimmunoassay for luteinizing hormone in human plasma or serum: physiological studies, *J. Clin. Invest.*, 46, 248, 1967.
171. **Taymor, M. L., Aono, T., and Pheteplace, C.**, Follicle-stimulating hormone and luteinizing hormone in serum during the menstrual cycle determined by radioimmunoassay, *Acta Endocrinol.* (Kbh.), 59, 298, 1968.
172. **Ross, G. T., Cargille, C. M., Lipsett, M. B., Rayford, P. L., Marshall, J. R., Strott, C. A., and Rodbard, D.**, Pituitary and gonadal hormones in women during spontaneous and induced ovulatory cycles, *Recent Progr. Horm. Res.*, 26, 1, 1970.
173. **Vande Wiele, R. L., Bogumil, J., Dyrenfurth, J., Ferin, M., Jewelewicz, R., Warren, M., Rizkallah, T., and Mikhail, G.**, Mechanism regulating the menstrual cycle, *Recent. Progr. Horm. Res.*, 26, 63, 1970.
174. **Wide, L., Nillius, S. J., Gemzell, C., and Roos, P.**, Radioimmunosorbent assay of follicle-stimulating hormone and luteinizing hormone in serum and urine from men and women, *Acta Endocrinol.* [Suppl.] (Kbh.), 174, 1973.
175. **Robyn, C., Delvoye, P., Nokin, J., Vekemans, M., Badawi, M., Perez-Lopez, F. P., and L'Hermite, M.**, Prolactin and human reproduction, in *Human Prolactin*, Int. Congr. Ser. No. 308, Pasteels, J. L. and Robyn, C., Eds., Excerpta Medica, Amsterdam, 1973, 167.
176. **Fukushima, M., Stevens, V. C., Garitt, C., and Vorys, N.**, Urinary FSH and LH excretion during the normal menstrual cycle, *J. Clin. Endocrinol. Metab.*, 24, 205, 1964.
177. **Rosemberg, E. and Keller, P. J.**, Studies on the urinary excretion of follicle-stimulating and luteinizing hormone activity during the menstrual cycle, *J. Clin. Endocrinol. Metab.*, 25, 1262, 1965.
178. **Schmidt-Elmendorff, H.**, The urinary excretion of FSH, LH, and total gonadotrophic activity during the normal menstrual cycle estimated by biological and immunological assay methods, *Acta Endocrinol.* [Suppl.] (Kbh.), 100, 106, 1965.
179. **Stevens, V. C.**, Discrepancies and similarities of urinary FSH and LH patterns as evaluated by different assay methods, *Acta Endocrinol.* [Suppl.] (Kbh.), 142, 338, 1969.
180. **Brown, P. S.**, Human urinary gonadotropins. II. In relation to the menstrual cycle, secondary amenorrhoea and the response to oestrogen, *J. Endocrinol.*, 18, 46, 1959.
181. **Rosemberg, E., Joshi, S. R., and Nwe, T. T.**, Daily urinary excretion of follicle-stimulating hormone during the menstrual cycle, *J. Clin. Endocrinol. Metab.*, 28, 1419, 1968.
182. **Rocca, D. and Albert, A.**, Daily urinary excretion of follicle-stimulating hormone during the menstrual cycle, *Proc. Mayo Clinic*, 42, 536, 1967.
183. **Persson, B. H. and McCormick, W. G.**, Patterns of urinary FSH excretion during ovulatory and anovulatory menstrual cycles, *Acta Endocrinol.* (Kbh.), 59, 573, 1968.
184. **McNeilly, A. S., Evans, G. E., and Chard, T.**, Observations on prolactin levels during the menstrual cycle, in *Human Prolactin*, Int. Congr. Ser. No. 308, Pasteels, J. L. and Robyn, C., Eds., Excerpta Medica, Amsterdam, 1973, 231.
185. **Ehara, Y., Siler, T., Vandenberg, G., Sinha, Y. N., and Yen, S. S. C.**, Circulating prolactin levels during the menstrual cycle: episodic release and diurnal variation, *Am. J. Obstet. Gynecol.*, 117, 962, 1973.
186. **Vekemans, M., Delvoye, P., L'Hermite, M., and Robyn, C.**, Evolution des taux sériques de prolactine au cours du cycle menstruel, *C. R. Acad. Sci. [D]* (Paris), 275, 2247, 1972.
187. **Delvoye P., Hasan, S. H., L'Hermite, M., Neuman, F., and Robyn, C.**, Relationships between serum levels of prolactin, oestrogen, LH, FSH and progesterone during menstrual cycle and treatment with steroids, *Acta Endocrinol.* [Suppl.] (Kbh.), 177, 133, 1973.
188. **Lequin, R. M.**, Discussion, in *Human Prolactin*, Int. Congr. Ser. No. 308, Pasteels, J. L. and Robyn, C., Eds., Excerpta Medica, Amsterdam, 1973, 221.
189. **Dierschke, D. J., Bhattacharya, A. N., Aktinson, L. E., and Knobil, E.**, Circhoral oscillations of plasma LH levels in the ovariectomized rhesus monkey, *Endocrinology*, 87, 850, 1970.
190. **Midgley, A. R., Jr. and Jaffe, R. B.**, Regulation of human gonadotropins: X. Episodic fluctuation of LH during the menstrual cycle, *J. Clin. Endocrinol. Metab.*, 33, 962, 1971.
191. **Yen, S. S. C., Tsai, C. C., Naftolin, F., Vandenberg, G., and Ajabor, L.**, Pulsatile patterns of gonadotropin release in subjects with and without ovarian function, *J. Clin. Endocrinol. Metab.*, 34, 671, 1972.
192. **Nankin, H. R. and Troen, P.**, Overnight patterns of serum luteinizing hormone in normal men, *J. Clin. Endocrinol. Metab.*, 33, 705, 1971.
193. **Boyar, R., Perlow, M., Hellman, L., Kapen, S., and Weitzman, E.**, Twenty-four hour pattern of luteinizing hormone secretion in normal men with sleep stage recording, *J. Clin. Endocrinol. Metab.*, 35, 73, 1972.
194. **Krieger, D. T., Ossowski, R., Fogel, M., and Allen, W.**, Lack of circadian periodicity of human serum FSH and LH levels, *J. Clin. Endocrinol. Metab.*, 35, 619, 1972.
195. **Naftolin, F., Yen, S. S. C., and Tsai, C. C.**, Rapid cycling of plasma gonadotrophins in normal men as demonstrated by frequent sampling, *Nature*, 236, 92, 1972.
196. **Rubin, R. T., Kales, A., Adler, R., Fagan, T., and Odell, W. D.**, Gonadotropin secretion during sleep in normal adult men, *Science*, 175, 196, 1972.
197. **Faiman, C. and Winter, J. S. D.**, Diurnal cycles in plasma FSH, testosterone and cortisol in men, *J. Clin. Endocrinol. Metab.*, 33, 186, 1971.
198. **Peterson, N. T., Midgley, A. R., Jr., and Jaffe, R. B.**, Regulation of human gonadotropins. III. Luteinizing hormone and follicle stimulating hormone in sera from adult males, *J. Clin. Endocrinol. Metab.*, 28, 1473, 1968.

199. **Burger, H. G., Brown, J. B., Catt, K. J., Hudson, B., and Stockigt, J. R.,** Physiological studies on the secretion of human pituitary luteinizing hormone and gonadal steroids, in *Protein and Polypeptide Hormones,* Int. Congr. Ser. No. 161, Margoulies, M., Ed., Excerpta Medica, Amsterdam, 1969, 412.
200. **Nillius, S. J. and Wide, L.,** Effects of oestrogen on serum levels of LH and FSH, *Acta Endocrinol.* (Kbh.), 65, 583, 1970.
201. **Slivka, J., Cargille, C. M., and Ross, G. T.,** An analysis for superimposed biological rhythms: application to FSH and LH in women, *J. Appl. Physiol.,* 30, 905, 1971.
202. **Nokin, J., Vekemans, M., L'Hermite, M., and Robyn, C.,** Circadian periodicity of serum prolactin concentration in man, *Br. Med. J.,* 3, 561, 1972.
203. **Nicoll, C. S.,** "Spinoff" from comparative studies on prolactin physiology of significance to the clinical endocrinologist, in *Human Prolactin,* Int. Congr. Ser. No. 308, Pasteels, J. L. and Robyn, C., Eds., Excerpta Medica, Amsterdam, 1973, 119.
204. **L'Hermite M. and Robyn, C.,** Prolactine hypophysaire humaine: détection radioimmunologique et taux au cours de la grossesse, *Ann. Endocrinol.* (Paris), 33, 357, 1972.
205. **Pujol-Amat, P., Gamissans, O., Calaf, J., Benito, E., Perez-Lopez, F. R., L'Hermite, M., and Robyn, C.,** Influence of L-Dopa on serum prolactin, human chorionic somatomammotrophin (HCS) and human chorionic gonadotrophin (HCG) during the last trimester of pregnancy, in *Human Prolactin,* Int. Congr. Ser. No. 308, Pasteels, J. L. and Robyn, C., Eds., Excerpta Medica, Amsterdam, 1973, 316.
206. **Friesen, H., Tolis, G., Shin, R., and Hwang, P.,** Studies of human prolactin: chemistry, radioreceptor assay and clinical significance, in *Human Prolactin,* Int. Congr. Ser. No. 308, Pasteels, J. L. and Robyn, C., Eds., Excerpta Medica, Amsterdam, 1973, 11.
207. **Parke, L. and Forsyth, I. A.,** Discussion, in *Human Prolactin,* Int. Congr. Ser. No. 308, Pasteels, J. L. and Robyn, C., Eds., Excerpta Medica, Amsterdam, 1973, 40.
208. **Cotes, P. M.,** Research standard A for human prolactin (in ampoules coded 71/222), in *Human Prolactin,* Int. Congr. Ser. No. 308, Pasteels, J. L. and Robyn, C., Eds., Excerpta Medica, Amsterdam, 1973, 97.
209. **Hunter, W. M. and Ganguli, P. C.,** The separation of antibody bound from free antigen, in *Radioimmunoassay Methods,* Kirkham, K. E. and Hunter, W. M., Eds., Churchill Livingstone, Edinburgh, 1971, 243.
210. **Pasteels, J. L.,** Tissue culture of human hypophyses: evidence of a specific prolactin in man, in *Lactogenic Hormones,* Wolstenholme, G. E. W. and Knight, J., Eds., Churchill Livingstone, Edinburgh, 1972, 269.
211. **Hwang, P., Guyda, H., and Friesen, H.,** Purification of human prolactin, *J. Biol. Chem.,* 247, 1955, 1972.

NEUTRAL STEROIDS

M. J. Levell

TABLE OF CONTENTS

I. 17-Oxosteroids . . . 129
A. Choice of Method . . . 130
B. Zimmerman Reaction . . . 130
C. Methodology . . . 130

II. Adrenocortical Function – Cortisol . . . 132
A. Plasma or Urine? . . . 133
B. Cortisol Production Rate . . . 134
C. Cortisol Metabolites . . . 134
D. Plasma Cortisol . . . 139
E. Urine Cortisol . . . 140

III. Interpretations . . . 143
A. "Correction for Creatinine" . . . 143
B. 17-OS and 17-OGS – Drug Interference . . . 144
C. 17-Oxosteroids in Urine . . . 144
D. 17-Oxogenic Steroids in Urine . . . 145
E. Urine Cortisol . . . 146
F. Plasma Cortisol . . . 146
G. Conclusions . . . 147

IV. Aldosterone . . . 147
A. Interpretation . . . 148

References . . . 148

This chapter includes 17-oxosteroids, methods for measuring cortisol and its metabolites, and aldosterone. The indications for aldosterone assays are so few, and the methods are so expensive, that it is likely that most hospital laboratories will send specimens away. Therefore, no method is described, but a section on interpretation of results is included.

I. 17-OXOSTEROIDS

For many years the assay of 17-oxosteroids in urine was the only direct way of measuring adrenal cortical steroid production which was routinely available for clinical use. Unfortunately, it is still used in situations in which measurement of cortisol or its metabolites would be more appropriate.

The 17-oxosteroids are a mixture of steroids containing 19 carbon atoms and with a ketone at C_{17}. They are largely derived from the adrenal cortex with small contributions from the ovaries and, in the male, the testes. Although they are often regarded as androgen metabolites, the major adrenal C_{19} steroid (dehydroepiandrosterone) has very little androgenic activity, whereas the highly potent androgen testosterone contributes less than 10% to the urinary 17-oxosteroids of the female. The 17-oxosteroid excretion is, therefore, only a poor reflection of androgenic status. Because of this inherent limitation, little effort has been directed toward improving the rather poor specificity[1] of 17-oxosteroid methods, nor would such effort be justified.

A. Choice of Method

All modern methods are based on the Zimmerman reaction preceded by hydrolysis and extraction of the steroid. Each of these operations can be performed in many ways. The best method is the easiest one that gives adequate precision.

Hydrolysis is necessary for releasing the free steroid from its water-soluble conjugates (largely sulfates and glucuronosides). It is best done by heating the urine with HCl, even though this causes some destruction and artifact formation. Thus, the conditions chosen represent the compromise between hydrolysis and destruction that gives optimum yields for normal urines. Some 17-oxosteroid sulfates are hydrolyzed only very slowly;[2] therefore, it would be theoretically possible for an excess excretion of these particular compounds to be missed.

The choice of solvent for extraction depends mainly on toxicity and boiling point. If the boiling point is too low, evaporation during the manipulations will concentrate the steroid and give a high result. If it is too high, evaporation of solvent will take a longer time and increase the risk of contamination. The polarity of the solvent affects the ratio of 17-oxosteroids to unwanted pigments, but not to a marked extent.[3]

Washing the extract with alkali removes estrogens and much of the pigment. With solid NaOH[3] almost colorless extracts may be produced. However, the major problem with the color reaction is not preformed color, which rarely contributes more than 1 to 2 mg/24 hr to the final value, but unwanted substances which generate color with the Zimmerman reagent. There is no evidence that using solid NaOH significantly helps this problem.

B. Zimmerman Reaction

The reaction between alkaline *m*-dinitrobenzene and activated methylene groups[4] was applied to steroids by Zimmerman.[5] The reagent ethanolic KOH[6] was always somewhat unsatisfactory owing to the difficulty of predicting when its usable life was ending. The use of quaternary ammonium bases[7,8] has greatly improved the method.

Different 17-oxosteroids yield compounds of different molar extinction coefficients so that the results must be expressed as color equivalents of a standard. Dehydroepiandrosterone is normally used as standard because of its easy availability.

The 17-oxosteroid chromophore has an extinction maximum around 520 nm. Urinary extracts contain chromogens other than 17-oxosteroids which produce an appreciable absorption at 520 nm. Various ways have been used to eliminate the interference from these "nonspecific chromogens."

Talbot et al.[9] measured the extinctions at 520 nm and 435 nm and calculated the contribution of nonspecific color by assuming that its absorption spectrum always showed a ratio of 0.6 at these two wavelengths. Allen[10] assumed that the extinction of the nonspecific color at 520 nm was the mean of the extinctions at two wavelengths equally spaced above and below 520 nm (commonly 440 nm and 600 nm). When the 17-oxosteroid content of the specimen is normal or high, both formulae give similar results. When the 17-oxosteroid content is low, relative to that of nonspecific chromogens, neither formula is reliable. Neither formula is designed to deal with the atypical colors which may result from using old reagents or from drug interference.

An alternative method of separating 17-oxosteroid color is to extract it from the Zimmerman reagent with ether when the extinction at 520 nm may be used without further correction.[7]

C. Methodology

1. Principle

17-Oxosteroid glucuronosides are hydrolyzed with HCl. The free 17-oxosteroids are extracted from the urine with chloroform and measured colorimetrically with the Zimmerman reaction.

2. Apparatus

1. Test tubes holding at least 30 ml with ground glass stoppers and suitable for centrifuge (e.g., 150 mm X 20 mm B19).
2. Centrifuge suitable for no. 1.
3. Shaking machine suitable for no. 1.
4. Test tubes for color reaction (e.g., 150 mm X 15 mm).
5. Spectrophotometer (e.g., Unicam® SP 600).
6. Filter funnels and Whatman no. 1 paper.
7. Water bath at 90 to 95°C.

3. Reagents

1. Chloroform. Test each new bottle before use by evaporating to dryness 4 ml chloroform with 0.2 ml DHA standard and carrying out the Zimmerman reaction. The solvent residues should not interfere significantly with the standard color.

2. HCl (AR quality).

3. NaOH (3 mol/l). Dissolve 12 g NaOH (AR quality) in water, cool, and make up to 100 ml with water.

4. Ethanol (AR quality). Purification procedures have been recommended; however, it is preferable to use a brand that does not require purifying.

5. 30% Ethanol – 3 vol ethanol mixed with 7 vol water.

6. *m*-Dinitrobenzene. BDH Ltd., Poole, "specially purified for the determination of 17-ketosteroids." Add 50 ml ethanol to 0.5-g crystals and shake mechanically until it has dissolved. Store at 4°C for no longer than 3 weeks.

7. Benzyltrimethylammonium hydroxide. BDH Ltd., Poole. Use 40% solution as supplied. Store at 4°C.

8. Zimmerman reagent. Mix 2 vol of reagent 6 with 2 vol of reagent 7. Use within 5 min.

9. Standard DHA. Dehydroepiandrosterone. Dissolve 16.5 mg in ethanol, warming if necessary, and make up to 100 ml. Store at 4°C. It is stable for several months.

4. Procedure

1. If the patient is an adult, dilute the 24-hr specimen of urine (collected without preservative) to 2,000 ml (unless the volume is already greater than this). Urines from children are better analyzed without prior dilution. Analyze as follows in duplicate. Record the original volume.

2. Pipette 5 ml urine into a stoppered test tube. Add 1.5 ml HCl and place in a boiling water bath for 10 min.

3. Cool. Add 6.0 ml chloroform. Stopper and shake mechanically for 5 min.

4. Centrifuge to break any emulsion that has formed. Suck off and discard the upper (aqueous) layer using a Pasteur pipette attached to a suction pump. Wash with 2 ml H_2O. Suck off upper (aqueous) layer.

5. Add 1.5 ml NaOH. Shake mechanically for 5 min. Suck off and discard the upper (aqueous) layer.

6. Filter the chloroform through a no. 1 paper into a test tube.

7. Pipette 4.0 ml of the filtrate into a clean stoppered test tube. Add two or three bauxite chips and evaporate to dryness in a water bath at 90 to 95°C.

8. Pipette 0.2 ml DHA standard into a clean stoppered test tube. Add two or three bauxite chips and evaporate to dryness in a water bath at 90 to 95°C.

5. Zimmerman Reaction

1. This is performed in batches, each batch consisting of 2 blanks (for which clean stoppered test tubes are required), 2 DHA standards, and up to 12 unknowns in duplicate. Check that each tube is quite free of solvent; the bauxite chips should rattle freely in the bottom.

2. Add 0.6 ml of Zimmerman reagent to each tube, starting and finishing with a blank. Rotate the tube to spread the reagent over the bottom half inch of the tube. Stopper the tubes and place in a water bath at 25°C in the dark for 1 hr.

3. Add 2.0 ml of 30% (v/v) aqueous ethanol and 5 ml of ether to each tube. Stopper the tubes and shake by hand for 10 sec.

4. Allow the phases to separate and transfer the ethereal (upper) layer with a Pasteur pipette to a 1-cm optical cell. (The cell must be stoppered unless the reading is to be made immediately.)

5. Measure the optical density of the blanks at 515 mμ against ether.

6. Select the lower blank and read the remaining tubes (tests and standards) against this blank.

6. Calculation

1. Average the readings of the standard.

2. For each tube calculate

$$\frac{\text{test}}{\text{standard}} \times 10 = \text{mg 17-oxosteroid per liter of urine}$$

(The standard tubes contain 33 μg. If the readings of test and standard are equal, 4 ml of the chloroform extract contains 33 μg steroid; hence, 6 ml contains 50 μg. This amount was in 5 ml of urine; hence, the concentration in the urine is 10 mg/l.)

3. Add together the duplicate values in mg/l.

to give the mean value in mg/2 l. (i.e., mg/24 hr).

If the urine was not made up to 2 l. (e.g., if the volume was greater than 2 l., or if it was from a child), multiply the value in mg/2 l. by the actual 24-hr volume in liters/2 to give mg/24 hr.

7. Reliability

The specificity of the method is unknown. Certainly the 17-oxosteroid excretion as measured is always greater, sometimes considerably so, than the sum of the known 17-oxosteroids measured separately by specific methods, but the reasons for the differences are not defined. Reproducibility is probably the most relevant characteristic of the 17-oxosteroid assay. In the normal range and above, the coefficient of variation of analyses done in the same batch is about 5%; below the normal range the proportionate error is greater. Between-batch variation is rather greater; coefficients of variation of between 4 and 14% are found for normal urines. Between-laboratory variation is, as usual, enormous.[11]

It is important not to place too much emphasis on control samples and differences of duplicate estimations since the biggest potential sources of error in the method may not show up by these means. Contamination of tubes, deterioration of reagents, and some drug metabolites in the urine may produce very substantial errors. Usually these will produce colors in the Zimmerman reaction which are different from the typical colors. It is, therefore, important to inspect the colors before extracting into ether. If some of the tubes give atypical colors, careful inspection and comparison of unknowns, control urines, standards, and blanks should reveal the likely cause.

Drug interference is dealt with under "Interpretations."

II. ADRENOCORTICAL FUNCTION – CORTISOL

There is a wide range of methods available for measuring cortisol and its metabolites. In order to make a choice, and particularly a choice between plasma and urine methods, some understanding of cortisol metabolism is required.

An adult woman usually produces 10 to 20 mg of cortisol per 24 hr, up to half of it during the latter part of the sleep period. Studies involving frequent sampling of blood have shown that, like many hormones; cortisol is secreted intermittently, the plasma concentration rising and falling many times during 24 hr. The concentration does not usually fall to zero between episodes of secretion; indeed, in the early morning the relatively frequent episodes result in a build-up of the concentration so that at, say, 9 a.m. the concentration is likely to be higher than in the late evening.[12-15]

Most of the cortisol in plasma is bound to a protein – corticosteroid binding globulin (CBG, transcortin). The total binding capacity of CBG averages 25 μg cortisol/100 ml plasma, so that when the plasma cortisol concentration is above 25 μg/100 ml, any cortisol in excess of this figure is not CBG bound. Even at concentrations below 25 μg/100 ml some of the cortisol is not CBG bound, the proportion varying somewhat with the total concentration but being about 10% of the total. The fact that the binding capacity of CBG is limited means that increases of total plasma cortisol represent disproportionate increases of non-CBG-bound cortisol. For example, a doubling of the total plasma cortisol concentration from 20 to 40 μg/100 ml represents an eightfold increase of non-CBG-bound cortisol. Methods for measuring plasma cortisol measure the total. However, it is the non-CBG-bound cortisol which is metabolically active.

There are two complications of the CBG-cortisol interaction, neither of which seriously affects the principles stated above. First, a proportion of the non-CBG-bound cortisol is bound to albumin. Second, the speed of association and dissociation of the CBG-cortisol complex is very rapid at 38°C. Thus, if there is a net loss of cortisol from the circulation to an organ (for example, the liver, where irreversible metabolism occurs) some CBG-cortisol will dissociate.[16]

Cortisol production is regulated by adrenocorticotrophic hormone (ACTH) from the basophil cells of the anterior lobe of the pituitary gland. ACTH is controlled by a hormone from the hypothalamus, corticotrophin releasing factor (CRF). The output of CRF (and, hence, cortisol indirectly) is controlled in a very complex manner by the interplay of a number of factors.

Most episodes of cortisol secretion are not clearly related to outside stimuli; consequently, they are sometimes known as "spontaneous" episodes. In addition, there are various stimuli – trauma, infection, hypoglycemia, even apprehension – that will cause secretory episodes. A high concentration of plasma cortisol inhibits spontaneous secretory episodes, but not the second type of episode, that caused by hypoglycemia or outside stress. A low concentration of cortisol seems to stimulate CRF secretion, at least early in the day. Thus, looking at the system rather crudely, there is a negative feedback regulation of cortisol with stress superimposed on this. Again, it is the non-CBG-bound fraction of cortisol whose concentration is regulated.

Some cortisol appears in the urine. The relationship between plasma concentration and rate of urinary excretion is consistent with glomerular filtration of the nonprotein-bound cortisol, followed by tubular reabsorption of about 80% of that filtered.[17] The rate of excretion of free cortisol is likely, therefore, to provide a good index of the nonprotein-bound fraction in the plasma.

Cortisol is removed from the plasma reversibly by many tissues and irreversibly by the liver. The reversible removal may involve a cycle of oxidation and reduction at C_{11}.[18] The irreversible metabolism of cortisol is complex, with at least ten urinary metabolites being formed in easily measurable quantities, and a host of others. From the analytical point of view, it is convenient to distinguish three major groups of metabolites.

Most of the cortisol undergoes reduction at $C_{3,4,5}$ to form 3-hydroxypregnane compounds (mostly 3α, 5β). These compounds are coupled to glucuronic acid or sulfuric acid to form water-soluble conjugates which are secreted in the urine. Of these reduced compounds, about 50% have undergone oxidation at C_{11}, that is, they are cortisone metabolites. About 30% have undergone reduction at C_{20} to yield 20α- or 20β-hydroxy compounds; these are α- and β-cortol (if they have an 11-hydroxyl group) or α- and β-cortolone (if they have an 11-oxo group).

A small proportion (5 to 15%) of cortisol loses its side chain to become C_{19} steroids with a 17-oxo group, that is, 17-oxosteroids.

The third route of metabolism involves the introduction of an additional hydroxyl group at C_6 and possibly other places. The resulting steroids are relatively water-soluble and are excreted without conjugation. Their water solubility leads to their poor extraction during analyses of cortisol metabolites so that spuriously low results may be obtained. This metabolic route usually affects only a small proportion of cortisol, but it becomes a major pathway when the cortisol production rate is high[19] and in patients treated with barbiturates[20] or anticonvulsants.[21]

By these various metabolic routes, the cortisol removed from the plasma is inactivated and rendered freely water-soluble. At least 95% of cortisol is ultimately excreted in the urine in normal women; however, if there is renal failure, more may appear in the feces. There is some doubt as to the exact location of all steps in the metabolic pathways, but the liver is still thought to play the major role. Thus, the rate of removal of cortisol from plasma, represented normally by a half-life of about 80 min, may be considerably slowed in chronic liver failure. Conversely, the increased hepatic hydroxylation of cortisol during barbiturate or anticonvulsant therapy leads to a shorter half-life.

A. Plasma or Urine?

The relationships discussed in the previous section are brought together in Figure 1. It will be seen that measurements of metabolites give some indication of adrenal activity, whereas plasma cortisol or urinary cortisol methods give an indication of the effective concentration of cortisol in the body. The diagram also shows that cortisol production and the plasma level are linked and to this extent plasma and urine cortisol methods measure adrenal activity and metabolite methods measure effective concentration.

Three of the relationships require further comment. The fact that adrenal secretion is episodic does not prevent metabolite methods from assessing adrenal activity because such methods are normally made on 24-hr urines, but it is important in the link between plasma cortisol and adrenal activity. Variations of the metabolic clearance rate (i.e., of plasma $t_{1/2}$) of cortisol will affect the steady-state relationship between cortisol production and plasma concentration, but they will not invalidate the use of changes of plasma cortisol concentration to assess changes of cortisol production rate because the cortisol $t_{1/2}$ is relatively constant in each individual. The concentration of CBG is relatively constant. However, changes can occur, for example, in pregnancy. When the CBG concentration is higher or lower than normal, the plasma cortisol concentration will be higher or lower without any change in the rate of production or removal of cortisol.

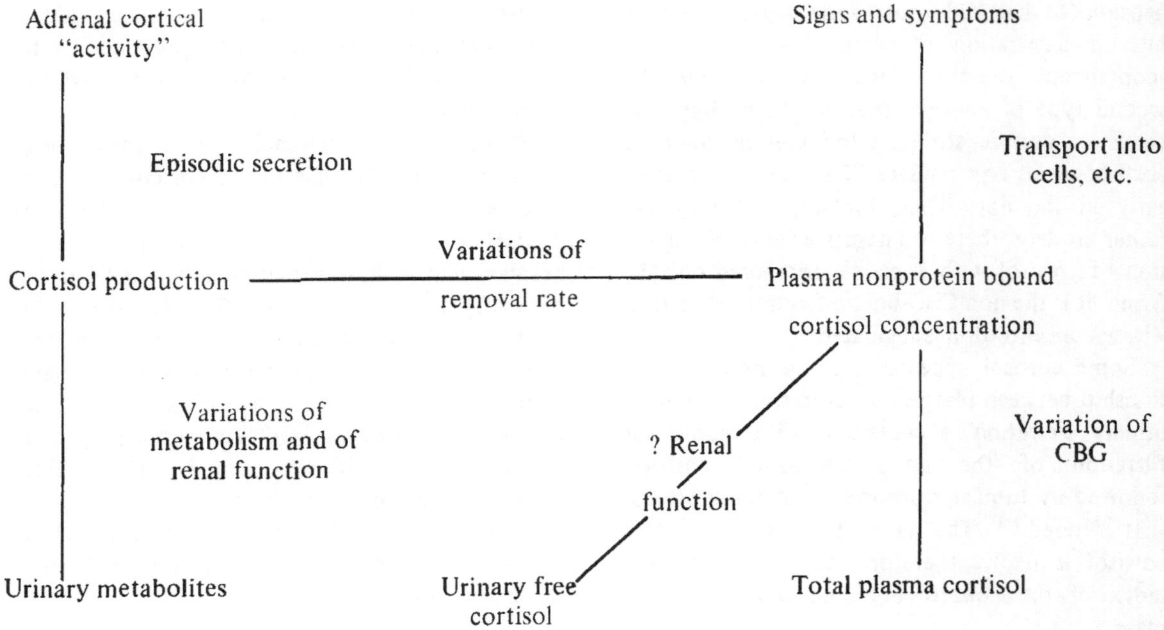

FIGURE 1. Relationships between cortisol production, its effect, and its measurement.

From the variety of factors disturbing the relationships between metabolite methods and plasma cortisol measurements it is clear that it would be naive to expect these to correlate; moreover, neither type of method gives unequivocal information on either the state of adrenal cortical activity or on the effective concentration of cortisol to which the tissues are exposed.

Fortunately, both types of measurement will detect *changes* of cortisol production. The most effective way of using either type of measurement for the diagnosis of diseases of the pituitary or adrenal is to make measurements in conjunction with maneuvers which stimulate or inhibit one or another gland to test the integrity of the control mechanisms. The choice between methods using 24-hr urines and plasma methods becomes largely a choice between more reliable, but more prolonged tests on the one hand and quicker but, in some cases, less reliable tests on the other hand. The measurement of urine cortisol appears to combine some of the better features of both approaches; it gives a measure of plasma cortisol but, because it is measured in 24-hr urines, it is not affected by transient stimuli.

B. Cortisol Production Rate (CPR)

Methods involving radioactive isotopes may be used to measure the average rate of cortisol production either over a period of 2 to 3 hr by using plasma measurements, or over a period of 1 to 2 days, using urinary measurements.

The urinary methods are more usual. Radioactive cortisol (0.5 to 1 μCi of [^{3}H]cortisol) is administered intravenously or orally. A portion of a cortisol metabolite is isolated from the 24- or 48-hr urine and both its mass and its ^{3}H content are measured. Cortisol production is calculated from the relationship:

$$\frac{\mu\text{g of metabolite}}{\mu\text{g of cortisol}} = \frac{^3\text{H in metabolite}}{^3\text{H administered}}$$

CPR measurements became widely accepted as a reference method against which the performance of other methods could be judged. More recently, however, the basic assumption that the administered [^{3}H]cortisol is metabolized in the same way as endogenous cortisol has been shown to be sometimes untrue.[22,23] Thus, these urinary methods can no longer be regarded as giving a true measure of the rate of cortisol production.

C. Cortisol Metabolites

These are excreted in the urine mainly in the

form of conjugates with sulfuric acid or glucuronic acid. The conjugates must be hydrolyzed before the free steroids can be extracted from the urine. Unlike many other steroids, cortisol metabolites are rapidly destroyed by acid or alkali and simple chemical hydrolysis, therefore, cannot be used.

Two solutions to this problem have been widely used. The enzyme β-glucuronidase splits the glucuronosides, which form the bulk of the conjugates. Methods based on the Porter-Silber reaction,[24] which is very popular in the U.S., use this approach. The alternative, proposed by Norymberski et al.,[25] oxidizes the conjugates to 17-oxosteroid conjugates, which may then be assayed by one of the 17-oxosteroid methods. The latter approach was introduced in the U.K. at about the same time that the Porter-Silber methods were introduced in the U.S. For many years each type of method had its adherents, but lately the Norymberski method (now considerably modified) has become the method of choice for most people.

The parallel development of the two methods produced a confusing clash of nomenclature. Norymberski originally called his method "17-ketogenic steroids." In 1955 the method was modified[26] to include all known 17-hydroxylated C_{17} steroids and it was renamed "17-hydroxycorticosteroids." This was unfortunate because the Porter-Silber methods were widely described as measuring 17-hydroxycorticosteroids (unjustifiably, because the methods measure only a proportion of such steroids). For many years the same name was used for the two quite different procedures. Finally, the matter was resolved by the recognition that nobody still used the original Norymberski method, so that the original name, 17-ketogenic steroid – or 17-oxogenic steroid as it has now become – could be used for the 1955 method and later methods using the same principle.

The diagram (Figure 2) shows that neither the Porter-Silber nor the Norymberski method is specific for cortisol metabolites and that the Norymberski method gives a wider coverage of metabolites.

There are also technical reasons for preferring the Norymberski method – it is quicker, fewer blanks are required, and there is less danger of other compounds interfering.

1. 17-Oxogenic Steroids – Choice of Method

The elegance of the original Norymberski method[25] lay in its finding simple experimental conditions for oxidizing 17-oxogenic steroid conjugates to 17-oxosteroids with sodium bismuthate. A disadvantage of the method was the need to measure 17-oxosteroids simultaneously to calculate the 17-oxogenic steroids by difference. In 1955 Appleby et al.[25] introduced the use of borohydride, a reagent which not only removed preexisting 17-oxosteroids, but, by reducing 20-oxo groups to 20-hydroxyls, included the side chain type 1 (Figure 2) which was excluded from the previous method. A second consequence of the conversion of type 1 to 2 and of type 3 to 4 (Figure 2) was that periodate could be used as oxidizing agent.[27] This was a great improvement since not all types of bismuthate were fully satisfactory.[28]

Although the original method incorporated a stage of acid hydrolysis after the oxidation, this is, in fact, unnecessary since glucuronoside moieties, present in the bulk of the metabolites, are also oxidized by bismuthate or periodate to formate. (Acid hydrolysis may appear to improve the yield but this is partly due to the formation of artifacts, especially $\Delta^{9(11)}$ steroids, which are more chromogenic in the Zimmerman reaction.) The addition of alkali after the oxidation quickly hydrolyzes the formates and avoids the necessity of an alkali wash later.

The choice of solvent for extraction is, as in the case of the 17-oxosteroid assay, largely a matter of convenience, although the solvent must be sufficiently polar to extract the 11-oxygenated 17-oxosteroids. Thus, CCl_4 is useless, but $CHCl_3$ is satisfactory.

The Zimmerman reaction has been considered above. The considerations which apply in the 17-oxosteroid method also apply here.

A number of simplifications of the periodate method were made by Metcalf;[29] the method described here is based on her method.

An aspect of the 17-oxogenic steroid assay which is not clearly resolved is the manner of expressing the results. The problem arises because different 17-oxosteroids produce different amounts of color in the Zimmerman reaction. (This problem is minimal in the 17-oxosteroid assay itself since the four major naturally occurring 17-oxosteroids have rather similar chromogenicities.) The major cortisol metabolites are

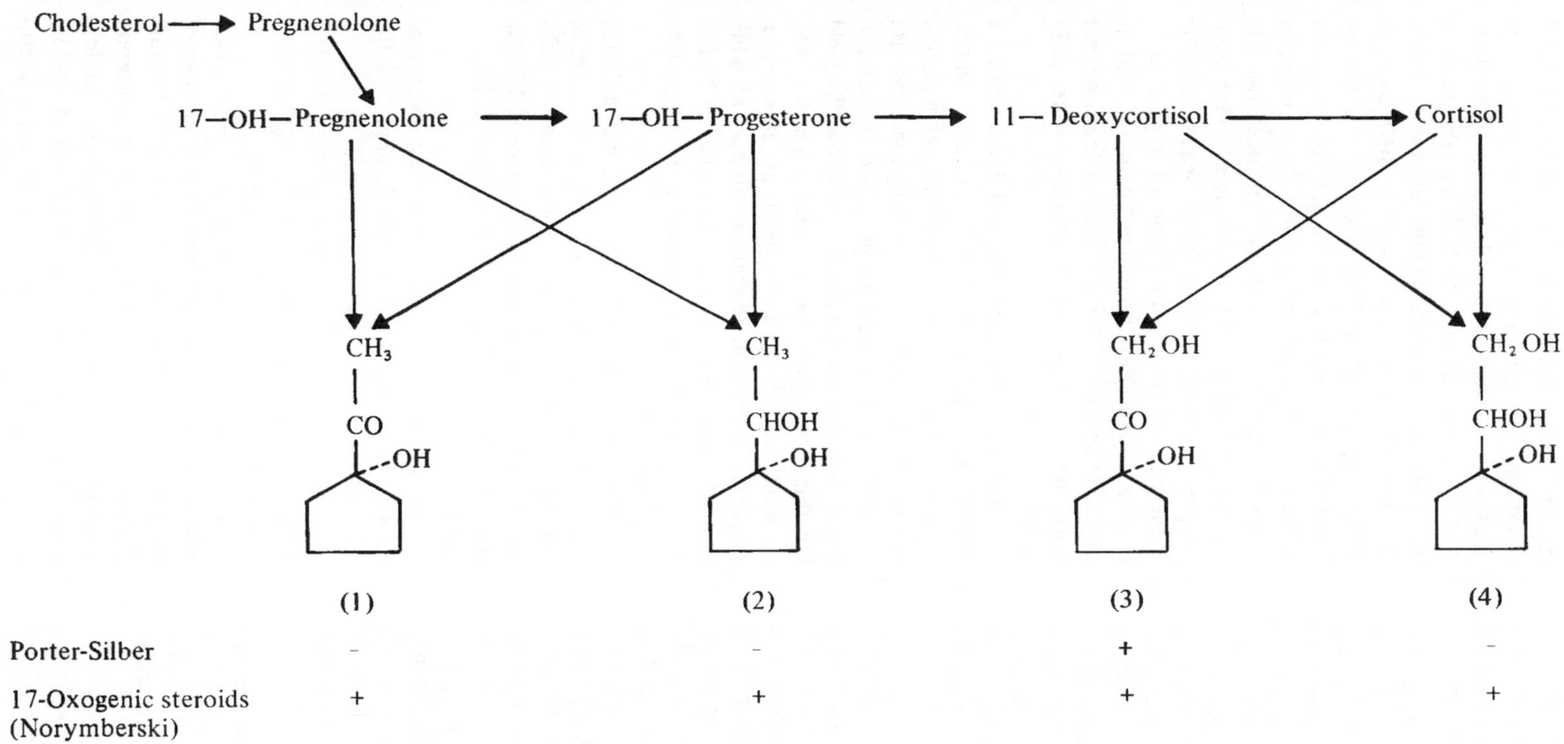

FIGURE 2. D-rings and side chains of 17-hydroxycorticosteroids showing relationships to biosynthetic pathway and to analytical methods. The numbers 1–4 of the side chains are referred to in the text.

converted in the 17-oxogenic steroid method to 11β-hydroxyetiocholanolone, which produces only about 75% of the color produced by dehydroepiandrosterone, the most commonly used standard. It has been proposed that results should be corrected for this difference in chromogenicity.[11] A further complication (unless moles are used) is the molecular weight change – should a correction be made for the fact that every 100 μg of 11β-etiocholanolone corresponds to 120 μg of cortisol metabolites? It is, of course, possible to use cortisol (or a cortisol metabolite) as a standard, taking it through the entire procedure. If this is done, clearly the corrections will be made automatically.

There are several arguments against applying corrections. Much of the literature and most, if not all, published normal ranges are based on uncorrected values. More compelling is the fact that with some pathological urines the correction would be quite inappropriate. For example, urine from a patient with untreated congenital adrenal hyperplasia yields largely etiocholanolone on oxidation. Perhaps the most important argument is that carrying out a correction might appear to invest the assay with more authority than it can, in fact, possess. Methods that measure groups of compounds according to some property which they share to a greater or lesser degree can only be expressed as equivalents. It is true that equivalents of 11β-hydroxyetiocholanolone will usually be less wrong than equivalents of dehydroepiandrosterone but, nevertheless, it seems better to accept that the figures are arbitrary and not burden the technician with extra arithmetic.

a. Interference from Glucose

Interference from glucose represents an important potential source of error. Glucose and other polyhydroxyl compounds compete with the steroids for the available periodate. A glucose concentration of more than 0.25 g/100 ml is liable to suppress the 17-oxogenic steroid result. If glucose is detectable by Clinitest®, action must be taken. Three methods are available. First, the amount of periodate may be increased.[30] This method suffers from the drawback that the glucose must be estimated to know how much extra reagent to add. For this reason it is not foolproof. The second method is to extract the conjugates, leaving the glucose behind. The urine is saturated with ammonium sulfate and extracted with 3 times its volume of 3:1 ether-ethanol.[31] This method is very effective but it is cumbersome. It is a useful technique to have available if one suspects interference from other highly polar compounds. The third and easiest way is to ferment out the glucose with yeast, remembering to correct for the dilution of the urine by the yeast.[32]

2. Methodology

1. Principle

17-Oxogenic steroid glucuronosides are converted by oxidation, followed by alkaline hydrolysis, to 17-oxosteroids, which are extracted into chloroform and assayed by the Zimmerman reaction. 17-Oxosteroids initially present in the urine are rendered nonchromogenic by treatment with borohydride.

2. Apparatus

1. Test tubes holding at least 30 ml with ground glass stoppers and suitable for centrifuging (e.g., 150 mm X 20 mm, B19).
2. Centrifuge suitable for no. 1.
3. Shaking machine suitable for no. 1.
4. Test tubes for color reaction (e.g., 150 mm X 15 mm).
5. Spectrophotometer (e.g., Unicam SP 600).
6. Filter funnels and Whatman no. 1 paper.
7. Water baths at 55°C and at 90 to 95°C.
8. Clinitest kit.

c. Reagents

1. NaOH (1 mol/l. [4 g/100 ml]; 0.1 mol/l. [0.4 g/100 ml]; 5 mol/l. [20 g/100 ml]). NaOH (AR quality) is made up to the appropriate volume with water.
2. Octan-2-ol.
3. Sodium borohydride: 10% (w/v) in 0.1 mol/l. NaOH. Prepare freshly each day.
4. Acetic acid (4 mol/l.); 25% (v/v) AR.
5. Sodium periodate: 10% (w/v) in water. Prepare freshly each day.
6. Chloroform. Test each new bottle before use by evaporating to dryness 4 ml chloroform with 0.2 ml DHA standard and carrying out the Zimmerman reaction. The solvent residues should not interfere significantly with the standard color.

7. Ethanol (AR quality). Purification procedures have been recommended; however, it is preferable to use a brand that does not require purifying.

8. 30% Ethanol: 3 vol/ethanol mixed with 7 vol water.

9. *m*-Dinitrobenzene. BDH Ltd., Poole, "specially purified for the determination of 17-ketosteroids." Add 50 ml ethanol to 0.5 g crystals and shake mechanically until it has dissolved. Store at 4°C for no longer than 3 weeks.

10. Benzyltrimethylammonium hydroxide. BDH Ltd., Poole. Use 40% solution as supplied. Store at 4°C.

11. Zimmerman reagent. Mix 2 vol of reagent 9 with 1 vol of reagent 10. Use within 5 min.

12. Standard DHA. Dehydroepiandrosterone. Dissolve 16.5 mg in ethanol, warming if necessary, and make up to 100 ml. Store at 4°C.

13. Standard cortisol (Cortisol, Steraloids Ltd., Croydon, U.K.): 0.32 mg/ml in ethanol. Store at 4°C.

14. Baker's yeast. Obtain fresh as required from the hospital kitchen or a local baker.

d. Procedure

1. If the patient is an adult, dilute the 24-hr specimen of urine (collected without preservative) to 2,000 ml (unless the volume is greater than this). Record (a) the original volume, and (b) the making of the dilution. Urine from children is analyzed better without dilution.

2. Test for reducing substances (Clinitest). If present, see note 1 below.

3. If necessary adjust the pH of a portion of urine to 7 to 8 with NaOH (1 mol/l.).

4. Pipette 5 ml of urine into each of two glass-stoppered tubes. Pipette 0.25 ml cortisol standard + 5 ml water into each of another two tubes. Each tube is treated as follows.

5. Add 1 drop of octanol and 0.5 ml borohydride. Place in a bath at 55°C for 45 min.

6. Without removing the tubes from the bath, add 0.5 ml acetic acid. Leave for 2 min.

7. Add 2 ml sodium periodate and leave in the 55°C bath for 10 min.

8. Add 0.5 ml NaOH (5 mol/l.). Remove from the bath and cool the tubes to room temperature.

9. Add 6.0 ml chloroform. Stopper and shake mechanically for 5 min.

10. Centrifuge to break any emulsion that has formed. Suck off and discard the upper (aqueous) layer.

11. Filter the chloroform through a no. 1 paper into a test tube.

12. Pipette 4.0 ml of the filtrate into a clean stoppered test tube. Add two or three bauxite chips and evaporate to dryness in a water bath at 90 to 95°C.

13. Pipette 0.2 ml DHA standard into a clean stoppered test tube. Add two or three bauxite chips and evaporate to dryness in a water bath at 90 to 95°C.

e. Zimmerman Reaction

1. This is performed in batches, each batch consisting of 2 blanks (for which 2 clean stoppered test tubes are required), 2 "cortisol" standards (see note 2), 1 DHA standard, and up to 12 unknowns. Check that each tube is quite free of solvent; the bauxite chips should rattle freely in the bottom.

2. Add 0.5 ml Zimmerman reagent to each tube, starting and finishing with a blank. Rotate the tube to spread the reagent over the bottom ½ in. of the tube. Stopper the tubes and place in a water bath at 25°C in the dark for 1 hr.

3. Add 2.0 ml 30% (v/v) aqueous ethanol and 5.0 ml ether to each tube. Stopper and shake for 10 sec.

4. Allow the phases to separate and transfer the ethereal (upper) layer with a Pasteur pipette to a 1-cm optical cell. (The cell must be stoppered unless the reading is to be made immediately.)

5. Measure the optical density of the blanks at 515 mμ against ether.

6. Select the lower blank and read the remaining tubes (tests and standards) against this blank.

f. Calculation

1. Average the readings of the DHA standard.

2. For each tube calculate

$$\frac{\text{test}}{\text{standard}} \times 10 = \text{mg 17-oxogenic steroid per liter of urine}$$

(The standard contains 33 μg. If the readings of test and standard are equal, 4 ml of chloroform extract contains the color equivalent of 33 μg steroid; hence, 6 ml contains 50 μg. This amount was in 5 ml of urine; hence, the concentration in the urine is 10 mg/l.)

3. Add together the duplicate values in mg/l. to give the mean value in mg/2 l., i.e., mg/24 hr.

If the urine was not made up to 2 l. (e.g., if the volume was greater than 2 l. or if it was from a child), multiply the value in mg/2 l. by the actual 24-hr volume in liters divided by two to give mg/24 hr.

Note that this result is expressed as mg color equivalents of DHA.

g. Notes

1. If glucose is present in the urine, remove it by fermentation. Suspend 1 to 2 g fresh baker's yeast in water in a graduated centrifuge tube, spin at a low speed, and pour away the cloudy supernatant. Repeat this washing. Add 15 ml of urine, suspend the yeast, and leave at room temperature overnight or until it is free of glucose. Centrifuge at higher speed than when washing the yeast. Note the volume of the packed yeast. Remove 2 X 5 ml supernatant for assay. Correct the final result for the dilution by the yeast by multiplying by

$$\frac{\text{volume of packed yeast} + 15}{15}$$

2. The cortisol standard is included as a check on the method. The amount used – 0.25 ml of 0.32 mg/ml, i.e., 0.08 mg – is oxidized to 0.067 mg of 17-oxosteroid. This is extracted into 6 ml chloroform of which 4 ml – 0.045 mg of steroid – is taken for colorimetry. The color equivalents are such that this gives approximately the same color as the 0.033 mg of DHA in the DHA standard. Thus, if the method is working properly, the ratio of cortisol standard to DHA standard should be constant and be between 0.9 and 1.0.

h. Reliability

The method has adequate reliability provided that the Zimmerman reaction is carried out with a reasonable degree of care, that new reagents are checked before use and not allowed to deteriorate, and that a careful watch is kept for evidence of contamination. A survey of ten laboratories showed within-laboratory coefficients of variation of 4 to 14% with a median of 7%.[11] Between-laboratory variation is enormous, but it is not clear to what extent this represents differences of methods.

It is a relatively easy matter to detect deterioration of the Zimmerman reagents because the blanks will become higher and finally all tubes, standards as well as unknowns, will be affected. Deterioration of the periodate will give poor recoveries of the standard cortisol and falling values for control urines. Deterioration of the borohydride will lead to high values of control urines because preexisting 17-oxosteroids will be measured as well as those formed from the 17-oxogenic steroids. It has been suggested that borohydride may be checked by assaying a solution containing 200 μg DHA; there should be at least 95% destruction.[30]

D. Plasma Cortisol

Most of the cortisol in plasma is protein bound. The physiological actions of the hormone depend mainly on the concentration of the nonprotein-bound fraction. In spite of this, it is usual to measure the total of protein bound and free. There are three reasons for doing so. First, it is technically very difficult to measure the fractions separately; second, the concentration of binding protein is relatively constant; and third, absolute values of plasma cortisol are of little clinical significance, the diagnostic use of the assay depending on the measurement of differences of concentration before and after various procedures.

Because the major concern is with changes rather than absolute values, precision is more important than accuracy or specificity. It is also appropriate for clinical purposes that the latter qualities should be sacrificed to speed and ease of operation. Thus, isotope derivative procedures (e.g., Fraser and James[33]), that offer the only way of getting accurate values of the cortisol concentration, are of use as reference methods and in many research situations, but are quite unsuitable for routine clinical use.

The earliest methods were based on the Porter-Silber reaction and enjoyed great popularity for many years, particularly in the U.S., perhaps because of the widespread use of urinary metabolite methods based on the same color reaction. Porter-Silber methods compare unfavorably with the two other methods available in sensitivity, specificity (particularly freedom from drug interference), and ease of working.

Fluorescence in ethanol-sulfuric acid provides the basis for the most popular group of methods. Most modern techniques are based on the method of De Moor et al.[34] The one to be described is that of Mattingly,[35] which was chosen as the recommended method by the Medical Research Council Working Party.[36]

The relative fluorescence intensities and shapes of the fluorescence spectra depend in a complex way on the concentration of the sulfuric acid and on the time and temperature of development as well as on the nature of the steroid. Under the conditions used, cortisol and corticosterone give approximately equal intensities with peaks near 520 nm, 20-dihydrocortisol gives a somewhat lower value, and all other cortisol metabolites give less than 10%. Because it assays both cortisol and corticosterone, the method is better described as "11-hydroxycorticosteroids," although it should be noted that there is a requirement for the Δ^4-3-oxo group as well as for an 11-hydroxyl.

Many other compounds fluoresce and, although the contribution of each one is small, their effects may add up to more than the fluorescence due to cortisol.[37] A correction procedure that involved reading the fluorescence twice[38] somewhat lowered the nonspecific fluorescence.[37] An automated version[39] of the Mattingly method has a better specificity[37] but has not been widely adopted. The poor specificity of the fluorescence methods remains their major disadvantage, particularly for any but established routine uses. Another disadvantage is the volume of plasma required (2 ml) is large by comparison with the requirement of the method which follows (although less than that required for most Porter-Silber methods).

The third type of method for plasma cortisol is that based on competitive protein binding.[40] Such methods are relatively free from drug interference. Ease of manipulation is similar to that of fluorescence methods. The major drawback is the cost of a liquid scintillation counter.

Protein binding methods use the same principle as radioimmunoassay -- labeled hormone is displaced from a specific protein by unlabeled hormone (either standards or unknowns). The amount of radioactivity remaining bound to protein is thus a measure of the amount of unlabeled hormone added. The advantage of the protein binding method over radioimmunoassay is that the binding protein is easily available; it is corticosteroid binding globulin which is present in plasma. Moreover, it need not be purified, so diluted pooled plasma is the key reagent. A number of methods have been published for cortisol by competitive protein binding, the main differences being in the manner by which free and bound label are separated. Florisil as an adsorbent of the free fraction has the advantage that its high density makes centrifuging unnecessary. The method to be described follows closely that of Barnes et al.[41]

E. Urine Cortisol

This is sometimes known as urine free cortisol to emphasize that it is the hormone itself rather than its metabolites which is being considered. The concentration of cortisol in urine is very similar to that in plasma; both the Mattingly method and the competitive protein binding method may be used with very little modification.

The specificity of the Mattingly method is poor when it is applied to urines; moreover, additional fluorogenic material may appear if urines are stored for more than 1 to 2 days. Despite these drawbacks, the method gives useful clinical results.[42,43]

Competitive protein binding appears superior to fluorescence methods for urine. The choice will, however, be dictated by availability of instruments, as in the case of the plasma cortisol methods.

1. Methodology – Fluorescence Method

a. Principle

Unconjugated steroids are extracted into dichloromethane, and from the dichloromethane into a fluorescence reagent. The fluorescence that develops is read in a simple fluorimeter.

b. Apparatus

1. Glass-stoppered tubes, 150 mm X 24 mm. Tubes for fluorimetry should be rinsed with ethanol and drained dry before use.

2. Fluorimeter. The Hg line at 436 nm is used. The fluorescence is read in the green (e.g., Chance filters – HA1 and OB10 primary and OGR1 and OY3 secondary).

3. Solid standard. A block of "Naton 136" (Thorn Electronics Ltd., Tolworth, Surrey, U.K.) machined to fit tightly into a cuvette.

c. Reagents

1. Dichloromethane. This is best purified before use by repeatedly extracting with one tenth its volume of concentrated sulfuric acid until the acid does not discolor, then washing with water, drying over Na_2SO_4, and distilling.

2. Sodium hydroxide (0.1 mol/l. [0.4 g/100 ml]).

3. Sulfuric acid (AR quality).

4. Ethanol (AR quality).

5. Fluorescence reagent. Add 7.5 vol of concentrated sulfuric acid to 2.5 vol of ethanol, keeping the mixture cool in a bath of ice water. This is best done by using a large conical flask so that the contents can be swirled round. The reagent should be colorless. In many situations it is most satisfactory to make up fresh reagent just before use; it will, however, keep for a few days.[36]

6. Cortisol stock standard (1 mg/ml in ethanol). Store at 4°C. It is stable for about 1 year.

7. Cortisol working standard (0.1 mg/100 ml in water). Dilute 0.1 ml of reagent 6 to 100 ml. Prepare freshly at least once a month and store at 4°C.

d. Procedure

The analysis is performed on either a 24-hr urine collected without preservative or heparinized plasma.

1. Into a series of tubes pipette either 2.0 of sample, 2.0 ml of water, or 2.0 ml working standard. Each batch should contain at least two blanks (preferably the first and last tubes) and two standards.

2. Add 15 ml dichloromethane, stopper, and shake for 1 min. Leave to stand for about 1 min, then suck off and discard the upper (aqueous) layer.

3. This step is only needed for urines and jaundiced sera. Add 2.0 ml NaOH, stopper, and shake for 1 min. Leave to stand for 1 min, then suck off and discard the upper layer.

4. Pipette 10 ml of the extract into a clean tube for fluorimetry.

5. Check that the fluorimeter is suitably set up. The fluorescence reagent should read zero. The solid standard should be set to what previous experience has shown to be a suitable value, for example, a value such that the cortisol standard (equivalent to 100 μg/100 ml) will give a scale reading of 100.

The remaining steps should be accurately timed.

6. Add 5 ml fluorescence reagent to the first tube. Stopper and shake by hand for 20 sec (timed!). Repeat for the remaining tubes at intervals determined by the operator's speed at step 8.

7. Suck off the upper (dichloromethane) layer. The interface is difficult to see – the angle at which it is viewed is critical.

8. Transfer the lower layer to a cuvette and read the fluorescence at exactly 12 min after the fluorescence reagent was added (step 6).

e. Calculation

The concentration of 11-hydroxycorticosteroids is given by

$$\frac{\text{unknown} - \text{blank}}{\text{standard} - \text{blank}} \times 100\ \mu g/100\ ml$$

If the specimen was of urine, the value in μg/24 hr should be calculated.

f. Reliability

Within-batch precision should not exceed a coefficient of variation of 10% at the middle of the normal range.

There are a number of recognized sources of error:

1. Some batches of dichloromethane cause suppression of the standards but not the unknowns; presumably the impurity is in some way neutralized by plasma.

2. The working standard can deteriorate rapidly.

3. The heparin used as anticoagulant for blood collection should be free of benzyl alcohol.[44]

4. If the plasma sample is less than 2 ml, it should not be diluted. Dilute plasma tends to give spuriously high results.

5. Spironolactone (Aldactone®) interferes giving very high values (100 to 200 μg/100 ml). Interference persists for up to a week after stopping therapy.

2. Methodology – Competitive Protein Binding

a. Principle

Unconjugated steroids are extracted from plasma or urine with dichloromethane. The concentration of cortisol in the extract is determined by measuring the displacement of [^{3}H]cortisol from a limited number of binding sites on corticosteroid binding globulin (CBG). Florisil® is used to remove, by adsorption, the unbound cortisol; the method depends on the fact that the equilibration of cortisol with CBG is rapid at 38°C and slow at 0°C.

b. Apparatus

1. Test tubes for extractions (100 mm X 15 mm with ground glass stoppers).
2. Test tubes for binding. Conical centrifuge tubes (Quickfit & Quartz, BC 24/C14T). Approximately 15-ml capacity with a C14 ground glass joint. These tubes should be as uniform as possible.
3. Liquid scintillation spectrometer and counting pots to fit it.
4. Water baths at 0°C, 45°C, and 55°C.
5. Shaking machine (Griffin & George Ltd.) mounted on a slope. The partitions are covered with 2 to 3 cm of sponge rubber. Tubes are inserted still in their wire test tube racks and shaken lengthwise along a path inclined at about 30° to the horizontal.
6. Stainless steel scoop machined to hold 80 mg of Florisil.

c. Reagents

1. Dichloromethane. Redistilled and stored in a brown bottle.
2. Pooled plasma. Leftovers from specimens received for analysis may be used provided their cortisol content is less than 20 μg/100 ml. Store in small containers (2 to 3 ml in each) at –15°C.
3. [^{3}H]Cortisol (Radiochemical Centre, Amersham, U.K.). At least 10 Ci/mmol. Prepare a working solution of 10 μCi/ml in ethanol.
4. Binding reagent. Use 1.25 ml pooled plasma diluted to 50 ml with water and with 0.25 ml [^{3}H]cortisol added. Prepare just before using.
5. Florisil (60/100 mesh). Wash with distilled water until it all settles out in 30 sec. Activate by heating to 600°C for 4 hr and store in a sealed container (see note 1).
6. Scintillator: 0.4% (w/v) Butyl PBD (Scintillator butyl-PBD, Ciba-Geigy Ltd., Duxford, Cambs., U.K.) in toluene (GPR grade) with 2% methanol (AR grade).
7. Cortisol standard (stock): 1 μg/100 ml in methanol (AR grade). Store at 4°C.
8. Working standards containing 0, 2, 5, 10, and 20 ng in 0.2 ml prepared by taking, respectively, 0, 1.0, 2.5, 5.0, and 10 ml of the stock standard and making each up to 100 ml with methanol. Store at 4°C.
9. Bauxite boiling chips.

d. Procedure

The analysis is performed on either a 24-hr urine collected without preservative or heparinized plasma.

1. Add 2.0 ml dichloromethane to 0.2 ml plasma or 0.4 ml urine. Shake for 1 min. Leave to separate out.
2. Pipette two 0.5-ml portions of dichloromethane into binding tubes. Add one or two boiling chips and boil dry on a water bath at 55°C.
3. Make up the tubes into batches. Each batch should contain a duplicate set of standards. The batch size may be up to 50 tubes but should not exceed the number of tubes which can be shaken simultaneously at step 7. The standards are prepared by pipetting 0.2 ml of the appropriate working standards into binding tubes and evaporating to dryness.
4. Make up the required volume of binding solution.
5. Add 1.0 ml binding solution to each tube beginning and ending with standards. Whirlmix for 1 to 2 sec. Place in a bath at 45°C for 10 min.
6. Cool in ice water for 10 min.
7. Add 80 mg Florisil to each tube. Stopper and shake vigorously for 1 min. The tubes should be handled quickly so that they do not warm up appreciably. The intensity of shaking must be standardized.
8. Return to the ice bath. Inspect the tubes and dislodge any Florisil particles trapped at the surface by tapping the tube. Leave for 30 min (see note 2).
9. Pipette 0.5 ml of supernatant into a counting pot completing the batch as quickly as possible. Add 8 ml of scintillator, screw the top on firmly, and shake for 3 to 4 sec. Place in the counter and carry out 1 min counts until a steady rate is found (see note 3).

e. Calculation

1. Plot a graph of counts per minute (cpm)

against ng cortisol. Read the unknowns off the graph (see note 4).

ng in the unknown X 2 = μg/100 ml plasma
or ng in the unknown = μg/100 ml urine

Urinary results are expressed in μg/24 hr.

If the unknown contains more than 20 ng, the assay must be repeated, taking a smaller portion of dichloromethane extract.

f. Notes

1. If the Florisil is insufficiently active, the calibration curve will be too flat between 10 and 20 ng. As its activity increases, not only will the curve become steeper, but the counts will diminish slightly at the zero point.

2. This was found to give better precision.[45]

3. There is a good deal of variability in the behavior of the count rate. It may fall to half its original value over several hours. Placing the pots in a deep freeze for 30 min sometimes speeds up the stabilization of the count rate.

4. The shape of the curve may be predicted from these considerations. The capacity of the CBG in the binding solution is approximately 6 ng. If the total of endogenous cortisol and added radioactive cortisol ($\leqslant$2 ng) is less than 6 ng, the calibration curve should be flat until the total cortisol present equals 6 ng. The curve should then fall to give half the zero binding after a further 6 ng.

In practice, in a series of calibration curves, the 2 ng standard was 80 to 90% of the zero standard and the cpm fell to half that of the zero standard at 7 to 8 ng.

If the curve appears not to be smooth, it may be useful, for checking, to convert it to a straight line. One way to do this is to graph the square root of the cpm against the logarithm of the mass of cortisol.

g. Reliability

Most of the error is between batches and is largely referable to uncertainty in the drawing of the calibration curve.

Coefficients of variation claimed for protein binding methods have varied from 2 to 8%. In routine practice, 10% is probably a reasonable value to accept from a control sample over a period of weeks.

III. INTERPRETATIONS

A. "Correction for Creatinine"

All urinary assays depend upon the completeness of 24-hr collections of urine. Most people find this a difficult procedure, and inadequate collection may well result if the importance of completeness has not been emphasized to the patient and her attendants (or, with some people, if it has been overemphasized!). Gross inadequacies are easy to detect from the volume and appearance of urine, bearing in mind that the lowest 24-hr volume from an adult compatible with normal renal function is 500 ml and with this volume the urine should be maximally concentrated.

Some advocate measuring creatinine in all 24-hr urines. It is unfortunate that confusion has been generated by the failure to realize that there are different possible reasons for doing this, only one of which is valid.

It has been suggested that creatinine values can be used to judge whether single 24-hr specimens from different patients are complete. Its use here depends on the normal range of creatinine, which in this laboratory is 0.8 to 1.5 g/24 hr – clearly much too wide to detect small losses of urine. Some would use creatinine assays to judge whether serial specimens from the same patient are complete. Here the crucial thing is the day-to-day variation within individuals, for which coefficients of variation of 5 to 13% have been found.[46] Thus, creatinine estimations will detect losses of 30% but not 20%. This may be a little better than looking at the urine.

Some poeple would go further and attempt to "correct" for inadequate collection by referring steroid results in suspect urines to creatinine excretion. To do so is to make the additional assumption that the diurnal rhythms of creatinine and steroids are identical. In fact, they are quite different.[47]

The use of creatinine measurements to compensate for failures of urine collection must be distinguished from their use as a basis for expressing steroid results when urine collections are believed to be complete. The excretions of both creatinine and adrenal steroids are dependent on body size and expressing steroid excretions as mg/g creatinine is a way of compensating for body

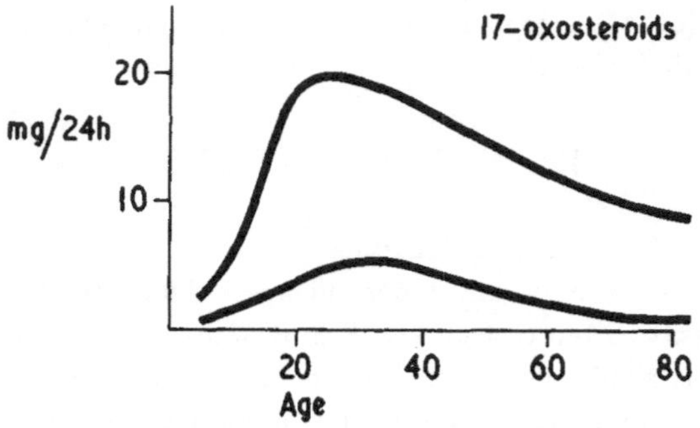

FIGURE 3. 17-Oxosteroids in normal women. The lines enclose 95% of observations. (Redrawn from Levell, M. J. et al., *J. Clin. Pathol.*, 10, 72, 1956. With permission.)

size differences. In practice, the normal ranges so produced have a similar width to those in which excretions are expressed as mg/24 hr, and serial estimations in the same patient showed a similar variation whether expressed as creatinine or time.[48] Referring steroid values to creatinine may, however, help in the differentiation of obesity and Cushing's syndrome.[49]

B. 17-Oxosteroids and 17-Oxogenic Steroids – Drug Interference

The Zimmerman reaction is influenced by a number of drugs. Many of these interfering effects have come to light as a result of casual observations rather than systematic studies; this and the varied pharmacological and chemical nature of the known drugs suggest that the list may well be incomplete.

Many of the drugs produce atypical, yellow colors in the Zimmerman reaction, which will at least warn the laboratory to disregard the result. It is, however, dangerous to rely on this; the only satisfactory approach is to ensure that patients are taken off all drugs preferably for at least a week, before collecting specimens.

The following have been reported[50] as causing an increased apparent 17-oxosteroid or 17-oxogenic steroid value:

Acetazolamide
Chlorpromazine
Hydralazine
Meprobamate
Methyprylone
Phenazopyridine
Spironolactone

The following cause a decrease:[50]

Chlordiazepoxide
Quinalbarbitone

Dexamphetamine and nalidixic acid also interfere.[11] Some other drugs reported as interfering[11] appear to be only minimal in their effects. Some radiocontrast media interfere with the 17-oxogenic steroid assay, possibly by competing for periodate.[51] Competition for periodate is always a potential source of interference; a method has been proposed for checking that excess is present.[52]

C. 17-Oxosteroids in Urine

The normal range in healthy ambulant children and adults is markedly age-dependent (Figure 3). Post-pubertal males have higher values than women, the greatest difference being in the 20 to 30-year age group when the male range is 3 to 4 mg/24 hr greater. Changes through the menstrual cycle are too small to influence the interpretation of pathological results. There is some increase in pregnancy. The 17-oxosteroid excretion is low in many chronic diseases. This is at least partly due to a decreased conversion of circulating androgens to their urinary metabolites.[53]

The main use of the 17-oxosteroid assay is in the differential diagnosis of virilism or hirsutism, when there may be increased excretion of 17-oxosteroids. Treatment with dexamethasone at a dose that will inhibit ACTH release (e.g., 0.5 mg 6-hourly) for 3 to 4 days will determine whether the increased androgen production is under ACTH control. If it is, the 17-oxosteroid excretion will fall to less than 5 mg/24 hr. If there is a source of androgens that is independent of ACTH, that is, a tumor of the adrenal or ovary, the 17-oxosteroid excretion will probably fall somewhat on dexamethasone (because the normal tissue will be suppressed) but the final value will be higher than 5 mg/24 hr and will reflect the output of the tumor. The interpretation when the value responds to dexamethasone involves other considerations. In congenital adrenal hyperplasia the 17-oxosteroid excretion is usually three to five times normal and the 17-oxogenic steroids are usually higher still. The results are so strikingly abnormal as to present little problem. Where a problem does arise is in the relatively large proportion of women with hirsutism in whom the 17-oxosteroids excre-

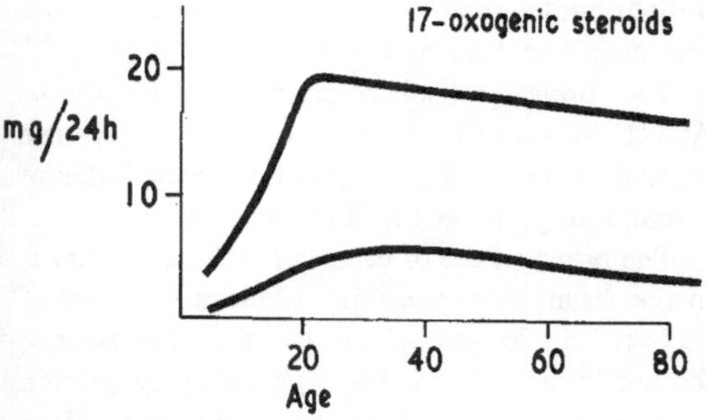

FIGURE 4. 17-Oxogenic steroids in normal women. The lines enclose 95% of observations. (Redrawn from Levell, M. J. et al., *J. Clin. Pathol.*, 10, 72, 1956. With permission.)

tion is typically 20 to 30 mg/24 hr, responding well to dexamethasone, and the 17-oxogenic steroids are normal. It has been shown that hirsute women with a raised 17-oxosteroid excretion and normal ovaries may benefit from estrogen-progestagen therapy.[54] Otherwise, little progress has been made toward defining therapeutic categories in this unfortunate group of patients.

The 17-oxosteroid excretion changes with the 17-oxogenic steroid excretion in Addison's disease and in those situations (ACTH-dependent Cushing's syndrome, hypopituitarism) in which the changes are due to altered ACTH secretion. The changes have very little diagnostic value, partly because the response of 17-oxosteroids to ACTH is less than that of 17-oxogenic steroids, and partly because of the nonspecific lowering found in many chronic diseases.

D. 17-Oxogenic Steroids in Urine

There is a good deal of variation in published normal ranges for females. All agree that there is a steady rise through childhood to a range of about 5 to 15 mg/24 hr at the age of 20 years. Some have reported a slight rise to a maximum at between 30 and 40 years with a gradual fall thereafter. Levell et al. found, on the contrary, no convincing evidence of a fall with age.[32] This discrepancy is puzzling and presumably reflects differences in the populations studied. Exercise, chronic illness, and body weight all influence the 17-oxogenic steroid excretion and are factors which might affect elderly "normals" more than younger ones. Figure 4 shows a normal range. Values in males are somewhat higher. This reflects their greater average size. There is an increase in pregnancy due largely to the excretion of 21-deoxycorticosteroids.[53]

1. Misleading Changes

In chronic liver failure the rate of cortisol metabolism may be reduced. This would cause a rise of plasma cortisol were it not for the negative feedback control. In the steady state the plasma cortisol is normal, the cortisol production rate is lower, and the 17-oxogenic steroid excretion is lower.

A similar mechanism operates in hypothyroidism in which the 17-oxogenic steroid excretion is often reduced.

Low values of 17-oxogenic steroids are found in many chronic diseases. This is probably also due to a lowered rate of metabolism of cortisol.

High excretions of 17-oxogenic steroids are found in patients receiving cortisone (or cortisol). When the steroid is given by mouth or intravenously it is metabolized in a similar (though not identical) fashion to the endogenous cortisol. Each 100 mg raises the 17-oxogenic steroid excretion by 40 to 50 mg. Intramuscular cortisone has a much smaller effect on the urinary 17-oxogenic steroids. This effect of cortisone or cortisol in raising the steroid excretion is partly offset by their inhibiting ACTH release, thus lowering the 17-oxogenic steroids derived from endogenous cortisol.

Any physiological situation in which ACTH is increased will increase the 17-oxogenic steroid excretion. After major surgery the excretion is elevated for 2 to 3 days. Physical activity may increase the plasma cortisol, as may apprehension, but these effects are unlikely to be sufficiently sustained to produce a marked effect on the 24-hr 17-oxogenic steroid excretion.

2. Significant Changes

Values of diagnostic significance may be found in diseases of the pituitary and adrenal.

In Cushing's syndrome, values are usually raised, but may be normal despite a raised cortisol production rate.[56] The 17-oxogenic steroids are also raised in simple obesity, although never to very high values. Thus, while a 17-oxogenic steroid excretion of over perhaps 30 mg/day points more to Cushing's syndrome than to obesity, less than half of patients with Cushing's syndrome will have such a value. It is not, therefore, a very efficient test. Of more value is the response to dexamethasone;* normals and obese patients usually suppress to less than 5 mg/24 hr on dexamethasone (0.5 mg 6-hourly), whereas patients with Cushing's syndrome show little or no fall. The higher dose of dexamethasone, 2.0 mg 6-hourly, distinguishes pituitary-dependent Cushing's from the others; the former group shows some fall. It should be noted that the metabolites of dexamethasone (and betamethasone) are not measured by the 17-oxogenic steroid method because the methyl group at C_{16} blocks the Zimmerman reagent.

In Addison's disease, values range from zero to the middle of the normal range. Patients who look Addisonian but who do not, in fact, have the disease also often have values in this range. It is, therefore, a poor screening test. The characteristic feature of Addison's disease is, of course, the failure to respond to corticotrophin. The 17-oxogenic steroids should rise to between 40 and 70 mg/24 hr by the second day of stimulation; a delayed rise indicates secondary hypoadrenalism.

In long-standing hypopituitarism, the 17-oxogenic steroids may be very low, although there may be no obvious signs of cortisol deficiency. This is at least partly due to the protective effect of thyroid failure, which usually accompanies ACTH failure, causing a slower metabolic clearance of the cortisol. As in the case of Addison's disease, a stimulation test is of more value than a single estimation. Metyrapone is used and the 17-oxogenic steroid assay may be used because, in a normal response, the increased excretion of 11-deoxycortisol metabolites exceeds the decreased excretion of cortisol metabolites. The 17-oxogenic steroids should rise 2 to 3 times if the test is conducted properly with careful attention to the timing of the doses of metyrapone.[59]

In patients with virilism or hirsutism, the 17-oxogenic steroids are usually normal. The exception is in congenital adrenal hyperplasia in which the 17-oxogenic steroid excretion may be 50 to 100 mg/24 hr, much of it being pregnanetriol.

E. Urine Cortisol

Normal ranges have not been defined as precisely as for 17-oxogenic steroids. In adults, the fluorescence method gave a range of 68 to 280 μg/24 hr in women, slightly more in men;[43] protein binding methods similar in principle to the one described here gave 0 to 100 μg/24 hr.[60,61]

The urine cortisol responds dramatically to ACTH in normal people; the value may rise tenfold or more. Thus, it provides a good index of response to prolonged ACTH tests.

The principal use of urine cortisol estimations is in the diagnosis of Cushing's syndrome. Values of 399 to 5,570 μg/24 hr by the fluorescence method[43] and 99 to 9,497 μg/24 hr by protein binding methods[62] have been reported. More significant is that obese patients fall in the normal range, and the true control group – patients thought to have Cushing's but in whom the disease was excluded – all had values below 106 μg/24 hr.[62]

The excretion of urine cortisol is low in those situations (Addison's disease, hypopituitarism) in which the plasma cortisol tends to be low, but because of the very low values which may also be encountered in normal people, single estimations have no diagnostic value in these situations.

F. Plasma Cortisol

The normal range is neither age- nor sex-dependent. At 9 a.m. most normals fall in the range 5 to 20 μg/100 ml. Values in this range may be encountered during much of the daytime, although toward evening high values become less frequent. By midnight it is unusual to have values over 10 μg/100 ml.

The protein binding method gives values 2 to 3 μg/100 ml less than the fluorescence method, but in both methods the absolute values are so dependent on detail of technique that normal ranges should be established locally (if they are to be used). Estrogens increase the concentration of corticosteroid binding globulin (CBG) and, therefore, of plasma cortisol. In late pregnancy the values of both these are about three times normal. The situation in women taking oral contraceptives is less clear. Ethinyl estradiol, at a dose of 50 μg per day, causes a 50% rise of both CBG[63] and plasma cortisol.[64] However, a number of reports suggest that women on oral contraceptives have values 2 to 3 times normal. The discrepancies may arise at least in part from the much greater usage, until recently, of preparations with a higher estrogen content.

*Details of ward procedures in pituitary-adrenal function tests are outside the scope of this chapter, but are available elsewhere.[57,58]

Smoking may increase the plasma cortisol acutely.[65]

Care must be taken to avoid stress artifacts. Some patients react swiftly to venipuncture. A particular problem is the plump adolescent with poor veins in whom Cushing's syndrome is being queried! It should be borne in mind that stress overrides the suppressive effect of dexamethasone.

Due to the episodic nature of cortisol secretion single samples are of very little diagnostic value. It is true that results are likely to be higher in Cushing's syndrome than in Addison's disease, but results found in these two diseases do, in fact, overlap. Insofar as normal subjects tend not to have secretory episodes in the late evening, a high value (over 10 to 15 μg/100 ml) at 11 to 12 p.m. is a pointer toward Cushing's syndrome, but a feeble one.

Plasma cortisol assays are of greatest value when used in function tests. Normal responses have been most clearly delineated with the fluorimetric method of analysis. In those situations in which they involve an increment, one would expect the protein binding method to give similar responses; where they involve an absolute value the protein binding method should be 2 to 3 μg/100 ml lower. It may be noted that for an increment to be significant at $P = 0.05$ it must exceed 2.7 times the error standard deviation of the method.

The normal responses to tetracosactrin (Synacthen®), insulin-induced hypoglycemia, lysine vasopressin, and pyrogen are roughly similar, lysine vasopressin providing a slightly weaker, and pyrogen a slightly stronger stimulus than the other two.[58] The response may be judged from the increment of plasma cortisol concentration or from the final value achieved; in some situations looking at both of these together may be helpful.[66] With all the tests, normal people show a rise of at least 7 μg/100 ml to a value of at least 20 μg/100 ml. Failure to respond must be treated with reservations. Thus, a failure to respond to tetracosactrin indicates either primary or secondary hypoadrenalism; a prolonged ACTH test must be carried out to distinguish these. Failure to respond to insulin-induced hypoglycemia, lysine vasopressin, or pyrogen tests indicates impairment of ACTH secretion only if the adrenal has been shown to be responsive to tetracosactrin.

Plasma cortisol measurements may be used to follow dexamethasone suppression; a value of less than 7 μg/100 ml at 9 a.m. after a single 2 mg dose at 11 to 12 p.m. on the previous evening excludes Cushing's syndrome. Failure to suppress in this test may be due to Cushing's syndrome, stress, or depression.[67]

G. Conclusions

It will be apparent from the foregoing sections that the process of interpretation is not simply a matter of values being high in this disease and low in that. It is rather an assessment of the technical reliability of a value and of the relative probabilities of such a value being associated with the disease under consideration or an alternative disease. It follows from this that the expression "checking that adrenal function is normal" has no useful meaning.

Furthermore, the diagnostic precision of cortisol measurements is enormously enhanced when they form part of dynamic function tests in which the integrity of the control mechanisms is challenged. Single assays can only be interpreted if there is an awareness of the factors such as stress which may produce large, but still physiological, changes.

IV. ALDOSTERONE

Aldosterone is produced by a different part of the adrenal cortex from that which produces cortisol and the androgens and is under the control of different trophic factors. In general, therefore, diseases causing increases or decreases of cortisol will not cause corresponding changes of aldosterone and vice versa. An exception to this rule is Addison's disease, in which the destruction of the adrenal cortex affects all its parts.

The production rate of aldosterone is only one hundredth that of cortisol and the plasma concentration one thousandth (it has no specific plasma binding protein). Simple chemical methods are not, therefore, applicable to its measurement. Until about 1970 very elaborate double isotope derivative techniques were used; since then, radioimmunoassay has become more usual.

The indications for aldosterone assay are so few and the methods so demanding that there is little justification for every hospital laboratory attempting to carry them out; the recommended method is to send the specimens to a centralized laboratory that has a high enough throughput to maintain adequate control of the technique.

A. Interpretation

In the interpretation of aldosterone assays, particular attention must be paid to the mechanisms that stimulate aldosterone production. This is difficult because the mechanisms are imperfectly understood and insofar as they are understood, they involve factors, such as plasma volume, which are not readily assessed.

The production rate of aldosterone and its concentration in plasma are increased by ACTH,[68] potassium,[68] and angiotensin II.[69] Angiotensin II is formed in the plasma from angiotensin I, which is generated in the plasma by the action of renin, a hormone from the kidney. There are a number of stimuli of renin production, a major one is sodium depletion. Thus, a low sodium intake or a course of diuretics will lower the sodium content of the extracellular fluid (though not the concentration because water will also be lost) stimulating renin which, via angiotensin, will increase aldosterone production.

This effect of sodium intake on aldosterone is part of a homeostatic feedback since aldosterone causes sodium retention.

Sodium status is the most important of the factors affecting aldosterone; thus, an increase of body sodium will suppress aldosterone even in the face of continuous ACTH stimulation. The dominant role of angiotensin II is further illustrated by the low aldosterone levels resulting from a primary failure of renin production.[70] Nevertheless, there may also be other mechanisms operating in parallel with angiotensin, linking sodium to aldosterone.[71]

The principal clinical use of aldosterone assays is in the diagnosis of primary aldosteronism. In this condition the plasma aldosterone, the aldosterone production rate, the "urinary aldosterone" (i.e., aldosterone + aldosterone-18-glucuronide) are typically 2 to 10 times the values seen on a normal sodium intake; that is, they are in or above the range seen in normal people on a low sodium intake. It is not unusual to find a high aldosterone production in a hypertensive, hypokalemic patient. In most cases the high aldosterone is not due to primary aldosteronism, but some other cause. The two most common causes are diuretic therapy and various sorts of renal damage; these should be excluded before carrying out aldosterone assays.

Primary aldosteronism may be due to either an adrenal tumor or to bilateral hyperplasia and it is desirable to make the distinction between them before surgery. Possible ways of approaching this very difficult problem have included computer-assisted quadric analysis,[72] analyses on adrenal venous blood,[73] and measurement of [^{131}I]cholesterol uptake by external scintillation counting.[74]

It is clear from the foregoing discussion that the interpretation of aldosterone results in hyperaldosteronism matches the difficulty of performing the analyses! This subject has recently been well reviewed.[75-77]

REFERENCES

1. **Ernest, I., Hakansson, B., Lehmann, J., and Sjogren, B.,** *Acta Endocrinol.,* 46, 552, 1964.
2. **Metcalf, M. G.,** *Steroids,* 17, 85, 1971.
3. **Drekter, I. J., Heisler, A., Scism, G. R., Stern, S., Pearson, S., and McGavack, T. H.,** *J. Clin. Endocrinol. Metab.,* 12, 55, 1952.
4. **Janovsky, J. V. and Erb, L.,** *Ber. Dtsch. Chem. Gesell.,* 19, 2155, 1886.
5. **Zimmerman, W.,** *Hoppe Seyler Z. Physiol. Chem.,* 233, 257, 1935.
6. **Callow, N. H., Callow, R. K., and Emmens, C. W.,** *Biochem. J.,* 32, 1312, 1938.
7. **James, V. H. T. and De Jong, M.,** *J. Clin. Pathol.,* 14, 425, 1961.
8. **Corker, C. S., Norymberski, J. K., and Thow, R.,** *Biochem. J.,* 83, 583, 1962.
9. **Talbot, N. B., Berman, R. A., and McLachlan, E. A.,** *J. Biol. Chem.,* 143, 211, 1942.
10. **Allen, W. M.,** *J. Clin. Endocrinol.,* 10, 71, 1950.
11. **Gray, C. H., Baron, D. N., Brooks, R. V., and James, V. H. T.,** *Lancet,* 1, 124, 1969.
12. **Hellman, L., Nakada, F., Curti, J., Weitzman, E. D., Kream, J., Roffwarg, H., Ellman, S., Fukushima, D. K., and Gallagher, T. F.,** *J. Clin. Endocrinol.,* 30, 411, 1970.
13. **Weitzman, E. D., Fukushima, D. K., Nogeire, C., Roffwarg, H., Gallagher, T. F., and Hellman, L.,** *J. Clin. Endocrinol.,* 33, 14, 1971.
14. **Kreiger, D. T., Allen, W., Rizzo, F., and Krieger, H. P.,** *J. Clin. Endocrinol.,* 32, 266, 1971.
15. **de Lacerda, L., Kowarski, A., and Migeon, C. J.,** *J. Clin. Endocrinol.,* 36, 227, 1973.

16. Paterson, J. Y. F., *J. Endocrinol.,* 56, 551, 1973.
17. Lindholm, J., *Scand. J. Clin. Lab. Invest.,* 31, 115, 1973.
18. Hellman, L., Nakada, F., Zumoff, B., Fukushima, D. K., Bradlow, H. L., and Gallagher, T. F., *J. Clin. Endocrinol.,* 33, 52, 1971.
19. Frantz, A. G., Katz, F. H., and Jailer, J. W., *J. Clin. Endocrinol.,* 21, 1290, 1961.
20. Berman, M. L. and Green, O. C., *Anesthesiology,* 34, 365, 1971.
21. Choi, Y., Thrasher, K., Werk, E. E., Sholiton, L. J., and Olinger, C., *J. Pharmacol. Exp. Ther.,* 176, 27, 1971.
22. Gallagher, T. F., Fukushima, D. K., and Hellman, L., *J. Clin. Endocrinol.,* 31, 625, 1970.
23. Levell, M. J., *Acta Endocrinol.,* 70, 89, 1972.
24. Peterson, R. E., Karrer, A., and Guerra, S. L., *Anal. Chem.,* 29, 144, 1957.
25. Norymberski, J. K., Stubbs, R. D., and West, H. F., *Lancet,* 1, 1276, 1953.
26. Appleby, J. I., Gibson, G., Norymberski, J. K., and Stubbs, R. D., *Biochem. J.,* 60, 453, 1955.
27. Few, J. D., *J. Endocrinol.,* 22, 31, 1961.
28. Diczfalusy, E., Plantin, L.-O., Birke, G., Ingall, S. C., and Norymberski, J. K., *Acta Endocrinol.,* 27, 275, 1958.
29. Metcalf, M. G., *J. Endocrinol.,* 26, 415, 1963.
30. Medical Research Council Committee on Clinical Endocrinology, *Lancet,* 1, 1415, 1963.
31. Edwards, R. W. H., Kellie, A. E., and Wade, A. P., *Mem. Soc. Endocrinol.,* 2, 53, 1953.
32. Levell, M. J., Mitchell, F. L., Paine, C. G., and Jordan, A., *J. Clin. Pathol.,* 10, 72, 1956.
33. Fraser, R. and James, V. H. T., *J. Endocrinol.,* 40, 59, 1968.
34. De Moor, P., Steeno, O., Raskin, M., and Hendrikx, A., *Acta Endocrinol.,* 33, 297, 1960.
35. Mattingly, D., *J. Clin. Pathol.,* 15, 374, 1962.
36. Medical Research Council Working Party, *Br. Med. J.,* 2, 310, 1971.
37. James, V. H. T., Townsend, J., and Fraser, R., *J. Endocrinol.,* 37, xxviii, 1967.
38. Spencer-Peet, J., Daly, J. R., and Smith, V., *J. Endocrinol.,* 31, 235, 1965.
39. Townsend, J. and James, V. H. T., *Steroids,* 11, 497, 1968.
40. Murphy, B. E. P., *J. Clin. Endocrinol.,* 27, 973, 1967.
41. Barnes, N. D., Joseph, J. M., Atherden, S. M., and Clayton, B. E., *Arch. Dis. Child.,* 47, 66, 1972.
42. Mattingly, D., Dennis, P. M., Pearson, J., and Cope, C. L., *Lancet,* 2, 1046, 1964.
43. Mattingly, D. and Tyler, C., *Br. Med. J.,* 4, 394, 1967.
44. Kendall, J. W., Egans, M. L., and Stott, A. K., *J. Clin. Endocrinol.,* 28, 1373, 1968.
45. Few, J. D. and Cashmore, G. C., *Ann. Clin. Biochem.,* 8, 205, 1971.
46. Chattaway, F. C., Hullin, R. P., and Odds, F. C., *Clin. Chim. Acta,* 26, 567, 1969.
47. Curtis, G. and Fogel, M., *Psychosom. Med.,* 32, 337, 1970.
48. Cryer, P. E. and Sode, J., *Clin. Chem.,* 16, 1012, 1970.
49. Streeten, D. H. P., Stevenson, C. T., Dalakos, T. G., Nicholas, J., Dennick, L. G., and Fellerman, H., *J. Clin. Endocrinol.,* 29, 1191, 1969.
50. Young, D. S., Thomas, D. W., Friedman, R. B., and Pestaner, L. C., *Clin. Chem.,* 18, 1041, 1972.
51. Nelson, J. C., Krueger, G. G., Wilcox, R. B., and Thompson, W. P., *J. Clin. Endocrinol.,* 28, 1515, 1968.
52. Abernethy, M. H. and Metcalf, M. G., *Clin. Chem.,* 16, 274, 1970.
53. Zumoff, B., Bradlow, H. L., Gallagher, T. F., and Hellman, L., *J. Clin. Endocrinol.,* 32, 824, 1971.
54. Hancock, K. and Levell, M. J., in preparation.
55. Appleby, J. I. and Norymberski, J. K., *J. Endocrinol.,* 15, 310, 1957.
56. James, V. H. T. and Caie, E., *J. Clin. Endocrinol.,* 24, 180, 1964.
57. Zilva, J. F. and Pannall, P. R., *Clinical Chemistry in Diagnosis and Treatment,* Lloyd-Luke, London, 1971, 109.
58. James, V. H. T. and Landon, J., *Hypothalamic-pituitary-adrenal Function Tests,* CIBA Laboratories Ltd., Horsham, 1971.
59. Cope, C. L., Dennis, P. M., and Pearson, J., *Clin. Sci.,* 30, 249, 1966.
60. Beardwell, C. G., Burke, C. W., and Cope, C. L., *J. Endocrinol.,* 42, 79, 1968.
61. Murphy, B. E. P., *J. Clin. Endocrinol.,* 28, 343, 1968.
62. Burke, C. W. and Beardwell, C. G., *Quart. J. Med.,* 42, 175, 1973.
63. Musa, B. U., Seal, U. S., and Doe, R. P., *J. Clin. Endocrinol.,* 25, 1163, 1965.
64. Bulbrook, R. D., Hayward, J. L., Herian, M., Swain, M. C., Tong, D., and Wang, D. Y., *Lancet,* 1, 628, 1973.
65. Kershbaum, A., Pappajohn, D. J., Bellet, S., Hirabayashi, M., and Shafiiha, H., *J.A.M.A.,* 203, 275, 1968.
66. Levell, M. J., Stitch, S. R., and Noronha, M., *Acta Endocrinol.,* 65, 608, 1970.
67. Carroll, B. J. and Davies, B., *Br. Med. J.,* 1, 789, 1970.
68. Williams, G. H. and Dluhy, R. G., *Am. J. Med.,* 53, 595, 1972.
69. Boyd, G. W., Adamson, A. R., Arnold, M., James, V. H. T., and Peart, W. S., *Clin. Sci.,* 42, 91, 1972.
70. Brown, J. J., Chinn, R. H., Fraser, R., Lever, A. F., Morton, J. J., Robertson, J. I. S., Tree, M., Waite, M. A., and Park, D. M., *Br. Med. J.,* 1, 650, 1973.
71. Editorial, *N. Engl. J. Med.,* 286, 100, 1972.
72. Aitchison, J., Brown, J. J., Ferriss, J. B., Fraser, R., Kay, A. W., Lever, A. F., Neville, A. M., Symington, T., and Robertson, J. I. S., *Am. Heart J.,* 82, 660, 1971.

73. Melby, J. C., Spark, R. F., Dale, S. L., Egdahl, R. H., and Kahn, P. C., *N. Engl. J. Med.,* 277, 1050, 1967.
74. Conn, J. W., Morita, R., Cohen, E. L., Beierwaltes, W. H., McDonald, W. J., and Herwig, K. R., *Arch. Intern. Med.,* 129, 417, 1972.
75. Cain, J. P., Tuck, M. L., Williams, G. H., Dluhy, R. G., and Rosenoff, S. H., *Am. J. Med.,* 53, 627, 1972.
76. Brown, J. J., Fraser, R., Lever, A. F., and Robertson, J. I. S., *Clin. Endocrinol. Metab.,* 1, 397, 1972.
77. Brown, J. J., Fraser, R., Lever, A. F., and Robertson, J. I. S., *Br. Med. J.,* 2, 391, 1972.

GAS CHROMATOGRAPHIC PROCEDURES FOR THE MEASUREMENT OF PREGNANEDIOL, ANDOSTERONE, ETIOCHOLANOLONE, AND DEHYDROEPIANDROSTERONE

M. G. Metcalf

TABLE OF CONTENTS

I. General Introduction . . . 151
- A. Introduction . . . 151
- B. Apparatus . . . 153
- C. Gas Chromatography . . . 153
- D. Reagents . . . 154
- E. Urine Samples . . . 155

II. The Measurement of Pregnanediol in Urine . . . 155
- A. Introduction . . . 155
- B. Procedure . . . 156
- C. Discussion . . . 157
- D. Results . . . 161

III. The Measurement of Androsterone and Etiocholanolone in Urine . . . 163
- A. Introduction . . . 163
- B. Procedure . . . 164
- C. Discussion . . . 165
- D. Results . . . 167

IV. The Measurement of Dehydroepiandrosterone in Urine . . . 169
- A. Introduction . . . 169
- B. Procedure . . . 171
- C. Discussion . . . 172
- D. Results . . . 173

Acknowledgments . . . 178

References . . . 178

I. GENERAL INTRODUCTION

A. Introduction

In 1958, Lovelock[1] described the argon ionization detector and McWilliam and Dewar[2] introduced the flame ionization detector. Two years later, Vanden Heuvel et al.,[3] using a thin film of silicone gum as a liquid phase, succeeded in separating steroids on a gas chromatographic column. These two discoveries made possible the use of gas chromatography for the estimation of steroids in blood and in urine. The two detectors, with sensitivities at least a thousand times that of existing equipment, were capable of responding to amounts of the order of 10 ng for the flame ionization detector and 1 pg for the argon ionization detector. The new silicone gums were stable at the temperatures necessary to volatilize steroids. Early difficulties associated with the breakdown of steroids at high temperatures were soon surmounted by the discovery of derivatives which were stable in the vapor phase. The development of steroid trimethylsilyl ethers,[4] trifluoroacetates, heptafluorobutyrates,[5] and, more recently, of the alkyl boronates,[6] has made practicable the measurement of most human steroids by gas chromatographic techniques.

In the analysis of steroids in urine, the gas chromatograph has been used primarily as a sensitive measuring device, with little use made of its potent separatory properties. Most gas chromatographic assays for urinary steroids are long and involved. Steroids are isolated in almost pure state and are converted to stable derivatives before injection into a gas chromatograph. In general, there has been little interest in the 1963 reports by Cox[7] and Turner et al.[8] that pregnanediol could be measured directly in toluene extracts of hydrolyzed urine. Their demonstration that at least one steroid could be analyzed by the examination of gas chromatograms of crude urinary extracts was, until recently, largely ignored.

Because of the importance of a prompt feedback of laboratory results in patient care, there is a continuing search for more rapid and more efficient ways of measuring substances of clinical relevance. The observations of Cox and Turner resulted in a major simplification in the assay of pregnanediol. Their discovery suggested that other urinary steroids might be amenable to measurement by a similar procedure, without the need for preliminary chromatographic purification of the assay sample and without the need for derivative formation.

To be amenable to analysis by this simplified type of procedure, steroids must satisfy certain requirements. First, and most important, they must volatilize without decomposition. This requirement eliminates from consideration all 17-hydroxycorticosteroids for which derivative formation is an essential preliminary to gas chromatographic analysis. It leaves only steroids such as pregnanediol and the urinary 17-oxosteroids, which possess no heat-labile glycol or dihydroxyacetone groups. The second prerequisite is that the steroids occur in urine at concentrations of at least 0.2 mg/l. Steroids that are to be measured directly in extracts of hydrolyzed urine must be excreted in sufficient amount to give gas chromatogram peaks which can be clearly differentiated from the nonspecific urinary background. A third requirement is that the steroids be separable from other urinary constituents by gas chromatography and that the separation be preceded only by simple and quantitative chemical manipulations. Last, it is desirable that a simple means of hydrolyzing the steroid conjugates exists. In urine, steroids occur predominantly as glucuronides or sulfates.[9] Satisfactory analytical procedures involve the quantitative hydrolysis of these urinary conjugates, both to convert the steroids to a form suitable for gas chromatography and to facilitate their removal from water-soluble contaminants by transfer to an organic solvent. Hydrolytic techniques described as nondestructive tend to be slow compared with procedures based on the use of hot acid.[10] In the context of this discussion, it is, therefore, advantageous if a steroid is resistant to acid attack, in addition to the attributes listed above.

A search for urinary steroids that conform with these conditions has led to the development of assays for etiocholanolone, dehydroepiandrosterone, pregnanediol, and for etiocholanolone and androsterone together. In each case the preparation of 40 samples for gas chromatography takes less than 4 hr, and each chromatographic analysis between 10 and 20 min. These assays are described below. It appears likely that in normal urine, few steroids other than these four are measurable by techniques of this degree of simplicity.

The urinary levels of this group of steroids have been measured in studies on many subjects of clinical interest. Pregnanediol is a major urinary metabolite of progesterone; the fluctuations in the secretion rate of progesterone that are found in menstruating women are reflected by similar changes in the excretion rate of pregnanediol. These changes may be used to distinguish ovulatory from anovulatory cycles. Conception is accompanied by a continuing rise in urinary pregnanediol levels, due first to the nonregression of the corpus luteum and later to the development of the placenta. In the later stages of pregnancy therefore, the urinary output of pregnanediol may be used as one of the indices of placental adequacy. The other three steroids in the group, androsterone, etiocholanolone, and dehydroepiandrosterone, are 11-desoxy-17-oxosteroids. They have their source both in the adrenal cortex and in the gonads, and have been measured in studies on growth, ovarian and testicular function, infertility, hirsutism, and virilism. It is claimed that women with breast carcinoma who respond favorably to endocrine ablation excrete them in greater amounts than women who fail to respond.[88] In an occasional patient they are end-products of malignant processes, and changes in their urinary levels

TABLE 1

Steroid Nomenclature

Androsterone (A)	5α-Androstan-3α-ol-17-one
Etiocholanolone (E)	5β-Androstan-3α-ol-17-one
Dehydroepiandrosterone (DHA)	5-Androsten-3β-ol-17-one
Progesterone (P)	4-Pregnen-3,20-dione
Pregnanediol (PD)	5β-Pregnan-$3\alpha,20\alpha$-diol
Epiandrosterone	5α-Androstan-3β-ol-17-one
Testosterone	4-Androsten-17β-ol-3-one
Androstenedione	4-Androsten-3,17-dione
5-Androstenediol	5-Androsten-$3\beta,17\beta$-diol
Allopregnanediol	5α-Pregnan-$3\alpha,20\alpha$-diol
Pregnanolone	5β-Pregnan-3α-ol-20-one
Pregnanetriol	5β-Pregnan-$3\alpha,17\alpha,20\alpha$-triol
5-Pregnenediol	5-Pregnen-$3\beta,20\alpha$-diol
5-Pregnenolone	5-Pregnen-3β-ol-20-one
Pregnanetriolone	5β-Pregnan-$3\alpha,17\alpha,20\alpha$-triol-11-one
11-Desoxy-tetrahydrocortisol	5β-Pregnan-3α-17α,21-triol-20-one
17-Oxogenic steroids (OGS)	17,20-Dihydroxysteroids + 17-hydroxy-20-oxosteroids

may be used to follow the results of surgery or of chemotherapy.

Table 1 gives details of steroid nomenclature relevant to this chapter.

B. Apparatus

The recommendations that follow are relevant to each of the assays described in this chapter.

1. Syringes

Glass-tipped syringes are fitted with stops (Figure 1) for the rapid dispensing of precise volumes (standard deviation = 0.0012 ml (10 trials) for a 1 ml syringe). Needles are omitted from syringes set to deliver volumes of more than 1 ml.

2. Glassware

Culture tubes (25 to 35 ml) with Teflon®-lined screw caps are used for hydrolysis and extraction, and conical centrifuge tubes (11 ml) for the concentration of extracts. All glassware used in these assays is cleaned by being soaked in detergent solution overnight, rinsed, soaked for not less that 5 min in methylated spirits, rinsed repeatedly with tap water, and twice with distilled water.

3. Test Tube Racks

Rectangular racks holding up to 40 test tubes are fitted with lids lined with foam plastic to facilitate manual shaking (Figure 1). Shallow racks in which the centrifuge tubes may be placed at an angle are used for the evaporation of extracts.

C. Gas Chromatography

1. Column Preparation

Glass columns (length 65 to 78 cm, internal diameter 3 mm) are cleaned thoroughly by successive immersions in benzene and concentrated sulfuric acid, followed by exhaustive rinsing with water, ethanol, and distilled water. After drying, the columns are filled with a solution of dimethyl dichlorosilane in toluene (5% v/v), left 15 min, drained, rinsed 4 times with toluene (5 ml), and then rinsed exhaustively with water, ethanol, and distilled water. After drying, the columns may be packed.

2. Preparation of Coated Supports

About 12 to 15 g of Gas Chrom Q is slurried with gentle degassing in a 2 g % (w/v) solution of SE30 (dehydroepiandrosterone assay) or of XE60 (all other assays) in chloroform. After 15 min, the support is filtered, left to dry for 1 hr at room temperature, and a further hour at 90 to 100°C.

3. Column Packing

The columns are packed with hot coated support under gentle suction and tapped repeatedly with a rubber-surfaced rod to ensure even

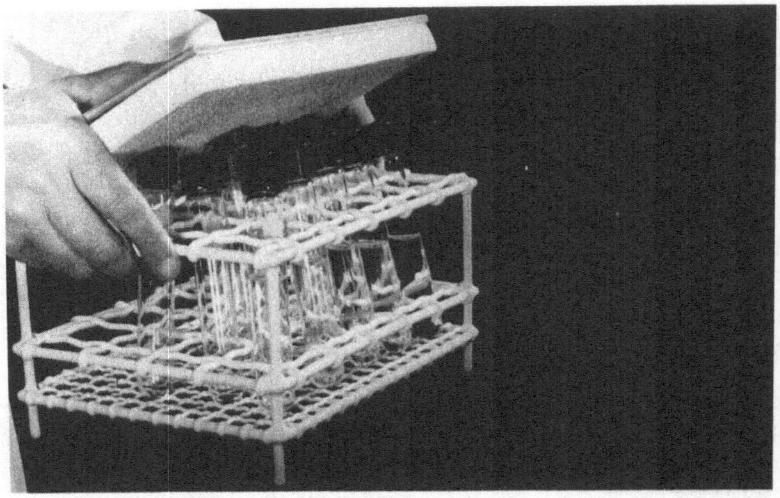

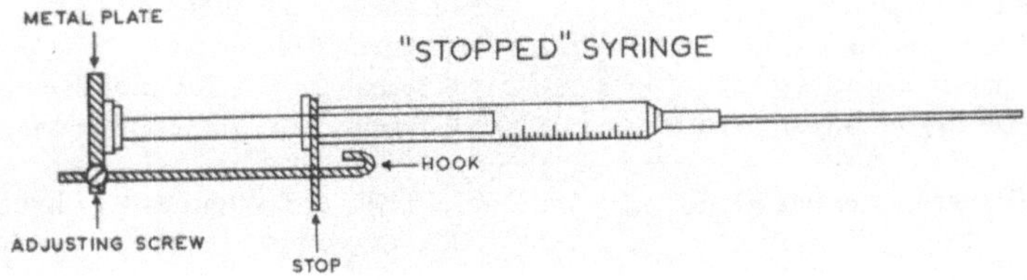

FIGURE 1. Extraction rack and lid (above). Diagram of a stopped syringe designed by Dr. A. C. Arcus, Princess Margaret Hospital, Christchurch, N.Z. (below).

packing. Care is necessary if damage to the particle surfaces is to be avoided.

4. Column Conditioning

Before use, the packed columns are conditioned in a slow stream of nitrogen at about 235°C. Conditioning is continued until the bleed of surplus liquid phase from the column falls to a low and constant level. This may take from as little as 4 hr for an SE30 column to 40 hr or more for an XE60 column. To prevent detector contamination, the column effluent is vented to air for the first few hours of conditioning.

5. Chromatographic Conditions

Measurements are made with a flame ionization detector. Both a Perkin Elmer® 800 and a Varian Aerograph 2700® have been used for gas chromatography. The operating conditions chosen have, in general, been those recommended by the makers, with optimization of gas flows achieved as outlined in standard texts.[11,12] Because column life is inversely related to oven temperature, it is usual to operate the instrument at as low a temperature as is consistent with reasonable analytical efficiency. Separation times of 10 to 15 min may be achieved if the oven is maintained at a temperature close to 235°C, the precise temperature chosen being a function of the column used and of carrier gas flow. The life of a column is variable, but usually of the order of 6 months. Columns are renewed when repeated injections of a sample containing two steroids give relative peak areas which differ repeatedly by more than ±10% from the mean value for all the injections. Satisfactory columns give symmetrical peaks as well as replicable results.

D. Reagents

In the following sections, it should be noted that the firms named are not necessarily the only agents supplying satisfactory reagents. Materials obtained from other firms may well be equally satisfactory.

Steroids and steroid conjugates were purchased from Ikapharm Ltd., Israel, or from Steraloids Inc., Pawling, N.Y., and used as received. Boric acid, hydrochloric acid, acetic acid, ethanol, and sodium acetate-$3H_2O$, all of analytical reagent grade, were obtained from British Drug Houses Ltd., as was the '4.5' indicator and the dimethyl dichlorosilane. Chloroform of anesthetic grade containing 1 to 2% ethanol was obtained in bulk from Industrial Chemical Industries Ltd. Before use it is distilled and the fraction boiling between 61 and 63° collected. Not more than 2 l. is prepared at any one time. The quality of the chloroform is checked by evaporating 5 ml to 50 μl and injecting a sample of the concentrate into a gas chromatograph. The absence of peaks other than the solvent front in the resultant gas chromatogram (Figure 7) indicates that the chloroform is of acceptable quality.

Both the silanized diatomaceous earth used as solid support for gas chromatography (Gas Chrom Q, 80/100 mesh) and the liquid phases, XE60 (cyano ethyl silicone polymer) and SE30 (methyl silicone polymer), were obtained from Applied Science Laboratories Inc. of Pennsylvania.

E. Urine Samples

Urine that cannot be assayed immediately is acidified with hydrochloric acid to a pH between 2 and 3, and stored at 4°C or less. Long-term storage is at -20°C.

II. THE MEASUREMENT OF PREGNANEDIOL IN URINE

A. Introduction

In normal women, three endocrine glands are capable of producing progesterone (4-pregnen-3,20-dione). They are the adrenal glands, the ovaries, and the placenta. Progesterone is an intermediate in the biosynthesis of cortisol and is secreted by the adrenal cortex in small and relatively unvarying amounts; it is an important ovarian hormone, formed after ovulation in the cells of the corpus luteum at a rate which varies characteristically around the menstrual cycle; it is also synthesized by the placenta. The progesterone-secreting potential of the placenta in late pregnancy exceeds by many times that of the mature corpus luteum, while that of the adrenal cortex is so small as to be negligible by comparison.

The increase in the secretion rate of progesterone that may be observed within a few days of follicle rupture is accompanied by a rise both in the blood levels of progesterone and in the rate of excretion of its urinary metabolites. Demonstration of a sustained increase in either quantity is usually regarded as evidence of ovulation. No increase occurs in anovulatory cycles. There is, therefore, considerable clinical interest in assays for progesterone in blood and for progesterone metabolites in urine.

The concentration of progesterone in the peripheral blood of menstruant women is low. It increases from less than 2 ng/ml[13,14] in the proliferative phase of the menstrual cycle to levels that rarely exceed 20 ng/ml at the height of the luteal phase. Such concentrations require analytical techniques of considerable sensitivity. These are, fortunately, now available. The discovery of the principles of saturation analysis early in 1960[15,16] and their application to the measurement of steroid hormones[17] have led to the development of assays for the measurement of progesterone in blood which have adequate sensitivity and specificity for following ovulatory changes in women. Many such assays, of varying degrees of technical complexity, have been reported.[13,14,18,19]

Changes in corpus luteal activity may also be followed by measuring the excretion rate of the urinary metabolites of progesterone. The excretion of pregnanediol (PD, 5β-pregnan-3α,20α-diol), quantitatively the most important urinary metabolite of progesterone, increases from less than 1 mg/24 hr in the proliferative phase of the menstrual cycle to levels as high as 10 mg/24 hr at the luteal peak.[20] No cyclical changes occur during anovulatory menstrual cycles. The close correlation between the excretion of pregnanediol and plasma progesterone levels[21] is somewhat surprising in view of the observation that only about 15% of administered progesterone appears in urine as pregnanediol.[22] Its existence means that it is as acceptable to use the urinary excretion of pregnanediol as an index of ovarian activity as it is to use plasma progesterone levels. Urine is abundantly available and requires no expert assistance for its collection. Furthermore, the recent demonstration that urinary pregnanediol to creatinine ratios correlate with pregnanediol excretion rates, and that the ratios change only slowly with time, means that it is no longer necessary to collect complete 24-hr specimens for studies of ovarian function. The measurement of the pregnanediol to creatinine ratio in small samples of urine provides the same information as is obtainable from the measurement of pregnanediol alone in 24-hr samples.[23]

Pregnanediol occurs in urine mainly as the glucuronide.[9] Before the advent of gas chromatography, urinary pregnanediol was assayed both gravimetrically and colorimetrically. These procedures were reviewed by Klopper et al. in 1955.[24] The more sensitive of them were liable to interference from urinary constituents and it was usual for colorimetry to be preceded by hydrolysis, extraction, oxidation, acetylation, and up to two paper chromatographic purifications.[24] Since 1962, when Cooper observed[25] that pregnanediol could be seen in gas chromatograms prepared directly from extracts of hydrolyzed pregnancy urine, many relatively simple assays for measuring pregnanediol in urine have been reported.[7,26-31] Typically, these assays involve hydrolysis, extraction, and gas chromatographic analysis, either of the free steroid[7,26,28,29,31] or of steroid derivatives such as the trimethyl silyl ether or diacetate.[27,32] Many recommendations as to assay conditions have been made, but only recently have systematic studies made it possible to establish these conditions with greater confidence.[28,29,33] It appears that the simplest of the gas chromatographic assays yet proposed[29] is adequate for following the changes accompanying corpus luteal development. This assay, together with the theoretical considerations on which it is based, will be described here.

B. Procedure

1. Principle

Urine is filtered to remove an interfering substance, hydrolyzed with acid, and extracted with chloroform containing internal standard. After neutralization, the extract is analyzed by gas chromatography using a liquid phase (XE60) which separates peaks due to pregnanediol, allopregnanediol, and pregnanolone from other urinary constituents.

2. Solutions

Pregnanediol (PD)	100 μg/ml in ethanol
Progesterone (P) internal standard	About 6 μg/ml in chloroform
Hydrochloric acid	6 *M* in water
Sodium hydroxide	1 *M* in water

3. Protocol

1. Tubes are set up as follows:

Sample tube	5.00 ml filtered urine* (X2)
Blank tubes	5.00 ml water (X1)
Standard tubes	0.20 ml pregnanediol in ethanol (100 μg/ml) + 5.00 ml water (X4)

It is convenient to prepare from 20 to 40 tubes at a time.

2. To each tube add 1.0 ml 6 *M* HCl. Stopper the tubes, shake to mix, and, without delay, heat them in a boiling water bath for 12.5 min.

3. Cool the tubes in cold water for 3 min.

4. To each tube add 4.00 ml of chloroform containing internal standard. Stopper the tubes and shake them vigorously for 1 min. Remove the aqueous phase from each by aspiration.

5. To each tube add 2 ml 1 *M* NaOH. Stopper the tubes and shake them vigorously for 1 min. Remove the aqueous phase by aspiration.

6. Using a syringe fitted with a 6-in. stainless steel needle (16 gauge), transfer about 3 ml of each extract to a pointed tube. Place the tubes at an angle in a hot water bath (60°C increasing to 100°C) until evaporation is complete.

7. Dissolve each residue in 50 μl chloroform. Inject 1 to 4 μl of the concentrate into a gas chromatograph fitted with an XE60 column.

4. Calculation

The ratio of the 'areas' (R_A) of the pregnanediol and progesterone peaks is measured and the concentration of pregnanediol calculated from that of the standard by the expression:

$$PD = \frac{R_{A(sample)}}{R_{A(standard)}} \times 4 \text{ mg/l.} \qquad \text{(Figure 2).}$$

If all gas chromatograms resembled the one illustrated in Figure 2, it would be advantageous to automate peak area measurement by use of an electronic integrator. Unfortunately, however, only chromatograms prepared from standard solutions have this appearance. In chromatograms prepared from urinary extracts, the two assay peaks are superimposed on an uneven background trace and on a complex of other peaks, not all of

*Assay samples are diluted with distilled water when (a) the 24-hr urine volume is less than 800 ml (1:1 dilution) and (b) the expected PD concentration exceeds 10 mg/l., (dilution to <10 mg/l.) p 6

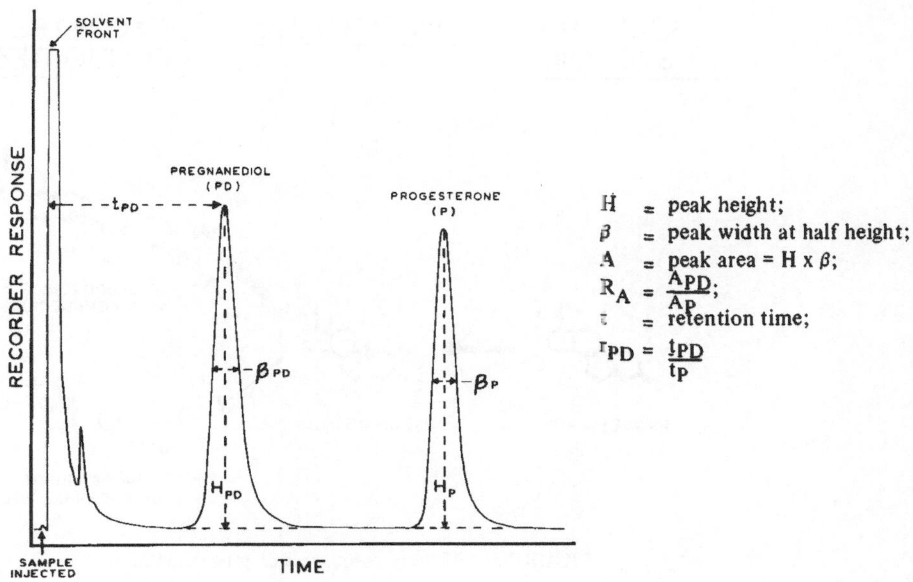

FIGURE 2. Calculation of pregnanediol concentrations from gas chromatogram data.

which are clearly separated from them (Figures 8 and 13). An approximation to the area of each assay peak must be made. In practice it has been found quicker to do this manually by drawing in a peak base and calculating the area, as shown above, than it is to allow for the contribution of the background trace and of neighboring peaks to an integrated signal.

Doubts as to the identity of a peak can be resolved by measuring its retention time relative to that of progesterone (r). For a peak to be identified as pregnanediol, it must have a relative retention time which is indistinguishable from that observed for the pregnanediol standards. For different XE60 columns this quantity has a value between 0.40 and 0.46. For any one column the variation is less and rarely exceeds ±2% of the mean value.

C. Discussion

A detailed justification of each of the steps recommended in the assay procedure has been reported elsewhere.[28,29,33] In what follows, only the more important considerations will be discussed.

1. Hydrolysis

Pregnanediol occurs in urine mainly as the 3α-glucuronide[9] and it is in this form that it must be analyzed (Figure 3). There are also reports[34] that considerable amounts of the 3α-sulfate may be excreted during pregnancy, although this finding has been disputed.[35] These conjugates are not volatile and they must be hydrolyzed prior to gas chromatographic analysis. The evidence presented in Figure 4 shows that both the 3α-glucuronide and the 3α-sulfate of pregnanediol are rapidly cleaved on exposure to aqueous solutions of hydrochloric acid at 100°C.

The urinary conjugates of pregnanediol hydrolyze in acid a little faster than aqueous solutions of pregnanediol-3α-glucuronide in similar conditions (Figure 4). This sort of evidence suggests two things: first, that urinary constituents do not inhibit acid hydrolysis and, second, that pregnanediol is excreted in urine either as the 3α-glucuronide or as conjugates which behave like it in acid.

There have been many reports of the instability of pregnanediol in the presence of hydrochloric acid.[10(p. 143),28,36] In a recent study, this effect has been confirmed and the rate of pregnanediol decomposition measured as a function of acid strength (Figure 5). The use of hydrochloric acid for hydrolysis is, therefore, suspect and will only be justified if conditions can be found in which

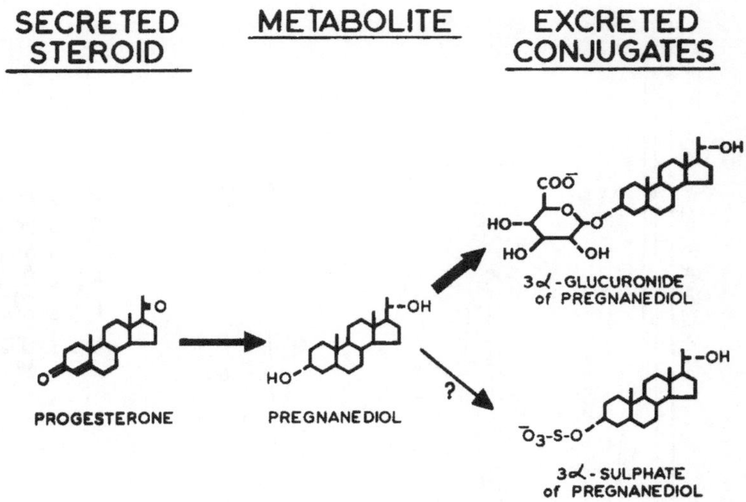

FIGURE 3. Urinary conjugates of pregnanediol.

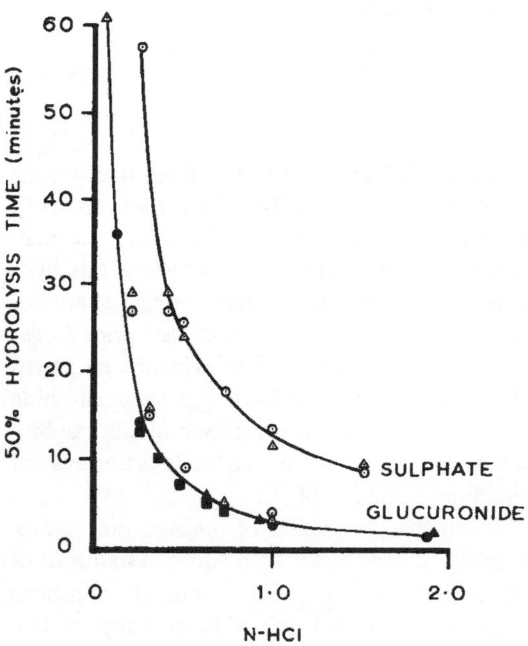

FIGURE 4. Relationship between acid concentration and the rate of hydrolysis at 100°C of the conjugates of pregnanediol (clear points represent results obtained in aqueous solution and shaded points those obtained in urine; changes in point shape indicate separate experiments). (From Metcalf, M. G., *Steroids,* 21, 193, 1973. With permission.)

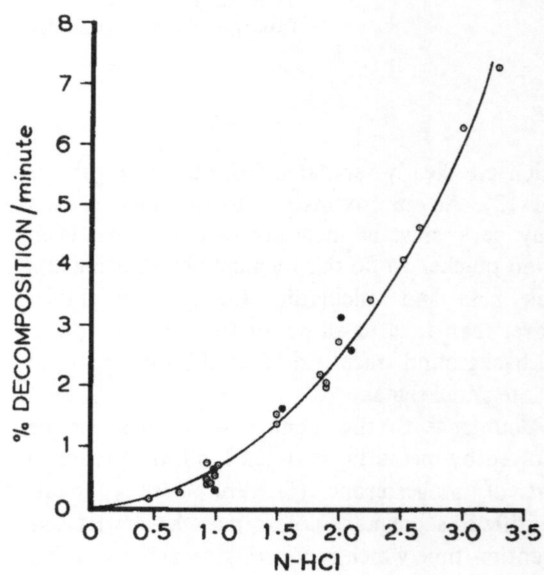

FIGURE 5. Pregnanediol decomposition at 100°C as a function of acid concentration (each point is the mean of 2 to 13 measurements; S.D. of the 47 measurements with decomposition rates <1%/min = 0.057%/min, of the 47 measurements with decomposition rates 1 to 4%/min = 0.16, and of the 11 measurements with decomposition rates >4%/min = 0.43%/min; total decomposition <50%). (From Metcalf, M. G., *Steroids,* 21, 193, 1973. With permission.)

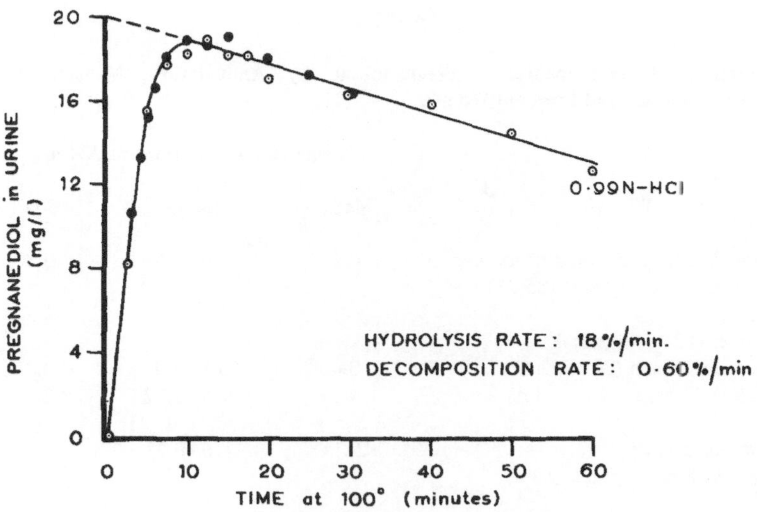

FIGURE 6. Effect of heat on the appearance and subsequent disappearance of pregnanediol from acidified pregnancy urine (third trimester pregnancy urine; changes in point shading indicate separate experiments; each point represents the mean of duplicate measurements, overall S.D. = 0.6 mg/l). (From Metcalf, M. G., *Steroids*, 21, 193, 1973. With permission.)

conjugate hydrolysis proceeds at a rate considerably in excess of steroid decomposition. Such conditions exist for pregnanediol and its urinary conjugates. In 1 *M* HCl at 100°C, the rate of formation of free pregnanediol in urine is about 20 times greater than its rate of decomposition.[33] The combined effects of these two processes (Figure 6) produces a plateau in the concentration of pregnanediol which extends from approximately 10 to 15 min from the onset of heating. The amount of pregnanediol released will be least sensitive to small changes in experimental conditions if a point midway between these two times is chosen for urine hydrolysis. The actual pregnanediol content of a urine specimen may be estimated by extrapolating the pregnanediol curve in Figure 6 to zero time. After 12.5 min in 1 *M* HCl at 100°C, about 90% of the available pregnanediol in urine is released.[29] These conditions offer a satisfactory compromise between speed and reliability. However, they are not unique. As the rates of both hydrolysis and decomposition are acid-dependent, it will be equally satisfactory to hydrolyze urine for longer periods with more dilute acid. The use of concentrations greater than 1 *M* is inadvisable because of the increasing instability of pregnanediol in stronger acid.

2. Filtration of Urine

All urine specimens are filtered or centrifuged before assay because of evidence that there is a substance in some urinary sediments which interferes with the gas chromatographic analysis of pregnanediol.[28] The source of interference appears to be completely removed by a single filtration through Whatman 4 paper. No loss of pregnanediol glucuronide from aqueous solution has been detected during this operation.

3. Stationary Phase

Because the cyanoethyl silicone polymer, XE60, has been found to give the clearest separation of pregnanediol from other urinary steroids, it is used in preference to the more usual SE30 and NPGA. This has an important consequence. It means that pregnanolone, with a retention time almost equal to that of pregnanediol, is also measured.[29] The inclusion of pregnanolone is not regarded as an objection since it, like pregnanediol, is an important metabolite of progesterone[37] and its concentration in urine also reflects progesterone secretion.[38] As the aim of the assay is to provide a simple index of corpus luteal activity, it is as acceptable to measure pregnanediol+pregnanolone as it is to measure pregnanediol. Pregnanolone added to urine is recovered quantitatively (Table 3). Its inclusion can be expected to increase assay results about 30%[34] above what would have been obtained for pregnanediol alone.

TABLE 2

Excretion of Pregnanediol + Pregnanolone by Anovulatory, Menstruant, Postmenopausal, and Pregnant Women

	Pregnanediol excretion (mg/24 hr)		
Women	Mean	Range	S.D.
Anovulatory, and with amenorrhea	0.41	0–1.5	0.35
22, aged 17 to 35 years (85 d*)			
Normal (19 to 45 years)			
18, prolif. phase (22 c, 59 d*)	0.40	0.1–0.9	0.20
37, luteal phase (44 c, 107 d)	4.67	1.5–12.2	2.20
Postmenopausal	<0.1	0–0.2	–
8, aged 28 to 63 years, (14 d)			
Pregnant			
10, 1st trimester (11 d)	11.63	4.3–28	8.23
9, 2nd trimester (27 d)	27.5	8–73	15
28, 3rd trimester (45 d)	64.5	16–125	29

*'c' represents the number of menstrual cycles and 'd' the number of days during which urine was collected.

(From Metcalf, M. G., *Clin. Biochem.*, 6, 4, 1973. With permission.)

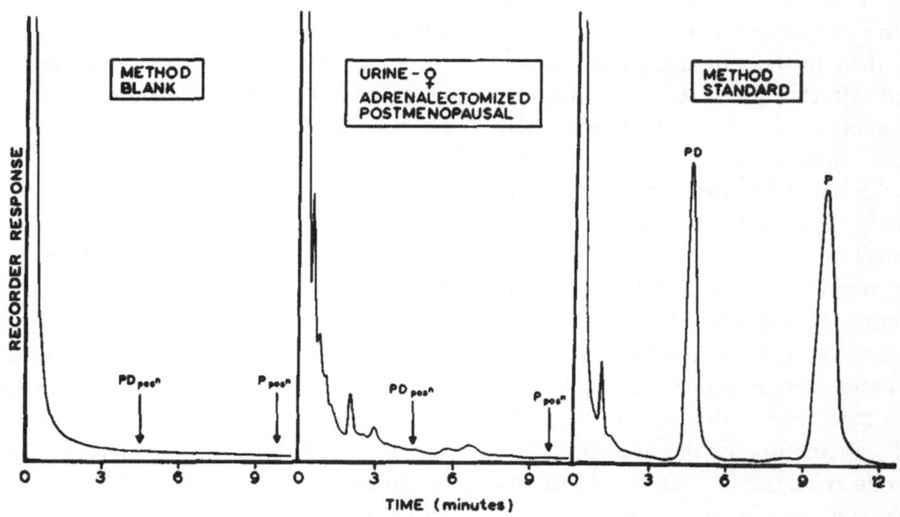

FIGURE 7. Gas chromatograms of a method blank, urine "blank," and method standard.

4. Specificity

Any method involving the gas chromatographic assay of crude urinary extracts raises the question of specificity. A gas chromatogram that is typical of those seen when unhydrolyzed urine or urine from women with minimal adrenocortical and gonadal activity is analyzed is shown in Figure 7. The pregnanediol position on the chromatogram is unoccupied. Accumulated evidence of this sort makes it unlikely that nonsteroidal constituents of urine interfere with the measurement of pregnanediol. No peak with a retention time near that of the internal standard, progesterone, has been detected, even in tracings prepared from late pregnancy urine. This is in accord with the evidence of Van der Molen,[39] who found that the excretion of progesterone in late pregnancy is less than 5 μg/day.

Of the urinary steriods tested, only pregnanetriol was found to interfere significantly with the assay.[29] Corticosteroids such as tetrahydrocortisol and 17-oxosteroids such as androsterone and dehydroepiandrosterone have been shown not to alter assay results. The extent of the interference from pregnanetriol means that there is one very real limitation on the use of the assay; it cannot be used to measure pregnanediol in urine collected from patients with congenital adrenal hyperplasia. The interference is not sufficient to affect more than trivially the estimation of pregnanediol in urine from normal subjects.

5. Standards

The results obtained from any assay critically depend on the substance chosen as the assay standard and on the way in which that standard is treated. Ideally, results should indicate the actual amount of a substance being analyzed; in practice, they are more often proportional to it. Direct equivalence is, in general, possible only when a single analytical species is to be measured and when assay standards and assay specimens are treated identically. In clinical chemistry, this combination of circumstances is frequently unattainable either because more than one substance is included in the assay or because the appropriate analytical standard is not available. An arbitrary choice of standard must be made, and the results depend on this choice.

In the assay for pregnanediol which is described here, at least two and possibly more urinary steroids are measured. The major species included are likely to be the 3*a*-glucuronides of pregnanediol and pregnanolone, with the former predominant. Their 3*a*-sulfates may also be present in measurable amounts, as may also the urinary conjugates of allopregnanediol. Free pregnanediol and pregnanolone are not excreted to any extent.[29] From the quantitative point of view, pregnanediol-3*a*-glucuronide is the obvious choice for assay standard. Unfortunately, however, other considerations prevail and pregnanediol must be used instead. The glucuronide of pregnanediol is only available for research purposes. Although the free steroid is made up in aqueous solution in an attempt to mimic urine, it cannot be expected that the results obtained by "hydrolyzing" a solution of pregnanediol will exactly match those obtained by hydrolyzing an equivalent amount of pregnanediol-3*a*-glucuronide.

Many workers use free pregnanediol as standard for their pregnanediol assays, but few treat urine samples and standard solutions in a comparable manner.[7,26,27,30] Their standards are not subjected to the whole assay procedure, as is recommended here. Instead, the urinary concentrations of pregnanediol are calculated by reference to a standard curve prepared by the direct gas chromatographic analysis of ethanolic solutions of pregnanediol and progesterone. In these assays no allowance is made for the pregnanediol lost during hydrolysis and extraction nor for the effect of day-to-day variations in assay conditions. The results of such assays can be expected to be increased by an amount which is proportional to the procedural loss of pregnanediol.

D. Results

A gas chromatogram prepared from a method blank is shown in Figure 7. Like tracings prepared from the urine of adrenalectomized postmenopausal women, it has no peaks in the pregnanediol and progesterone positions. Chromatograms prepared from the urine of normal menstruant women change in characteristic fashion round the menstrual cycle, with the pregnanediol peak rising high above the urinary background as the luteal phase advances (Figure 8). Conception and the onset of placental progesterone secretion are accompanied by the increasing dominance of the pregnanediol peak over all other urinary peaks.

The excretion of pregnanediol and pregnanolone by normal and infertile women is shown in Table 2. Ovulatory cycles can be readily distinguished from anovulatory cycles. Pregnancy was associated with a progressive increase in the urinary levels of pregnanediol until almost full term. The figures obtained in the pregnancy series are somewhat higher than those reported elsewhere,[7,26,27,30] partly because of the inclusion of pregnanolone in the measurements and partly because of the way in which the pregnanediol standards are treated.

The accuracy of the method can be seen in Table 3. Quadruplicate pregnanediol standards were analyzed on eight different occasions. The standard deviation of the 32 measurements from their respective means was 0.07 mg/l. for the 4.0

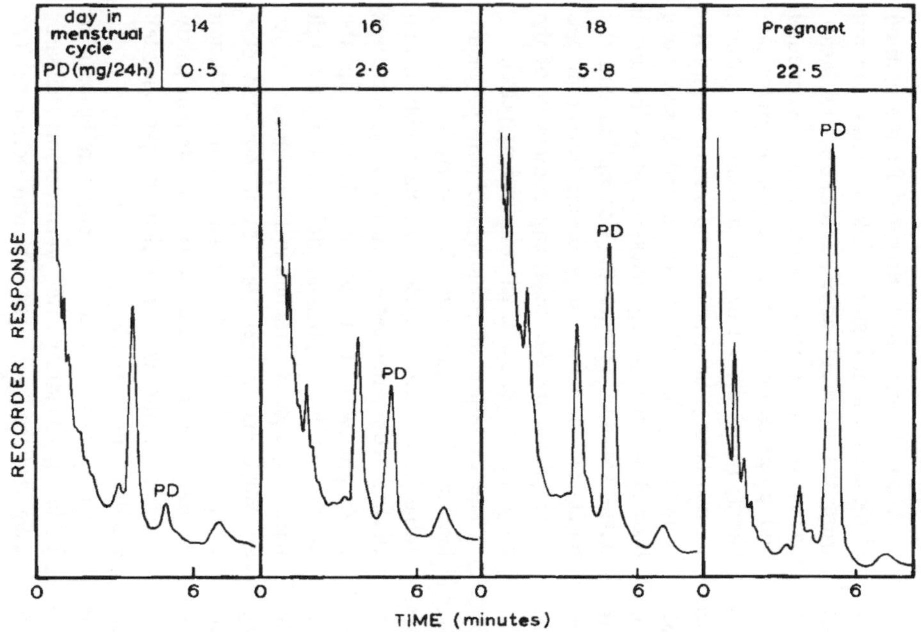

FIGURE 8. Gas chromatograms of proliferative and luteal phase urines and of pregnancy urine. (From Metcalf, M. G., *Am. J. Obstet. Gynecol.*, Nov. 1973. With permission.)

TABLE 3

The Recovery of Pregnanediol Added to Urine, of Pregnanediol Measured in Urine at Different Dilutions, and of Pregnanolone Added to a Standard (4 mg/l.) Solution of Pregnanediol

Pregnanediol added (mg/l.)	Pregnanediol found (mg/l.)	% Recovery
0	3.11	–
2.5	5.73	105
5.0	7.95	97
7.5	10.50	99
Urine dilution (urine:water)		
1:0	12.70	–
1:1	6.67	105
1:2	4.48	105
Pregnanolone added (mg/l.)		
0	4.00	–
0.5	4.58	102
1.0	5.05	101
1.5	5.61	102
2.0	6.25	104
2.0	6.11	102

Pregnanediol experiment – each figure is the mean of quadruplicates; S.D. of all measurements from their respective means, 0.24 mg/l. Pregnanolone experiment – each figure is the mean of 3 to 4 estimations; S.D. of the 25 measurements from their respective means, 0.10 mg/l.

(From Metcalf, M. G., *Clin. Biochem.*, 6, 4, 1973. With permission.)

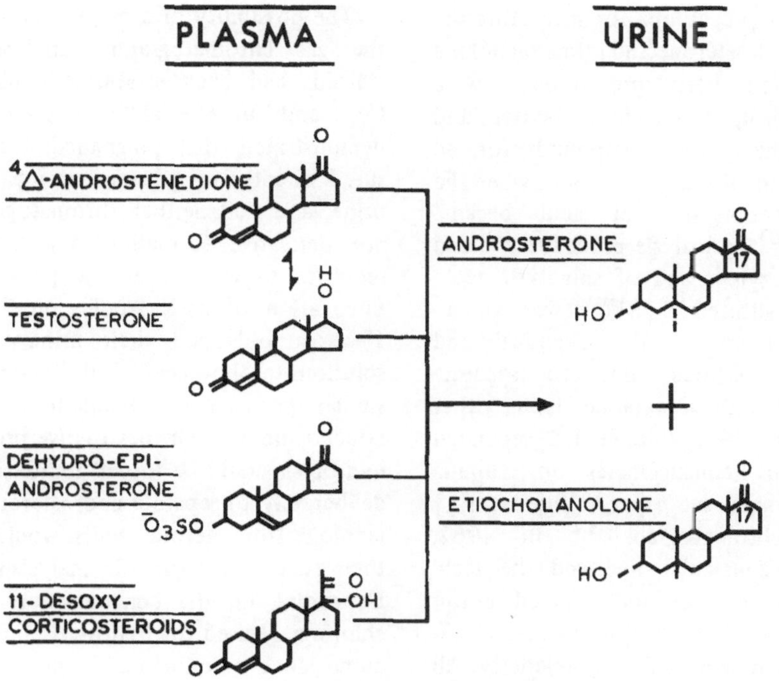

FIGURE 9. Major precursors of androsterone and etiocholanolone.

mg/l. solution. Thirty urine samples containing between 1 and 10 mg/l. pregnanediol (mean 3.88 mg/l.) were analyzed in duplicate over a 6-month period. The overall standard deviation, calculated from the means of duplicates, was 0.12 mg/l.

III. THE MEASUREMENT OF ANDROSTERONE AND ETIOCHOLANOLONE IN URINE

A. Introduction

Androsterone (A, 5α-androstan-3α-ol-17-one) and etiocholanolone (E, 5β-androstan-3α-ol-17-one) are 17-oxosteroids which differ structurally only in shape. Apart from its pyrogenic properties,[40] etiocholanolone is reported to have little biological activity. Androsterone is a weak androgen.[41] Although the two steroids have low hormonal activity, their excretion reflects changes in the plasma levels of steroids which are potent androgens. In normal women, the urinary conjugates of androsterone and etiocholanolone are formed as a result of both adrenal and ovarian activity.[42a,42b] They are metabolic products of the 11-desoxycorticosteroids and dehydroepiandrosterone-sulfate secreted by the adrenal gland; they are also metabolites of the 4-androstenedione and testosterone which are secreted both by the adrenal glands and, as estrogen precursors, by the ovaries (Figure 9). If the adrenocortical contribution is suppressed with an agent such as dexamethasone, the urinary levels of androsterone and etiocholanolone are a function of ovarian activity[43] and, in particular, of the production rate of ovarian-based androgens.

Most of the androsterone and etiocholanolone in blood circulates as the 3α-sulfate of androsterone.[42,44] Relatively little is carried as the free steroid and the total 3α-glucuronide level probably does not exceed 5 μg/100 ml of plasma. Etiocholanolone sulfate, unlike its 5α-epimer, is present in trace quantities. Only androsterone-sulfate, with plasma concentrations of 10 to 60 μg/100 ml,[45] is reasonably abundant. The distribution of the two steroids in urine differs considerably from that in blood. In urine, androsterone and etiocholanolone appear in roughly comparable amounts.[46] They are excreted mainly as conjugates of β-glucuronic acid, with almost all the androsterone and more than 75% of the etiocholanolone occurring in this form.[47] The remainder is conjugated with sulfuric acid. Total androsterone and etiocholanolone excretions for normal women are variously reported to lie between 1 and 10 mg/24 hr.[43,48,49]

Since 1931, when Butenandt[50] first isolated and later identified androsterone in the urine of normal men, many procedures for measuring the urinary levels of androsterone and etiocholanolone have been reported. Early procedures[51] were based on acid hydrolysis, solvent extraction, and quantitation as the purple compounds formed with alkaline *m*-dinitrobenzene.[52] Later, when the destructive effects of hot acid became apparent,[10(p. 150)] hydrolysis procedures, based on the use of enzymes and of solvolytic techniques, were substituted.[53] There was an increasing sophistication in the chemical and chromatographic methods used to separate androsterone and etiocholanolone from other urinary steroids. In 1941, Wolfe et al.[54] separated them as steroidal hemisuccinates on alumina columns. Other groups used paper,[55] thin-layer,[56] and column chromatography.[57] In 1962, Vestergaard and Claussen[58] reviewed the technique of gradient elution and claimed better separations of the three major 11-desoxy-17-oxosteroids than had been achieved previously. All of these assays were, however, time-consuming and all depended on the efficiency of the preparatory chromatographic separations for their success.

The announcement in 1960 by Vanden Heuvel et al.[3] of the separation of a group of steroids by gas chromatography introduced a new phase in the development of methods for measuring androsterone and etiocholanolone in urine. Reports of assays based on the new techniques appeared the following year.[59,60] These and subsequent assays were reviewed by Wotiz and Clark in 1966[27(p. 167)] and again by Eik-Nes in 1968.[49(p. 290)] It was apparent that the main new feature of these procedures was the use of the gas chromatograph as a sensitive measuring device. Little use was made of its potent resolving powers and few attempts appear to have been made to shorten or simplify the preparative work-up which had been an essential preliminary to colorimetric analysis. In a typical assay,[61] gas chromatography was preceded by thin-layer chromatography and conversion of the steroids to trimethylsilyl ethers. Later procedures have, in general, been variants of this, although a novel approach reported by Beale et al. in 1971[62] did not include a preliminary chromatographic separation. In this assay, a series of differential oxidations and chemical transformations was used to separate the sulfates and glucuronides of androsterone and etiocholanolone from other urinary steroids. Quantitation was by gas chromatography of the steroid formates.

The possibility of a major technical advance in the gas chromatographic analysis of urinary steroids had been established independently by Cox[7] and Turner et al.[8] as long ago as 1963. They demonstrated that pregnanediol could be measured directly in toluene extracts of hydrolyzed urine, and that neither chromatographic clean-up nor derivative formation was a necessary prerequisite to gas chromatography. Following the publication of basic studies on the behavior of 17-oxosteroids and their urinary conjugates in solutions of hot acid,[63] it became clear that a similar approach was possible for androsterone and etiocholanolone. The destructive properties of acid hydrolysis could be turned to advantage and used deliberately to separate androsterone and etiocholanolone from steroids which would interfere with their gas chromatographic analysis if not removed. Depending on the concentration of the hydrochloric acid used for hydrolysis, either etiocholanolone alone or etiocholanolone and androsterone together could be isolated. The feasibility of measuring one or both in extracts of acid-hydrolyzed urine was proven and assays of a new order of simplicity developed.[64] The procedure for measuring androsterone + etiocholanolone is described below.

B. Procedure

1. Principle

Androsterone (A), etiocholanolone (E), and dehydroepiandrosterone (DHA) are stable at high temperatures and appear together after gas chromatography on XE60, well separated from other urinary components. When urine is acidified to 3 *N* with HCl and heated for 10 min at 100°C, their conjugates are completely hydrolyzed, dehydroepiandrosterone is entirely destroyed, and 80% of the androsterone and all of the etiocholanolone remain. Gas chromatography of a chloroform extract gives a measure of E + 0.8 A.

2. Solutions

Androsterone + etiocholanolone – (50 μg A + 50 μg E)/ml in ethanol

Progesterone, internal standard – about 10 μg/ml in chloroform

Hydrochloric acid – 31.5 to 32.5% (w/w)

Sodium Hydroxide – 1 *M* in water

3. Protocol

1. Tubes are set up as follows:

Sample tubes	4.00 ml filtered urine* + 0.25 ml ethanol (X2)
Blank tubes	4.00 ml water + 0.25 ml ethanol (X1)
Standard tubes	0.25 ml A+E in ethanol + 4.00 ml water (X4)

It is convenient to prepare from 20 to 40 tubes at a time.

2. To each tube add 1.0 ml concentrated HCl. Stopper the tubes, shake to mix, and, without delay, place them in a boiling water bath for 10 min.

3. Cool the tubes in cold water for 3 min.

4. To each tube add 4.00 ml of chloroform containing internal standard. Stopper the tubes and shake them vigorously for 1 min. Remove the aqueous phase from each by aspiration.

5. To each tube add 2 ml 1 *M* NaOH. Stopper the tubes and shake them vigorously for 1 min. Remove the aqueous phase by aspiration.

6. Using a syringe fitted with a 6-in. needle, transfer about 3 ml of each extract to a pointed tube. Place the tubes at an angle in a hot water bath (60°C increasing to 100°C) until evaporation is complete.

7. Dissolve each residue in 50 μl chloroform. Inject 1 to 4 μl of the concentrate into a gas chromatograph fitted with an XE60 column.

4. Calculation

The ratio of the 'areas' of the A + E and progesterone peaks is measured and the concentration of A+E calculated from that of the standard by the procedure outlined in Figure 2.

$$A{+}E = \frac{R_{A(sample)}}{R_{A(standard)}} \times 6.25 \text{ mg/l.}$$

The A+E peak on the gas chromatograms prepared from urinary extracts should have a retention time relative to progesterone, which is indistinguishable from that observed for the standards. For different XE60 columns the relative retention time of A+E has been found to lie between 0.31 and 0.35.

C. Discussion

The experimental reasons for choosing the recommended assay conditions have been described in detail elsewhere.[63,64] Only the more important aspects of these studies will be commented on here.

1. Hydrolysis

Androsterone, etiocholanolone, and dehydroepiandrosterone react very differently upon exposure to hot acid, dehydroepiandrosterone breaking down rapidly, etiocholanolone being resistant to acid attack, and androsterone showing an in-between type of behavior (Figure 10). This differential response means that conditions exist in which dehydroepiandrosterone can be completely removed from a solution of all three steroids, leaving only androsterone and etiocholanolone. Because dehydroepiandrosterone interferes with the gas chromatographic analysis of androsterone and etiocholanolone, this phenomenon makes acid the preferred medium for hydrolyzing their urinary conjugates.

In the conditions recommended for acid hydrolysis (10 min at 100°C in 3 *N* HCl), the mean loss of androsterone observed was 21.2% (28 measurements on 5 occasions, S.D. = 5.2%). This compared with a loss of less than 2% for etiocholanolone. The standard used in the assay is made up of equal amounts of androsterone and etiocholanolone. It can be inferred, therefore, that the loss of steroid from the standard solutions during the period of acid hydrolysis will be of the order of 10%. Urine, with A to E ratios varying from 0.4 to 1.8,[64] will be subjected to hydrolysis losses varying between 6 and 13%.

Most androsterone and etiocholanolone in urine is conjugated to glucuronic acid and only a small proportion to sulfuric acid. Measurements done on the hydrolysis rates of these conjugates show that the glucuronides cleave more readily than the sulfates in acid solution.[63] These data and evidence such as that summarized in Figure 11 can be used to demonstrate that at 100°C in 3 *N* HCl, the hydrolysis of the urinary conjugates of A+E is complete within 10 min.

The combined effects of the destructive and hydrolytic properties of hydrochloric acid are shown in Figure 12. Results such as these may be

*Urine samples are diluted with distilled water when (a) the 24-hr urine volume is less than 800 ml (1:1 dilution or (b) the expected A+E concentration exceeds 12 mg/l or the expected 17-oxosteroid concentration exceeds 40 mg/l.

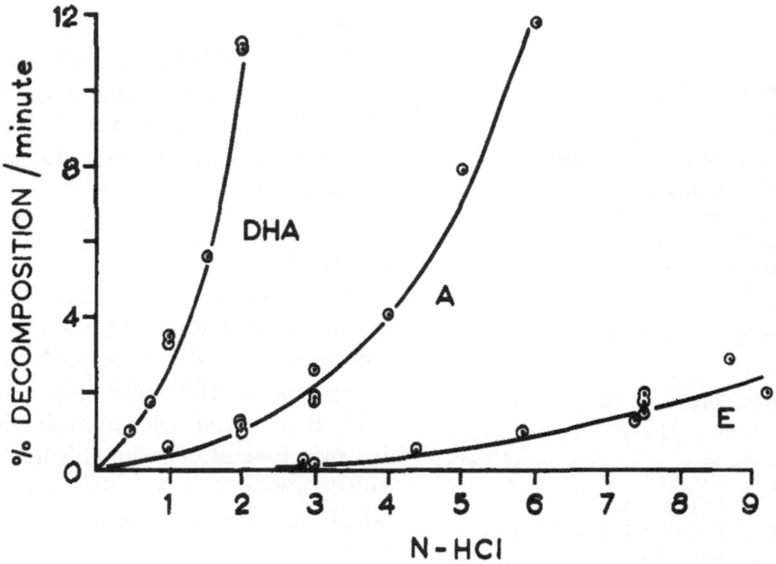

FIGURE 10. Decomposition of androsterone, etiocholanolone, and dehydroepiandrosterone at 100°C as a function of HCl concentration (each point is the mean of 2 to 12 observations; S.D. of all the measurements from their respective means, 0.22%/min; total decomposition always <40%). (From Metcalf, M. G., *Clin. Biochem.*, 3, 271, 1970. With permission.)

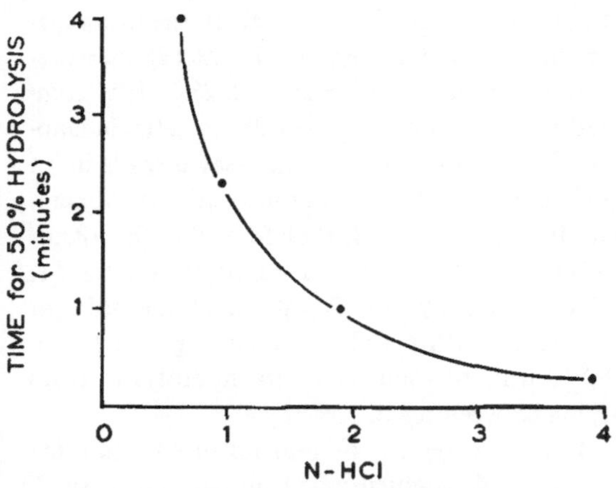

FIGURE 11. The effect of acid concentration on the rate of hydrolysis of urinary A+E at 100°C. (From Metcalf, M. G., *Clin. Biochem.*, 3, 271, 1970. With permission.)

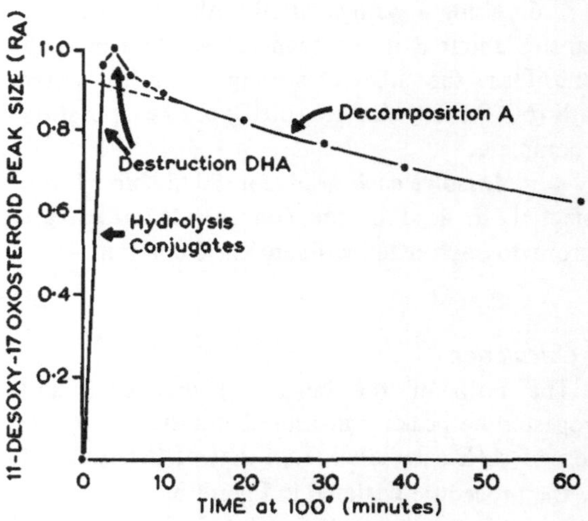

FIGURE 12. Appearance and disappearance of androsterone, etiocholanolone, and dehydroepiandrosterone in urine acidified to 3 *N* with HCl and heated to 100°C. (Each point is the mean of triplicates: S.D. of all measurements from their respective means = 0.017.) (From Metcalf, M. G., *Clin. Biochem.*, 3, 271, 1970. With permission.)

used to estimate the actual A+E content of a urine sample. Extrapolation of the decay curve to zero time shows that in this experiment the recovery of urinary A+E after hydrolysis for 10 min in 3 *N* HCl was 94% of the theoretical maximum.

2. Stationary Phase

Of the liquid phases tested, only XE60 separated the acid-decomposition products of androsterone from the parent compound. The failure of both SE30 and QF1 in this respect means that neither can be used for the assay. With XE60 as liquid phase, a clear separation of A+E from

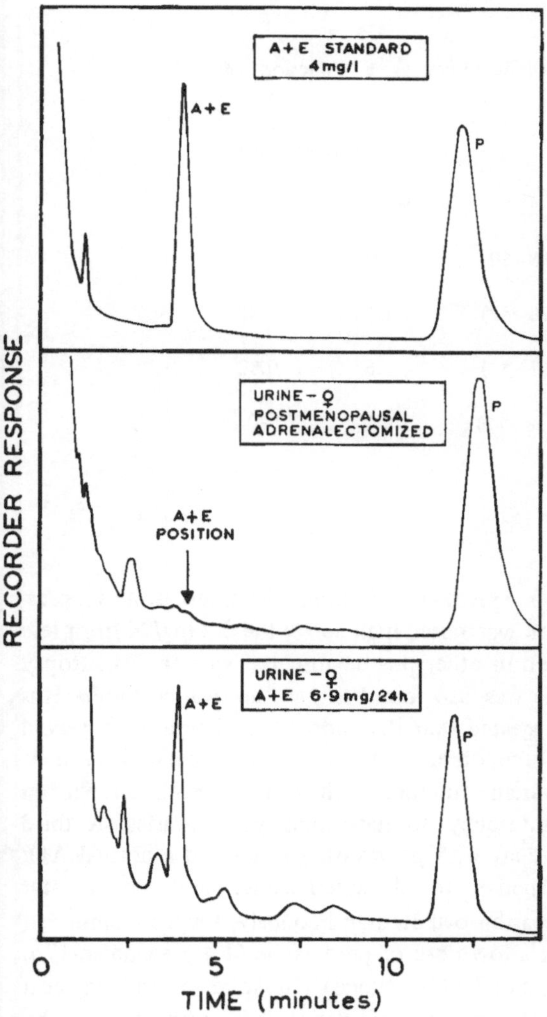

FIGURE 13. Gas chromatograms of extracts prepared from a method standard, a normal urine, and urine from a bilaterally adrenalectomized woman aged 65 years.

other urinary substituents is usually achieved (Figure 13). Gas chromatograms prepared from the urine of adrenolectomized hypogonadal patents have no peaks in either the A+E or the internal standard positions.

3. *Specificity*

In this assay, as in the one described for pregnanediol, interference from nonassay substances is a possibility which must be examined with care. The risk of nonsteroidal constituents being measured as A+E has been discounted. Potential interference from urinary steroids is more difficult to evaluate with certainty, and the possibility of spurious results from this source should always be kept in mind.

Dehydroepiandrosterone, with a retention time close to that of A+E, is the major urinary steroid likely to cause interference with the measurement of A+E. In this assay all dehydroepiandrosterone is destroyed during acid hydrolysis and none of its decomposition products have retention times near that of A+E. It has been shown not to interfere with the assay.[64] Another unlikely source of interference is cortisol and its urinary metabolites, and no A+E could be detected in the urine of a postmenopausal woman without adrenal glands while she was taking 80 mg of cortisol daily.

Of the many steroids which have been tested, only pregnanetriol 11β-androsterone, 11-desoxytetrahydrocortisol and pregnanetriolone interfered with the measurement of A+E. In practice, interference will not matter unless the ratio of interfering steroids to A+E in a urine sample is appreciable. For urine from normal women, it has been calculated that the total interference from these four steroids is unlikely to exceed 3.5%.[64] The effect is, therefore, trivial. Only urine from patients with congenital adrenal hyperplasia due to a 21-hydroxylase deficiency presents a problem. These patients excrete enormous amounts of pregnanetriol. Unless this pregnanetriol is removed prior to hydrolysis, A+E cannot be measured in their urine by the simple procedure described here. Few other patients excrete disproportionate amounts of interfering steroids, and the method with this one exception appears generally applicable.

D. Results

A gas chromatogram that is typical of those obtained when urine from a normal woman is assayed may be seen in Figure 13. Tracings prepared from a method standard and from urine collected from a patient unable to synthesize steroids have no peaks in either the A+E or the internal standard positions.

The excretion of A+E by women reflects both adrenocortical and ovarian activity and the loss of either set of glands is accompanied by a decreased output of urinary A+E (Table 4). Eight oophorectomized women and 16 women who were postmenopausal excreted significantly less A+E than women with intact ovaries. These women had normal adrenocortical function, as judged by their excretion of 17-oxogenicsteroids (7 to 19 mg/24 hr; mean, 11.5 mg). Similarly, a group of hypoadrenal women, who were judged to have normal

TABLE 4

The Excretion of Androsterone and Etiocholanolone as a Function of Adrenocortical and Ovarian Activity

Endocrine status	No.	(Age/years)	A + E (mg/24 hr) Mean	S.D.
Normal adrenal and ovarian function	66	27 (16–50)	5.0	2.07
Normal adrenal function with hypogonadism	24	56 (28–82)	1.1	0.58
Normal ovarian function with hypoadrenalism	21	27 (17–41)	1.6	0.62
Hypogonadism and hypoadrenalism	12	53 (28–72)	<0.1	(0–0.2)

ovarian function by reason of their cyclical excretion of pregnanediol, had a reduced excretion of A+E. Two of these women had undergone bilateral adrenalectomy and 19 had had their adrenocortical function suppressed for 3 or more days with dexamethasone (1,4-pregnadien-9α-fluoro-16α-methyl-11β,17α,21-triol-3,20-dione) during which time their excretion of 17-oxogenicsteroids fell to a mean of 31% (S.D. ±10%) of pretreatment levels. A small group of women with defects in both the adrenal glands and the ovaries (three with panhypopituitarism, two postmenopausal women with Addison's disease and two whose adrenal function was suppressed with dexamethasone, and five adrenalectomized women who were either postmenopausal or who had no ovaries) excreted almost no A+E.

Useful information on the etiology of hirsutism in women may be gained by considering the effects of adrenocortical suppression. After 3 or more days suppression with dexamethasone (1 mg given at midnight, 0.5 mg on rising, and 0.5 mg at noon), the excretion of A+E dropped to a steady level. Menstruant women whose main presenting symptom was facial hirsutism responded to dexamethasone like normal women and excreted ⩽3 mg/24 hr (Figure 14). By contrast, the A+E excretion of four women with secondary amenorrhea and symptoms of virilization, in addition to their hirsutes, did not suppress in a normal manner. One of these women subsequently had an ovarian tumor removed and responded normally to dexamethasone thereafter. Another, a woman of 62 years, had gross symptoms of androgen excess. Her A+E excretion during dexamethasone suppression was raised high above the 0.2 mg/24 hr or less seen in other postmenopausal women. At autopsy she was also found to have an ovarian tumor. It is suggested that the androgenic symptoms observed in the other two patients were related to their ovarian function. They were both shown on culdoscopy to have polycystic ovaries. A third patient with polycystic ovaries had a normal A+E response to dexamethasone, and on a later occasion ovulated and conceived while maintained on a low dose of prednisone (1,4-pregnadien-17α,21-diol-3,11,20-trione). In spite of the physical appearance of her ovaries, it is presumed that the precipitating cause of infertility in this patient was associated with some disturbance in adrenocortical function.

A remarkable picture was presented by a well-feminized young woman who complained only of acne and facial hirsutism. Her daily excretion of A+E, consistently two to three times higher than any level previously observed, did not fall below 35 mg/24 hr during prolonged treatment with suppressive doses of dexamethasone. The condition was subsequently found due to the presence of a large dehydroepiandrosterone-secreting tumor in the right adrenal gland. The abnormal urinary levels of A+E reflected the enormous outpouring of DHA-SO_4, a finding which was consistent with her physical appearance and the relative lack of androgenic symptoms.

The accuracy of the A+E assay is shown in Table 5.

Quadruplicate standards containing 6.25 mg/l

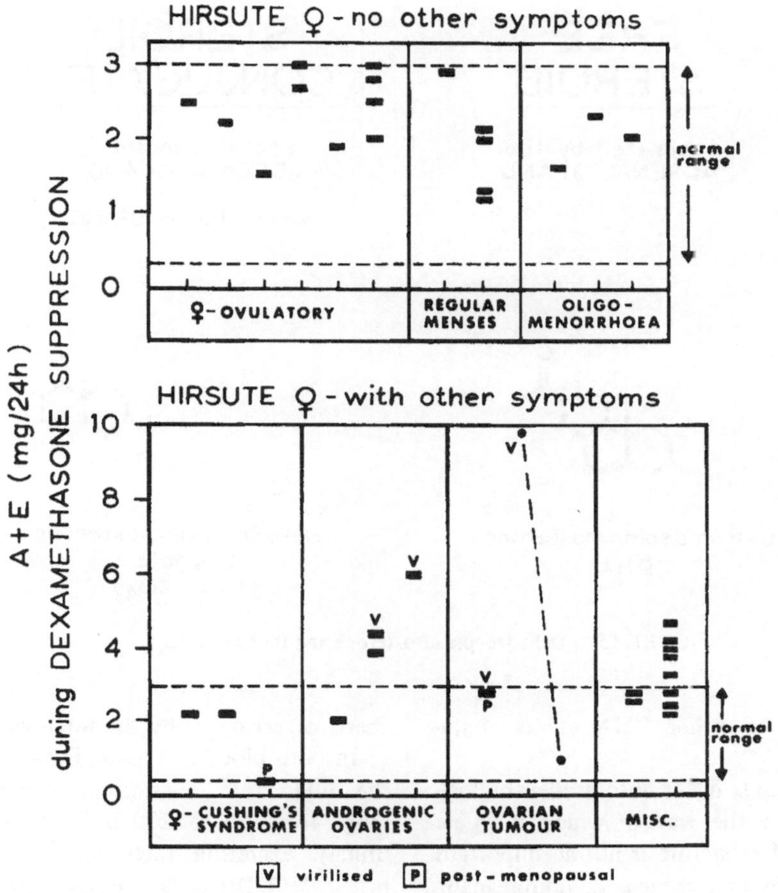

FIGURE 14. The excretion of A+E by hirsute women during adrenocortical suppression with dexamethasone.

TABLE 5

Recovery of A+E Added to Urine*

A+E added (mg/l.)	A+E found (mg/l.)	% Recovered
0	1.69	–
1.61	3.11	88
3.22	4.78	96
4.82	6.42	98
6.43	7.94	97

*Each figure is the mean of triplicates; S.D. of all measurements from their respective means = 0.3 mg/l. (From Metcalf, M. G., *Clin. Biochem.*, 3, 271, 1970. With permission.)

A+E were analyzed on seven occasions. The standard deviation of the measurements from their respective means was 0.14 mg/l. Forty urine samples containing 1 to 10 mg/l A+E (mean concentration 4.34 mg/l) were analyzed in duplicate. The overall standard deviation, calculated from the means of duplicates, was 0.14 mg/l. As expected, the between assay variability is greater than the within assay variability. When 31 urine samples were analyzed in duplicate on two different occasions, the standard deviation, calculated from the differences between the pairs of mean values, was 0.32 mg/l.

IV. THE MEASUREMENT OF DEHYDROEPIANDROSTERONE IN URINE

A. Introduction

Dehydroepiandrosterone (DHA, 5-androsten-3β-ol-17-one) is a 17-oxosteroid which, as the 3β-sulfate (Figure 15), is secreted in large amounts by the cells of the adrenal cortex. In women, the production rate of the conjugate is reported to lie between 7 and 20 mg/24 hr[65-67] and to exceed the rate reported for the free steroid.[67] The ovaries are not thought to be a quantitatively

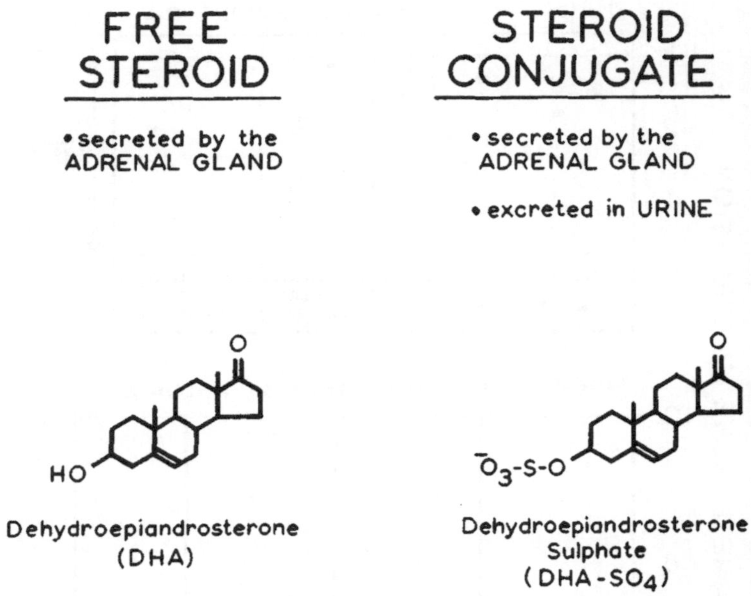

FIGURE 15. Dehydroepiandrosterone and its 3β-sulfate.

significant source of either DHA or its conjugate.[65,68,69]

DHA and DHA-SO_4 differ both in their biological activity and in the way in which they are metabolized. The free steroid is not an important source of androgens or estrogens in nonpregnant women. In 1969, Baird and co-workers[70] estimated that in normal women only about 6% of the androstenedione and testosterone, and a considerably smaller fraction of the subsequently formed estrone and estradiol, originated from the DHA in peripheral blood. Somewhat surprisingly, in view of its interconvertibility with DHA, DHA-SO_4 has even less biological activity than the free steroid.[65] The phenomenon may be due to the extensive binding of the conjugate, but not of the free steroid, by serum proteins. The absence of DHA-SO_4 does not prevent maturation, and women with Addison's disease are often fertile.[65] In an experiment in which a small group of hypogonadal individuals was given large doses of DHA-SO_4 for prolonged periods, only minimal signs of androgenic and estrogenic activity were observed and these changes were transitory.[71a] Altogether, there is little evidence so far to indicate an essential role for either DHA or DHA-SO_4 in maintaining the hormonal status of nonpregnant women. Only in pregnancy have these compounds been shown to be important. DHA-SO_4 is an essential precursor for the biosynthesis of estrogens by the fetal-placental unit.[71b]

In both blood and urine, DHA occurs mainly as the sulfate.[65,72] Plasma levels in normal women range from 50 to 200 μg/100 ml[42b,65,73] and urinary excretion rates from 0.1 to 8 mg/24 hr.[65,74,75] DHA-SO_4, unlike DHA, is excreted in feces as well as in urine;[76] and up to half of the radioactivity from an injected dose has been shown to be eliminated by this route. The major urinary metabolites of DHA and DHA-SO_4 are the glucuronides of androsterone and etiocholanolone and, to a lesser extent, DHA-SO_4 itself.[77] It should be noted that in experiments on a group of normal women about one fifth of an infusion of DHA-SO_4 was excreted in this way, but that only from 0.3% to 9% of the initial dose appeared as DHA-SO_4.[77] It is unlikely, therefore, that there will be any close relationship between plasma and urinary levels of DHA-SO_4. Only gross changes in production rates, such as occur in the presence of a DHA-secreting tumor of the adrenal gland or when adrenocortical activity is stimulated or suppressed, are likely to be accompanied by unequivocal changes in urinary excretion rates. The measurement of DHA-SO_4 levels in urine is invaluable on those rare occasions when a patient presents with an adrenal tumor associated with a hypersecretion of DHA. On other occasions such measurements may be less useful. The introduction of an assay for the analysis of DHA-SO_4 in urine is justified in the context of this chapter because any laboratory able to measure pregnanediol or androsterone + etiocholanolone by rapid gas chromatographic techniques is also able to measure DHA with a minimum of extra effort.

More than 20 years elapsed between the first isolation of DHA from urine in 1934[78] and the publication of a satisfactory method for its quantitation. The development of suitable analytical techniques had to await the completion of basic studies on the nature of the DHA conjugates in urine and on the development of satisfactory methods for hydrolyzing them. The results of these studies, reviewed by the Lieberman group in 1954,[79] showed that DHA is excreted almost exclusively as the sulfate, that it is unstable in acid solution, and that its sulfate may be hydrolyzed either at room temperature in the presence of an organic solvent such as diethyl ether (solvolysis[10]) or at elevated temperatures in an aqueous solution buffered to neutral or near-neutral pH (buffer hydrolysis). The latter technique is selective and cleaves only the 3β-sulfates of $^5\Delta$-steroids; 3α-sulfates and the sulfates of saturated steroids are untouched. It was used by Fotherby, who published[80] what was possibly the first reliable assay for DHA in human urine. In this assay, urine is heated for 6 hr at 100°C without adjustment of pH. The resultant DHA is extracted, isolated by column chromatography, and measured colorimetrically as the mauve complex it forms with acetic acid and furfural.

Since 1959, many assays have been described for DHA. These have typically involved conjugate solvolysis for 2 to 5 days at room temperature, followed by chromatographic purification and measurement of the DHA by the Zimmerman,[52] Oertel,[81] or Pettenkofer[80] reactions or, more recently, by derivative formation and gas chromatographic analysis.[49,82-84] None of these assays is remarkable for the speed with which samples can be analyzed. In all of them, analysis is preceded by the chromatographic isolation of DHA, a step which is both slow and potentially nonquantitative. While paper or thin-layer chromatography may have been an essential preliminary to colorimetric measurement, it is not obvious that it is necessary before gas chromatographic analysis; nor is it obvious that derivative formation is an indispensible preliminary. It is usual to convert DHA to a derivative such as the trimethyl silyl ether so that it can be measured separately from the more abundant androsterone and etiocholanolone. If, however, the buffered hydrolysis technique described by Fotherby is used, it is possible to hydrolyze DHA-SO_4 without affecting the urinary conjugates of androsterone and etiocholanolone. Only DHA will be extracted from the aqueous phase.

In the assay described below,[74] DHA-SO_4 is hydrolyzed in conditions that do not cleave the conjugates of androsterone and etiocholanolone. The procedure is sufficiently selective to obviate the need for chromatographic pre-purification and derivative formation. DHA is measured by the direct gas chromatographic analysis of extracts of hydrolyzed urine.

B. Procedure

1. Principle

Urine at pH 4.5 is heated for 3 hr at 100°C to hydrolyze DHA-SO_4. The DHA formed is extracted into chloroform containing internal standard, washed with boric acid to remove 5-pregnenetriol, washed with water to remove substances which shorten column life, and concentrated for chromatographic analysis.

2. Solutions

DHA-SO_4 (3β-sulfate of DHA)	20 μg/ml in water
progesterone, internal standard	About 4 μg/ml in chloroform
boric acid	0.5 *M* in water
1 *M* acetate buffer, pH 4.5	8.85 ml acetic acid + 13.6 g sodium acetate, in 100 ml water
Sodium hydroxide	5 *M* in water

3. Protocol

1. Tubes are set up as follows:

Sample tubes	10.0 ml filtered urine (X2)
Blank tube	10.0 ml water (X1)
Standard tubes	1.00 ml DHA-SO_4 in water (20 μg/ml) + 9.0 ml water (X4)

It is convenient to prepare from 20 to 40 tubes at a time.

2. To each tube add 1 drop of BDH 4.5 indicator. Swirl each solution in turn on a vortex mixer and add 5 *M* NaOH dropwise till the first appearance of a grey color.

3. To each tube add 1 ml 1 *M*-acetate buffer. Stopper the tubes, shake to mix, and heat them in a boiling water bath for 3 hr. Cool.

4. To each tube add 4.00 ml of chloroform containing internal standard. Stopper the tubes and shake them vigorously for 1 min. Remove the aqueous phase by aspiration.

5. To each tube add 2 ml boric acid solution. Stopper the tubes and shake them vigorously for 1 min. Remove the aqueous phase by aspiration.

6. To each tube add 2 ml distilled water. Stopper the tubes and shake them vigorously for 1 min. Remove the aqueous phase by aspiration.

7. Transfer about 3 ml of each extract to a pointed tube. Warm the extracts until evaporation is complete.

8. Dissolve each residue in 50 μl chloroform. Inject 1 to 4 μl of the concentrate into a gas chromatograph fitted with either an SE30 or an XE60 column.

4. Calculation

The ratio of the 'areas' (R_A) of the dehydroepiandrosterone and progesterone peaks is measured and the concentration of DHA calculated from that of the standard by the procedure illustrated in Figure 2.

$$\text{DHA-SO}_4 = \frac{R_{A(\text{sample})}}{R_{A(\text{standard})}} \times 2 \text{ mg/l.}$$

$$\text{DHA} = \text{DHA-SO}_4 \times 0.677 \text{ mg/l.}$$

C. Discussion

A detailed description of the experimental evidence that determined the choice of assay conditions has been reported elsewhere.[74] Only the more important aspects of these studies will be discussed here.

1. Hydrolysis

The extreme fragility of DHA in hot acid[63] precludes the use of acid to hydrolyze its urinary conjugates. DHA is excreted almost entirely as the 3β-sulfate and 'buffer' hydrolysis, a procedure which is specific for the 3β-sulfates of $^5\Delta$-steroids, cleaves it with a minimum release of other steroids. The rate of hydrolysis of urinary DHA at 100°C is pH dependent, being very slow in neutral solution and increasing as the pH falls. At a pH of 4.5, hydrolysis is complete within 3 hr (Figure 16). In these conditions DHA is stable and no damage occurs if heating is prolonged for 6 hr. The use of more acid solutions to shorten the time necessary for hydrolysis is not practicable because of the increasing instability both of DHA and of the urinary conjugates of androsterone and etiocholanolone at lower pH values.[63]

2. Specificity

The degree of selectivity of the hydrolysis procedure used in this assay is important. Androsterone and etiocholanolone have almost the same retention times as DHA (Table 6). If any splitting of their urinary conjugates occurs, they will interfere with the gas chromatographic analysis of DHA. It is claimed that only 3β-sulfates of $^5\Delta$-steroids hydrolyze in the recommended conditions. Experiments undertaken to test these claims confirmed their truth so far as androsterone and etiocholanolone were concerned. When aqueous solutions of their 3α-glucuronides and 3α-sulfates were buffered at pH 4.5 and heated for 3 hr, no androsterone or etiocholanolone was released.[74] The possibility that the conjugates of epiandrosterone (5α-androstan-3β-ol-17-one) might hydrolyze in these conditions was also examined, and it was found that although the 3β-glucuronide was stable, about 8% of the 3β-sulfate split. The effect was not sufficient to alter the results obtained when a $DHA\text{-}SO_4$ standard was analyzed with and without the addition of 2 mg/l. of epiandrosterone-sulfate. It is fortunate, however, that the daily excretion of this conjugate by normal women is thought to be small compared with that of $DHA\text{-}SO_4$.[85] Interference from this source can probably be ignored.

Steroidal $^5\Delta$-3β-sulfates likely to be found in urine in amounts comparable with $DHA\text{-}SO_4$[86,87] are those of 5-pregnenediol (5-pregnen-3β,20α-diol) and 5-pregnenelone (5-pregnen-3β-ol-20-one), the 16-oxygenated $^5\Delta$-steroids, 16-oxo-androstenediol (5-androsten-3β, 17β-diol-16-one) and 16-hydroxy-DHA (5-androsten-3β, 16α-diol-17-one), and 5-pregnenetriol (5-pregnen-3β, 17α,20α-triol). These conjugates are likely to hydrolyze in the assay conditions. The first four,

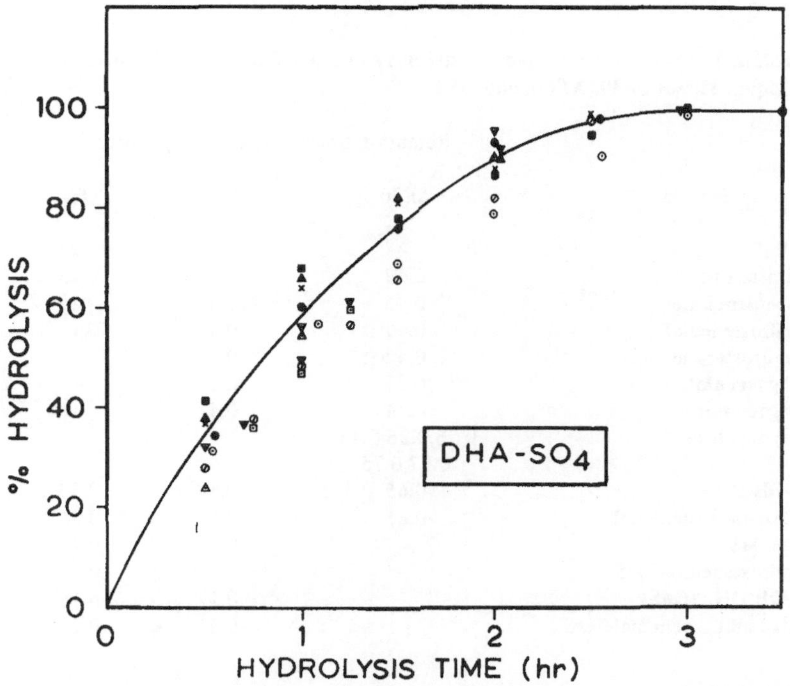

FIGURE 16. The rate of hydrolysis of DHA-SO_4 at pH 4.5 and 100°C. (Clear points represent measurements on DHA-SO_4 in aqueous buffer; shaded points represent measurements on urinary DHA-SO_4; changes in point shape indicate separate experiments; all points are the result of single measurements; on the 5 occasions when hydrolysis was continued beyond 3 hr, there was no further change in the amount of DHA released and hydrolysis was assumed to be complete.) (From Metcalf, M. G., *Clin. Biochem.*, 4, 241, 1971. With permission.)

with retention times on SE30 and XE60 which differ considerably from that of DHA, are unlikely to interfere with its measurement (Table 6). A peak with a retention time close to that of the two 16-oxysteroids can be seen in some urinary chromatograms (Figure 18). 5-Pregnenetriol, unlike the other $^5\Delta$-steroids listed, decomposes during gas chromatography and one of the decomposition products has a retention time indistinguishable from that of DHA. Fortunately, this source of interference is easily removed by washing the chloroform extract with a dilute solution of boric acid.[74] The 3β-sulfate of 5-androstenediol (5-androsten-3β, 17β-diol), which also has a retention time close to that of DHA, has not been detected in urine collected from normal subjects, although it has been reported in the urine of patients with adrenal carcinomata.[86] It has been reassuring to observe little other than peaks due to DHA and the trimethysilyl ether of DHA in gas chromatograms of trimethylsilylated extracts of urine (Figure 17). It had been shown earlier that there was no 16α-OH-DHA in these urine samples.

3. Stationary Phase

As can be seen from Figure 18, the liquid phase SE30 provides a better separation of DHA from other urinary substituents than does XE60. In practice, XE60 columns may be used when urine containing DHA in concentrations above 1 mg/l. is to be analyzed. With both columns care should be taken to check that the relative retention time of the peak thought to be DHA has the same value as the DHA peak on the standard tracings.

The possibility that nonsteroidal substituents of urine might interfere with the assay is made unlikely by data such as those shown in Figure 18. No DHA could be detected in the urine of eight women whose adrenocortical function had been suppressed for 3 or more days with dexamethasone.

D. Results

Gas chromatograms of method blanks resemble those seen in the pregnanediol assay (Figure 7) and are not dissimilar to the tracings obtained when urine from dexamethasone-suppressed patients is

TABLE 6

Retention Times of Some 3β-Hydroxysteroids and Their Trimethylsilyl Ethers on the Liquid Phases SE30, XE60, and QF1

	Retention time relative to progesterone		
Steroid	SE30	XE60	QF1
DHA	0.47	0.34	0.23
Androsterone	0.49	0.32	0.25
Etiocholanolone	0.45	0.33	0.25
5-Androstenediol	0.49	0.30	0.16
Epiandrosterone	0.48	0.36	–
5-Pregnenediol	0.83	0.46	–
5-Pregnenolone	0.74	0.47	–
5-Pregnenetriol	0.26,0.36,0.46, 0.67,0.75	–	–
16α-OH-DHA	0.65	0.68	0.33
16-Oxo-androstenediol	0.65	0.69	0.34
DHA-TMS	–	0.17	0.15
5-Androstenediol-TMS	–	–	0.06
16α-OH-DHA-TMS	–	0.17	0.16
16-Oxo-androstenediol-TMS	–	0.21	0.20

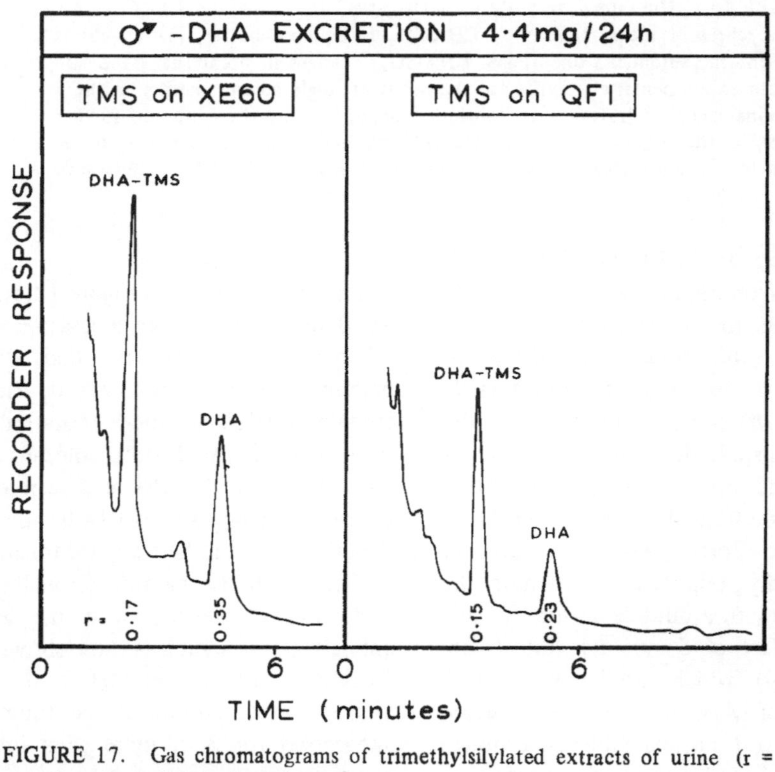

FIGURE 17. Gas chromatograms of trimethylsilylated extracts of urine (r = retention time relative progesterone).

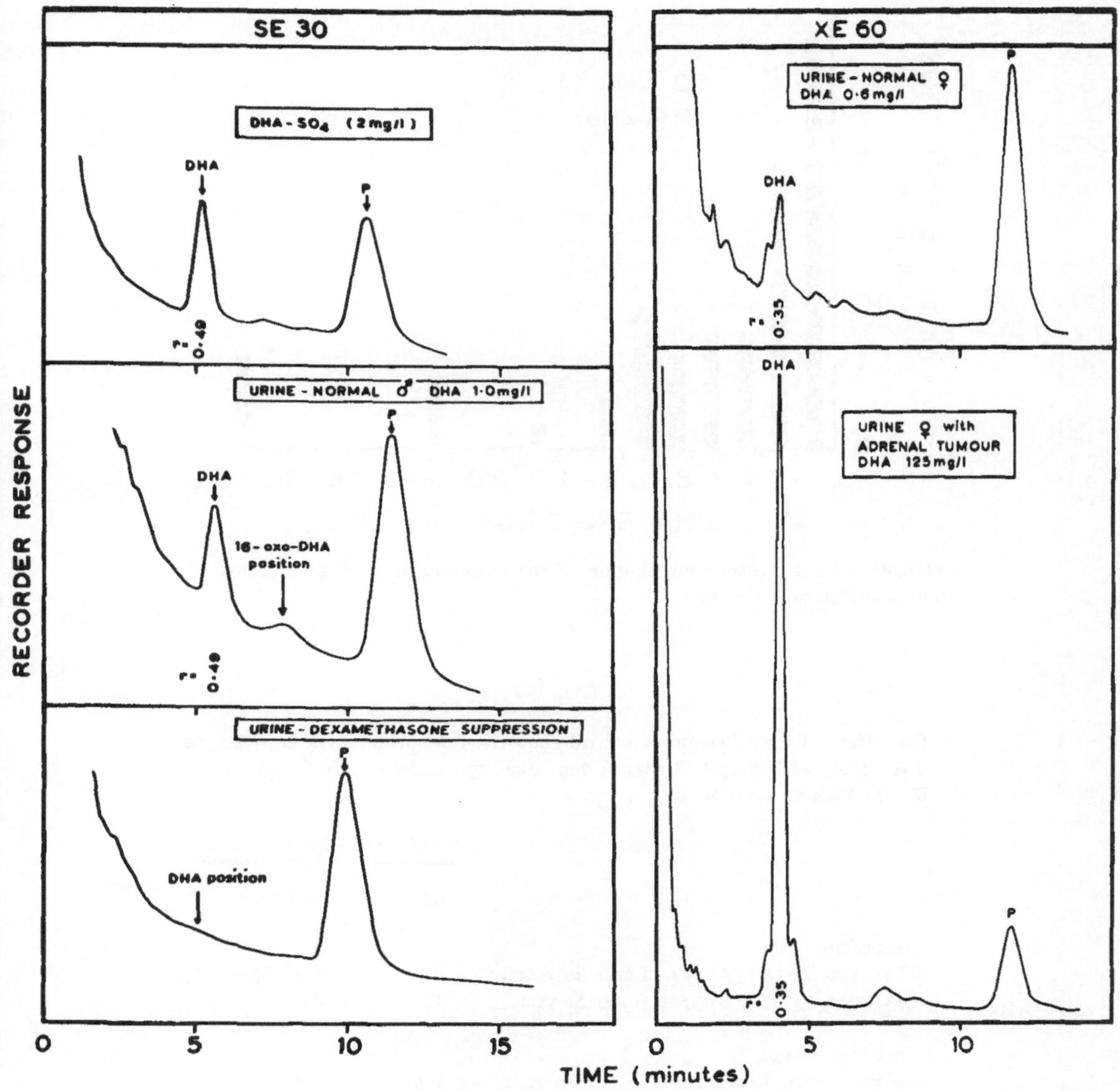

FIGURE 18. Gas chromatograms of urinary and standard extracts on the liquid phases SE30 and XE60.

analyzed (Figure 18). Tracings typical of those seen when urine containing different amounts of DHA is processed are seen in Figure 18. The limit on the sensitivity of the assay is about 0.2 mg/l. Concentrations less than this barely show above the urinary background and cannot be quantitated. In some tracings prepared from the urine of normal individuals, a prominent peak can be seen midway between the DHA and progesterone peaks. This peak has a retention time close to or equal to that of the 16-oxosteroids, 16α-OH-DHA, and 16-oxo-androstenediol.

The 24-hr excretion of DHA-SO_4 (measured as DHA) by normal menstruant women has a skew distribution (Figure 19), 70% excreting <1 mg/24 hr and only 6% >4 mg/24 hr. In the same women, the basal excretion of DHA was found not to correlate with adrenocortical activity as judged by 17-oxogenic steroid output.[74] This phenomenon was not unexpected in view of the reported inefficient and variable clearance of injected DHA-SO_4 through the kidneys.[77] A similar discordance was seen when DHA and estrogen excretions were compared. It is apparent that the amount of DHA appearing in the urine of normal subjects does not reflect day-to-day changes in adrenal and gonadal function and probably bears only the

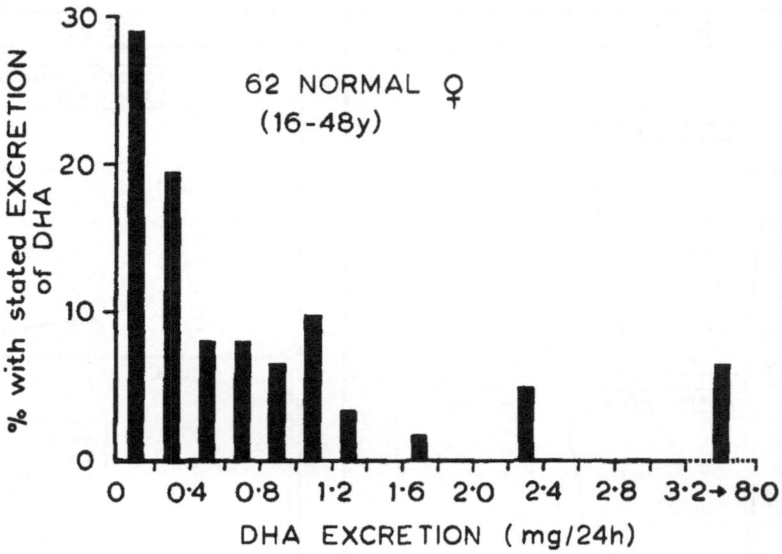

FIGURE 19. Frequency distribution of the excretion of DHA by a group of normal menstruant women.

TABLE 7

The Effect of Age, Starvation, and Adrenocortical Suppression (Dexamethasone Given Daily at Midnight (1 mg), 6 a.m. (0.5 mg), and Noon (0.5 mg) for ≥ 3 Days) on DHA Excretion

	DHA excretion (mg/24 hr)	
	Mean	Range
Effect of age		
62 women age 16 to 48 years (mean 26 years)	1.0	0–8.0 (Figure 19)
10 women age 52 to 61 years (mean 56 years)	0.1	0–0.6
Effect of starvation		
9 obese women, basal state	0.6	0–1.9
3rd fast day	0.1	0–0.2
Effect of dexamethasone		
12 women, basal state	1.8	0.2–7.1
suppressed	0.1	0–0.2

most tenuous relationship to DHA and DHA-SO_4 secretion rates.

Three conditions in women are associated with the disappearance of DHA from urine. These conditions are advanced age, food deprivation, and adrenocortical suppression (Table 7). Although no correlation between age and DHA excretion was observed in menstruant women, few women over the age of 50 years had any detectable DHA in their urine using the techniques described above. Even when their excretion of 17-oxogenic steroids was increased sixfold during adrenocorticotrophic hormone stimulation, two women, aged 51 and 57 years, excreted less than 0.1 mg/24 hr DHA. In the menstruant group, adrenocortical suppression and starvation were associated with a prompt decrease in urinary DHA levels (Table 7). The effects appeared within 2 to 3 days in each case.

In younger women, adrenocortical stimulation often causes a rise in the urinary levels of

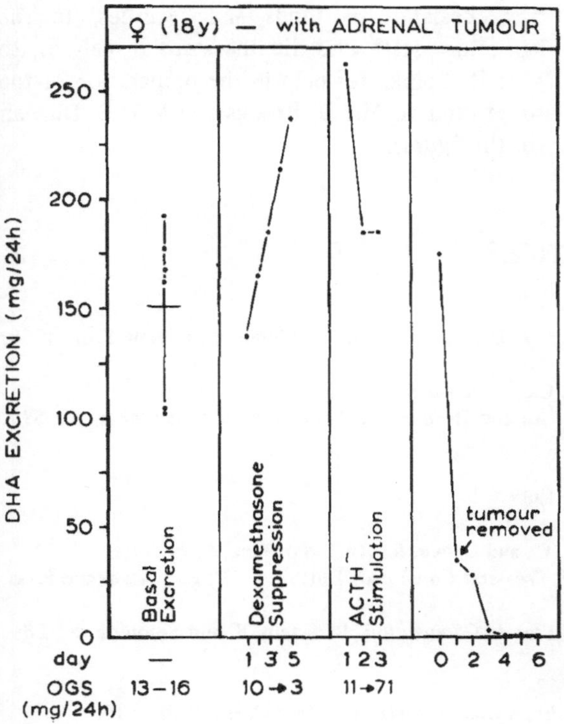

FIGURE 20. Effect of changes in adrenocortical activity and of tumor removal on the excretion of DHA by a patient with a DHA-SO_4-secreting adenoma.

DHA-SO_4. Five of a group of nine women aged 22 to 38 years (mean 29 years), excreted increased amounts of DHA in response to adrenocorticotrophic hormone (ACTH, 25 I.U. infused during 8 hr on 2 successive days; urine collected on the second day). Only two of them had increments greater than 1 mg/24hr (viz., 5 and 13 mg/24hr). The extent of the increase did not, in this small series, appear invariably to be related to the basal excretion of DHA, and it bore no relationship to adrenocortical activity as judged by 17-oxogenic-steroid output.

Large increases in urinary levels of DHA-SO_4 are sometimes associated with the presence of an adrenal tumor. The young woman aged 18 years with a DHA-SO_4-secreting tumor of the adrenal gland, who was referred to previously, excreted grossly elevated amounts of DHA (Figures 18 and 20). Her daily output of urinary DHA-SO_4 (measured as DHA) varied between 100 and 250 mg and was unaffected either by adrenocortical suppression with dexamethasone, or by adrenocortical stimulation with ACTH. The patient's only presenting symptoms were acne and marked facial hirsutism. She was normally feminized and showed no signs of virilization. After the removal of a large adrenal tumor, her excretion of DHA fell to 2.3 mg by the second day after operation and to undetectably low levels by the third and subsequent days (Figure 20).

TABLE 8

Recovery of DHA-SO_4 Added to Urine, and of DHA-SO_4 in Urine at Different Dilutions

DHA-SO_4 added (mg/l.)	DHA-SO_4 found (mg/l.)	% Recovery
0	0.47	–
0.4	0.84	92.5
0.8	1.24	96.2
1.2	1.62	95.7
1.6	2.09	100.7
2.0	2.47	99.7
Urine dilution (urine:water)		
1:0	0.99	–
1:1	0.47	94.8

*Each figure is the mean of 3 measurements; S.D. of all measurements from their respective means, 0.05 mg/l. (From Metcalf, M. G., *Clin. Biochem.*, 4, 241, 1971. With permission.)

Accuracy was measured by assaying DHA-SO_4 in urine containing added conjugate and in urine at two dilutions. The results, shown in Table 8, demonstrate that recoveries are adequate for clinical studies. Quadruplicate DHA-SO_4 standards were hydrolyzed and assayed for DHA on six different occasions. The standard deviation of all the results from their group means was 0.04 mg/l. DHA. Standard deviations calculated from the results of duplicate assays on urine samples were 0.04 mg/l. in the 0 to 1 mg/l. range (31 samples with a mean DHA concentration of 0.36 mg/l.) and 0.09 mg/l. in the >1 mg/l. range (30 samples with a mean DHA concentration of 2.79 mg/l.). Between-assay replication is less satisfactory. When 31 urine samples containing 0.1 to 9.0 mg/l. DHA (mean 1.74 mg/l.) were measured in duplicate on 2 different occasions, the standard deviation calculated from the differences between the pairs of mean values was 0.2 mg/l. Although DHA concentrations of 0.1 mg/l. can, in general, be distinguished from the urinary background, concentrations of less than 0.4 mg/l. cannot be measured with acceptable precision.

ACKNOWLEDGMENTS

I am most grateful to the North Canterbury Hospital Board and the Medical Research Council, New Zealand for funds and facilities, to the laboratory staff who did the steroid analyses, to Mrs. R. Spinks for help in the preparation of the script, and to Mr. D. Brooks and Miss B. Duncan for the figures.

REFERENCES

1. **Lovelock, J. E.,** *J. Chromatogr.*, 1, 35, 1958.
2. **McWilliam, I. G. and Dewar, R. A.,** in *Gas Chromatography*, Desty, D. H., Ed., Academic Press, New York, 1958, 142.
3. **Vanden Heuvel, W. J. A., Sweeley, C. C., and Horning, E. C.,** *J. Am. Chem. Soc.*, 82, 3481, 1960.
4. **Luukkainen, T., Vanden Heuvel, W. J. A., Haahti, E. O. A., and Horning, E. C.,** *Biochim. Biophys. Acta*, 52, 599, 1960.
5. **Clark, S. J. and Wotiz, H. H.,** *Steroids*, 2, 535, 1963.
6. **Brooks, C. J. W. and Harvey, D. J.,** *Biochem. J.*, 114, 15, 1969.
7. **Cox, R. I.,** *J. Chromatogr.*, 12, 242, 1963.
8. **Turner, D. A., Seegar-Jones, S. E., Sarlos, I. J., Barnes, A. C., and Cohen, R.,** *Anal. Biochem.*, 5, 99, 1963.
9. **Jayle, M. F. and Pasqualini, J. R.,** in *Glucuronic Acid, Free and Combined*, Dutton, G. J., Ed., Academic Press, London, 1966, 533.
10. **Bradlow, H. L.,** in *Chemical and Biological Aspects of Steroid Conjugation*, Bernstein, S. and Solomon, S., Eds., Springer-Verlag, New York, 1970, 131.
11. **Knox, J. H.,** *Gas Chromatography*, Methuen, London, 1962.
12. **McNair, H. M. and Bonelli, E. J.,** *Basic Gas Chromatography*, Varian Aerograph, Walnut Creek, Calif., 1967.
13. **Johansson, E. D. B.,** *Acta Endocrinol.*, 61, 592, 1969.
14. **Swain, M. C.,** *Clin. Chim. Acta*, 39, 455, 1972.
15. **Ekins, R. P.,** *Clin. Chim. Acta*, 5, 453, 1960.
16. **Yalow, R. S. and Berson, S. A.,** *J. Clin. Invest.*, 39, 1157, 1960.
17. **Murphy, B. E. P.,** *Nature*, 201, 679, 1964.
18. **Lipsett, M. B., Doerr, P., and Bermudez, J. A.,** *Acta Endocrinol.*, Suppl. 147, 155, 1970.
19. **Kutas, M., Chung, A., Bartos, D., and Castro, A.,** *Steroids*, 20, 697, 1972.
20. **Townsend, S. L., Brown, J. E., Johnstone, J. W., Adey, F. D., Evans, J. H., and Taft, H. P.,** *J. Obstet. Gynaecol. Br. Commonw.*, 73, 529, 1966.
21. **Sobota, J. T. and Kirton, K. T.,** *Obstet. Gynecol.*, 35, 752, 1970.
22. **Dorfman, R. I. and Ungar, F.,** *Metabolism of Steroid Hormones*, Academic Press, New York, 1965, 602.
23. **Metcalf, M. G.,** *Am. J. Obstet. Gynecol.*, accepted for publication.
24. **Klopper, A., Michie, E. A., and Brown, J. B.,** *J. Endocrinol.*, 12, 209, 1955.
25. **Cooper, J. A., Abbot, J. P., Rosengreen, B. K., and Claggett, W. R.,** *Am. J. Clin. Pathol.*, 38, 388, 1962.
26. **Barrett, S. A. and Brown J. B.,** *J. Endocrinol.*, 47, 471, 1970.
27. **Wotiz, H. H. and Clarke, S. J.,** in *Gas Chromatography in the Analysis of Steroid Hormones*, Plenum Press, New York, 1966, 129.
28. **Metcalf, M. G.,** *Anal. Biochem.*, 25, 510, 1968.
29. **Metcalf, M. G.,** *Clin. Biochem.*, 6, 4, 1973.
30. **Evans, C., Fahmy, D., Morgan, C. A., Pearce, M. A., and Cooke, I. D.,** *Clin. Chim. Acta*, 38, 25, 1972.
31. **Rogers, M. and Chamberlain, J.,** *Clin. Chim. Acta*, 39, 439, 1972.
32. **Van der Molen, H. J.,** in *Gas Phase Chromatography of Steroids*, Eik-Nes, K. B. and Horning, E. C., Eds., Springer-Verlag, New York, 1968, 204.
33. **Metcalf, M. G.,** *Steroids*, 21, 193, 1973.
34. **Chang Shen, N., Katzman, M. B., Francis, F. H., and Kinsella, R. A.,** *J. Lab. Clin. Med.*, 61, 174, 1963.
35. **Crepy, O., Judas, O., Rulleau-Meslin, F., and Jayle, M. F.,** *Bull. Soc. Chim. Biol.*, 44, 327, 1962.
36. **Metcalf, M. G.,** *Clin. Chem.*, 15, 24, 1969.
37. **Peterson, R. E.,** in *The Human Adrenal Cortex*, Christy, N. P., Ed., Harper and Row, London, 1971, 118.
38. **Vela, B. A. and Acevedo, H. F.,** *Steroids*, 14, 499, 1969.
39. **Van der Molen, H. J.,** *J. Clin. Endocrinol.*, 28, 1361, 1968.
40. **Driessen, O., Voute, P. A., and Vermeulen, A.,** *Acta Endocrinol.*, 57, 177, 1968.
41. **Fieser, L. F. and Fieser, M.,** in *Steroids*, Reinhold, New York, 1959, 519.

42a. **Solomon, S. and Bhavnani, B. R.,** in *Chemical and Biological Aspects of Steroid Conjugation*, Bernstein, S. and Solomon, S., Eds., Springer-Verlag, New York, 1970, 326.

42b. **Migeon, C. J.,** *Am. J. Med.*, 53, 606, 1972.

43. **Metcalf, M. G.,** *Clin. Biochem.*, 5, 19, 1972.

44. Bongiovanni, A. M. and Cohn, R. M., in *Chemical and Biological Aspects of Steroid Conjugation,* Bernstein, S. and Solomon, S., Eds., Springer-Verlag, New York, 1970, 416.
45. Jänne, O., Vhiko, R., Sjövall, J., and Sjövall, K., *Clin. Chim. Acta,* 23, 405, 1969.
46. Johnsen, S. G., *Acta Endocrinol.,* 57, 595, 1968.
47. Jayle, M. F. and Pasqualini, J. R., in *Glucuronic Acid, Free and Combined,* Dutton, G. J., Ed., Academic Press, London, 507, 1966.
48. Bulbrook, R. D., Hayward, J. L., and Spicer, C. C., *Lancet,* 2, 395, 1971.
49. Eik-Nes, K. B., in *Gas Phase Chromatography of Steroids,* Eik-Nes, K. B. and Horning, E. C., Eds., Springer-Verlag, New York, 1968, 298.
50. Butenandt, A. and Tscherning, K., *Z. Physiol.,* 229, 167, 1934.
51. Callow, N. H., *Biochem. J.,* 33, 559, 1939.
52. Zimmerman, W., *Z. Physiol. Chem.,* 233, 257, 1935.
53. Dobriner, K., *J. Clin. Invest.,* 32, 940, 1953.
54. Wolfe, J. K., Fieser, L. F., and Friedgood, H. B., *J. Am. Chem. Soc.,* 63, 582, 1941.
55. Bush, I. E., *Biochem. J.,* 50, 370, 1952.
56. Miller, H. and Durant, J. A., *Clin. Chim. Acta,* 14, 475, 1966.
57. Jones, J. K. N. and Stitch, S. R., *Biochem. J.,* 53, 679, 1953.
58. Vestergaard, P. and Claussen, B., *Acta Endocrinol.,* Suppl. 64, 39, 1962.
59. Cooper, J. A. and Creech, B. G., *Anal. Biochem.,* 2, 502, 1961.
60. Haahti, E. O. A., Vanden Heuvel, W. J. A., and Horning, E. C., *Anal. Biochem.,* 2, 182, 1961.
61. Kirschner, M. A. and Lipsett, M. B., *J. Clin. Endocrinol.,* 23, 255, 1963.
62. Beale, R. N., Croft, D., and Taylor, R. F., *Steroids,* 18, 619, 1971.
63. Metcalf, M. G., *Steroids,* 17, 85, 1971.
64. Metcalf, M. G., *Clin. Biochem.,* 3, 271, 1970.
65. Roberts, K. D. and Lieberman, S., in *Chemical and Biological Aspects of Steroid Conjugation,* Bernstein, S. and Solomon, S., Eds., Springer-Verlag, New York, 1970, 242.
66. Wieland, R. G., Levy, R. P., Katz, D., and Hirschmann, H., *Biochim. Biophys. Acta,* 78, 566, 1963.
67. MacDonald, P. C., Chapdelaine, A., Gonzales, O., Gurpide, E., Vande Wiele, R. L., and Lieberman, S., *J. Clin. Endocrinol.,* 25, 1557, 4, 1569, 1965.
68. Longhino, N., Tajic, M., Vedris, M., Jankovic, D., and Drobnjak, P., *Acta Endocrinol.,* 59, 644, 1968.
69. Laatikainen, T. and Vihko, R., *J. Steroid Biochem.,* 2, 173, 1971.
70. Baird, D. T., Horton, R., Longcope, C., and Tait, J. F., *Recent Progr. Horm. Res.,* 25, 611, 1969.
71a. Drucker, W. D., Blumberg, J. M., Gandy, H. M., David, R. R., and Verde, A. L., *J. Clin. Endocrinol. Metab.,* 35, 48, 1972.
71b. France, J. T., *Aust. N.Z. J. Obstet. Gynaecol.,* 10, 132, 1970.
72. Garnham, J. R., Bulbrook, R. D., and Wang, D. Y., *Eur. J. Cancer,* 5, 239, 1969.
73. Hampl, R. and Starka, L., *Clin. Chim. Acta,* 34, 77, 1971.
74. Metcalf, M. G., *Clin. Biochem.,* 4, 241, 1971.
75. Faucette, N. E. and Cawley, L. P., *Clin. Biochem.,* 4, 287, 1971.
76. Hellstrom, K., *Acta Endocrinol.,* 60, 501, 1969.
77. Baulieu, E. E., Corpéchot, C., Dray, F., Emiliozzi, R., LeBeau, M-C., Mauvais-Jarvis, P., and Robel, P., *Recent Progr. Horm. Res.,* 21, 411, 1965.
78. Butenandt, A. and Dannenbaum, H., *Hoppe Seyler Z. Physiol. Chem.,* 229, 192, 1934.
79. Lieberman, S., Mond, B., and Smyles, E., *Recent Progr. Horm. Res.,* 9, 113, 1954.
80. Fotherby, K., *Biochem. J.,* 73, 339, 1959.
81. Oertel, G. W. and Eik-Nes, K. B., *Anal. Chem.,* 31, 98, 1959.
82. Jungmann, R. A., Calvary, E., and Schweppe, J. S., *J. Clin. Endocrinol.,* 27, 355, 1967.
83. Solomon, D., Strummer, D., and Nair, P. P., *Am. J. Clin. Pathol.,* 48, 295, 1967.
84. Brooksbank, B. W. L. and Gower, D. B., *Acta Endocrinol.,* 63, 79, 1970.
85. Uozumi, T., Manabe, H., Tanaka, H., Hamanaka, Y., Kotoh, K., Suzuki, K., and Matsumoto, K., *Acta Endocrinol.,* 61, 17, 1969.
86. Wilson, H., Lipsett, M. B., and Ryan, D. W., *J. Clin. Endocrinol.,* 21, 1304, 1961.
87. Jänne, O. and Vihko, R., *Steroids,* 14, 235, 1969.
88. Miller, H., Durant, J. A., Jacobs, A. G., and Allison, J. F., *Br. Med. J.,* 1, 147, 1967.

TESTOSTERONE AND METABOLITES IN WOMEN USING RADIOIMMUNOASSAY TECHNIQUES

R. G. Wieland, E. M. Zorn, M. C. Hallberg, and B. Furst

TABLE OF CONTENTS

I. Appraisal 181

II. Methodology 182
A. Principle of Radioimmunoassay 182
B. Radioimmunoassay for Combined Testosterone and Dihydrotestosterone 182
1. Principle 182
2. Apparatus 183
3. Reagents 183
4. Procedure 183
5. Calculations 184
6. Accuracy and Precision 185
C. Assay for Fractionated Testosterone and Dihydrotestosterone 185
1. Principle 185
2. Apparatus 185
3. Reagents 185
4. Procedure 185
5. Calculations 186
6. Accuracy and Precision 187
D. Interpretation 187

III. Appendix 188
A. T-3-BSA Preparation 188
B. T-3-BSA Antibody Preparation 188

References 189

I. APPRAISAL

Assay of testosterone in female serum is preferable under clinical conditions to assay of the hormone in urine. This is because the urinary metabolite assayed, testosterone glucuronoside, is derived from both testosterone and prehormones,[1] some of the latter being converted and conjugated by one passage through the liver. Thus, it has been demonstrated that a significant amount of androstenedione, a very important testosterone prehormone, can be converted directly to testosterone glucuronoside.[1] This amount, therefore, does not circulate as free testosterone and gives a false elevation of the testosterone production rate, using the urinary excretion technique in the woman. Amother possible source of error is the secretion of testosterone conjugates such as has been indicated by spermatic vein studies in males.[2]

Technological advances have resulted in the reporting of various techniques for the determination of serum testosterone. These include double isotope derivative methods,[3] gas-liquid chromatography,[4] and competitive protein binding using testosterone binding globulin.[5] Such methods are accurate and sensitive enough to measure testosterone in the serum of women and pre-pubertal males. However, in the authors' experience the radioimmunoassay, utilizing a first antibody generated in rabbits against testosterone-3-bovine serum albumin (T-3-BSA) and a second antibody of goat anti-rabbit γ-globulin, has proved to be the most reproducible technique for the assay of serum testosterone.[6,7] Efforts to reduce the incubation time have resulted, in the authors' hands, in less sensitivity and accuracy of the assay.

II. METHODOLOGY

A. Principle of Radioimmunoassay

Radioimmunoassay (RIA) is sometimes described as displacement analysis. An excess of antigen is added to a constant and limited quantity of antibody. Two forms of the antigen are used.[8,9] The concentration of the labeled form is constant, while that of the unlabeled form is variable. The two forms of the antigen are assumed to compete equally for antibody sites according to the law of mass action. This leaves the unbound antigens free of antibody in amounts inversely proportional to that of the antigen-antibody complexes formed. Once equilibrium is obtained, the antibody bound antigen is separated from the unbound (free) antigen. The separation has been accomplished by several methods: electrophoresis;[8] using adsorbents to adsorb the free antigen[10] (charcoal and Florisil®); salting out the antigen-antibody complex with ammonium sulfate;[11] adsorbing covalently the antibody to an insoluble matrix;[12] or polymerizing the antibody[13] to render it insoluble but still reactive. Commonly used is the double antibody method[14] for which the first antibody is the antigen for the second antibody. This latter technique is the procedure to be described in this chapter.

The equilibrium of the reaction[15] may be described by the classic equilibrium equation.

$$Ag + Ab \underset{k_1}{\overset{k}{\rightleftharpoons}} AgAb$$

$$K = \frac{k}{k_1} \qquad (1)$$

Ag = antigen
Ab = antibody

Therefore, $[Ag]\,[Ab] = k_1\,[AgAb]$ dissociation equation and $k\,[Ag]\,[Ab] = [AgAb]$ association equation. At equilibrium $k_1\,[Ag][Ab] = k_1\,[AgAb]$. Rearrangement and substitution in Equation 1 gives

$$K[Ag]\,[Ab] = [AgAb] \text{ equilibrium equation} \qquad (2)$$

The ratio of the bound (B) to the free (F) antigen which exists at equilibrium is

$$B/F = \frac{[AgAb]}{[Ag]} \qquad (3)$$

if

Ag_i = initial or total antigen
Ab_i = initial or total antibody

the following definitions follow

$$[Ag] = [Ag_i] - [AgAb] \qquad (4)$$

$$[Ab] = [Ab_i] - [AgAb] \qquad (5)$$

By substitution of 4 and 5 in Equation 3 and rearrangement and substitution of Equation 1, one obtains

$$(B/F)^2 + B/F\,(1+K[Ag_i] - K[Ab_i]) - K[Ab_i] = 0 \qquad (6)$$

This is the equation of a hyperbola, which is the classic form for the expression of the dose response curve for RIA. However, a more readable form is found in the log-linear plot of the results which may be expressed mathematically as

$$[Ag_i] = Ag_{i_o}\, e^{(\text{slope} \times B/F)}$$
$$[Ag_{i_o}] = \text{intercept}$$

Others have graphed the results on a log-logit plot in attempts to linearize the curve.[16] However, the high and low ends of the curve, which represent the extremes of the antigen concentrations, do not fit the equation of a hyperbola due to the multivalent nature of the antibody. This inherent error is carried over into the log-logit calculations.

B. Radioimmunoassay for Combined Testosterone (T) and Dihydrotestosterone (DHT)

1. Principle

Serum, with added radioactive antigen (T-1,2-^{3}H), is incubated with T-3-BSA antibody and the resulting antigen-antibody complex is precipitated by incubation with a second antibody (anti-rabbit γ-globulin). After centrifugation, the amount of radioactivity in the supernatant solution is compared with that of a standard curve to determine the concentration of testosterone in the serum.

2. Apparatus

1. Analytical balance.

2. Extraction tubes, approximately 10-ml capacity with glass stoppers or plastic caps.

3. Disposable glass culture tubes, 10 X 75 mm (Tek-tube, Abbot Scientific Products Division, 820 Mission St., South Pasadena, Calif.).

4. Hamilton microsyringes: 100 μl and 10 μl.

5. Automatic pipette (#60422. BBl. Division of BioQuest, Cockeysville, Md.).

6. "Fibrin" pipette tips (Elkay Products, Inc., 95 Grand St., Worcester, Mass.).

7. Thermostatically controlled hot water bath.

8. Refrigerated centrifuge (International Equipment Model PR-2 with trunnion carriers #381).

9. Scintivials (VP-1 Isolab Inc. Drawer 4350, Akron, Ohio).

10. Liquid scintillation counter (Model LS-100, Beckman Instruments, Inc., Fullerton, Calif.).

11. Log-logit graph paper (W. Heffer & Sons, Ltd., 26 King St., Cambridge, CB1 1LN, England).

3. Reagents

1. Diethyl ether. Laboratory grade. A new can of ether is opened each day or redistilled before use.

2. Compressed nitrogen, water pumped.

3. PBSA, 0.01 *M,* pH 7.25.

Solution A: 8.765 g NaCl, 1.38 g $NaH_2PO_4 \cdot H_2O$, and 1 g sodium azide are dissolved in distilled water and diluted to 1 liter. Solution B: 8.765 g NaCl, 1.42 g Na_2HPO_4, and 1 g sodium azide are dissolved in distilled water and diluted to 1 liter. Solutions A and B are then mixed to give a pH of 7.25. The exact volumes of each depend on the distilled water used; in the authors' laboratory approximately 1 vol of solution A is added to 9 vol of solution B.

4. Testosterone (T) standards (m.p. 153 to 154°C). Stock solution: 1.0 mg testosterone/ml absolute ethanol. Working standard: 10 ng testosterone/100 μl ethanol. With a 10 μl Hamilton syringe, 10 μl stock solution is added to 95% ethanol in a 100-ml volumetric flask and diluted to the mark with 95% ethanol. Both the stock solution and the working standard are kept under refrigeration. Parafilm is wrapped around the glass stoppers to minimize the dangers of evaporation.

5. 1% HSA in PBSA: 1 g human serum albumin is dissolved in 100 ml PBSA.

6. Tracer T-1,2-^{3}H, 0.05 ng T/100 μl PBSA (40-60 Ci/mmol, New England Nuclear, Boston, Mass.). When received, T-1,2-^{3}H is diluted to 10 ml with absolute ethanol. Then 100 ml of tracer solution containing 0.05 ng T/100 μl is prepared by adding the appropriate amount of T-1,2-^{3}H to PBSA in a 100-ml volumetric flask and diluting to the mark with PBSA.

7. T-3-BSA antibody (see Appendix for antibody preparation). The optimum dilution of the antibody varies with each preparation; therefore, it is necessary to check each antibody. A dilution is chosen which gives a maximum precipitation of 70 to 80% (zero tubes) and with which 1.0 ng of T gives a precipitation of 65 to 75% of the maximum. The concentrated antibody is thawed and diluted with PBSA just before use.

8. Goat anti-rabbit γ-globulin (Antibodies, Inc., Davis, Calif.). This antibody is diluted 1:50 with a mixture of PBSA and EDTA just before use (9 vol PBSA plus 1 vol 0.1 *M* EDTA).

9. Dioxane counting solution: 100 g naphthalene and 5 g 2,5-diphenyloxazole are dissolved in 1 liter of a high grade dioxane.

10. The purity of the nonradioactive T and DHT and T-1,2-^{3}H and DHT-1,2-^{3}H is checked regularly. The unlabeled steroids are checked by GLC-SE 30 + XE 60 1% on Anakron ABS, with a hydrogen flame detector; the labeled steroids are checked by thin-layer chromatography, as described below.

4. Procedure

a. Extraction of Sera

First, 2 ml serum is pipetted into extraction tubes, then 5 ml fresh ether is added and the tubes are stoppered and shaken vigorously for at least 1 min. The serum is permitted to settle and 4.0 ml (2.0 ml twice) of the supernatant ether (equivalent to 1.6 ml serum) is transferred to disposable glass culture tubes and evaporated to dryness in a water bath at 35 to 40°C under a stream of nitrogen. After the ether extract is dried, care must be taken to wash down the sides of the tubes with more ether so that all of the extracted testosterone is at the very bottom of the tube. All unknown sera are extracted and assayed in duplicate. If high values are anticipated, less serum is used. Volumes of less than 1 ml are diluted to 1.0 ml with PBSA before extraction.

b. Preparation of Standards

In an attempt to equalize the amount of ether-extracted material other than testosterone in the tubes containing standards and unknowns, an extract of serum is combined with the standards: 12 ml of a pool of normal female sera is extracted with 30 ml ether. Then 1.0 ml of the ether extract (equivalent to 0.4 ml serum) is transferred to each of 16 disposable glass culture tubes and dried under nitrogen in a 35 to 40°C water bath. Next, 4.0 ml of the ether extract (equivalent to 1.6 ml serum) is pipetted into each of two tubes for the assay of the pool used.

Using a 100-µl Hamilton syringe, 100 µl of the working standard is diluted to 1.0 ml with 95% ethanol. A standard curve ranging from 0.2 to 6 ng testosterone is prepared by delivering in duplicate 20, 40, 70, and 100 µl of the dilute working standard and 20, 40, and 60 µl of the undiluted working standard to the tubes containing the ether extract of 0.4 ml of normal female serum, leaving two tubes with no added testosterone. The tubes are again dried under nitrogen and the sides washed down with ether to bring all the material to the bottom of the tubes. Two tubes with neither ether extract nor added testosterone are included in the assay as zero tubes.

c. Assay

With the automatic pipette and pipette tips 100 µl 1% HSA is added to each tube. This is followed by the addition of 100 µl tracer T-1,2-^{3}H and 100 µl dilute T-3-BSA antibody. The tubes are mixed, placed in a 40°C water bath for 15 min, left at room temperature for 15 to 30 min, and refrigerated at 4°C. After 2 days of refrigeration, 100 µl goat anti-rabbit γ-globulin is added to each tube. The tubes are left at 22°C for 30 min and refrigerated at 4°C for 2 days. The tubes are then centrifuged for 30 min at 4°C at 800 g, placed in an ice bath, and, using the automatic pipette, 150 µl of the supernatant solution is transferred to scintivials containing 10 ml dioxane counting solution. The samples are counted in the scintillation counter to a 1.5% counting error.

5. Calculations

The percentage of the maximum precipitation is calculated for each standard and unknown tube as follows:

$$\frac{\text{A} - \text{cpm of standard or unknown}}{\text{B}} \times 100 = \text{\% of maximum precipitation}$$

where

$$\text{A} = \frac{150}{400} \times (\text{cpm of } 100\ \mu\text{l tracer})$$

$$\text{B} = \text{A} - \text{cpm of zero tubes} = \text{maximum precipitation.}$$

The percentage of maximum precipitation of the standard tubes is plotted on semilogarithmic paper, resulting in a sigmoid curve. The amount of testosterone in 1.6 ml of the female pool used with the standards is then read from the standard curve and divided by 3 to give the amount of testosterone in the 0.4 ml serum in the standard tubes. (If x = ng T in 0.4 ml female pool, then any point on the standard curve is equal to ng + x and 1.6 ml female pool is equal to 4x. Therefore, 4x = ng + x, and $x = \frac{ng}{3}$.)

The standard curve is then replotted with this correction. The concentration of testosterone in the unknown samples can then be read from the corrected standard curve. The results are expressed as ng/100 ml serum, and calculated as follows:

$$\frac{\text{ng} \times 100}{1.6} = \text{ng/100 ml}$$

The percentage of maximum precipitation of the standards may also be plotted on the log-logit paper to give a straight line. However, because of the large errors at the upper and lower parts of the curve, the line must be drawn using only the points that fall between about 20 and 80% of the maximum precipitate. The nanogram amount of testosterone in the unknown samples can be read directly from this line or the semilogarithmic curve can be corrected from this line, thereby normalizing the curve and reducing some of the experimental errors of pipetting.

6. Accuracy and Precision

1. A pool of female sera was assayed 15 times, giving a mean value of 72 ± 2.5 (S.E.) ng/100 ml. The interassay variation calculated from the standard error was 3.5%.

2. Four 2-ml samples of a pool of female sera with 1.0 ng added testosterone were assayed as described. Recoveries of testosterone were 103.3, 99.4, 100.7, and 96.2%.

3. A pool of male sera was assayed in triplicate using volumes of 0.2, 0.4, and 0.6 ml of serum, yielding values of 548, 546, and 527 ng/100 ml, respectively.

C. Assay for Fractionated Testosterone and Dihydrotestosterone

1. Principle

A method is described for the chromatographic separation and subsequent radioimmunoassay of testosterone and dihydrotestosterone in female sera. The method used for chromatography is that of Demetriou and Austin.[17] The hormones are assayed using the method described in this chapter.

2. Apparatus

1. Chromatographic equipment consists of a rectangular glass tank (Arthur H. Thomas Co.), supporting 9 x 8 in. glass plates, and stainless steel clips.

2. The needle-syringe elution technique utilizes glass syringes (1 ml), bent needles (22 gauge), polyethylene bottles (240 mm), and glass funnels (45 mm).[17]

The tops of the polyethylene bottles are cut off so that the bottles can slip over the funnels and a hole is cut into the bottom of the bottles to hold the syringes. The needles are bent to form a hook and attached to the syringes. A section of ChromAR® applied to the hook of the needle is inside the bottle and hanging over the funnel.

3. Shortwave length UV lamp (Ultra Violet Products, Inc. UVS-12).

3. Reagents

1. ChromAR 500 (silicic acid/glass fiber, Mallinckrodt Chemical Works, St. Louis, Mo.).

2. Tracer DHT-1,2-^{3}H, 0.05 ng DHT/100 μl PBSA (40 to 60 Ci/mmol) (New England Nuclear, Boston, Mass.). DHT-1,2-^{3}H is diluted in the same manner as the testosterone tracer.

3. Chromatographic system. Benzene and ethyl acetate (80:20 v/v).

4. Dihydrotestosterone standards (m.p. 178 to 182°C). A stock solution of 1.0 mg DHT/ml absolute ethanol is diluted to make a working standard of 10 ng DHT/100 μl ethanol as in the preparation of testosterone standard.

4. Procedure

a. Preparation of Chromatographic Paper

First, 8 X 8-in. sheets of ChromAR 500 are ruled into five lanes with a soft pencil, leaving about 0.5 cm at the edges. Four lanes for sample migration are 4 cm wide and a center lane for a T marker is about 3.2 cm wide. A line is marked across the lanes 3 cm from the bottom of the sheet to indicate the origin for application of the sample and another line is marked 15 cm higher to indicate the solvent front. In order to decrease the amount of nonspecific interfering substances, the sheets are soaked in two changes of ethyl acetate for several days, clipped to glass plates, and dried in an oven at 40°C. The blank sheets are then placed upright in the tank so that the sheets hang free and are developed twice with the chromatographic system. One extra sheet must be prepared for each assay so that blank sections can be eluted for use with the standards. This sheet must be developed four times to correspond to the treatment of the sample sheets. Sheets not used immediately should be wrapped in aluminum foil and protected from exposure to air.

b. Extraction

Because of the low concentration of DHT in female serum, 4.0 ml serum must be used for each sample, or 8 ml for duplicate determinations. Prior to extraction, 100 μl tracer T-1,2-^{3}H and 100 μl tracer DHT-1,2-^{3}H (approximately 12,000 dpm) are added to each unknown sample for internal recovery determinations. For recovery standards, 100 μl of each tracer is pipetted directly into scintivials containing 10 ml counting fluid. The samples are extracted three times with 10 ml ether and the combined extracts are evaporated to dryness in single disposable culture tubes, as in the assay for combined T and DHT.

c. Chromatographic Separation

The dried extracts of 4.0 ml serum are dissolved in a mixture of 0.02 ml dichloromethane and 0.09 ml 95% ethanol and incubated in the water bath for 10 min at 40°C. Each sample is transferred in repeated 5-μl portions to the middle 2 or 3 cm of the origin line in the ChromAR sheet sample lanes. Then 5 μl of the stock T standard solution (1 μg/μl) is applied as a single spot in the center lane for a marker. When dry, the chromatogram is clipped to a glass plate and developed in the chromatography jar for a distance of 15 cm, dried in a 40°C oven, and redeveloped as before.

d. Elution

The T. marker in the center lane is located under ultraviolet light in a dark room and the top of the spot marked with a pencil. A 2 X 4 cm section in each sample lane is marked extending 2 cm from the top of the T spot toward the origin for the elution of T. A 2 X 4 cm section 1.5 to 3.5 cm above the top of the T spot is marked for the elution of DHT in each lane. Another 2 X 4 cm section below the origin is used as a blank. The sections are excised with scissors and hung on the needles of the elution units with forceps. About 3.5 ml ethyl acetate is allowed to drip over the ChromAR sections through the needles and syringes, and the eluates are collected in glass culture tubes placed under the funnels.

The eluates are dried under nitrogen in the usual manner. The dried eluates are redissolved in 1.0 ml ethyl acetate and 100 μl is transferred to scintivials for counting and determination of recoveries. The remaining 0.9 ml is evaporated to dryness, again washing down the sides of the tubes to make sure that all of the eluted material is in the bottom of the tubes, and used for the assay.

e. Preparation of Standards

The stock and working solutions of T standards are those used in the assay for combined T and DHT. The standard curves for T and DHT are prepared as previously described using tubes containing the dried eluates of blank ChromAR sections.

f. Assay

The separate assays for T and DHT are essentially the same as the assay for combined T and DHT. The automatic pipette and pipette tips are used in adding the reagents. First 100 μl 1% HSA is added to all tubes in both assays including two control tubes for each assay. In the T assay 100 μl tracer T-1,2-^{3}H is added to all standard, blank, and control tubes. Then, 50 μl tracer is added to the unknown tubes with 50 μl PBSA to yield an equal assay volume. In the DHT assay tracer DHT-1,2-^{3}H is used in the same manner in place of T-1,2-^{3}H. Next, 100 μl of dilute T-3-BSA antibody is added to all tubes in both assays. After incubation as in the assay for combined T and DHT, 100 μl goat anti-rabbit γ-globulin is added to all tubes except the controls which instead receive 100 μl PBSA. The assay is then completed as before.

5. Calculations

The percentage of maximum precipitation of the standard tubes is plotted on semilogarithmic paper resulting in a sigmoid curve or on log-logit paper to linearize the curve. If

c = cpm control tubes
z = cpm zero tubes
s = cpm standard tubes

then the percentage of maximum precipitation of the standard tubes is equal to $(\frac{c-s}{c-z})$ X 100. The maximum precipitation of the unknown tubes must be calculated individually because the tubes contain varying amounts of recovery tracer. If

t = cpm 100 μl tracer
r = cpm recovery tube
u = cpm unknown tubes
m = maximum cpm of unknown tubes
mp = maximum precipitate of unknown tubes

then the fractions of T and DHT recovered in the chromatographic separation are equal to $\frac{10r}{t}$. The maximum cpm of the unknown tubes is equal to $(\frac{150}{400})(9r + \frac{t}{2})$ and the maximum precipitate is equal to $(m)(\frac{c-z}{c})$. The percentage of maximum precipitation is then equal to $(\frac{m-u}{mp})$ 100. The value in ng can then be read from the standard curve and expressed in ng/100 ml serum as follows:

$$\text{ng/100 ml} = (\frac{ng}{4})(\frac{t}{10r})\,100$$

The blank tubes are not used in the calculations, but serve as ChromAR controls.

6. Accuracy and Precision

A pool of female sera was assayed several times giving mean values of 52 ng/100 ml for T and 30 ng/100 ml for DHT. The chromatography does not separate DHT from androstenedione and dehydroepiandrosterone. These steroids, however, have negligible cross-reactivity with our antiserum.

D. Interpretation

The antiserum to testosterone-3-bovine serum albumin that the authors utilize has approximately 53% cross-reaction with dihydrotestosterone in the female testosterone range. Dihydrotestosterone arises by the peripheral conversion of testosterone and by direct secretion.[18] The androstenediols and androst-5-en,3β,17β-diol, and other circulating androgens cross-react to a very insignificant extent. The concentration of dihydrotestosterone is approximately 33%[19] of the circulating testosterone level in normal females, but one study of hirsute patients showed somewhat more variation.[19] Chromatographic separation of testosterone from dihydrotestosterone can be accomplished prior to radioimmunoassy using thin-layer chromatography as described above or by the use of a microcolumn of Al_2O_3 as described by Furuyama and associates.[20] The present authors have found that the required accuracy demands the use of an internal standard with all samples and elution of the supporting medium with the standards. In the clinical practice of evaluating females with hirsutism with or without virilism, the authors have generally performed the assay on unchromatographed serum and have reported it as testosterone plus dihydrotestosterone (T + DHT).[21]

The authors have found an elevated level of T + DHT in 50% of the patients studied with hirsutism. However, many of these individuals may have another abnormal biochemical finding – a decreased level of steroid-binding β-globulin (SBβG).[22,23] This abnormality may explain the finding of a normal serum testosterone level, but an elevated testosterone production rate[24] in certain of these individuals. Such a combination (normal testosterone but decreased SBβG) could result in an increased testosterone metabolic clearance rate.[25] Thus, some evaluation of protein binding of testosterone (measurement of SBβG, calculation of apparent free testosterone index, or 17β-hydroxysteroid index) is of additional help in the evaluation of abnormalities of androgen production. The authors found the assay of the dihydrotestosterone precipitation index (DHT-PI), which is the ratio of SBβG bound DHT-^{3}H to free DHT-^{3}H, as described by Heyns and DeMoor,[26] to be easily performed and quite helpful in their laboratory. Episodic pulsing of testosterone in the human male, apparently unrelated to similar pulsing of gonadotropins of higher magnitude, has been recently reported.[27] Similar pulsing of testosterone and of androstenedione in the female (adrenal and ovarian) is likely, but currently unproven. This could also explain random normal testosterone levels in patients with hirsutism.

The sources of circulating testosterone in the female have been recently reviewed.[28] The peripheral conversion of androstenedione of both adrenal and ovarian origin accounts for 50% of the production rate in the blood. The major part of the balance is most likely from ovarian and adrenal secretion of testosterone. Finally, the peripheral conversion of dihydroepiandrosterone of adrenal origin accounts for 5%. The blood production rate of testosterone in the female is about 0.3 mg daily or about 5% of the male testosterone production rate. The mean peripheral testosterone concentration in the adult female is 40 ng/100 ml, whereas the concentration of T + DHT as performed in the present authors' laboratory is 57 ± 12 ng/100 ml.

The most significant clinical use of serum testosterone determination in the female occurs in the evaluation of female hirsutism with or without virilization. Hair on the upper lip or on the sideburn area is difficult to evaluate. Hair on the chin, neck, lower abdomen, medial aspect of thighs, buttocks, or on the thorax merits investigation. The true virilizing lesions (adrenal tumor, adrenal enzymatic defects, and ovarian tumor) account etiologically for less than 4% of hirsute females in the senior author's clinical experience. The vast majority of females with hirsutism have an unexplained excessive testosterone production of ovarian and/or adrenal origin.[28] Virilization of the external genitalia rarely occurs in these individuals. Although menstrual disturbances with polycystic ovaries frequently occur in these individuals, the finding of polycystic ovaries does not necessarily implicate them as the source of excessive androgen.

A group of 28 patients with hirsutism evaluated in the authors' clinic had a mean T + DHT level of 83 ± 26 ng/100 ml ($p < 0.01$ as compared to controls).[21] One half of the patients had levels two standard deviations or greater above the controls. Androgen suppression with prednisone, 5 mg at bedtime, resulted in a decrease in the level to 68 ± 24 ng/100 ml. The suppression of circulating androgens with glucocorticoids does not necessarily represent a reduction in the adrenal output since there is evidence to support the hypothesis that glucocorticoids produce a systemic rather than organ-specific suppression in androgen production.[29] Testosterone levels with virilizing lesions tend to be higher than in simple hirsutism.

Patients with this problem frequently have decreased SBβG as reflected by a decreased DHT-PI. This is more striking in patients with oligomenorrhea and polycystic ovaries (0.49 ± 0.27)[21] than in hirsute patients with regular menses (0.92 ± 0.34). The normal DHT-PI is 1.22 ± 0.28. This latter biochemical abnormality is not improved with chronic androgen suppression, but is improved by exogenous estrogen .

Data on testosterone and SBβG in pregnancy indicate testosterone levels similar to nonpregnant individuals, but increased SBβG values. Hyperthyroidism also raises SBβG levels. The menopause is associated with decreased SBβG levels but a normal female testosterone concentration.

III. APPENDIX

A. T-3-BSA Preparation

The stimulating antigen is T-3-BSA. The antigen is prepared according to the method of Erlanger et al.[30] with minor modifications by Chen et al.[6] Testosterone-3-oxime is prepared first, using 4.9 g testosterone and 5.3 g carboxymethoxylamine hemihydrochloride (K & K lab #4159) dissolved in a liter of absolute ethanol. This solution is made alkaline by the addition of 100 ml of NaOH (50 g/l. H_2O) and refluxed for 90 min. At this point the reaction mixture is reduced to a small volume by removal of most of the ethanol on a flash evaporator. To the resulting concentrate 350 ml H_2O is added. This is extracted three times with 250 ml diethyl ether, discarding the ether phase. The remaining alkaline aqueous phase is acidified slowly by the addition of concentrated HCl until a precipitate forms. A high yield may be obtained by allowing the precipitation to continue overnight at 4°C. The precipitate is collected by centrifugation, discarding the supernatant solution. The precipitate is extracted three times with 150 ml diethyl ether and the ether phases pooled and washed with 300 ml H_2O. The ether extract is made anhydrous with Na_2SO_4 and dried under a stream of nitrogen in a warm water bath. The crude testosterone-3-oxime (T-3-oxime) powder is then recrystallized from benzene using ligroin as a precipitating solvent. The resulting T-3-oxime has a m.p. of 179 to 181°C with a UV absorption maximum at 248 nm in 0.05 *M* Tris buffer (pH 8.6).

For the preparation of T-3-BSA, 1.1 g of T-3-oxime and 0.75 ml of *N*-tributylamine are dissolved in 30 ml of 1,4-dioxane and cooled to 10°C. To this 0.40 ml isobutyl chlorocarbonate is added. The mixture is allowed to react for 20 min at 4°C (cracked ice and water bath). This is then added to a cooled solution of 4.2 g bovine serum albumin in 200 ml dioxane:water 1:1 (v/v). Then 4.2 ml of 1 *N* NaOH is added with stirring (magnetic stirring bar); gas is evolved. The reaction mixture is kept cool (4 to 10°C) and after 1 hr an additional 2 ml of 1 *N* NaOH is added. The reaction is allowed to continue for an additional 2 hr with stirring and cooling. The reaction mixture is then dialyzed overnight against running water. To the impermeate, 1 *N* HCL is added (several drops) to bring the pH to 4.5 (indicator paper). A precipitate begins to form and is allowed to continue for 24 hr at 4°C. The precipitate is collected by centrifugation in a refrigerated centrifuge. The pellet is suspended in 100 ml water and then made soluble by the addition of 5% aqueous $NaHCO_3$. The clear solution is then lyophilized, yielding T-3-BSA. T-3-BSA produced in the authors' laboratories has a maximum absorption at 240 to 250 nm against BSA as a blank. Calculations based on absorption indicated that this preparation has about 27 steroid residues per protein molecule.

B. T-3-BSA Antibody Preparation

First, 10 mg of T-3-BSA is dissolved in 10 ml physiological saline which is then mixed with 10 ml Freund's complete adjuvant. Female rabbits are immunized in the following fashion:

Day	1	10	20*	30	37	44	51	58	65*	79	93*	107	121*	135
Ag (mg)	1	1	1	1	0.5	0.5	0.5	0.5	0.5	0.5	0.5	0.5	0.5	0.5
Vol (ml)	2	2	2	2	1	1	1	1	1	1	1	1	1	1

*Blood drawn and antibody formation determined. After day 135 the animals are injected monthly with 0.5 mg in 1 ml as boosters.

The serum is drawn from the animals every month and stored frozen. The antibody may be used as it is or purified with Rivanol® (6,9-diamino-2-ethoxyacridinelactate) which renders the albumins insoluble. Next, 1 ml of undiluted serum is added to 4 ml of 0.4% Rivanol, mixed, allowed to stand for 15 min, and then centrifuged. The supernatant solution is transferred to another centrifuge tube and 100 mg Norit A charcoal added and mixed. This is centrifuged after 15 min. The supernatant solution is treated with charcoal in the same manner at least two more times or until it is clear. The antibody is now at a 1:4 aqueous dilution and may be stored as frozen aliquots at −20°C (up to at least 1 year) until needed for the assay.

REFERENCES

1. **Korenman, S. G. and Lipsett, M. B.,** *J. Clin. Invest.,* 43, 2125, 1964.
2. **Ekstrom, B., Vorys, N., Neri, A., Taylor, J. N., and Wieland, R. G.,** *Am. J. Med. Sci.,* 255, 202, 1968.
3. **Rivarola, M. A. and Migeon, C. J.,** *Steroids,* 7, 193, 1966.
4. **Wotiz, H. H. and Clark, G. J.,** *Meth. Biochem. Anal.,* 18, 357, 1970.
5. **Hallberg, M. C., Zorn, E. M., and Wieland, R. G.,** *Steroids,* 12, 241, 1968.
6. **Chen, J. C., Zorn, E. M., Hallberg, M. C., and Wieland, R. G.,** *Clin. Chem.,* 17, 581, 1971.
7. **Wieland, R. G., Hallberg, M. C., Zorn, E. M., Klein, D. E., and Luria, S. L.,** *Fertil. Steril.,* 23, 779, 1972.
8. **Yalow, R. S. and Berson, S. A.,** *Diabetes,* 9, 254, 1960.
9. **Grodsky, G. M. and Forsham, P. H.,** *J. Clin. Invest.,* 39, 1070, 1960.
10. **Law, K. S., Gottlieb, C. W., and Herbert, V.,** *Proc. Soc. Exp. Biol. Med.,* 123, 126, 1966.
11. **Mayes, D. and Nugent, C. A.,** *J. Clin. Endocrinol.,* 28, 1169, 1968.
12. **Wide, L. and Porath, J.,** *Biochim. Biophys. Acta,* 130, 257, 1966.
13. **Isojima, S., Naka, O., Koyama, K., and Adachi, H.,** *J. Clin. Endocrinol.,* 31, 693, 1970.
14. **Skom, J. H. and Talmage, D. W.,** *J. Clin. Invest.,* 37, 783, 1958.
15. **Potts, J. T., Jr., Sherwood, L. M., O'Riordan, J. L. H., and Aurbach, G. D.,** *Adv. Intern. Med.,* 13, 183, 1967.
16. **Rodbard, P., Bridson, W., and Rayford, P. L.,** *J. Lab. Clin. Med.,* 74, 770, 1969.
17. **Demetriou, J. A. and Austin, F. G.,** *Clin. Chem.,* 16, 111, 1970.
18. **Wieland, R. G., Chen, J. C., Zorn, E. M., Webster, K. D., and Tang, P. H.,** *Prog. Centr. Soc. Clin. Res.,* 44, 45, 1971.
19. **Ito, T. and Horton, R.,** *J. Clin. Endocrinol.,* 31, 362, 1970.
20. **Furuyama, G., Mayes, D. M., and Nugent, C. A.,** *Steroids,* 16, 415, 1970.
21. **Wieland, R. G., Zorn, E. M., and Hallberg, M. C.,** *Am. J. Obstet. Gynecol.,* 117, 983, 1973.
22. **Vermeulen, A., Verdonck, L., Van der Straeten, M., and Arie, N.,** *J. Clin. Endocrinol.,* 29, 1470, 1969.
23. **Rosenfield, R. L., Ehrlich, E. N., and Cleary, R. E.,** *J. Clin. Endocrinol.,* 34, 92, 1972.
24. **Bardin, C. W. and Lipsett, M. B.,** *J. Clin. Invest.,* 46, 891, 1967.
25. **Corvol, P. and Bardin, C. W.,** *Biol. Reprod.,* 8, 277, 1973.
26. **Heyns, W. and DeMoor, P.,** *Steroids,* 18, 709, 1971.
27. **Wieland, R. G., Hallberg, M. C., Koepke, K. A., and Zorn, E. M.,** *Fertil. Steril.,* 24, 644, 1973.
28. **Kirschner, M. A. and Bardin, C. W.,** *Metabolism,* 21, 667, 1972.
29. **Kirschner, M. A. and Jacobs, J. B.,** *J. Clin. Endocrinol.,* 33, 199, 1971.
30. **Erlanger, B. F., Borek, F., Bieser, S. M., and Lieberman, S.,** *J. Biol. Chem.,* 228, 713, 1957.

HORMONES AND BREAST CANCER

D. Y. Wang and M. C. Swain

TABLE OF CONTENTS

I. Introduction . . . 192

II. Urinary 11-Deoxy-17-ketosteroids (11-DKS) and 17-Hydroxycorticoids (17-OHCS) . . . 193
- A. Advanced Breast Cancer . . . 193
- B. Early Breast Cancer . . . 193
- C. Preclinical Phase of Breast Cancer . . . 193
- D. Steroid Excretion as Parameters for Predicting Response to Treatment of Women with Advanced Breast Cancer . . . 194
- E. Steroid Assays and Prognosis of Women with Early Breast Cancer . . . 196
- F. Method for the Measurement of Urinary Androsterone, Etiocholanolone, Dehydroepiandrosterone, and 11-Deoxy-17-ketosteroids . . . 196
- G. Method for the Determination of Urinary 17-Hydroxycorticoids . . . 198

III. 11-Deoxy-17-ketosteroids in Plasma . . . 199
- A. The Established Disease . . . 199
- B. Studies on the Production Rate of 11-Deoxy-17-ketosteroids . . . 199
- C. Plasma 11-Deoxy-17-ketosteroids and Etiology of Breast Cancer . . . 200
- D. Method for the Measurement of Plasma Dehydroepiandrosterone Sulfate and Androsterone Sulfate . . . 202

IV. Plasma 17-Hydroxycorticosteroids . . . 203
- A. Method for the Estimation of Plasma Cortisol . . . 204

V. Estrogens and Progesterone . . . 204
- A. Introduction . . . 204
- B. Early Breast Cancer . . . 206
- C. Advanced Breast Cancer . . . 206
- D. The Preclinical Phase of Breast Cancer . . . 206
- E. Estrogens and the Prediction of Response to Endocrine Therapy . . . 207
- F. The Assay of Estradiol and Estrone in Plasma . . . 208
- G. The Assay of Progesterone in Plasma . . . 208

VI. Binding of Steroid Hormones to Plasma Proteins . . . 209
- A. Method for Determining Plasma Protein Binding . . . 209

VII. Protein Hormones . . . 210
- A. Gonadotrophins . . . 210
- B. Growth Hormone and Prolactin . . . 211
- C. Thyroid Hormones . . . 212

VIII. Conclusion . . . 213

Acknowledgments . . . 213

References . . . 213

I. INTRODUCTION

There is a large body of literature describing the close involvement of hormones in the development of the normal breast. Both protein and steroid hormones play an intimate role in the growth and lactational activity of this organ.[1-7] It is logical, therefore, to consider that hormones may play as important a part in the etiology of malignant growth as they do in normal growth.

The estrogens were the first hormones to be recognized as being involved in breast cancer. Beatson[8] in 1896 demonstrated that the surgical removal of ovaries from two premenopausal women with advanced breast cancer produced a marked regression of the tumors. Results obtained by later workers[9] showing that the development of breast tumors in some strains of mice could be prevented by ovariectomy of the immature animal suggested that there might be common factors in the etiology of breast cancer in man and in the mouse. Laccasagne[10] showed that in a strain of mice in which the female developed spontaneous mammary tumors, the administration of estrogens to the males produced tumors in the mammary system. This work suggested that estrogens acting alone or in conjunction with other natural compounds in the body could be carcinogenic. Although estrogens have been shown to be carcinogenic in certain strains of rats and mice they do not appear to have this effect in all animals.[11]

There is no conclusive evidence of a carcinogenic action of progesterone alone although it may act as a co-carcinogen with other compounds. The induction of mammary tumors in rats by progesterone and estrogens has been demonstrated,[12] but the data are somewhat equivocal in the case of tumor induction by these compounds in mice.[13-15] However, Boot[16] has reported that progesterone may act as a co-carcinogen with prolactin in mice.

Pituitary isografts have been found to induce mammary tumors in experimental animals.[16,17] Such isografts secrete mainly prolactin and a minor amount of growth hormone because the pituitary tissue is no longer under the regulatory control of the hypothalamus. Normally prolactin release is regulated in a negative sense by the hypothalamus through its secretion of Prolactin Inhibitory Factor, while growth hormone is regulated by the trophic hormone Growth Hormone Releasing Factor. The carcinogenic activity of prolactin was confirmed by the administration of this hormone to mice that subsequently developed breast tumors. Prolactin was also found to be necessary for the growth of DMBA*-induced mammary tumors in the Sprague Dawley rat in a series of elegant experiments by Pearson et al.[18] Although the present evidence strongly implicates prolactin in the etiology of breast tumors in rodents, the data on growth hormone are equivocal. This is surprising since growth hormone has a high intrinsic lactogenic and mammogenic activity.

The interest in androgens arose from the possibility that they could act as antitumor agents, because androgens are antiestrogenic and estrogens are carcinogenic. The administration of androgens to patients with advanced breast cancer can bring about a remission in the disease but the remission is not permanent and only about 25% of the patients benefit from such treatment.[19-21] Furthermore, surgical removal of the adrenal glands and the concomitant loss of adrenal androgen secretion can also bring about a remission in about a quarter of the patients so treated.[22-24] More recently, measurements of urinary androgen metabolites have produced interesting reports claiming that the amount of urinary 11-deoxy-17-ketosteroids is related to the etiology and clinical course of the disease.[25-33]

This brief introduction shows that hormones are intimately involved in the etiology and biological behavior of breast cancer in man and in laboratory animals. In the remainder of this chapter we will describe in detail current work concerning the etiology and clinical course of the disease in man.

For most of the hormones mentioned, the method of their estimation has been set out at the end of the appropriate section. These assay methods are the ones presently employed in our laboratory and have been described to show the type of methodology used to obtain results in this field. No description of the radioimmunoassay of protein hormones is given; these assays are thoroughly dealt with by other authors in this book.

*7,12-dimethylbenz(a)anthracene

II. URINARY 11–DEOXY-17-KETOSTEROIDS (11-DKS) AND 17-HYDROXYCORTICOIDS (17-OHCS)

Of all the hormones studied in women with breast cancer those most extensively investigated have been the urinary metabolites of androgen and corticoid steroids. These have been found to be related to the etiology and clinical course of the disease.

A. Advanced Breast Cancer

There is general agreement that the excretion of 11-DKS in women with advanced breast cancer is lower than that for normal women.[34–37] Cameron and his co-workers have reported no difference in the excretion of urinary etiocholanolone or androsterone in women whose disease was generally disseminated, but they did find subnormal excretion of urinary androgen metabolites in women with locally advanced breast cancer.[38]

B. Early Breast Cancer

There has been some controversy concerning whether women with early breast cancer (at the time of mastectomy) excrete subnormal amounts of urinary 11-DKS. Bulbrook and his colleagues[39] have reported that such patients ten days after mastectomy excreted subnormal amounts of urinary androsterone and etiocholanolone. The excretion of 17-OHCS was slightly, although not significantly, raised. The subnormal excretion of 11-DKS has been confirmed by several workers,[34,40,41] but Cameron and his associates[38] could find no difference between the excretion of urinary androsterone and etiocholanolone in patients with early breast cancer and normal women. The confusion probably arises because of differences in urine collection, since Bulbrook et al. studied urine from women 10 days after mastectomy, whereas Cameron et al. studied patients before mastectomy. Data obtained in our laboratory indicate that the levels of urinary 11-DKS are significantly higher 1 day before mastectomy than 10 days after mastectomy (unpublished results).

The reasons for this lowered excretion of urinary 11–DKS after mastectomy are not known. However, it should be borne in mind that from the moment a diagnosis of breast cancer is made, a number of uncontrolled variables affect the results of any investigation made on the endocrine status of such patients. Among these are emotional stresses caused by diagnosis and entry to hospital, effects of drugs, surgery, and illness.[42–47]

C. Preclinical Phase of Breast Cancer

Because of the abnormal excretion of these metabolites in the established disease a prospective study was set up in 1961 to test the hypothesis that these abnormalities preceded the clinical onset of the disease. Urine was collected from 5,000 ostensibly normal, healthy women living on the island of Guernsey, and the excretion of urinary 11-DKS and 17-OHCS of women who subsequently developed breast cancer compared with that of normal controls. The first results of this study, published in 1967,[48] showed that women who developed breast cancer tended to excrete subnormal amounts of urinary etiocholanolone when compared with controls matched for age, height, weight, menstrual cycle, and parity. If the excretion of urinary 17-OHCS was taken into account, the women who subsequently developed breast cancer appeared to have steroid abnormalities which were "multi-directional." This implied that several patterns of abnormal steroid excretion were associated with a high risk of breast cancer.

A later report[49] based on 27 cases of subsequent breast cancer and 1,506 controls was published in 1971. This showed that subnormal excretion of urinary androsterone and etiocholanolone was associated with a high risk of breast cancer, and that women excreting less than 800 μg of etiocholanolone/24 hr had approximately five times the risk of developing breast cancer compared to those excreting between 1 and 2 mg/24 hr (Figure 1).

The average excretion of etiocholanolone was subnormal in women who subsequently developed breast cancer between 35 and 55 years of age. The abnormality was independent of the time before diagnosis and was not a late event in the history of the disease since subnormal excretion was observed up to 9 years before clinical diagnosis. It was also observed that several women who subsequently developed breast cancer excreted more than 2 mg of etiocholanolone/24 hr, which is in keeping with the concept of multidirectional abnormalities proposed in the earlier report.[48] The urinary excretion of 17-OHCS alone does not

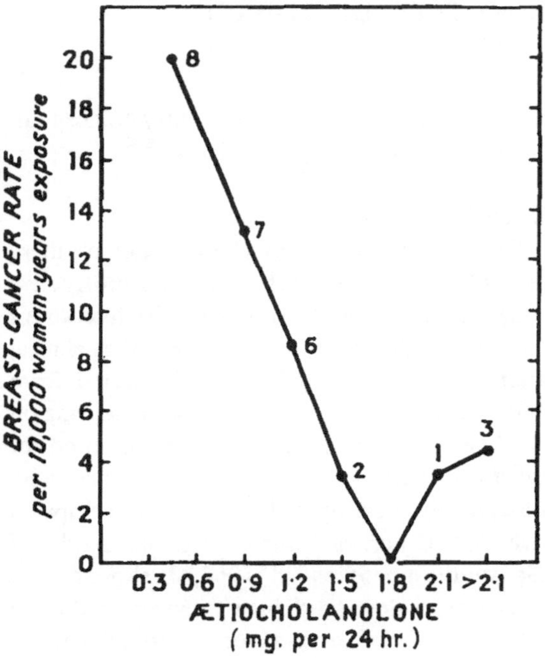

FIGURE 1. Relationship between the subsequent breast cancer rate and urinary etiocholanolone excretion. The figures shown on the curve refer to the number of subjects who subsequently developed breast cancer. (From Bulbrook, R. D., Hayward, J. L., and Spicer, C. C., *Lancet*, ii, 395, 1971. With permission.)

appear to be related to risk of developing breast cancer.

It has been shown that annual examination by palpation and mammography is effective in detecting breast cancer at an early stage (i.e., where there is no evidence of metastases). Furthermore, women in this study whose tumors had been detected by palpation or mammography had a better rate of survival than a control population.[50,51] Although there has been criticism[52] of the method of analysis of the data, the number of deaths in the screened group of patients was 30% lower than in a nonscreened group at 5 years. To provide such a screening service on a national basis would mean incurring heavy expenditure; the advantage in reducing costs is obvious if only women with a high risk of breast cancer are examined.

In view of this evidence, our laboratory hopes to investigate the feasibility of using the levels of urinary 11-DKS in conjunction with breast examination by palpation and mammography in the early detection of the disease. Also, it might be possible to identify those women with fast-growing tumors that become apparent between visits to the screening clinic.[50,51]

If measurement of urinary 11-DKS levels in women with high risk and large-scale screening of populations are envisaged, then much quicker and simpler methods of analysis will be needed. The reader is referred to the papers of Moore[53,54] in which a semiautomated method for measuring urinary 11-DKS is outlined.

D. Steroid Excretion as Parameters for Predicting Response to Treatment of Women with Advanced Breast Cancer

The only surgical treatment that may be effective in metastatic breast cancer is either adrenalectomy with oophorectomy or hypophysectomy. However, only about 25% will benefit from such surgery[22,55-57] and it was realized early in the history of this treatment that some means of predicting the response to endocrine ablation was desirable. Such a predictor would obviate unnecessary suffering for those patients unlikely to benefit from these forms of endocrine surgery.

In 1957 Allen, Hayward, and Merivale reported[58] on the urinary excretion of 11-oxy-17-ketosteroids and 11-deoxy-17-ketosteroids of women undergoing either adrenalectomy with oophorectomy or hypophysectomy. They noted that if the amounts of these metabolites were expressed as a simple ratio, those patients that responded to treatment had a ratio of greater than one and resembled the ratio found in normal women. Conversely, those patients who had a ratio of less than one tended not to respond. Although the methodology of these workers was justifiably criticized and not confirmed by two sets of workers[59,60] it suggested an approach to the problem of finding a predictor of response.

In subsequent papers by Bulbrook, Greenwood, and Hayward,[61,62] using more reliable methods of analysis, it was reported that the mean urinary 17-OHCS level was higher and mean etiocholanolone concentration lower in the group of women who failed to respond to adrenalectomy or hypophysectomy compared to those women who responded. These quantities of urinary 17-OHCS and etiocholanolone were then used to obtain a discriminant function, the formula of which was

$$80 - 80 \times 17\text{-OHCS (mg/24 hr)} + \text{etiocholanolone } (\mu g/24 \text{ hr})$$

This function was so designed that women in this

series who failed to respond tended to have a discriminant function which was negative while those that responded tended to have a positive discriminant function. The discriminant function was calculated using the data from 41 patients.[62] An inspection of the formula of this function shows that the major contribution is made by the amount of etiocholanolone excreted, high etiocholanolone levels giving positive values and low levels giving negative values. In any investigation of this type it is important that the method of prediction be evaluated in a prospective study. Hayward and Bulbrook in 1968 reported[63] their results on a subsequent series of 151 patients and showed that the claims of their earlier papers were essentially substantiated. The success rate among the 81 patients with positive discriminants was more than twice that found for those with negative discriminants. Nevertheless, it should be borne in mind that only 35% of patients with positive discriminants responded to endocrine ablation while if the decision to operate had rested solely upon the value of the discriminant function, 16% of those patients with negative discriminants would have been denied an objective remission. Thus, in practical terms, the use of the discriminant function alone as a routine guide as to whether to operate is far from perfect.

However, if the previous history of the patient is taken into account, the discriminant function in conjunction with menopausal status and length of time between mastectomy and recurrence of the disease (free period)[63] was found to be of value in the management of patients with advanced breast cancer. In a group of women with negative discriminants, with a free period of less than 2 years or within 6 years of the menopause, the response rate was only 5%. Hayward and Bulbrook[63,64] suggest, therefore, that on the basis of these figures such patients should be rejected for ablative surgery. Since a third of their patients fell within this category it was felt that the discriminant function together with these variables should prove a useful aid in the management of the advanced disease.

The measurement of etiocholanolone presents difficulties for most routine laboratories, and the report of Miller and his colleagues[65] should be noted as a major contribution in simplifying the measurement of the discriminant function. These workers have shown that a simple ratio of urinary 11-DKS to 17-OHCS was directly related to the discriminant function. Patients having ratios of greater than 0.16 had positive discriminants while ratios of less than 0.13 had negative discriminants. Subsequently, Thomas et al.[66,67] found that the success rate to endocrine ablation of patients with ratios above 0.17 was comparable to those with positive discriminants while ratios of 0.17 or less had a rate near that of those with negative discriminants.* Since the discovery of Few[68] that urinary 11-DKS can be separated from the 11-oxy-17-ketosteroids by a simple partition with isopentane, the measurement of the ratio should prove within the scope of most laboratories.

The predictive value of urinary 17-OHCS and 11-DKS has been confirmed by other workers. Juret et al.[69] reported that response to implantation of ^{90}Y in the pituitary was correlated with the amount of androsterone and etiocholanolone excreted. Kumaoka et al.[37] have shown that the response of Japanese patients to adrenalectomy is related to 11-DKS excretion and have used assay values to select patients, thereby obtaining not only a substantial but a sustained increase in remission rates. Fotherby et al.[70] have examined the relationship between urinary steroid excretion and response to treatment and reported essentially similar results. In America, Wilson and Moore[71] have used a discriminant function based on urinary 17-OHCS and 11-DKS after ACTH stimulation and the length of the free period. They claim that they were able to predict with 80% certainty those patients that would or would not respond to adrenalectomy. Sarfaty et al.[72] in Australia have reported that the excretion of urinary 11-DKS or etiocholanolone glucuronide was significantly lower in women who did not respond to adrenalectomy compared to those that did respond. Ghosh et al.[73] recently reported that the excretion of urinary corticosteroid sulfates was significantly above normal in women with metastatic breast cancer. These authors intend to investigate whether the measurement of urinary corticoids can be used as a parameter of prognosis.

Although there is general agreement that urinary steroid excretion can be used as a predictor of response, it should be noted that there are dissentors.[59,74,75]

*The difference between the ratios is due to the fact that Miller et al. expressed urinary 17-OHCS as cortisol whereas Thomas et al. expressed their results in terms of dehydroepiandrosterone.

E. Steroid Assays and Prognosis of Women with Early Breast Cancer

Since the discriminant function was able to predict with a certain degree of accuracy the response to ablative surgery it was natural to investigate what value it might have as a parameter of prognosis in the early disease. In 1964, Bulbrook et al.[76] reported their results on an 8-year follow-up of women whose discriminant function had been measured after mastectomy. This survey showed that approximately half the patients with early breast cancer had negative discriminants when measured 10 days after mastectomy. Patients with negative discriminants had 3 times the rate of recurrence and mortality at 8 years compared with those patients who had positive discriminants. Thus, of all the chemical measurements, this would seem at present to be the best in determining prognosis in the early disease (Figure 2).

F. Method for the Measurement of Urinary Androsterone, Etiocholanolone, Dehydroepiandrosterone, and 11-Deoxy-17-ketosteroids

Since the early report in 1957 by Allen et al.[58] of the predictive value of urinary steroid measurements, the growth of literature in this field has taken place concurrently with advances in methodology. Consequently, methods which were used in earlier studies on urinary 11-DKS and 17-OHCS have been modified or superceded by others. Because of these alterations over the years it is desirable to give a brief historical outline of these developments.

In 1957 Kellie and Wade[77] described their method for measuring urinary androsterone, etiocholanolone, and dehydroepiandrosterone by separation of the hydrolyzed urinary neutral steroid conjugates on alumina columns. This method was used in our own laboratory for the early work on the discriminant function.

The advent of gas-liquid chromatography and derivatives of androsterone, etiocholanolone, and dehydroepiandrosterone suitable for chromatography led to the development of more rapid methods for their estimations.[78-82] Our present method for measuring these metabolites is as follows. The 24-hr urine sample is collected without preservative and the pH adjusted to 6.0 to 7.0 with 50% acetic acid or 5 *N* sodium hydroxide. The volume is made up to 2 l with distilled water. The urine can be stored at −20°C. It is important that if stored urine is used it should be thoroughly mixed after being thawed.

1. Extraction

Urine aliquots (20 ml) in duplicate are put into glass-stoppered tubes containing ammonium sulfate (10 g) and the salt allowed to dissolve. A

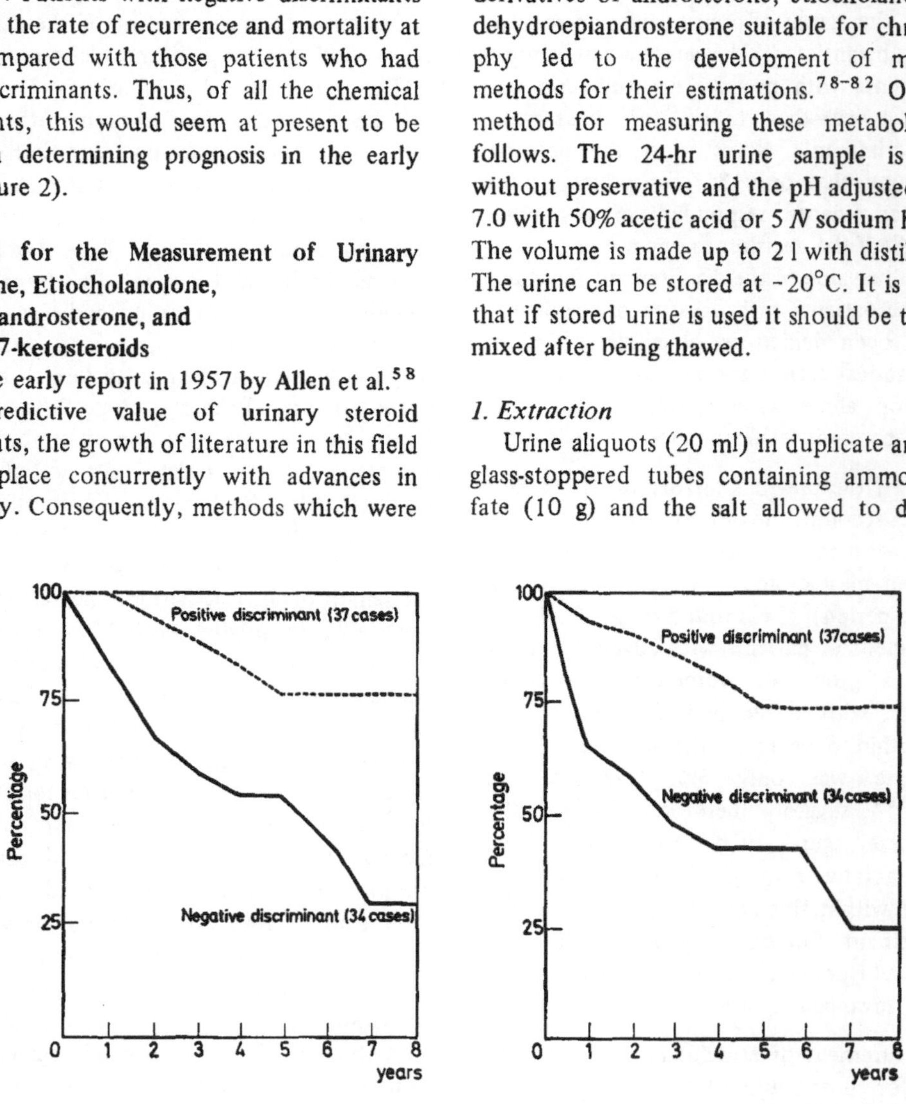

FIGURE 2. Relationship between the discriminant function and the rate of recurrence and survival. Figure 2a (left) shows the proportion free from recurrence and Figure 2b (right) the proportion surviving up to 8 years after mastectomy in patients with positive and negative discriminants. (From Hayward, J. L., *Recent Results in Cancer Research*, Springer-Verlag, Berlin, 1970, 24. With permission.)

mixture (2:1, v/v) of chloroform : *n*-propanol (20 ml) is added, and the tubes shaken for 2 min and centrifuged at 600 g for 10 min. The upper aqueous layer and interface are removed and anhydrous sodium sulfate (0.5 g) is added to the organic layer. The tubes are shaken for 2 min and the sodium sulfate allowed to settle. The clear organic phase is pipetted (10 ml) into glass-stoppered tubes, placed into a water bath at 45°C, and the solvent evaporated with a flow of nitrogen.

2. Hydrolysis

The steroid glucuronides are hydrolyzed enzymically by adding sodium acetate-acetic acid buffer (2 ml, 0.5 *M*, pH 4.0) containing β-glucuronidase (10,000 units) prepared from patella vulgata. The tubes are then incubated in a water bath at 62°C for 90 min. After incubation sodium chloride (2 g), water (8 ml), and 4 *N* sulfuric acid (0.2 ml) are added, the tubes shaken, and their contents transferred to 25-ml separating funnels. The emptied tubes are rinsed with ethyl acetate (10 ml) and, after shaking, the ethyl acetate is transferred to the appropriate separating funnel. The enzymically hydrolyzed steroids and sulfate conjugates are extracted into the ethyl acetate and the organic phase transferred to 25-ml glass-stoppered tubes, care being taken to avoid any contamination with the aqueous phase. These tubes are then incubated at 62°C for 1 hr to hydrolyze the sulfate conjugates.

After cooling, heptane (20 ml) is added and the contents of the tubes transferred to 50-ml separating funnels, washed twice with 1 *N* sodium hydroxide (3 ml), and twice with water (3 ml). The organic phase is evaporated to dryness under air at 45°C in 50-ml B25 glass-stoppered tubes.

3. Extraction of 11-DKS Fraction

The steroids are dissolved in a drop of ethanol and water (2 ml) and a drop of 1 *N* sodium hydroxide added. After shaking, this is transferred to a 25-ml separating funnel. The emptied tubes are rinsed with isopentane (10 ml), which is also transferred to the separating funnel and shaken for 2 min. The isopentane layer, which contains the 11-DKS, is transferred to 20-ml B14 glass-stoppered tubes and evaporated to dryness under nitrogen at 30°C.

To the dried tube and standard tube containing 40 μg of dehydroepiandrosterone, Zimmerman reagent is added, and optical density measured at 440, 520, and 600 nm. After correcting the 520-nm optical density using the Allen correction (viz., corrected $E_{520} = E_{520} - \frac{1}{2}[E_{440} + E_{600}]$), the quantity of urinary 11-DKS is calculated as follows:

$$\text{11-DKS (mg/24 hr)} = \frac{\text{corrected E test}}{\text{corrected E standard}} \times 40 \times \frac{2{,}000}{20} \times \frac{1}{1{,}000}$$

The results are expressed as mg/24 hr using dehydroepiandrosterone as a standard.

For a thorough understanding of the principle and practice of the Zimmerman reaction the reader is referred to Chapter 5.

4. Measurement of Androsterone, Etiocholanolone, and Dehydroepiandrosterone by Gas-liquid Chromatography

Three urine aliquots (20 ml) are taken, and 2 of these are processed by the method described above and total 11-DKS determined. The third aliquot is processed as above but instead of measuring the total 11-DKS with Zimmerman reagent, the steroids in this tube are converted to their chloromethyldimethylsilyl ether derivatives and chromatographed on a gas-liquid chromatogram, which is achieved in the following manner.

5. Preparation of the Chloromethyldimethylsilyl Ether Derivatives

Hexane (1 ml) is added to 2 tubes after which diethylamine (0.1 ml) is put into 1 tube and chloromethyldimethylchlorosilane (0.2 ml) is added to the other. The contents of the tubes are mixed, centrifuged (600 g for 5 min), and 0.1 ml of the supernatant added to each sample tube. Included in each batch is one tube containing 100 μg each of androsterone, etiocholanolone, and dehydroepiandrosterone. The tubes are stoppered and sealed with Parafilm® and incubated at 37°C for 30 min. After incubation the tubes are evaporated to dryness under a flow of nitrogen, and hexane (0.1 ml) added to each tube to dissolve the steroid derivatives. After centrifugation (at 600 g for 5 min) the hexane layer is removed and 5 μl run on a gas-liquid chromatogram. A 5-ft (i.d. 4 mm) glass column is used for chromatography, packed with "Supasorb" (British Drug Houses, Poole, England) (100-120 mesh), and coated with 1.5% XE60 as stationary phase. Temperature is 215°C with a flow of nitrogen of 60 ml/min. The areas of the peaks are measured by triangulation.

6. Calculation

The amounts of androsterone, etiocholanolone, and dehydroepiandrosterone are calculated using (1) the peak areas, (2) the total amount of 11-DKS present measured by Zimmerman, and (3) the relative chromogenicity of the Zimmerman derivatives of androsterone, etiocholanolone, and dehydroepiandrosterone.[81,82] From the standard, factors F_A, F_E, and F_D are obtained by dividing each peak area by the peak area of etiocholanolone (by definition $F_E = 1.00$). If the peak areas of androsterone, etiocholanolone, and dehydroepiandrosterone for the unknown are A_A, A_E, and A_D, respectively, then the corrected area for response of the detector is A_AF_A, A_EF_E, and A_DF_D. The sum of these corrected areas is S, i.e.,

$$S = A_AF_A + A_EF_E + A_DF_D$$

and the ratios

$$\frac{A_AF_A}{S} \text{ and } \frac{A_DF_D}{S}$$

calculated.

Since the total 11-DKS (T) is known, then the quantity of androsterone $= \frac{A_AF_A}{S} \times T$ and the quantity of dehydroepiandrosterone $= \frac{A_DF_D}{S} \times T$.

Because the color response of etiocholanolone in the Zimmerman reaction is different from dehydroepiandrosterone and androsterone, the amount of this steroid present needs correction and the amount of etiocholanolone $= \frac{A_EF_E \times T}{S + 0.12\,A_E}$.

7. Example

F_A, F_E, and F_D = 0.90, 1.00, and 1.10, respectively. Peak areas of unknown sample A_A, A_E, and A_D = 1.0, 2.0, and 3.0, respectively. The corrected areas A_AF_A, A_EF_E, and A_DF_D are 0.9, 2.0, and 3.3, respectively. $S = A_AF_A + A_EF_E + A_DF_D = 6.2$ and the ratios $\frac{A_AF_A}{S} = \frac{0.9}{6.2} = 0.145$, $\frac{A_DF_D}{S} = \frac{3.3}{6.2} = 0.532$

If total 11-DKS (T) is 10 mg/24 hr, then:

Quantity of androsterone = 0.145 x 10 = 1.45 mg/24 hr
Quantity of dehydroepiandrosterone = 0.532 x 10 = 5.32 mg/24 hr
Quantity of etiocholanolone $= \frac{2.0 \times 10}{6.2 + 0.12} \times 20$ = 3.11 mg/24 hr

G. Method for the Determination of Urinary 17-Hydroxycorticoids

The 17-OHCS were originally determined using a method based on that described by Few.[83] More recently the suggestions of Metcalf[84] that higher temperatures be used for the reduction and oxidation steps have been adopted, thereby shortening the time taken to estimate urinary 17-OHCS.

A 24-hr urine specimen is collected and treated in exactly the same way as described previously for the estimation of 11-DKS. Two aliquots (5 ml) of urine are pipetted into glass-stoppered tubes and 0.5 ml of freshly prepared 10% (w/v) sodium borohydride in 0.1 *N* sodium hydroxide added. The tubes are placed into a water bath at 55°C for 15 min, then 25% acetic acid (0.25 ml) is added and the tubes left at 55°C for a further 15 min.

To each tube is added freshly prepared 10% (w/v) sodium periodate (2 ml) in distilled water,[85] the pH is adjusted to 6.7 using narrow-range indicator paper, and the tubes are placed into a water bath at 55°C for 15 min. Sodium hydroxide (5 *N*, 0.25 ml) is added and the tubes incubated at 37°C for 15 min, after which they are cooled. Redistilled ethylenedichloride (10 ml) is added, the tubes are shaken for 15 min to extract the steroids, and centrifuged at 600 g for 10 min, and the top layer discarded by suction. To each tube is then added freshly prepared 5% (w/v) sodium dithionite (2.5 ml) in 2 *N* sodium hydroxide and the tube shaken for 2 min. After centrifugation at 600 g for 10 min, the top layer is removed by suction and the ethylenedichloride washed with water (2.5 ml). After further centrifugation and removal of the water layer the ethylenedichloride is filtered through a Whatman® No. 4 filter paper 10.5 cm in diameter. 5 ml of the extract is placed into a 15-ml B14 glass-stoppered tube, a bumping granule added, and the extract boiled to dryness in a boiling water bath.

The amount of 17-OHCS in each tube is then determined colorimetrically using the Zimmerman reaction in exactly the way outlined previously. The results are calculated using the formula:

$$17\text{-OHCS (mg/24 hr)} = \frac{\text{corrected E test}}{\text{corrected E standard}} \times 40 \times \frac{2{,}000}{2.5} \times \frac{1}{1{,}000}$$

III. 11-DEOXY-17-KETOSTEROIDS IN PLASMA

A. The Established Disease

Because of the interest aroused by the reports of abnormal excretion of urinary 11-deoxy-17-ketosteroids in women with breast cancer it was natural that investigations should be made to determine whether these abnormalities were reflected in the blood concentration of 11-DKS. Such investigations have shown a general agreement that the levels of blood 11-DKS in patients with advanced breast cancer are subnormal.

Deshpande et al.[86] found a tendency for plasma 11-DKS to be lower in these patients, compared to normal women, although the difference was not statistically significant.

More recent studies on the major component of blood 11-DKS, namely dehydroepiandrosterone sulfate (DS), have shown that the concentration of this steroid is reduced in patients with advanced breast cancer.[87] Investigations in our laboratory have revealed that not only is the level of DS subnormal, but that the concentration of plasma androsterone sulfate (AS) is also significantly decreased in women with advanced breast cancer.[88,89]

Bénard and his colleagues[90] have reported greatly increased concentrations of unconjugated 11-DKS in over half the subjects with breast cancer studied. This finding was not confirmed in our laboratory[91] and in view of the very large production rates which would be required to maintain such high concentrations of unconjugated 11-DKS, the results of Bénard et al. must be viewed with caution.

There is no general agreement as to the levels of plasma 11-DKS in women with early breast cancer. Deshpande et al.[86] reported that the level of total plasma 11-DKS in these patients was not significantly different from that found in normal women. Brownsey et al.,[87] measuring plasma DS, reported similar findings.

These conclusions are unexpected since urinary 11-DKS are significantly depressed in the early disease[39] and plasma and urinary 11-DKS levels are significantly correlated.[86] The discrepancy seems to be due to the type of patients studied since both Brownsey et al.[87] and Deshpande et al.[86] studied blood taken from women before mastectomy, while Bulbrook et al.[39] investigated urine specimens taken from women after mastectomy.

We have found that the concentrations of plasma DS and AS are significantly reduced after mastectomy. The results of this study (Figure 3) show that the concentrations of plasma DS and AS are significantly subnormal after mastectomy, whereas the plasma DS and AS before surgery are not. Also, we have observed a highly significant depression in urinary 11-DKS after mastectomy (unpublished results). Thus, the importance of defining clearly the patients who are being studied cannot be overemphasized.

There has been very little work done on the possibility of using plasma steroid levels as an index of prognosis or response to treatment. Preliminary results[91] from our laboratory suggest that plasma DS levels might be useful as an index of prognosis in the early disease. Plasma DS was measured after mastectomy and compared with a regression line of these concentrations on age for normal women. Patients whose plasma DS concentrations lay below this regression line tended to have a shorter free period than those patients whose levels were above the regression line (see Figure 4). It is stressed that these are preliminary results based on a short follow-up period.

B. Studies on the Production Rate of 11-Deoxy-17-ketosteroids

Investigations in this laboratory have shown that the amount of plasma DS and urinary excretion of 11-DKS are highly correlated in normal women, women with early breast cancer (before or after mastectomy), and patients with advanced breast cancer (Figure 5). Since all these correlations are statistically indistinguishable, it implies that the production rates of 11-DKS are related to urinary 11-DKS excretion and hence one might expect the production rate to be subnormal in women with advanced breast cancer. It is appropriate, therefore, to survey the literature for data on the production rates of C19 neutral steroids in women with breast cancer.

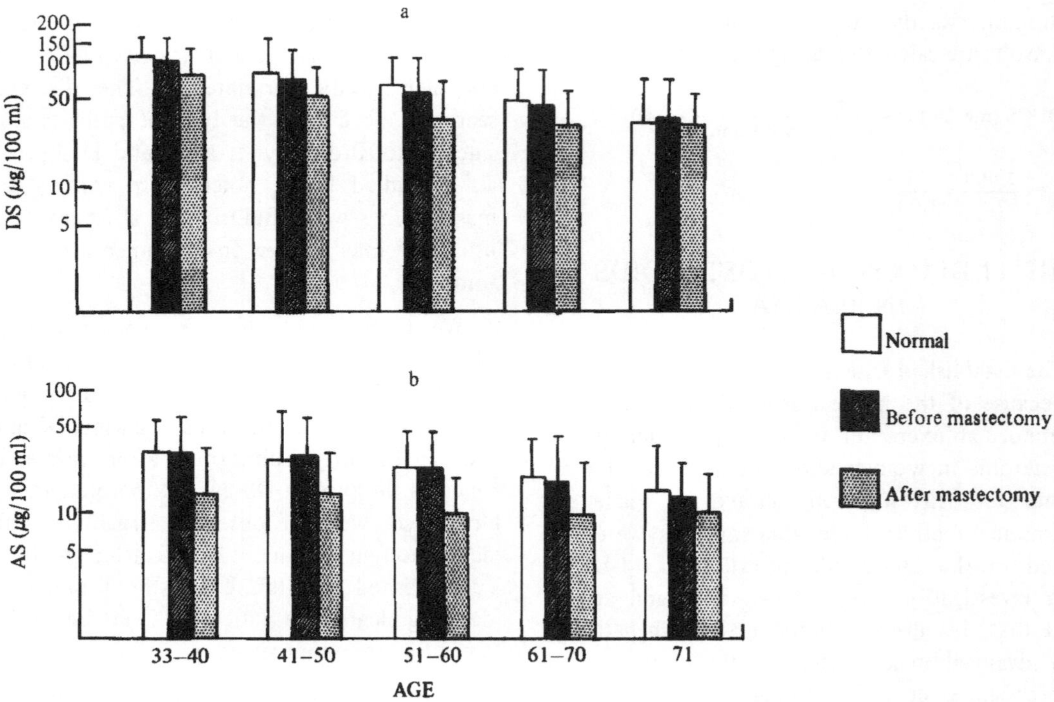

FIGURE 3. Plasma levels of DS and AS in normal women and patients with early breast cancer before and after mastectomy. The mean levels (+ S. D.) of plasma (a) DS and (b) AS for various age groups are shown for normal women and patients with early breast cancer before mastectomy and after mastectomy. The total number of subjects studied was 211 normal women and 202 women with breast cancer. There is no statistical difference between the mean level of DS or AS in normal women and patients before surgery. The mean levels of DS and AS in normal women are significantly higher ($P<0.001$) than those in patients after surgery in all age categories except for the oldest age group (71 years).

Early work showed that there was a direct correlation between the secretion rate of dehydroepiandrosterone and urinary excretion of 11-DKS.[92] However, this study was questioned when it was realized that not only dehydroepiandrosterone but also DS were adrenal secretory products.[93]

Although attempts have been made to measure the secretion rates of dehydroepiandrosterone and DS simultaneously no satisfactory models have been devised to make the measurements possible.[94-96] At present only production rates (i.e., no distinction being made between peripheral conversion and glandular secretion) can be measured in practice. Poortmen et al.[96] have reported that the urinary production rate of dehydroepiandrosterone and DS are significantly subnormal in patients with breast cancer. Deshpande et al. have reported results of a more direct approach; these workers infused tritiated pregnenolone into human adrenals *in situ*. They reported that the ratio of radioactive dehydroepiandrosterone and cortisol synthesized from pregnenolone was related to the urinary discriminant function.[97]

There has been a report that the subnormal excretion of urinary 11-DKS in women with advanced breast cancer is not due to a lowered production of steroid precursors but to a change in their metabolism.[47] This finding is difficult to reconcile with the data in Figure 5 and also with the report of lowered urinary production rates of dehydroepiandrosterone and DS.[76]

C. Plasma 11-Deoxy-17-ketosteroids and Etiology of Breast Cancer

As mentioned earlier, a prospective study has shown that most women have abnormally low levels of urinary 11-DKS before clinical diagnosis of breast cancer. Since there is a highly significant correlation between urinary and plasma 11-DKS it would be expected that subnormal levels of plasma 11-DKS might occur also in these women. Although a further prospective study on the island of Guernsey has been set up to determine whether

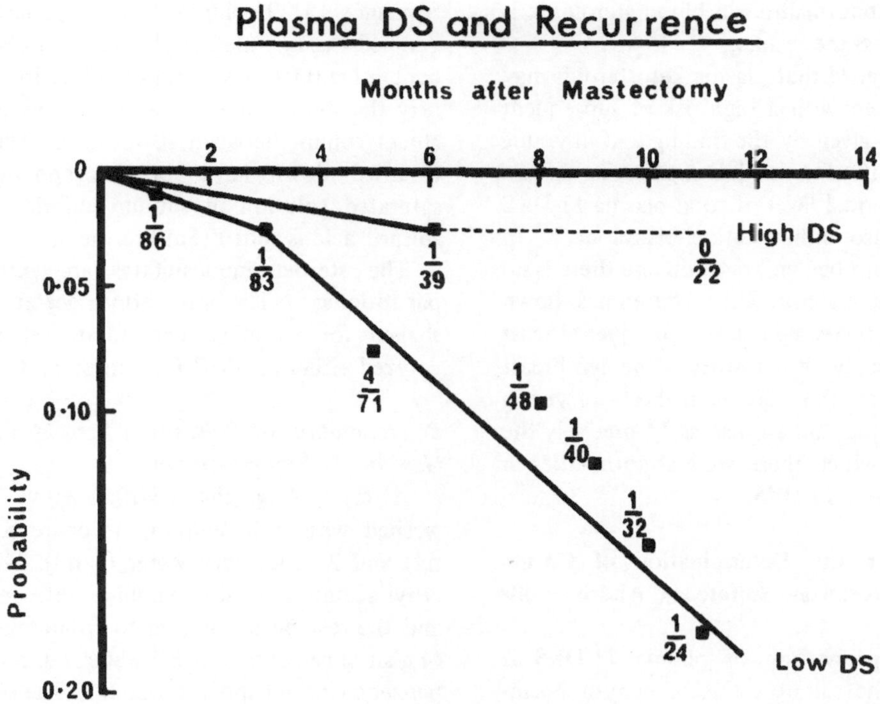

FIGURE 4. Plasma DS concentration and recurrence rate. The regression line of plasma DS concentration on age of normal women was used to divide patients with early breast cancer into two groups, one having higher than normal DS and the other lower than normal DS. The numerator in the fractions refers to the number of patients whose disease has recurred, and the denominator to the number of patients followed for that period of time and who were free of recurrence.

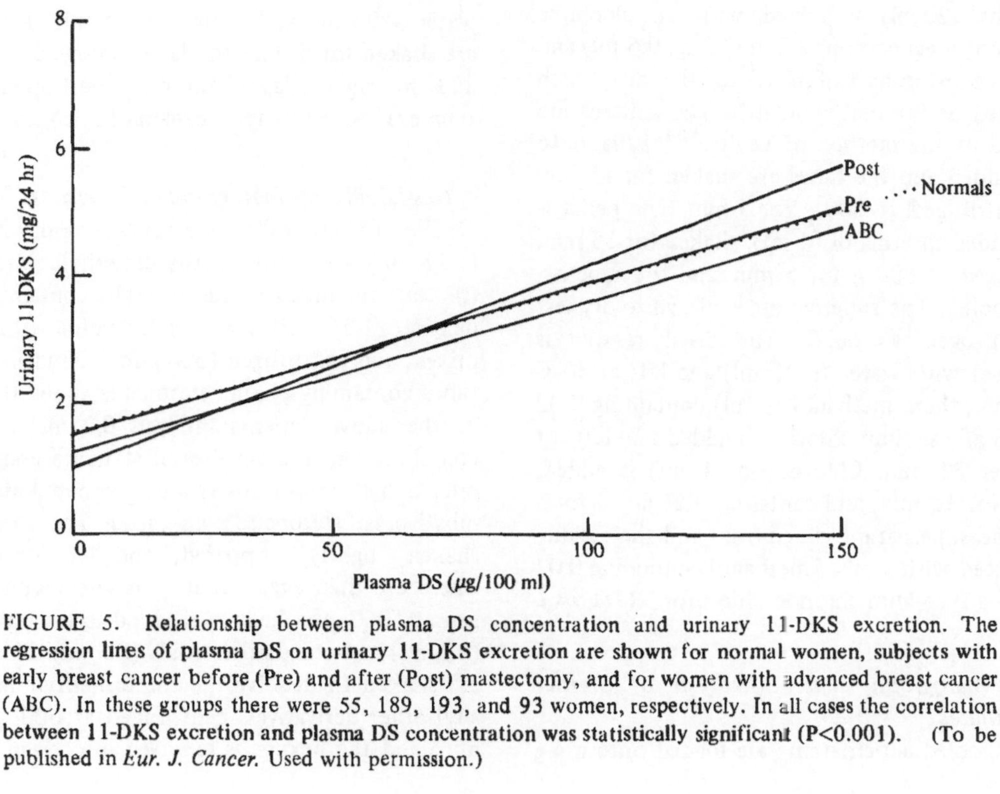

FIGURE 5. Relationship between plasma DS concentration and urinary 11-DKS excretion. The regression lines of plasma DS on urinary 11-DKS excretion are shown for normal women, subjects with early breast cancer before (Pre) and after (Post) mastectomy, and for women with advanced breast cancer (ABC). In these groups there were 55, 189, 193, and 93 women, respectively. In all cases the correlation between 11-DKS excretion and plasma DS concentration was statistically significant ($P<0.001$). (To be published in *Eur. J. Cancer.* Used with permission.)

there are any abnormalities in blood steroids it is too early to assess the results.

Indirect evidence that plasma ketosteroids may be low in women with a high risk of subsequent breast cancer is given by the finding that in young women (20 to 40 years) with benign breast disease there is a subnormal level of total plasma 11-DKS and DS and also urinary etiocholanolone.[98] In older women with benign breast disease there is no such demonstrable abnormality. Warren has shown that there is an increased risk of subsequent breast cancer in women with a history of benign breast lesions and that this risk is highest in young women with benign breast disease,[99] precisely the same group in which there were abnormalities in urinary and plasma 11-DKS.

D. Method for the Determination of Plasma Dehydroepiandrosterone Sulfate and Androsterone Sulfate

The major proportion of plasma 11-DKS is composed of the sulfate esters of dehydroepiandrosterone (DS) and androsterone (AS). The method employed at present for the measurement of the steroids uses gas-liquid chromatogrpahy,[100] although a competitive protein binding method is available.[101]

1. Extraction of Steroid Conjugates

Plasma (2 ml) is mixed with an alcoholic solution of pregnenolone sulfate (5 μg, 0.5 ml) and dihydrotestosterone sulfate (5 μg, 0.5 ml) which are added as internal standards. The sulfates are prepared by the method of Levitz.[102] Ethanol (4 ml) is added and the tubes are shaken for 15 min and centrifuged at 600 g for 5 min. The pellet is resuspended in ethanol (4 ml), shaken for 15 min, centrifuged at 600 g for 5 min, and the supernatants pooled. The supernatant is blown to dryness with nitrogen at 60°C. The dried residue is moistened with water (0.05 ml) and left at 50°C for 5 min; then, methanol (1 ml) containing 0.02 *M* (1.16 g/l) sodium chloride is added and left for a further 30 min. Chloroform (1 ml) is added, shaken for 10 min, and centrifuged at 600 g for 5 min. The supernatant is removed and the residue reextracted with 1 ml of methanol containing 0.02 *M* (1.16 g/l) sodium chloride:chloroform (1:1 v/v).

2. Chromatography and Solvolysis of Steroid Monosulfates

The pooled supernatants are loaded onto a 4-g Sephadex LH20 column (1 cm i.d.) and chromatographed using methanol:chloroform (1:1 v/v) containing 0.01 *M* sodium chloride. In our laboratory the steroid monosulfates are contained in the eluant volume between 20 to 58 ml. This fraction is dried down in a rotary evaporator and a saturated solution of sodium chloride in 0.1 *N* sulfuric acid is added (5ml) to the dried residue.

The steroid monosulfates are extracted by partitioning twice into ethyl acetate (10 ml, shaken for 15 min). The sulfate esters are solvolyzed either at 50°C for 4 hr or 37°C overnight.

3. Preparation of Solvolyzed Steroid Sulfates for Gas-liquid Chromatography

After cooling, the acidified ethyl acetate is washed with 20% sodium carbonate solution (5 ml) and 3 times with water (5 ml). The washed ethyl acetate is dried down under nitrogen at 50°C and the residue transferred to an alumina column (4 g alumina containing 8% water, i.d. 5 mm) using benzene (0.5 ml and 2 x 0.25 ml). The steroids are chromatographed with benzene containing 1% ethanol (v/v) and eluant volume between 10 and 23 ml is collected. This volume is checked for each new batch of alumina. After evaporation under nitrogen at 50°C the 11-DKS are extracted into isopentane. This is achieved by dissolving the steroids in ethanol (0.05 ml) and then adding water (0.95 ml) and isopentane (5 ml). The tubes are shaken for 5 min, the layers allowed to settle, and the organic layer removed. The isopentane is then evaporated under a stream of nitrogen.

4. Gas-liquid Chromatography of Steroids

The contents of a tube containing dry hexane (1.25 ml) and chloromethyldimethylchlorosilane (0.2 ml) are added to another tube containing dry hexane (1.25 ml) and diethylamine (0.1 ml), mixed, and centrifuged (600 g for 10 min). To the tubes containing plasma steroids is added 0.15 ml of the above supernatant and 0.2 ml to tubes containing equal quantities of standard androsterone, dehydroepiandrosterone, pregnenolone, and dihydrotestosterone (25 μg each). The tubes are shaken, tightly stoppered, and left overnight. These are then evaporated to dryness with nitrogen at 50°C, and hexane is added to the unknown (0.1 ml) and standard steroids (0.5 ml). The tubes are shaken to dissolve the chloromethyldimethylsilyl ether derivatives, centrifuged at 600 g for 5 min, and the hexane is removed into clean tubes.

Approximately 10 μl of plasma steroid extract or 2 μl of standards is introduced by solid injection into the gas chromatogram and chromatographed on a 5-ft (i.d. 4 mm) column of Supasorb (100-120 mesh) coated with 1.5% XE60. The oven temperature is 215°C and the nitrogen flow is 60 ml/min.

5. Calculation

The calculation is based on (1) the amount of the internal standards added, (2) the fact that the loss of these through the method is the same as for DS and AS in the plasma samples, and (3) the ratio of the peak heights on the gas chromatogram of the external standard, containing equal quantities of dehydroepiandrosterone, androsterone, dihydrotestosterone, and pregnenolone.

6. Example

Assume that the amount of dihydrotestosterone added to plasma is 5 μg (it should be remembered that dihydrotestosterone sulfate is added and that this figure is obtained after correcting for the sulfate moiety). Also, if the ratios

$$\frac{\text{peak height of androsterone of external standard}}{\text{peak height of dihydrotestosterone of external standard}} = 1.2$$

and

$$\frac{\text{peak height of dehydroepiandrosterone of external standard}}{\text{peak height of dihydrotestosterone of external standard}} = 2.0$$

and if the corresponding ratios for the unknown sample are 0.36 and 0.40, respectively, then the amount of androsterone in the unknown is

$$\frac{0.36}{1.2} \times 5 = 1.5\ \mu\text{g/2 ml plasma} = 75\ \mu\text{g/100 ml plasma}$$

Similarly, the amount of dehydroepiandrosterone in the unknown is

$$\frac{0.40}{2.0} \times 5 = 1.0\ \mu\text{g/2 ml} = 50\ \mu\text{g/100 ml plasma}$$

It should be noted that answers are expressed in terms of the unconjugated steroid. The calculation for the other internal standard, pregnenolone sulfate, is identical to the above.

It has been stated that the loss of internal standard is the same as that for the steroids in the sample. Although this is true for the majority of samples analyzed, we have experienced some samples in which this has not occurred and as a safeguard we have used two internal standards. Thus the results using pregnenolone sulfate as an internal standard should be the same as those for dihydrotestosterone sulfate.

A further safeguard, which is applicable to all methods, is the use of the quality control. In our laboratory, plasma from a quality control pool is included in every batch of analyses. Since the coefficient of variation of the precision of this method is about 7.5%, then, using 95% confidence limits, all batches in which the quality control varies more than 15% from the mean are rejected. The mean value of the quality control is obtained by performing multiple estimations on the pool.

IV. PLASMA 17-HYDROXYCORTICOSTEROIDS

Although there is a diurnal variation in plasma levels of 17-OHCS[103] the concensus of results would suggest that there is an increased concentration in women with breast cancer. Schubert et al.[104] in 1961 reported elevated resting levels of plasma cortisol in women with advanced breast cancer, which were associated with a marked response to ACTH administration. Benard and his colleagues in 1962 reported a raised mean concentration of plasma 17-OHCS in women with breast cancer although there was great variability between patients.[105] Deshpande et al.[86] have confirmed that the mean plasma levels of 17-OHCS are elevated, although Beck et al.[106] did not.

In the series of Deshpande et al.[86] over half the patients with advanced breast cancer had abnormally high levels of plasma 17-OHCS. However, these high concentrations were not reflected by correspondingly elevated excretion of urinary 17-OHCS. Unlike normal women there was no relationship between plasma and urinary 17-OHCS. Beck et al.[106] reported similar findings.

The situation in the advanced disease may be

similar to that found in pregnancy and in women taking steroidal oral contraceptives, when there is an increased protein binding resulting in elevated plasma cortisol levels.[107,108] Whether this occurs in this case is uncertain since there are reports that the mean level of cortisol binding is elevated,[109,110] countered by another report of a lower mean binding.[111] In the latter there was also a larger variance of the binding in advanced breast cancer cases compared to normal women.

A possible consequence of high plasma cortisol levels is a change in immune response; MacKay et al.[112] found that plasma cortisol was lower in women with breast cancer who had a positive response to tuberculin than in those with a negative response.

A. Method for the Estimation of Plasma Cortisol

The concentration of plasma cortisol has been measured in our laboratories by fluorimetry,[113] colorimetry,[114] and by double isotope derivative[115] techniques. The present method used is based on competitive protein binding,[116,117] and it is this method that will be described.

1. Extraction of Plasma

Plasma (1ml) is added to ethanol (9 ml in a ground glass-stoppered tube, the tube is shaken for 1 min, and the contents filtered through a Whatman No. 1 filter paper. The estimations are performed in duplicate and aliquots (1 ml) of filtrate, equivalent to 0.1 ml plasma, are transferred to 2-ml disposable plastic cups and evaporated to dryness in a stream of air at 40°C. If high concentrations of plasma cortisol are anticipated, aliquots of 0.5 ml are also taken.

2. Preparation of Standard Curve

An ethanolic cortisol solution (100 ng/ml) in volumes of 0, 50, 100, 150, and 200 μl are pipetted, in duplicate, into 2-ml disposable plastic cups and evaporated to dryness in a stream of air at 40°C.

3. Preparation of Corticosteroid Binding Globulin (CBG) Solution

The source of CBG is freshly obtained normal human plasma. Over the range of the method (0 to 30 ng) a 5% dilution in water was found to be optimal. However, the dilution of CBG should be assessed for each batch of plasma used. The method of assessment is to use the concentration of plasma that gives 50% binding at zero concentration of cortisol. The CBG solution is prepared by evaporating to dryness 4 μC; ^{3}H-cortisol in ethanol in a 100-ml volumetric flask. To this is added 5 ml normal human plasma and distilled water to 100 ml.

4. Reaction with CBG Solution

CBG solution (1 ml) is added to each of the dried standards and plasma extracts and the cups are well shaken. The samples are warmed to 45°C for 1 hr and cooled in an ice water bath for 10 min.

A mixture of equal quantities of fuller's earth and calcium carbonate (12 mg) is added to each sample. The calcium carbonate is nonabsorbent and added as bulk to enable the absorbent, fuller's earth, to be measured with greater precision. After shaking for 5 min on an automatic shaker the cups are centrifuged at 600 g for 2 min and returned to the ice bath. Aliquots (0.5 ml) of the supernatant are transferred to counting vials and Bray's scintillation fluid (5 ml) added.

5. Radioactive Counting and Calculations

Samples are counted twice to a preset number of counts and the mean time taken to accumulate these counts calculated. The degree of quenching was found to be about 50%, and there was no significant variation in quenching between samples when tested either by internal standardization using ^{3}H-cortisol or external standardization using a γ-source (radium). Therefore, the counts were not corrected for quenching.

A standard curve of the time required to obtain a fixed number of counts against the amount of cortisol present is plotted. This gives a straight line over the range 0 to 30 ng of cortisol. The concentration of cortisol in the plasma specimens can then be read from the standard curve.

V. ESTROGENS AND PROGESTERONE

A. Introduction

There is no conclusive evidence that estrogens or progesterone or both produce malignant tumors in man, although some workers have claimed that estrogens may be carcinogenic[118-120] and have described cases where patients treated with estrogens have subsequently developed breast cancer. This view has been endorsed by Hertz[121] in a review on the possible effects of long-term administration of oral contraceptives. However, if this were the case one might expect that the incidence of breast cancer would have increased with the increasing use of estrogenic compounds as contraceptives and for the relief of menopausal symptoms. Recent papers[122,123] suggest that this is not so, and that the administration of hormonal contraceptives may provide some protection against benign breast disease.[123] This study relates to a time when oral contraceptives had been in use for 7 years, and the possibility of a prolonged latent period for the development of tumors cannot be discounted. Incidentally, although some premenopausal patients with breast cancer will respond to oophorectomy,[124–126] some postmenopausal patients benefit from the administration of large amounts of estrogen.[19,127]

An etiological role for the ovary is implied in much of the data obtained from epidemiological studies. It has been shown that there is a reduced risk of breast cancer in women whose menarche occurs above the age of 16 years and in women whose menopause is induced surgically[128], while there is an increased risk of developing breast cancer in those women in whom the menopause occurs naturally at a late age.[129]

During pregnancy the breast is subjected to gross increases in hormonal stimuli, but this appears to give a protective effect since there is a reduction in risk of breast cancer associated with pregnancy, particularly in the case of pregnancy at an early age.[130] Lactation, however, appears to have little effect on the risk of developing breast cancer.[131]

Early work on the assessment of endocrine status of patients with breast cancer was hampered by the lack of methods for the estimation of the small amounts of steroid hormones found in urine and blood. Using semiquantitative biological assays three groups of workers[132–134] reported that the estrogen excretion of breast cancer patients was essentially similar to that of normal women. The sensitive chemical assay of Brown[135] has made possible the repetition of these early experiments on a larger scale. However, the results obtained are far from conclusive. Three groups of workers[136–138] have reported that patients with breast cancer excrete abnormal amounts of urinary estrogen metabolites, and while two other groups have reported an increased excretion,[139,140] two further groups found no difference between the breast cancer patients and control groups.[141,142]

The results of Lemon et al.[137] showed that estriol formed a smaller proportion of the total estrogen excretion in patients with breast cancer than in control cases. Lemon[143] subsequently suggested that estriol might impede the action of estradiol at the cellular level and that the protective effect of pregnancy might be due to a large and continuous infusion of estriol from the fetoplacental unit. It has been postulated that the estrogen profile of pregnancy, characterized by a high ratio of estriol to the other estrogen fractions, may be of etiological significance and that the estrogen profile of young women may be indicative of the risk of subsequent breast cancer.[144]

In a study of young North American women and young Asian women[145] it has been shown that the estriol quotient (estriol/estrone + estradiol) is higher in the Asian women who are less at risk than the North American women.[146] The differences exist in both the follicular and luteal phases of the menstrual cycle and are more marked in the early years of reproductive life. Further evidence to support this finding is given in the reports by Gross[147] and Briggs.[148] Gross has studied the endocrine environment of Yemenite and European women, and Briggs has studied three ethnic groups of women, African, European, and Indian, living in Lusaka, Zambia. In both cases the higher estriol quotients were associated with a lower risk of breast cancer.

Data on the urinary excretion of pregnanediol suggest that there is a diminished progestational stimulus in patients with breast cancer.[149] This is in agreement with the findings of Grattarola.[150] In a histological study of the endometrium of women with breast cancer he found many cases in which there was no evidence of luteal stimulation, and in cases where ovulation had occurred there was an apparent diminution of corpus luteum activity.

More recently the development of competitive protein binding assays and radioimmunoassays has made possible the estimation of estrogens in small samples of blood, and simplified the assay of progesterone. To date, the concentrations of these hormones in blood have not been studied extensively in patients with breast cancer, but a preliminary report of investigations carried out in these laboratories is given below.

B. Early Breast Cancer

The assessment of the estrogenic status of premenopausal women is complicated by the wide variations between individual subjects and by the fluctuation of these hormones during the menstrual cycle. In an attempt to minimize variation due to the latter factor, the following study has included only those patients who were expected to be in the luteal phase of the menstrual cycle. The control group consisted of healthy women, none of whom had any history of malignant disease at any site and who were also expected to be in the luteal phase of the menstrual cycle.

In the case of postmenopausal women, at least 5 years had elapsed since the last menstrual period and the control group again consisted of healthy women without previous history of malignancy at any site.

All the hospital patients were awaiting mastectomy and blood samples were collected on the day before operation. None of the patients had any evidence of visceral or skeletal metastases and neither they nor the control subjects were receiving any form of steroid medication. Plasma estradiol, estrone, and progesterone were estimated by the methods to be described at the end of this section, and the results are shown in Table 1. It can be seen that there is no evidence to suggest that the patients with breast cancer are subjected to an abnormal estrogenic stimulus.

C. Advanced Breast Cancer

The criteria that were applied in the selection of patients and controls in Section V.B, "Early Breast Cancer," have been applied similarly in the selection of patients with advanced breast cancer but in this case all the patients had evidence of metastatic disease. The premenopausal patients were awaiting oophorectomy and postmenopausal patients were to be treated with estrogens or androgens. Plasma estradiol, estrone, and progesterone were estimated in samples collected prior to treatment and the results are shown in Table 2.

There are no significant differences in the levels of plasma estradiol and estrone between the control group and the advanced breast cancer patients in the postmenopausal women. In the premenopausal group the mean plasma levels of estrone are comparable in the controls and in the cancer patients but the mean level of estradiol is higher in the cancer patients, although the differences are not statistically significant. The mean plasma level of progesterone is significantly lower in the advanced breast cancer cases than in the control groups ($P<0.05$). As yet it is unclear whether these differences have any prognostic significance but it is hoped to extend the study to include more patients and to accumulate follow-up data.

D. The Preclinical Phase of Breast Cancer

A prospective study has been set up to assess the role of the blood hormones in the etiology of breast cancer, and blood samples have been collected from 3,500 ostensibly normal women living on the island of Guernsey. Subsequently a diagnosis of carcinoma of the breast has been made in five of these women. Plasma estradiol was estimated in the samples taken from these women before the clinical appearance of a carcinoma and also in samples from women who have been matched as closely as possible for age, height, weight, parity, time of collection of sample, and, where applicable, the day of the menstrual cycle.

TABLE 1

Plasma Concentrations (± S.E.M.) of Estrone, Estradiol, and Progesterone in Women with Early Breast Cancer and in a Control Group

	Premenopausal women				Postmenopausal women		
	Estrone (pg/ml)	Estradiol (pg/ml)	Progesterone (ng/ml)	Number of subjects	Estrone (pg/ml)	Estradiol (pg/ml)	Number of subjects
Control group	72± 8	86±14	11.6±2.0	14	15±1	8±1	20
Early breast cancer patients	69±11	94±17	11.3±1.2	22	22±4	12±3	13

TABLE 2

Plasma Concentrations (±S.E.M.) of Estrone, Estradiol, and Progesterone in Women with Advanced Breast Cancer and in a Control Group

	Premenopausal women				Postmenopausal women		
	Estrone (pg/ml)	Estradiol (pg/ml)	Progesterone (ng/ml)	Number of subjects	Estrone (pg/ml)	Estradiol (pg/ml)	Number of subjects
Control group	72± 8	86±14	11.6±2.0	14	15±1	8±1	20
Advanced breast cancer patients	65±14	151±43	5.5±1.3*	15	17±8	8±3	15

*Significantly different from control group ($P < 0.05$)

Preliminary results obtained suggest that there is no evidence of an increased estrogenic stimulus in the preclinical phase of the disease.[151]

The relationship between benign breast disease and carcinoma of the breast has been discussed in an earlier part of this chapter, and in view of this relationship it seemed appropriate to include benign breast disease in a section dealing with the preclinical phase. The endocrine status of women with benign breast disease is unclear, and there have been few reports on studies of this subject. An early study by Bucher and Geschickter[152] suggested that although urinary estrogen excretion appeared normal in all their patients, there was subnormal excretion of pregnanediol in the luteal phase by some of the patients. This was supported by Nathanson[153] who reported that endocrine dysfunction and menstrual abnormalities were found frequently in patients with benign diseases of the breast. Taylor[154] claimed that both estrogen and pregnanediol excretion by these patients was essentially normal and that there did not appear to be evidence of ovarian dysfunction. However, all these studies were carried out on very small numbers of patients and for this reason may be considered equivocal. More recent studies[149,155] on larger groups of patients and controls support the view that there are no gross endocrine differences between the groups as measured by urinary excretion of estrogens and pregnanediol.

In an attempt to assess the hormonal stimulus to the breast by endogenous hormones, the blood concentrations of estradiol and progesterone have been measured in patients with a variety of benign lesions and in a control group matched for age and day of the menstrual cycle.[156] The results suggest that there are no significant differences between the two groups. In a study of the histology of specimens obtained at biopsy from the patients with benign breast disease, there were several cases showing marked epithelial proliferation, the lesion considered to carry the greatest risk of subsequent malignancy.[157,158] The estradiol and progesterone concentrations in these patients were distributed evenly throughout the normal range of values. While further work is required it seems unlikely that this lesion is associated with abnormalities in the plasma concentrations of the female sex hormones.

E. Estrogens and the Prediction of Response to Endocrine Therapy

Several groups of workers have attempted to predict the response of patients with advanced breast cancer to endocrine ablation using the values obtained postoperatively of the urinary excretion of estrogen metabolites.[62,159,161] However, none of these groups was able to show any significant differences between those patients who responded to treatment and those who did not.

Although there do not appear to be marked abnormalities in plasma estrogen concentrations and the urinary excretion of estrogen metabolites in patients with breast cancer, it is possible that there may be alterations in the response to hormonal stimuli of the breast tissue. Hormone responsive tissues such as uterus, vagina, and breast contain specific proteins that have a high affinity for estrogens (estrogen receptors), and these appear to play a major role in hormone action. The first indication that the specific uptake of estrogens by tumor tissue might be of prognostic importance was given in work by Folca, Glascock, and Irvine.[162] They administered tritium-labeled hexoestrol to patients before adrenalectomy, and measured the uptake of labeled hexoestrol in breast tumor tissue and in muscle. There was a greater response to oophorectomy and adrenalectomy in those patients in whom there was a relatively greater accumulation of tritium in tumor tissue than in muscle.

More recent studies[163] have shown that estrogen receptors are not found in all breast tumor tissue and that those patients whose tumors lack the receptor proteins are unlikely to respond to endocrine ablation. Of the patients whose tumors do contain estrogen receptors, most, but not all, will respond to ablative therapy. Further studies using the in vitro approach have been reported.[164-166] These workers have also found that many, but not all, tumor tissue specimens contain estrogen receptors and that this high affinity binding is seldom seen in biopsy specimens of benign breast lesions. As yet, detailed follow-up studies are not available.

Terenius[167] has demonstrated the presence of a specific receptor for progesterone in human and rat mammary tissue. This may be found in the presence or absence of the estrogen receptor.

While these studies on the specific binding proteins appear to have some value in the prediction of response to endocrine ablation, further data are required before the full clinical significance of such tests can be assessed.

F. The Assay of Estradiol and Estrone in Plasma

The method described below is a modification of the method of Korenman, Perrin, and McCallum.[168]

1. Sample Preparation

^{3}H-Estradiol and ^{3}H-estrone (2,000 cpm) are added to plasma (4 ml) and allowed to equilibrate for 15 min. The plasma is extracted with dichloromethane (2x2½ vol), and the combined dichloromethane extracts are washed with distilled water (1x1/10 vol) and evaporated to dryness. The dried plasma extracts are dissolved in benzene:methanol (85:15, v/v) applied to Sephadex LH20 columns (1 g, i.d. 0.9 cm, height 7.0 cm) and eluted with the benzene:methanol mixture. The estrone (4.0 to 5.5 ml) and estradiol (6.5 to 8.5 ml) fractions are collected and evaporated to dryness. They are then dissolved in 0.4 ml tris-hydrochloric acid buffer (0.01*M*, pH 7.4), an aliquot is taken for assessment of procedural losses, and duplicate aliquots (0.1 ml) are taken for assay.

2. Assay of Estrone and Estradiol

Standard solutions containing 0, 20, 50, 100, and 200 pg estrone or estradiol in tris-hydrochloric acid buffer (0.1 ml) are pipetted into disposable plastic tubes. To these and the unknown samples is added ^{3}H-estrone or ^{3}H-estradiol (10,000 cpm in 0.1 ml tris-hydrochloric acid buffer) and rabbit uterine cytosol (0.05 ml). The tubes are incubated at room temperature for 20 min and in an ice water bath for 20 min. The protein-bound steroid and free steroid are separated by the addition of dextran-coated charcoal (0.5% charcoal, 0.05% dextran, 0.5 ml). The tubes are shaken and returned to the ice water bath for a further 10 min. After centrifugation at 12,000 g for 3 min, aliquots of the supernatant (0.5 ml) are transferred to counting vials and scintillation liquid (10 ml) is added to each sample.

3. Radioactive Counting

Those samples in which the recovery is to be estimated are counted for 20 min each and those in which the estrone and estradiol are assayed are counted until 10,000 counts have accumulated. When the time required to accumulate 10,000 counts is plotted as the ordinate with picograms estrone or picograms estradiol as the abscissa the resultant graph is a straight line over the range 0 to 200 pg. The amount of estrone or estradiol in the unknown samples is read from the standard curve and corrections for procedural losses are applied.

G. The Assay of Progesterone in Plasma

The method described is a competitive protein binding method which uses diluted guinea-pig serum as the specific binding protein.[169]

1. Preparation of Protein Binding Solution

^{3}H-Progesterone (4 μCi) is evaportated to dryness in a 100-ml volumetric flask. Guinea-pig serum (0.17 ml) obtained from blood collected 1 day postpartum is added and the solution diluted to 100 ml with tris-hydrochloric acid buffer (0.01 *M*, pH 7.4).

2. Extraction of Plasma Samples

Hexane (2.5 ml) is added to plasma samples (0.5 ml) in glass test tubes. The tubes are shaken mechanically for 10 min and centrifuged at 2,000 g for 5 min. A 1-ml aliquot of the hormone extract is used for the assay of plasma samples obtained from women in the luteal phase of the menstrual cycle. The hexane extracts are transferred to conical glass centrifuge tubes. Standards containing 0,. 1, 2, 3, 4, and 5 ng progesterone are pipetted into similar tubes and these and the hexane extracts of plasma are evaporated to dryness.

3. Competitive Protein Binding Assay

Protein binding solution (0.5 ml) is added to each tube, the contents are mixed on a vortex mixer, and the stoppered tubes placed in a water bath at 37°C for 15 min. The tubes are mixed again and transferred to an ice bath for 15 min. Dextran-coated charcoal (0.2 ml) is added, and the contents of the tubes are mixed again and allowed to stand in the ice bath for 5 min. The tubes are centrifuged at 2,000 g for 10 min at 4°C. Aliquots of the supernatant (0.5 ml) are transferred to counting vials and scintillation liquid (10 ml) is added. All samples are counted until at least 10,000 counts have accumulated. A graph is plotted with nanograms progesterone in the standards as the abscissa and counts per minute bound in the supernatant as the ordinate. The amount of progesterone in the plasma samples is read from the standard curve and the values obtained corrected for the volume of the hexane extract used and for recovery during the extraction procedure.

VI. BINDING OF STEROID HORMONES TO PLASMA PROTEINS

It hås been mentioned that there is an increased plasma level of cortisol in pregnant women and in women taking oral contraceptives compared with that in normal women. This is not reflected in the urinary excretion of 17-OHCS, and, in fact, in women on oral contraceptives there is a subnormal level of urinary 17-OHCS. Most or all of the rise in plasma cortisol concentration is due to increased protein binding of cortisol.[108,170,171]

These findings indicate that the possibility of changes in protein binding of the hormones in question should not be overlooked in any investigation of the concentration of blood hormones in women with breast cancer, and this is especially true if only a nonbound hormone is physiologically active.

Although cortisol binding appears to be abnormal in women with advanced breast cancer,[109,110] studies on the binding of the following hormones have revealed no abnormalities: estradiol-17β, estriol, dehydroepiandrosterone, androsterone, etiocholanolone, dehydroepiandrosterone sulfate, androsterone sulfate, testosterone, androstenedione, and progesterone.[172]

Although the mean percentage binding of each steroid varied widely, 98 (androsterone sulfate) to 64% (estriol), there were no differences in protein binding for any of these steroids between the plasma from normal women, subjects with benign breast disease, and patients with early or advanced breast cancer.[172] This was confirmed by multiple equilibrium dialysis[173] (see Section VI.A.3, "Multiple Equilibrium Dialysis").

A. Method for Determining Plasma Protein Binding

1. Equilibrium Dialysis

Plasma is diluted with 4 times its volume of 0.9% sodium chloride solution. The diluted plasma (2 ml) is put into a sac made from 9/16-in Visking Nojax® casing and dialyzed at 37°C in 6 ml of physiological saline in which the appropriate radioactively labeled steroid (30,000 cpm) is dissolved. This solution is prepared by adding the radioactive compound in ethanol to a vessel and evaporating to dryness; saline is then added and the vessel is shaken for 15 min. The solution is then transferred to another vessel before pipetting into the flasks used for dialysis. The equilibrium dialysis vessels are gently shaken in a water bath at 37°C for 18 hr. This incubation time was sufficient to ensure that equilibrium had been reached for all steroids examined in our laboratory.

2. Determination of Radioactivity and Calculations

At the end of the period of dialysis the outside surfaces of the sacs are quickly rinsed with water and blotted with paper tissue, and the contents are emptied into test tubes. Duplicate 0.1-ml aliquots of plasma solutions and 0.4 ml of the saline solutions are transferred to counting vials and phosphor is added (15 ml). The phosphor is prepared from 7 g 2,5-diphenyloxazole (PPO) and 600 mg 1,4-bis-2-(4-methyl-5-phenyloxazoyl)-benzene(dimethyl POPOP) dissolved in 500 ml ethoxyethanol and 500 ml toluene. The vials are twice counted for 10 min and the mean counts calculated. Quenching is determined by the addition of a constant amount of an internal standard (^{3}H-toluene).

The percentage binding is calculated using the formula

$$1 - \frac{(S-C)\,(T_p-P)}{4\,(Ts-S)\,(P-C)} \times 100$$

where C is the background counts, P and S are counts for plasma and saline solutions, respectively, and Tp and Ts are counts for plasma and saline solutions after the addition of internal standard, respectively. These counts are all expressed for the same length of time.

3. Multiple Equilibrium Dialysis

This is a useful technique[173] in which the binding property of plasma can be compared by allowing the different plasmas, contained in separate dialysis sacs, to compete for the radioactive steroid in the surrounding saline. The method employed is identical to that outlined above except that instead of equilibrating one sac, two or more sacs are equilibrated together. Also, the volume of saline is increased to cover the extra sacs. For a single dialysis sac the percentage bound $(P) = \frac{Cp}{Cs + Cp} \times 100$ where Cp and Cs are the counts bound per milliliter of plasma and counts per milliliter in the saline outside the dialysis sac, respectively.

In the case of two dialysis sacs, if P′ and P″ are percentage binding for two different plasmas, and

$$P' = \frac{C'p}{C's + C'p} \times 100$$

and

$$P'' = \frac{C''p}{C''s + C''p} \times 100$$

therefore

$$\frac{P'(C's + C'p)}{100} = C'p$$

and

$$\frac{P''(C''s + C''p)}{100} = C''p$$

Rearranging,

$$C'p\,(100-P') = P'C's$$
$$C''p\,(100-P'') = P''C''s$$

therefore,

$$\frac{C'p\,(100-P')}{C''p\,(100-P'')} = \frac{P'}{P''}$$

since C′s = C″s being the counts in the saline or

$$\frac{C'p}{C''p} = \frac{P'}{P''} \cdot \frac{(100-P'')}{(100-P')}$$

The usefulness of this procedure is illustrated by the following example: if P′ = 98% and P″ = 98.5%, then $\frac{C'p}{C''p} = 0.75$. This implies that a 0.5% difference in percentage binding leads to a 25% difference in the number of counts bound to the two plasma solutions.

VII. PROTEIN HORMONES

A. Gonadotrophins

Investigations of the relation between the excretion of urinary gonadotrophins and response to treatment of breast cancer in postmenopausal women have led to a confused picture. Loraine et al.,[174] in 1957, measured urinary gonadotrophin excretion in 47 postmenopausal women with advanced breast cancer before treatment with stilbestrol. The response to this treatment was classified as "worse," "no apparent change," or "improved." They reported that patients classified as "worse" had a higher mean level of excretion than that of any other group of patients or 37 women with noncancerous diseases. In 1958 Boyland et al.[175] also reported that patients with high concentrations of urinary gonadotrophins did not respond to pituitary gland irradiation. However, Segaloff et al.[176] in 1954 reported results that were in disagreement with those described above. Hayward et al.[62] estimated urinary gonadotrophins in 41 patients with advanced breast cancer and found no significant difference between patients who were responsive or unresponsive to adrenalectomy or hypophysectomy, although they did observe that patients who responded to endocrine ablation tended to have higher urinary gonadotrophin levels. Martin[177] reported the results of 166 assays of urinary gonadotrophins and found a significantly lower excretion in women who were responsive to hormonal therapy. Beck et al.[106] reported that an attempt to correlate preoperative levels of gonadotrophin with response to treatment was inconclusive because of the large overlap in concentrations. It did appear, however, that preoperative levels were in the normal or elevated range in patients in whom subsequent remission occurred, as compared to those in whom no remission was seen. In all these studies the urinary gonadotrophin excretion was measured by bioassay, and the findings can be criticized, either because of the methodology, treatment, or assessment of response to treatment.[64]

Results in our laboratory using radioimmunoassay methods for measuring plasma luteinizing hormone (LH) and follicle stimulating hormone (FSH) in postmenopausal women have revealed no significant differences in mean levels between women with early breast cancer, before and after mastectomy, compared to those in normal healthy control women. Furthermore, preliminary results show no significant changes in LH or FSH levels due to mastectomy or correlation between these levels and length of free period. Thus plasma concentration of LH or FSH would not appear to have any prognostic value in the early disease.

B. Growth Hormone and Prolactin

Both growth hormone and prolactin have a profound effect on the growth and development of the mammary gland and lactogenesis.[1-7,178-183] The inability to separate prolactin from human pituitaries, using methods that were successful in isolating prolactin from pituitaries of other species, led to a controversy over whether human prolactin actually existed.[184-187] This argument was reinforced by the high lactogenic activity of human growth hormone as tested by the pigeon crop assay.[188-191] It is only recently that this argument has been resolved by the isolation of human prolactin.[192] Growth hormone was one of the first protein hormones to be measured by radioimmunoassay, while a method for human prolactin has only recently become available.[193,194] It was soon evident that the levels of growth hormone in humans varied widely depending on the metabolic state of the subject. Various stimuli such as exercise, fasting, and stress affected the secretion of this hormone,[195] and for a comparative study strictly controlled conditions must be applied. Greenwood et al.[196] used an insulin sensitivity test on 10 patients with breast cancer before and 7 days after mastectomy and found higher resting levels of growth hormone at both times compared with normal subjects. The response of these patients to insulin-induced hypoglycemia was normal. A second series of 21 patients with breast cancer was tested with oral glucose and plasma cortisol and growth hormone measured. Anomalous rises in cortisol were observed in 11 patients and in growth hormone in 3 patients. Suppression of growth hormone was not observed after oral glucose in the remainder of the patients.

Pearson et al.[197] examined the effect of oral glucose on plasma growth hormone and insulin levels in 23 women with metastatic breast cancer. There was an anomalous increase in growth hormone in 9 patients. Diabetic-type glucose tolerance curves were observed in 12 patients; also, these patients had delayed insulin secretion and the levels were lower than in noncancer subjects, suggestive of decreased pancreatic insulin reserve. Both Greenwood et al.[196] and Pearson et al.[197] conclude that alterations in carbohydrate metabolism in women with breast cancer cannot be explained on the basis of abnormalities in growth hormone secretion alone.

There is much evidence that prolactin is important in mammary carcinogenesis in laboratory rodents.[12,16,18,198] Using mammary tumors induced in Sprague Dawley rats by DMBA, Pearson et al.[18] have shown that prolactin alone is required for the growth of the tumor. However, in the Fisher rat bearing the R3230 AC tumor, increased prolactin secretion was associated with tumor regression.[199]

The remission observed in some 30% of patients after hypophysectomy has been assumed to be due to the removal of anterior pituitary hormones, growth hormone, and prolactin. Sebum secretion is elevated in patients with breast cancer[200,201] without a concomitant increase in androgen secretion,[202] which implies that prolactin secretion could be increased. It has been reported that injection of ovine prolactin appeared to stimulate the growth of breast cancer as judged by the increased excretion of urinary calcium.[203] However, the evidence, direct and indirect, against prolactin being of prime importance in the etiology or clinical course of human breast cancer is considerable. The epidemiological findings indicate that pregnancies at an early age, which would involve a large increase in blood prolactin concentration, afford a lifelong reduction in risk of breast cancer. In countries where prolonged lactation is common, the incidence of breast cancer is low.[130,204,205]

In the advanced disease, stalk section may lead to remission[206] in spite of the fact that prolactin levels are increased.[207] Furthermore, patients with breast cancer may benefit from becoming pregnant.[208,209] Ergot alkaloids, which inhibit prolactin release,[210] have been found to be ineffective in the treatment of advanced breast cancer.[211]

Measurements of plasma prolactin levels in women with breast cancer have led to contradictory findings. Murray et al.[212] found that the mean prolactin level of 24 patients with metastatic breast cancer was significantly higher than the mean level of 23 hospitalized controls. However, Boyns et al.[213] found no significant differences in the mean plasma prolactin levels between 10 hospitalized patients, 12 women with benign breast diseases, 27 patients with primary breast cancer, and 41 patients with metastatic breast cancer. It should be mentioned that both sets of investigators used heterologous assays to measure human prolactin. Murray et al.[212] used ^{125}I-labeled porcine prolactin and antiovine prolactin serum. Boyns et al.[213] used ^{125} I-labeled ovine prolactin and a rabbit antiserum, originally raised against human growth hormone, which had a low cross reaction to human prolactin. The first group expressed their results in porcine prolactin equivalents and the second group in terms of a W.H.O. standard serum(71/222). Neither employed a method of choice, namely a homologous assay such as that devised by Sinha et al.[194] However, unless large numbers of patients are studied, it is difficult to see how any significant differences in plasma prolactin levels between normal women and patients with breast cancer are to be detected since the variations in these levels found by Boyns et al.[213] were about 100-fold between lowest and highest titer.

It may well be that to obtain meaningful results prolactin will need to be measured under strictly controlled conditions similar to that described for human growth hormone. Such a test could involve using a tranquilizer such as chlorpromazine or after administration of thyrotropin releasing factor, both of which stimulate prolactin secretion.[214,215]

C. Thyroid Hormones

There have been various reports in the literature of the relationship between thyroid function and breast cancer. In 1952 Repert[216] reported that the incidence of thyroid disease was 10 times greater than expected in 306 women with breast cancer. Subsequent reports of Loeser[217] and Wynder et al.[146] supported these findings when they observed that breast cancer was more common in hypothyroid patients and that hyperthyroidism was more prevalent in a control group of women compared to women with breast cancer.

The possibility of thyroid hormones being involved in the etiology of the disease has been investigated by Bulbrook et al.,[218,219] who compared urinary steroid excretion in British and Japanese women. They found that the ratio of androsterone to etiocholanolone was higher in Japanese women aged 40 to 59 than in comparable British women. Since thyroxine influences the metabolism of androgens in favor of androsterone at the expense of etiocholanolone,[220] it was argued that Japanese women appeared to have a higher circulating level of thyroid hormones than British women. This, therefore, may be one reason for the incidence of breast cancer in Japan being lower than that in Britain.

In the established disease there is some evidence that although thyroid function appears to be normal in women with early breast cancer, a decreased thyroid function[221] and increased serum protein-bound iodine[222] has been reported in the advanced disease. Sommers[223] reported thyroid atrophy in patients with disseminated breast cancer.

It has been known for some time that galactorrhea occurs in a small number of patients with primary hypothyroidism.[224,225] Forsyth et al.[226] in 1971 reported their study on one of these patients and showed that the galactorrhea was associated with high plasma prolactin levels and that the prolactin concentration could be reduced with thyroxine therapy. The various reports that thyrotropin releasing factor stimulates the secretion of prolactin[207,215,227,228] gives a link between thyroid function and breast development. These advances have, therefore, opened up a new and interesting line of research in the field of breast cancer and thyroid-linked hormones.

VIII. CONCLUSION

Although a remarkable quantity of data has been accumulated from endocrinological studies on women with breast cancer, it is a sobering thought that there has seldom been agreement between the results of studies. This lack of agreement may be partly the result of conclusions being drawn from work on small series of patients, and partly due to the fact that patients within a series may represent a very heterogeneous population in terms of clinical status, age, weight, menstrual status, etc. In some cases there is the further complication that the methods used have not been adequately evaluated. For these reasons endocrine studies have contributed little to the clinical management of the disease.

In this chapter we have dealt mainly with hormones individually, but it should be remembered that hormones may act synergistically or antagonistically. Ideally, a complete study of all hormones should be undertaken together with measurements of other parameters, e.g., viral and immunological. Such a multidisciplinary approach might yield valuable results not only in the treatment of the established disease, but also in the recognition of its preclinical phase.

Acknowledgments

We would like to thank Drs. R. D. Bulbrook and B. S. Thomas and Mr. J. W. Moore for their advice and helpful criticism in the preparation of this chapter.

REFERENCES

1. **Folley S. J.,** *The Physiology and Biochemistry of Lactation,* Charles C Thomas, Springfield, Ill., 1956.
2. **Lyons, W. R.,** *Proc. R. Soc. Lond. Ser. B Biol. Sci.,* 149, 303, 1958.
3. **Lyons, W. R., Li, C. H., and Johnson, R. E.,** *Recent Prog. Horm. Res.,* 14, 219, 1958.
4. **Cowie, A. T.,** in *Milk: The Mammary Gland and Its Secretion,* Kon, S. K. and Cowie, A. T., Eds., Academic Press, New York, 1961, 163.
5. **Cowie, A. T. and Folley, S. J.,** in *Sex and Internal Secretions,* Vol. 1, Young, W. C., Ed., Bailliere, Tindall and Cox, London, 1961, 590.
6. **Jacobsohn, D.,** in *Milk: The Mammary Gland and Its Secretion,* Kon, S. K. and Cowie, A. T., Eds., Academic Press, New York, 1961, 127.
7. **Topper, Y. J.,** *Recent Prog. Horm. Res.,* 26, 287, 1970.
8. **Beatson, G. T.,** *Lancet,* 2, 104, 162, 1896.
9. **Lathrop, A. E. C. and Loeb, L.,** *J. Cancer Res.,* 1, 1, 1916.
10. **Laccasagne, A.,** *C. R. Hebd. Seances Acad. Sci.,* 195, 630, 1932.
11. **Noble, R. L.,** in *The Hormones,* Vol. 5, Pincus, G., Thimann, K. V., and Astwood, E. B., Eds., Academic Press, New York, 1964, 559.
12. **Riviere, M. R., Chouroulinkov, I., and Guerin, M.,** *C. R. Seances Soc. Biol. Fil.,* 155, 2102, 1961.
13. **Heiman, J.,** *Am. J. Cancer,* 39, 172, 1940.
14. **Burrows, H. and Hoch-Ligeti, C.,** *Cancer Res.,* 6, 608, 1946.
15. **Symconidis, A.,** *Acta Unio Int. Contra Cancrum,* 6, 163, 1948.
16. **Boot, L. M.,** *Int. J. Cancer,* 5, 167, 1970.
17. **Muhlbock, O. and Boot, L. M.,** in *CIBA Foundation Symposium on Carcinogenesis,* Wolstenholme, G. E. W. and O'Connor, M., Eds., J. and A. Churchill, London, 1959, 83.
18. **Pearson, O. H., Llerena, O., Llerena, L., Molina, A., and Butler, T.,** *Trans. Assoc. Am. Physicians Phila.,* 82, 225, 1969.
19. Council on Drugs, Subcommittee on Breast and Genital Cancer, *J.A.M.A.,* 172, 1271, 1960.
20. **Prudente, A.,** *Surg. Gynecol. Obstet.,* 80, 575, 1945.
21. Co-operative Breast Cancer Group, *J.A.M.A.,* 188, 1069, 1964.
22. Joint Committee on Endocrine Ablative Procedures in Disseminated Mammary Carcinoma, *J.A.M.A.,* 175, 787, 1961.
23. **Fracchia, A. A., Randall, H. T., and Farrow, J. H.,** *Surg. Gynecol. Obstet.,* 125, 747, 1967.

24. Prohaska, J., in *Major Endocrine Surgery for the Treatment of Cancer of the Breast in Advanced Stages*, Dargent, M. and Romieu, C., Eds., Simep Editions, Lyon, France, 1967, 37.
25. Hayward, J. L. and Bulbrook, R. D., *Acta Unio Int. Contra Cancrum*, 18, 890, 1962.
26. Bulbrook, R. D. and Hayward, J. L., *Acta Unio Int. Contra Cancrum*, 18, 893, 1962.
27. Bulbrook, R. D., *Vitam. Horm.*, 23, 329, 1965.
28. Hayward, J. L. and Bulbrook, R. D., *Cancer Res.*, 25, 1129, 1965.
29. Bulbrook, R. D. and Hayward, J. L., *Cancer Res.*, 25, 1135, 1965.
30. Bulbrook, R. D., *Proc. 13th Annu. Clin. Conf. Cancer*, Houston, Year Book Medical Publishers, Chicago, 1970, 51.
31. Bulbrook, R. D., in *Advances in Steroid Biochemistry and Pharmacology*, Vol. 1, Briggs, M. H., Ed., Academic Press, New York, 1970, 387.
32. Bulbrook, R. D., *Oncology (Basel)*, 3, 367, 1970.
33. Bulbrook, R. D., *J. Natl. Cancer Inst.*, 48, 1039, 1972.
34. Marmorston, J., *Ann. N.Y. Acad. Sci.*, 125, 959, 1966.
35. Bulbrook, R. D., Hayward, J. L., Spicer, C. C., and Thomas, B. S., *Lancet*, ii, 1235, 1962.
36. Juret, P., in *Prognostic Factors in Breast Cancer*, Forrest, A. P. M. and Kunkler, P. B., Eds., E. and S. Livingstone, Edinburgh, 1968, 393.
37. Kumaoka, S., Sakauchi, N., Abe, O., Kusama, M., and Takatani, O., *J. Clin. Endocrinol. Metab.*, 28, 667, 1968.
38. Cameron, E. H. D., Griffiths, K., Gleave, N., Stewart, H. J., Forrest, A. P. M., and Campbell, H., *Br. Med. J.*, 4, 768, 1970.
39. Bulbrook, R. D., Hayward, J. L., Thomas, B. S., and Spicer, C. C., *Lancet*, ii, 1238, 1962.
40. Bacigalupo, G. and Lingk, H., *Arch. Geschwulstforsch.*, 32, 95, 1968.
41. Gutierrez, R. M. and Williams, R. J., *Proc. Natl. Acad. Sci. U.S.A.*, 59, 938, 1968.
42. Chou, C. Y. and Wang, C. W., *Chin. J. Physiol.*, 14, 151, 1939.
43. Hardy, J. D., Richardson, E. M., and Dohan, F. C., *Surg. Gynecol. Obstet.*, 96, 448, 1953.
44. Thorn, G. W., Jenkins, D., and Laidlaw, J. C., *Recent Prog. Horm. Res.*, 8, 171, 1953.
45. Katz, J. L., Ackman, P., Rothwax, Y., Sachar, E., Weiner, H., Hellman, L., and Gallagher, T. F., *Psychosomatics*, 22, 1, 1970.
46. Tanaka, H., Manabe, H., Koshiyama, K., Hamanaka, Y., Matsumoto, K., and Uozumi, T., *Acta Endocrinol.*, (Kbh.), 65, 1, 1970.
47. Zumoff, B., Bradlow, L. H., Gallagher, T. F., and Hellman, L., *J. Clin. Endocrinol. Metab.*, 32, 824, 1971.
48. Bulbrook, R. D., and Hayward, J. L., *Lancet*, i, 519, 1967.
49. Bulbrook, R. D., Hayward, J. L., and Spicer, C. C., *Lancet*, ii, 395, 1971.
50. Shapiro, S., Strax, P., and Venet, L., *J.A.M.A.*, 215, 1777, 1971.
51. Venet, L., Strax, P., Venet, W., and Shapiro, S., *Cancer*, 28, 1546, 1971.
52. Feinleib, M. and Zelen, M., *Arch. Environ. Health*, 19, 412, 1969.
53. Moore, J. W., *Clin. Chim. Acta*, 39, 532, 1972.
54. Moore, J. W., *Clin. Chem.*, 19, 706, 1973.
55. Atkins, H. J. B., Falconer, M. A., Hayward, J. L., and MacLean, K. S., *Lancet*, i, 489, 1957.
56. Atkins, H. J. B., Falconer, M. A., Hayward, J. L., MacLean, K. S., Schurr, P. H., and Armitage, P., *Lancet*, i, 1148, 1960.
57. Hayward, J. L., in *Clinical Evaluation in Breast Cancer*, Hayward, J. L. and Bulbrook, R. D., Eds., Academic Press, New York, 1966, 131.
58. Allen, B. J., Hayward, J. L., and Merivale, W. H. H., *Lancet*, i, 496, 1957.
59. Hobkirk, R. and Forrest, A. P. M., *Lancet*, i, 636, 1957.
60. Plantin, L. O., Birke, G., Diczfalusy, E., Frankson, C., Hellstrom, J., Hultberg, S., and Westman, A., in *Endocrine Aspects of Breast Cancer*, Currie, A. R., Ed., E. and S. Livingstone, Edinburgh, 1958, 224.
61. Bulbrook, R. D., Greenwood, F. C., and Hayward, J. L., *Lancet*, i, 1154, 1960.
62. Hayward, J. L., Bulbrook, R. D., and Greenwood, F. C., *Mem. Soc. Endocrinol.*, 10, 144, 1961.
63. Hayward, J. L. and Bulbrook, R. D., in *Prognostic Factors in Breast Cancer*, Forrest, A. P. M. and Kunkler, P. B., Eds., E. and S. Livingstone, Edinburgh, 1968, 383.
64. Hayward, J. L., *Recent Results in Cancer Research*, Springer-Verlag, Berlin, 1970, 24.
65. Miller, H., Durant, J. A., Jacobs, A. G., and Allison, J. F., *Br. Med. J.*, i, 147, 1967.
66. Thomas, B. S., Bulbrook, R. D., and Hayward, J. L., *Br. Med. J.*, iii, 523, 1967.
67. Thomas, B. S., Bulbrook, R. D., Durant, J. A., Miller, H., and Ross, D. M., *Clin. Biochem.*, 2, 311, 1969.
68. Few, J. D., *J. Endocrinol.*, 41, 213, 1968.
69. Juret, P., Hayem, M., and Fleisler, A., *J. Chir.*, 87, 409, 1964.
70. Fotherby, K., Sellwood, R. A., and Burn, J. I., *Br. J., Surg.*, 55, 868, 1968.
71. Wilson, R. E. and Moore, F. D., in *Prognostic Factors in Breast Cancer*, Forrest, A. P. M. and Kunkler, P. B., Eds., E. and S. Livingstone, Edinburgh, 1968, 399.
72. Sarfaty, G., Tallis, M., and Pitt, P., *Med. J. Aust.*, in press, 1973.
73. Ghosh, P. C., Lockwood, E., and Pennington, G. W., *Br. Med. J.*, i, 328, 1973.
74. Sim, A. W., Hobkirk, R., Stewart, H. J., Blair, D. W., and Forrest, A. P. M., *Br. J. Cancer*, 14, 460, 1960.

75. **Ahlquist, K. A., Jackson, A. W., and Stewart, J. C.,** *Br. Med. J.*, i, 217, 1968.
76. **Bulbrook, R. D., Hayward, J. L., and Thomas, B. S.,** *Lancet*, i, 945, 1964.
77. **Kellie, A. E. and Wade, A. P.,** *Biochem. J.*, 55, 315, 1953.
78. **Thomas, B. S. and Bulbrook, R. D.,** *J. Chromatogr.*, 14, 28, 1964.
79. **Thomas, B. S.,** in *Gas-Liquid Chromatography of Steroids in Biological Fluids*, Lipsett, M., Ed., Plenum Press, New York, 1965, 1.
80. **Thomas, B. S., Eaborn, C., and Walton, D. R. M.,** *Chem. Commun.*, 2, 408, 1966.
81. **Thomas, B. S.,** in *Androgens in Normal and Pathological Conditions*, Vermeulen, A., Ed., Excerpta Medica Foundation, Amsterdam, 1966, 49.
82. **Thomas, B. S. and Walton, D. R. M.,** *J. Endocrinol.*, 41, 203, 1968.
83. **Few, J. D.,** *J. Endocrinol.*, 22, 33, 1961.
84. **Metcalf, M. G.,** *J. Endocrinol.*, 26, 415, 1963.
85. **Thomas, B. S.,** *J. Clin. Endocrinol. Metab.*, 25, 710, 1965.
86. **Deshpande, N., Hayward, J. L., and Bulbrook, R. D.,** *J. Endocrinol.*, 32, 167, 1965.
87. **Brownsey, B., Cameron, E. H. D., Griffiths, K., Gleave, E. N., Forrest, A. P. M., and Campbell, H.,** *Eur. J. Cancer*, 8, 131, 1972.
88. **Wang, D. Y., Swain, M. C., Hayward, J. L., and Bulbrook, R. D.,** *Recent Results Cancer Res.*, 39, 177, 1972.
89. **Wang, D. Y. and Herian, M.,** *Acta Endocrinol.* [Suppl.] (Kbh.), 177, Abstr. No. 30, 1973.
90. **Bénard, H., Bourdin, J.-S., Saracino, R., and Seeman, A.,** *Eur. J. Cancer*, 4, 141, 1968.
91. **Garnham, J. R., Bulbrook, R. D., and Wang, D. Y.,** *Eur. J. Cancer*, 5, 239, 1969.
92. **Bulbrook, R. D., Hayward, J. L., and Salokangas, R. A. A.,** *J. Endocrinol.*, 26, i, 1963.
93. **Van de Wiele, R. L., MacDonald, P. C., Gurpide, E., and Lieberman, S.,** *Recent Prog. Horm. Res.*, 19, 275, 1963.
94. **Wang, D. Y.,** in *The Human Adrenal Gland and Its Relation to Breast Cancer. First Tenovus Workshop*, Griffiths, K. and Cameron, E. H. D., Ed., Alpha Omega Alpha Publishing, Cardiff, Wales, 1969, 71.
95. **Loras, B. and Migeon, C. J.,** *Steroids*, 7, 459, 1966.
96. **Poortman, J., Thijssen, J. H. H., and Schwarz, F.,** *J. Clin. Endocrinol. Metab.*, 37, 101, 1973.
97. **Deshpande, N., Jensen, V., Bulbrook, R. D., and Doouss, T. W.,** *Steroids*, 9, 393, 1967.
98. **Brennan, M. J., Bulbrook, R. D., Deshpande, N., Wang, D. Y., and Hayward, J. L.,** *Lancet*, i, 1076, 1973.
99. **Warren, S.,** *Surg. Gynecol. Obstet.*, 71, 257, 1940.
100. **Wang, D. Y., Bulbrook, R. D., Thomas, B. S., and Friedman, M.,** *J. Endocrinol.*, 42, 567, 1968.
101. **Andre, C. M. and James, V. H. T.,** *Clin. Chim. Acta*, 43, 295, 1973.
102. **Levitz, M.,** *Steroids*, 1, 117, 1963.
103. **Doe, R. P., Vennes, J. A., and Flink, E. B.,** *J. Clin. Endocrinol. Metab.*, 20, 253, 1960.
104. **Schubert, K., Bacigalupo, G., and Frankenberg, G.,** *Arch. Geschwulstforsch.*, 17, 108, 1961.
105. **Bénard, H., Bourdin, J. S., Saracino, R. T., and Seeman, A.,** *Ann. Endocrinol.*, 23, 15, 1962.
106. **Beck, J. C., Blair, A. J., Griffiths, M. M., Rosenfeld, M. W., and McGarry, E. E.,** *Proc. Can. Cancer Res. Conf.*, 6, 3, 1966.
107. **Slaunwhite, W. R., Jr. and Sandberg, A. A.,** *J. Clin. Invest.*, 38, 384, 1959.
108. **Bulbrook, R. D., Hayward, J. L., Herian, M., Swain, M. C., Tong, D., and Wang, D. Y.,** *Lancet*, i, 628, 1973.
109. **Sandberg, A. A., Nelson, D. H., Glenn, E. M., Tyler, E. H., and Samuels, L. T.,** *J. Clin. Endocrinol. Metab.*, 13, 1445, 1953.
110. **Jensen, V., Deshpande, N., Bulbrook, R. D., and Doouss, T. W.,** *J. Endocrinol.*, 42, 425, 1968.
111. **Bell, E., Bulbrook, R. D., and Deshpande, N.,** *Lancet*, ii, 395, 1967.
112. **MacKay, W. D., Edwards, M. H., Bulbrook, R. D., and Wang, D. Y.,** *Lancet*, ii, 1001, 1971.
113. **Mattingly, D.,** *J. Clin. Pathol.* (Lond.), 15, 374, 1962.
114. **Deshpande, N. and Bulbrook, R. D.,** *J. Endocrinol.*, 28, 289, 1964.
115. **James, V. H. T. and Fraser, R.,** *J. Endocrinol.*, 34, xvi, 1966.
116. **Murphy, B. E. P.,** *Nature*, 201, 679, 1964.
117. **Murphy, B. E. P.,** *J. Clin. Endocrinol. Metab.*, 27, 973, 1967.
118. **Allaben, G. R. and Owen, S. E.,** *J.A.M.A.*, 112, 1933, 1939.
119. **Parsons, W. H. and McCall, E. F.,** *Surgery*, 9, 780, 1940.
120. **Haagensen, C. D.,** in *Diseases of the Breast*, W. B. Saunders, Philadelphia, 1956.
121. **Hertz, R.,** *Cancer*, 24, 1140, 1969.
122. **Fechner, R. E.,** *Cancer*, 26, 1204, 1970.
123. **Vessey, M. P., Doll, R., and Sutton, P. M.,** *Br. Med. J.*, iii, 719, 1972.
124. **Lewison, E. F.,** *Obstet. Gynecol. Surv.*, 17, 769, 1962.
125. **Lewison, E. F.,** *Cancer*, 18, 1558, 1965.
126. **Nissen-Meyer, R.,** *Acta Radiol.*, Suppl. 249, 1965.
127. Co-operative Breast Cancer Group, *Cancer Chemotherapy Rep.*, 41, Suppl. 1, 1964.
128. **MacMahon, B. and Cole, P.,** in *Current Problems in the Epidemiology of Cancer and the Lymphomas*, Grundmann, E. and Tulinius, H., Eds., Springer-Verlag, Berlin, 1972, 185.
129. **Feinleib, M.,** *J. Natl. Cancer Inst.*, 41, 315, 1968.

130. MacMahon, B., Cole, P., Lin, T. M., Lowe, C. R., Mirra, A. P., Ravnihar, B., Salberg, E. J., Valaoras, V. G., and Yuasa, S., *Bull. W.H.O.* 43, 209, 1970.
131. MacMahon, B., Feinleib, M., *J. Natl. Cancer Inst.*, 24, 733, 1960.
132. Ross, M. and Dorfman, R. I., *Cancer Res.*, 1, 52, 1941.
133. Taylor, H. C., Mecke, F. E., and Twombley, G. H., *Cancer Res.*, 3, 180, 1943.
134. Nathanson, I. T., *Cancer Res.*, 3, 132, 1943.
135. Brown, J. B., *Biochem. J.*, 60, 185, 1955.
136. Bacigalupo, G. and Schubert, K., *Eur. J. Cancer*, 2, 75, 1966.
137. Lemon, H. M., Wotiz, H. H., Parsons, L., and Mozden, P. J., *J.A.M.A.*, 196, 112, 1966.
138. Schweppe, J. S., Jungman, R. A., and Lewin, I., *Cancer*, 20, 155, 1967.
139. Brown, J. B., in *Endocrine Aspects of Breast Cancer*, Currie, A. R., Ed., E. and S. Livingstone, Edinburgh, 1958, 197–208.
140. Nissen-Meyer, R. and Sanner, T., *Acta Endocrinol.* (Kbh.), 44, 334, 1963.
141. Jull, J. W., Shucksmith, H. S., and Bonser, G. M., *J. Clin. Endocrinol. Metab.*, 23, 433, 1963.
142. Hellman, L., Fishman, J., Zumoff, B., Cassouto, J., and Gallagher, T. F., *J. Clin. Endocrinol. Metab.*, 27, 1087, 1967.
143. Lemon, H. M., *Cancer*, 23, 781, 1969.
144. Cole, P. and MacMahon, B., *Lancet*, i, 604, 1969.
145. MacMahon, B., Cole, P., Brown, J. B., Aoki, K., Lin, T. M., Morgan, R. W., and Woo, N–C., *Lancet*, ii, 900, 1971.
146. Wynder, E. L., Bross, I. J., and Hirayama, T., *Cancer*, 13, 559, 1960.
147. Gross, J., in *Proceedings of the First Breast Cancer Task Force Working Conference* (Williamsburg, Va.), National Cancer Institute, National Institutes of Health, U. S. Department of Health, Education, and Welfare, 1973.
148. Briggs, M., *Lancet*, i, 324, 1972.
149. Marmorston, J., Crowley, L. G., Myers, S. M., Stern, E., and Hopkins, C. E., *Am. J. Obstet. Gynecol.*, 92, 447, 1965.
150. Grattarola, R., *Cancer*, 17, 1119, 1964.
151. Bulbrook, R. D., Wang, D. Y., and Swain, M. C., in *Prolactin and Carcinogenesis*, Boyns, A. R. and Griffiths, K., Eds., Alpha Omega Alpha Publishing, Cardiff, Wales, 1972, 143.
152. Bucher, N. L. R. and Geschickter, C. F., *J. Clin. Endocrinol. Metab.*, 1, 58, 1941.
153. Nathanson, I. T., *Surgery*, 16, 108, 1944.
154. Taylor, H. C., *Surg. Gynecol. Obstet.*, 74, 326, 1942.
155. Marmorston, J., Crowley, L. G., Myers, S. M., Stern, E., and Hopkins, C. E., *Am. J. Obstet. Gynecol.*, 92, 460, 1965.
156. Swain, M. C., Hayward, J. L., and Bulbrook, R. D., *Eur. J. Cancer*, 9, 553, 1973.
157. Kern, W. H. and Brooks, R. N., *Cancer*, 24, 668, 1969.
158. Black, M. M., Barclay, T. H. C., Cutler, S. J., Hankey, B. F., and Asire, A. J., *Cancer*, 29, 338, 1972.
159. Strong, J. A., Brown, J. B., Bruce, J., Douglas, M., Klopper, A. I., and Loraine, J., *Lancet*, ii, 955, 1956.
160. Irvine, W. T., Aitken, E. H., Rendleman, D. F., and Folca, P. J., *Lancet*, ii, 791, 1961.
161. McAllister, R. A., Sim, A. W., Hobkirk, R., Stewart, H., Blair, D. W., and Forrest, A. P. M., *Lancet*, i, 1102, 1960.
162. Folca, P. J., Glascock, R. F., and Irvine, W. T., *Lancet*, ii, 796, 1961.
163. Jensen, E. V., Block, G. E., Smith, S., Kyser, K., and De Sombre, E. R., *Natl. Cancer Inst. Monogr.*, 34, 55, 1971.
164. Johansson, H., Terenius, L., and Thoren, L., *Cancer Res.*, 30, 692, 1970.
165. Feherty, P., Farrer-Brown, G., and Kellie, A. E., *Br. J. Cancer*, 25, 697, 1971.
166. Hahnel, R., Twaddle, E., and Vivian, A. B., *Steroids*, 18, 681, 1971.
167. Terenius, L., *Eur. J. Cancer*, 9, 291, 1973.
168. Korenman, S. G., Perrin, L. E., and McCallum, T. P., *J. Clin. Endocrinol. Metab.*, 29, 879, 1969.
169. Swain, M. C., *Clin. Chim. Acta*, 39, 455, 1972.
170. Bulbrook, R. D. and Hayward, J. L., *Lancet*, ii, 1033, 1969.
171. Sandberg, A. A., Rosenthal, H., and Slaunwhite, W. R., in *Proc. Sec. Int. Congr. Horm. Steroids*, Martini, L., Fraschini, F., and Motta, M., Eds., Excerpta Medica, Amsterdam, 1967, 707.
172. Wang, D. Y. and Bulbrook, R. D., *Eur. J. Cancer*, 5, 247, 1969.
173. Wang, D. Y. and Bulbrook, R. D., *J. Endocrinol.*, 39, 405, 1967.
174. Loraine, J. A., Strong, J. A., and Douglas, M., *Lancet*, ii, 575, 1957.
175. Boyland, E., Godsmark, B., Greening, W. P., Rigby-Jones, P., Stevenson, J. J., and Abul-Fadl, M. A. M., in *Endocrine Aspects of Breast Cancer*, Currie, A. R., Ed., E. and S. Livingstone, Edinburgh, 1958, 170.
176. Segaloff, A., Gordon, D., Carabasi, R. A., Horwitt, B. N., Schlosser, J. V., and Murison, P. J., *Cancer*, 7, 758, 1954.
177. Martin, F. I. R., *Br. Med. J.*, ii, 351, 1964.
178. Nandi, S., *Science*, 128, 772, 1958.
179. Nandi, S., *Proc. Soc. Exp. Biol. Med.*, 108, 1, 1961.
180. Hallowes, R. C., Wang, D. Y., Lewis, D. J., *J. Endocrinol.*, 57, 253, 1973.
181. Hallowes, R. C., Wang, D. Y., Lewis, D. J., Strong, C., and Dils, R., *J. Endocrinol.*, 57, 265, 1973.
182. Wang, D. Y., Hallowes, R. C., Smith, R. H., Amor, V., and Lewis, D. J., *J. Endocrinol.*, 52, 349, 1972.

183. Wang, D. Y., Hallowes, R. C., Bealing, J., Strong, C. R., and Dils, R., *J. Endocrinol.*, 53, 311, 1972.
184. Wilhelmi, A. E., *Can. J. Biochem.*, 39, 1659, 1961.
185. Tashjian, A. H., Jr., Levine, L., and Wilhelmi, A. E., *Endocrinology*, 77, 1023, 1965.
186. Apostolakis, M., *Vitam. Horm.*, 26, 197, 1968.
187. Bewley, T. A. and Li, C. H., *Science*, 168, 1361, 1970.
188. Lyons, W. R., Li, C. H., and Johnson, R. E., in 43rd Meeting of the Endocrine Society, Abstr. 7, 1961.
189. Chadwick, A., Folley, S. J., and Gemzell, C. A., *Lancet*, ii, 241, 1961.
190. Ferguson, K. A. and Wallace, A. L. C., *Nature*, 190, 632, 1961.
191. Forsyth, I. A., in *Growth Hormone*, Pecile, A. and Muller, E. E., Eds., Excerpta Medica, Amsterdam, 1968, 364.
192. Lewis, U. J., Singh, R. N. P., and Seavey, B. K., in *Prolactin and Carcinogenesis*, Boyns, A. R. and Griffiths, K., Eds., Alpha Omega Alpha Publishing, Cardiff, Wales, 1972, 4.
193. Glick, S. M., Roth, J., Yalow, R. S., and Berson, S. A., *Nature*, 199, 784, 1963.
194. Sinha, Y. N., Selby, F. W., Lewis, U. J., and Vanderlaan, W. P., *J. Clin. Endocrinol. Metab.*, 36, 509, 1973.
195. Pecile, A. and Muller, E. E., Eds., *Growth Hormone*, Excerpta Medica, Amsterdam, 1968.
196. Greenwood, F. C., James, V. H. T., Meggitt, B. F., Miller, J. D., and Taylor, P. H., in *Prognostic Factors in Breast Cancer*, Forrest, A. P. M. and Kunkler, P. B., Eds., E. and S. Livingstone, Edinburgh, 1968, 409.
197. Pearson, O. H., Llerena, O., Samaan, N., and Gonzalez, D., in *Prognostic Factors in Breast Cancer*, Forrest, A. P. M. and Kunkler, P. B., Eds., E. and S. Livingstone, Edinburgh, 1968, 421.
198. Meites, J., Cassell, E., and Clark, J., *Proc. Soc. Exp. Biol. Med.*, 137, 1225, 1971.
199. Hilf, R., Bell, C., Goldenberg, H., and Michel, I., *Cancer Res.*, 31, 1111, 1971.
200. Krant, M. J., Brandrup, C. S., Greene, R. S., Pochi, P. E., and Strauss, J. S., *Nature*, 217, 463, 1968.
201. Burton, J. L., Cunliffe, W. J., and Shuster, S., *Br. Med. J.*, i, 665, 1970.
202. Wang, D. Y., Bulbrook, R. D., Guillebaud, J., and Lewis, A., *Eur. J. Cancer*, 8, 381, 1972.
203. McCalister, A. and Welbourne, R. B., *Br. Med. J.*, i, 1669, 1962.
204. Hwang, P., Guyda, H., and Friesen, H., *Proc. Natl. Acad. Sci. U.S.A.*, 68, 1902, 1971.
205. MacMahon, B., Lin, T. M., Lowe, C. R., Mirra, A. P., Ravnihar, B., Salber, G. J., Trichopoulos, D., Valaora, V. G., and Yuasa, S., *Bull. W. H. O.*, 42, 185, 1970.
206. Ehni, G. and Eckles, N. E., *J. Neurosurg.*, 16, 628, 1959.
207. Bowers, C. Y., Friesen, H. G., Hwang, P., Guyda, H. J., and Folkers, K., *Biochem. Biophys. Res. Commun.*, 45, 1033, 1971.
208. Bond, W. H., in *The Treatment of Carcinoma of the Breast*, Jarrett, A. S., Ed., Excerpta Medica, 1967, 24.
209. Peters, M. V., in *Prognostic Factors In Breast Cancer*, Forrest, A. P. M. and Kunkler, P. B., Eds., E. and S. Livingstone, Edinburgh, 1968, 65.
210. Besser, G. M. and Edwards, C. R. W., *Br. Med. J.*, ii, 280, 1972.
211. European Breast Cancer Group, *Eur. J. Cancer*, 8, 155, 1972.
212. Murray, R. M. L., Mozaffarian, G., and Pearson, O. H., in *Prolactin and Carcinogenesis*, Boyns, A. R. and Griffiths, K., Eds., Alpha Omega Alpha Publishing, Cardiff, Wales, 1972, 158.
213. Boyns, A. R., Cole, E. N., Griffiths, K., Roberts, M. M., Buchan, R., Wilson, R., and Forrest, A. P. M., *Eur. J. Cancer*, 9, 99, 1973.
214. Turkington, R. W., *J. Clin. Endocrinol. Metab.*, 34, 247, 1972.
215. Jacobs, L. S., Snyder, P. J., Wilber, J. F., Utiger, R. D., and Daughaday, W. H., *J. Clin. Endocrinol. Metab.*, 33, 996, 1971.
216. Repert, R. W., *J. Mich. St. Med. Soc.*, 51, 1315, 1952.
217. Loeser, A. A., *Br. Med. J.*, ii, 1380, 1954.
218. Bulbrook, R. D., Thomas, B. S., and Utsunomiya, J., *Nature*, 201, 189, 1964.
219. Bulbrook, R. D., Thomas, B. S., Utsunomiya, J., and Hamaguchi, E., *J. Endocrinol.*, 38, 401, 1967.
220. Hellman, L., Bradlow, H. L., Zumoff, B., Fukushima, D. K., and Gallagher, T. F., *J. Clin. Endocrinol. Metab.*, 19, 936, 1959.
221. Edelstyn, G. A., Lyons, A. R., and Welbourne, R. B., *Lancet*, i, 670, 1958.
222. Carter, A. C., Feldman, E. B., and Schwarz, H. L., *J. Clin. Endocrinol. Metab.*, 20, 477, 1960.
223. Sommers, S. C., *Lab. Invest.*, 4, 160, 1955.
224. Hennes, A., Wajchenberg, B. L., and Ulhoa-Cintra, A. B., *Port. Med.*, 44, 693, 1960.
225. Bayliss, P. F. C. and Van't Hoff, W., *Lancet*, ii, 1399, 1969.
226. Forsyth, I. A., Besser, G. M., Edwards, C. R. W., Francis, L., and Myres, R. P., *Br. Med. J.*, iii, 225, 1971.
227. Hershman, J. M., Kojima, A., and Friesen, H. G., *J. Clin. Endocrinol. Metab.*, 36, 497, 1973.
228. Gautvik, K. M., Weintraub, B. D., Graeber, C. T., Maloof, F., Zuckerman, J. E., and Tashjian, A. H., *J. Clin. Endocrinol. Metab.*, 36, 135, 1973.

ENZYMES IN CARCINOMA

S. Winsten

TABLE OF CONTENTS

I. General Introduction . . . 219
II. Adenosine Deaminase . . . 220
III. Alkaline Phosphatase . . . 221
IV. Alkaline Phosphatase Isoenzymes . . . 223
V. Copper Resistant Serum Acid Phosphatase . . . 224
VI. Gamma-glutamyl Transpeptidase . . . 226
VII. Lactic Acid Dehydrogenase . . . 228
VIII. Lactic Acid Dehydrogenase Isoenzymes . . . 229
IX. Phosphohexoseisomerase . . . 231
References . . . 234

I. GENERAL INTRODUCTION

The simple diagnostic test for cancer has been an elusive Holy Grail that the Galahads of science have pursued for many years. Several procedures have initially shown great promise, but almost all have fallen by the wayside as more meticulous experimentations have disclosed a lack of specificity and an inability to differentiate between normal and oncologic biochemistry. Clinical enzymology has been one of the areas most assiduously investigated. Many biochemists, including the author, felt that because of the nature of the enzyme action, early detection of cancer might be accomplished by the investigation of enzyme activity in body fluids. It was the hope of most clinical enzymologists that there would be a sufficient increment in enzyme levels at the site of origin of the carcinoma so that the changes could be easily detected in the body fluids. Unfortunately, this is not true. The conversion of normal progeny to malignant cells does not sufficiently alter the enzyme milieu so that those changes can be detected in body fluids. Most malignant growths are not characterized by a *de novo* elaboration of new enzyme pathways; in fact, the anaplastic tumors actually lose their ability to produce enzymes rather than increase enzyme activity. Even in the neoplasm in which enzyme increases have been noted, a large number of neoplastic cells are often required to overcome the dilution factor of the body fluid. Therefore, the detection of occult carcinoma by enzyme analysis appears to be a futile pursuit.

Leaving aside the impossible dream, however, enzyme determinations do play a role in the diagnosis and treatment of carcinoma. Although this text is concerned with the biochemistry of women, any discussion of this topic must, like the women liberationists, cross sex lines. This chapter cannot be limited to carcinoma specifically affecting female anatomical structures, since the biochemistry of malignancy and especially metastatic disease does not discriminate between the sexes. The author will, therefore, be forced to describe, because of their relationship to metastatic disease, such routine enzyme techniques as alkaline phosphatase and lactic acid dehydrogenase. The discussion will also include an analysis and a description of specific isoenzyme methods which may be used to pinpoint more specifically the location of the primary carcinoma or the metastatic region.

The enzyme analysis of vaginal fluid has been assiduously studied for the detection of female neoplasms of the urogenital system. Recent evidence has indicated that this area of investigation as currently pursued certainly has not provided any valuable information which could not be obtained by other means. However, we will include a description of an enzyme analysis in vaginal fluid where applicable, so that investigators interested in reopening this area will be able to follow a well-defined technique. Currently the author's laboratory is assaying enzyme levels of tissues obtained from patients with breast carcinoma. Although it is recognized that we are still in the preliminary stages of our investigation, a description of the author's technique will be included so that biochemists interested in this area may have the benefit of his experience.

Whenever possible, in addition to presenting workable methodology, the available clinical experience in the area of treatment and detection of neoplasm will be enumerated. It is hoped that a clinician using these tests faced with the problems of oncology will find his unenviable task lightened and his patients' often dark future brightened.

II. ADENOSINE DEAMINASE

A. Principle

Adenosine deaminase (adenosine aminohydrolase, E.C. 3.5.4.4.) catalyzes the following reaction

Adenosine → inosine + ammonia

Ammonia formed during 30 min of incubation at 37°C in the presence of adenosine in a phosphate buffer at pH 7.4 is estimated colorimetrically using a modified phenol-alkaline hypochlorite procedure. The enzyme activity is expressed in terms of μg ammonia-nitrogen formed per hr per ml serum at 37°C.

B. Specimen

The assay is run on serum. Since the enzyme appears to be unstable at high concentrations and there is an increase in endogenous ammonia in blood at room temperatures, it is advisable to place the sample tube on ice immediately after collection and freeze the separated serum sample as soon as possible at −20°C.

C. Reagents

All reagents should be prepared in deionized distilled water or bottled water for parenteral use to minimize interference from ammonium ions.

1. Stock sodium phosphate solutions
   Stock solution A – dissolve 13.9 g monobasic sodium phosphate ($NaH_2PO_4 \cdot 7H_2O$) in water and make up to 500 ml. Stock solution B – dissolve 26.7 g $Na_2HPO_4 \cdot 7H_2O$ or 35.9 g $Na_2HPO_4 \cdot 12H_2O$ in water and make up to 500 ml. Keep both solutions refrigerated. Warm to dissolve any precipitate before use.
2. Working phosphate buffer, 0.1 *M*, pH 7.4
   Mix 19 ml of stock solution A with 81 ml of stock solution B. Adjust the pH to 7.4 on meter. (If pH is above 7.4 add A, if below 7.4 add B.) Make up to 200 ml with water. Store in the refrigerator.
3. Buffered substrate
   Dissolve 250 mg adenosine (Sigma Chemical Co.) in 50 ml of 0.1 *M* phosphate buffer pH 7.4. Keep refrigerated, discard after 15 days. (If the test is run occasionally, prepare 5 mg adenosine/ml only as required.)
4. Phenol reagent
   Dissolve 5 g of phenol and 25 mg sodium nitroprusside in water and make up to 100 ml.
5. Alkaline hypochlorite reagent
   Dissolve 2.5 g sodium hydroxide in water, add 5 ml of sodium hypochlorite solution (laboratory grade, containing at least 5% available chlorine), and make up to 100 ml.
6. Ammonium sulfate, stock solution

Dissolve 70.7 mg ammonium sulfate in 0.1 *N* H_2SO_4 and make up to 100 ml with the same solution. This solution is equivalent to 15 mg ammonia-nitrogen/100 ml.

7. Ammonium sulfate standard

Dilute 1 ml of ammonium sulfate stock solution to 10 ml with water. This solution is equivalent to 1.5 mg ammonia-nitrogen/100 ml. Prepare fresh with each determination.

The phenol reagent, alkaline hypochlorite reagent, and ammonium sulfate stock solutions are available commercially from Hyland Division, Travenol Laboratories, Inc., Costa Mesa, Calif.

D. Procedure

1. Warm the working phosphate buffer solution to 37°C to ensure complete solution of the precipitate formed during refrigeration.

2. Pipette 1 ml phenol reagent in four tubes marked "control," "test," "standard," and "blank."

3. Pipette 0.2 ml of buffered substrate into a 12 X 75 mm test tube. Add 0.4 ml serum, mix on a vortex mixer, immediately withdraw 0.1 ml of the mixture, and add to the tube marked "control."

4. Incubate the 12 X 75 mm tube containing serum and buffered substrate for exactly 30 min in a 37°C water bath. Cover the tube with parafilm to minimize evaporation.

5. At the end of 30 min, withdraw 0.1 ml of the mixture and add to the tube marked "test."

6. Add 0.1 ml of ammonium sulfate standard to the tube marked "standard" and 0.1 ml of water to the tube marked "blank."

7. After at least 3 min, pipette 1 ml of alkaline hypochlorite reagent into all 4 tubes. Mix and allow to stand for at least 45 min.

8. Read the absorbance of the blank, standard, control, and test at 660 nm in a suitable colorimeter or spectrophotometer, adjust to zero with water.

E. Calculation

$$\frac{Abs_T - Abs_C}{Abs_S - Abs_B} \times 45.5 = \text{adenosine deaminase units/ml serum}$$

where

Abs_T = absorbance of "test"
Abs_C = absorbance of "control"
Abs_S = absorbance of "standard"
Abs_B = absorbance of "blank"

The factor used in the calculations is arrived at as follows:

$$\frac{Abs_T - Abs_C}{Abs_S - Abs_B} \times 1.5 = \mu g \text{ ammonia-nitrogen formed/0.1 ml reaction mixture}$$

using (0.066 ml serum/30 min) in the above analysis this gives

$$\frac{Abs_T - Abs_C}{Abs_S - Abs_B} \times \frac{1.5 \times 2}{0.066} = \mu g \text{ ammonia-nitrogen formed/hr/ml serum}$$

$$\frac{Abs_T - Abs_C}{Abs_S - Abs_B} \times 45.5 = \text{units/ml serum}$$

F. Interpretation

The normal range for adenosine deaminase as determined in the author's laboratory is 5 to 20 units/ml/hr serum. Elevated serum adenosine deaminase has been reported in carcinoma. The usefulness of this parameter in the evaluation of cancer is equivocal. There is no question that in acute leukemia the level may reach five times the upper limit of normal.[2] This is one of the few enzymes in which a nucleoside such as adenosine is used as a substrate. It would appear that reinvestigation of this enzyme and an attempt to determine whether there are isoenzyme forms would appear to have some value in the study of neoplasms.

III. ALKALINE PHOSPHATASE

A. Principle

Many substrates and buffers have been used to measure alkaline phosphatase (E.C. 3.1.3.1.). They have ranged from the beta glycerol phosphate and the veronal buffer of Bodansky to *p*-nitrophenol phosphate and tris and ethanolamine buffers. We have had recent experience with sodium thymolphthalein phosphate and an amino methyl propanol buffer. The principle of technique can be summarized as follows. Serum alkaline phosphatase hydrolyzes sodium thymolphthalein phosphate to produce thymolphthalein and phosphate. The concentration of the dye thymolphthalein can be measured directly after increasing the pH with the introduction of an alkaline solution. Not only is there a development of a blue color, which is proportional to the concentration of enzyme, but the addition of alkali halts the enzyme activity.[3,4]

B. Specimen

An unhemolyzed serum sample should be obtained. Plasma should not be used since EDTA, oxalate, and citrate inhibit the alkaline phosphatase reaction. The enzyme is stable for 8 hr at 25°C and for 7 days at 4°C. Frozen specimens (-20°C) will be stable for at least 2 months.

C. Reagents

1. Magnesium chloride, 0.3 *M* ($MgCl_2 \cdot 5H_2O$)

   Dissolve 6.009 g of magnesium chloride in water and dilute to 100 ml. This is stable, when refrigerated, for at least 3 months.

2. Brij-35, 20%

   Dissolve 20 g of Brij-35 in a beaker containing approximately 50 ml of water. Gentle warming may be necessary. Cool and dilute to 100 ml. This is stable at room temperature for 3 months.

3. Hydrochloric acid, 6 *N*

   Carefully dilute 50 ml of concentrated HCl to 100 ml with water.

4. Concentrated buffer: 2-amino-2-methyl-1-propanol, pH 10.15

   To 32.2 g of 2-amino-2-methyl-1-propanol, add sufficient 6 *N* HCl to adjust the pH to approximately 10.30 to 10.35 (at 25°C). This should be done in 100 ml volumetric flask. Add 1 ml of $MgCl_2$ (Reagent 1) and 10 ml Brij-35 (Reagent 2) and bring the volume within 5 ml of the 100-ml mark with water. With a 0.1-ml pipette add sufficient 6 *N* HCl so that the pH is brought to 10.25 and then bring the level of the buffer to 100 ml with water.

5. Diluted buffer: 10 ml of this concentrated buffer pH 10.25 buffer when diluted with 90 ml of water will give a pH of 10.15 at 25°C. This should be checked. Place the diluted buffer in a tightly sealed bottle and keep refrigerated. This buffer is stable for 1 month when refrigerated.

6. Buffered substrate: Sodium thymolphthalein monophosphate in aminomethyl propanol buffer, 3 mg/ml

   Dissolve 0.3 g of sodium thymolphthalein monophosphate in approximately 85 ml of water. Add 10 ml of concentrated buffer (Reagent 4) and dilute to 100 ml with water. The final pH should be 10.15 at 25°C. Keep refrigerated in a tightly stoppered bottle. This is stable for at least 2 weeks.

7. Alkali reagent, 0.1 *N* sodium hydroxide (NaOH) 0.1 *M* sodium carbonate (Na_2CO_3)

   Dissolve 4 g of sodium hydroxide and 10.6 g of anhydrous sodium carbonate in water, dilute to 1 liter. The solution is stable indefinitely when stored at 4°C.

8. Thymolphthalein stock standard, 0.01 *M*

   Dissolve 430.5 mg of thymolphthalein in 70% *n*-propanol and dilute to 100 ml with propanol. This is stable indefinitely when refrigerated.

All reagents are available as an alkaline phosphatase reagent set from the Worthington Biochemical Company.

D. Standardization

1. Working standard: 1 ml, 0.10 μmol of thymolphthalein

   Dilute 1 part of stock standard (Reagent 7) with 9 parts of dilute buffer.

2. Set up six cuvettes of 1 cm path length as follows and read against a reagent blank at 590 nm.

Cuvette no.		1	2	3	4	5	6
Standard	ml	0.00	0.20	0.40	0.60	0.80	1.00
Buffer (diluted)	ml	1.00	0.80	1.60	0.40	0.20	0.00
Alkali	ml	5.1	5.1	5.1	5.1	5.1	5.1
I.U./liter		0	20	40	60	80	100

3. One I.U. is defined as 1 micromol of thyolphthalein formed per minute per liter of serum at 37°C.

4. Experiments with several sera have indicated that with the Gilford 300N Spectrophotometer®, the curve is linear to 600 I.U. liter. It is, therefore, possible to choose a standard value and calculate activity on the basis of that value. Using the Worthington Kit, the author's laboratory has chosen the 50 I.U. liter standard for calculation as follows:

$$\frac{\text{Absorbance of unknown}}{\text{Absorbance of standard}} \times 50 = \text{activity of unknown in I.U./liter}$$

E. Procedure

1. A reagent blank and at least one standard must be run with each assay batch. The standard should have a value of 50 I.U. liter; its preparation is described above.

2. Label tubes for each serum sample and the reagent blank. Into each tube pipette 1.0 ml buffered substrate and incubate for 5 min at 37°.

3. Add 0.1 ml distilled water to the tube marked reagent blank.

4. At 10-sec intervals add 0.1 ml of each serum with an accurate micropipette to the appropriate tubes.

5. A second reagent blank (Cuvette 1 in the calibration series) should be set up. (Note this does not undergo stage 2 above.)

6. Exactly 10 min after the addition of serum, add 5.0 ml of alkali reagent to each tube at the same time intervals.

7. Stopper and mix by inversion and read against the reagent blank that has undergone stage 2 at 590 nm. Check that this reagent blank absorbance does not exceed the stage 5 reagent blank by more than 0.300. If it does, prepare fresh reagents. The blue color formed is stable for at least 4 hr.

F. Interpretation

The normal range is 15 to 45 I.U. liter with no obvious difference between the sexes. Using this technique, Roy has observed a coefficient of variation of approximately 14% and the author observed a day-to-day variability with serum controls of about 8%.[5] Due to choice of wavelength (590 nm), any interference with bilirubin is essentially eliminated.

Elevated alkaline phosphatases are observed in many conditions including neoplasms. In women, increases in alkaline phosphatase levels have been reported with bone and hepatic metastases to primary breast carcinoma. Elevated alkaline phosphatases in anicteric serum are often the first indication of metastatic spread. With the above technique it is not possible to determine the tissue origin of the elevated alkaline phosphatase. Some investigators have used isoenzyme separation of alkaline phosphatase to attempt to determine the site of metastatic involvement. A relatively simple technique for isoenzyme separation will be described in the next section.

IV. ALKALINE PHOSPHATASE ISOENZYMES

A. Principle

Alkaline phosphatase (E.C. 3.1.3.1.) is determined in serum before and after incubation with 2.5 *M* urea. The residual activity after urea treatment is an indication of the alkaline phosphatase isoenzyme pattern of the serum, since the isoenzyme originating in bone is more susceptible to denaturation by urea than the isoenzyme from liver.[6]

B. Reagents

1. Urea, 5 *M*

Prepared fresh. Dissolve 30 g of urea in 81 ml of water and dilute to 100 ml.

2. All other reagents are similar to those described in the alkaline phosphatase technique in this chapter.

C. Procedure

1. Prepare three tubes as follows:

Sample	Urea blank (ml)	Untreated (ml)	Urea treated (ml)
Serum	–	0.05	0.05
Water	0.05	0.05	–
Urea	0.05	–	0.05

2. Add the urea last and incubate at 37°C for 15 min.

3. At the completion of the incubation period, add 0.5 ml thymolphthalein monophosphate (Reagent 5 in alkaline phosphatase technique) and proceed to estimate the alkaline phosphatase in each of the tubes as described previously.

D. Calculation

The urea blank value should be zero. Calculate the alkaline phosphatase activity in the untreated and treated serum as described in the alkaline phosphatase method. Express the results in terms of residual activity as calculated by the following formula:

$$\text{Percent residual activity} = \frac{\text{Alkaline phosphatase in urea treated serum}}{\text{Alkaline phosphatase in untreated serum}}$$

E. Interpretation

It is obvious from Table 1 that the technique is not ideal. There is a large gray zone and repeated experimentation in the present author's laboratory both with isoenzyme separation by electrophoresis and phenylalanine inhibition or heat inactivation has not resulted in measurable improvement in the interpretation of the results. Separation of alkaline phosphatase activity in prepubertal males and females appears to be a futile endeavor, since during this period of bone growth only the bone alkaline phosphatase can be determined by this technique. Furthermore, this procedure should only be performed on sera with elevated alkaline phosphatase, since normal levels of alkaline phosphatase in nonpregnant females are invariably intestinal in origin. During the latter stages of normal pregnancy, the alkaline phosphatase isoenzyme is urea resistant.

TABLE 1

Interpretation of Results: Alkaline Phosphatase

Residual activity	Tissue origin of alkaline phosphatase
0–25%	Bone
25–35%	Questionable bone or liver
35–65%	Liver
65% and over	Intestinal or placental

V. COPPER RESISTANT SERUM ACID PHOSPHATASE

A. Principle

Acid phosphatase (E.C. 3.1.3.2.) reacts with diphenyl phosphate, producing phenol and phosphate. The amount of phenol liberated can be determined by the reaction of phenol with Folin-Ciocalteau reagent. If a 0.2 nm copper ion is added, erythrocytic acid phosphatase is inhibited.[7]

B. Sample

Serum should be collected and rapidly removed from the clot. For copper resistant serum acid phosphatase, storage of separated serum at −20°C is adequate.

C. Reagents

1. Acetate buffer, pH 4.9

   Dissolve 27.2 g sodium acetate trihydrate in 500 ml of water. Add 6.6 ml of glacial acetic acid and dilute to 1,000 ml with water.

2. Phenol reagent (Folin-Ciocalteau). This reagent can be purchased commercially or prepared as described below.

   Add 100 g sodium tungstate ($Na_2WO_4 \cdot 2H_2O$), 25 g sodium molybdate ($Na_2MoO_4 \cdot 2H_2O$) and 700 ml of water to a 2-l. round-bottom flask.

When all the solids are dissolved, add 50 ml of 85% orthophosphoric acid and 100 ml concentrated hydrochloric acid. Connect the round-bottom flask to a reflux condenser fitted with ground glass points and reflux gently for approximately 10 hr. At the end of this refluxing period, add 150 g of lithium sulfate (Li_2SO_4) 50 ml of water, and 20 drops of bromine. Boil the flask for 15 min in a fume hood without the reflux condenser to remove excess bromine. Cool and dilute to 1 liter with water. Filter if necessary. The reagent should be yellow without a greenish tint. Store in refrigerator. If the reagent turns green during storage, it can be reoxidized to the yellow color with the bromine treatment.

3. Dilute phenol reagent
   Dilute 1 vol of stock phenol reagent with 2 vol of water.
4. Disodium phenylphosphate,1%
   Dissolve 1 g of disodium phenylphosphate in 100 ml of water.
5. Copper sulfate reagent, 0.26%
   Dissolve 262 mg of copper sulfate ($CuSo_4 \cdot 5H_2O$) in 100 ml of water.
6. Sodium carbonate reagent, 20%
   Dissolve 20 g sodium carbonate (Na_2CO_3, anhydrous) in 100 ml of water.

D. Procedure

1. First, 25-ml centrifuge tubes are prepared as shown in the following table:

	Blank (ml)	Serum control (ml)	Serum test (ml)
Acetate buffer	8.8	8.8	8.8
Copper sulfate solution	0.2	0.2	0.2
Phenyl-phosphate	1.0	1.0	1.0
Serum	–	0.5	0.5
Water	0.5	–	–
Phenol Reagent	–	4.5	–

Stopper the tubes, incubate for 60 min, then add

Diluted phenol reagent	4.5	–	4.5

2. Mix and centrifuge the tubes, and transfer 10 ml of the supernatant to 1-cm cuvettes.

3. To each cuvette add 2.5 ml of 20% sodium carbonate and read the serum cuvette aagainst the blank cuvette at 660 nm after 1 hr.

4. The μmoles of phenol in the incubated and nonincubated tubes are determined from a calibration curve. The difference between the tubes represents the phenol released during 1-hr incubation.

5. Calibration curve

a. Prepare a 0.1 *N* HCl solution. Dilute 8 ml of concentrated HCl to 1 liter with water.

b. Phenol standard (mg/ml)
   Dissolve 100 mg of phenol in 100 ml of water. Since phenol has a molecular weight of 94, this solution is equal to 106 μmol/ml.

c. The following 6 standard tubes are prepared:

Standard #	Stock standard (ml)	0.1 *N*/HCl (ml)	μmol phenol/100 ml
1	0	10	0
2	0.1	9.9	1
3	1	9	10
4	2	8	20
5	5	5	50
6	10	0	100

d. Then 4.5 ml of phenol reagent and 2.5 ml of 20% Na_2CO_3 are added to each tube and the tubes, 2 to 6, are read against tube #1 at 660 nm after 1 hr incubation at room temperature.

Activity of serum acid phosphatase = amount of phenol liberated in the incubated tube in μmol/100 ml/hr – amount of phenol in the nonincubated tube in μmol/100 ml/hr.

E. Interpretation

The normal female range for this enzyme is 0 to

24 μmol/100 ml serum of acid phosphatase activity. The mean in females appears to be approximately 12 μmol/100 ml.

The coefficient of variation for this method is approximately 8%, well within the limits of enzyme assays. Although serum acid phosphatase is primarily used to detect prostatic carcinoma in males, several investigators have reported that serum contains a mixture of phosphomonoesterases derived from several varieties of cells. In fact, it appears to be a ubiquitous enzyme, although the role in mammalian biochemistry is still obscure. The technique described in this section apparently lends itself to detection of metastatic breast cancer in females. In a study in 1956, Reynolds et al.[6] reported that of 70 cases of metastatic carcinoma in breast, 52 had levels exceeding the upper limits of normal. In addition, Lemon and Reynolds have reported a correlation between regression of osseous metastases, decreases in hypercalcemia, and a decrease in serum acid phosphatase.[7] It would appear that a reinvestigation of this enzyme in conjunction with some of the enzymes indicating visceral metastasis such as gamma-glutamyl transpeptidase might provide a more rational approach to the chemotherapy of breast carcinoma than is currently available.

VI. GAMMA–GLUTAMYL TRANSPEPTIDASE

A. Principle

Gamma-glutamyl transpeptidase (GGTP) activity can be estimated in serum using L-gamma-glutamyl-*p*-nitroanilide as a substrate and estimating the enzymatic release of *p*-nitroaniline employing the Bratton-Marshall reaction.[8]

B. Specimen

Unhemolyzed serum may be kept for at least 8 hr at 25°C and up to 2 weeks at 4°C without any appreciable loss of activity.

C. Reagents

1. Hydrochloric acid, 0.1 *N*
   Add 0.8 ml concentrated hydrochloric acid to approximately 50 ml water and make up to 100 ml.
2. Tris HCl buffer, 0.1 *M*, pH 8.1, containing glycylglycine, 0.1 *M*
   Dissolve 1.21 g tris (hydroxymethyl) aminomethane in approximately 30 ml water. Add 1.32 g glycylglycine and mix until dissolved. Add 0.1 *N* hydrochloric acid gradually with constant stirring until a pH of 8.1 is obtained as measured on a pH meter. Make up to 100 ml.
3. Saturated substrate solution
   Add 25 mg L-gamma-glytamyl-*p*-nitroanilide to 10 ml tris HCl buffer containing glycylglycine. Heat for 5 min in a 56°C water bath and shake vigorously to dissolve as much as possible. Cool and filter. Prepare fresh for each test.
4. Acetic acid, 10% (v/v)
   Dilute 10 ml glacial acetic acid to 100 ml with water.
5. Sodium nitrite solution, 0.1% (w/v)
   Dissolve 100 mg sodium nitrite in water and make up to 100 ml. Store in a refrigerator.
6. Ammonium sulfamate solution, 1.0% (w/v)
   Dissolve 1 g ammonium sulfamate in water and make up to 100 ml. Store in a refrigerator.
7. *N*-naphthylamine dihydrochloride (NED) solution, 0.05% (w/v)
   Dissolve 50 mg NED in water and make up to 100 ml. Store in an amber-colored bottle in a refrigerator.
8. *p*-Nitroaniline, stock solution, 1 μmol/ml
   Dissolve 138 mg *p*-nitroaniline in water and make up to 1 liter.
9. *p*-Nitroaniline working standard, 0.2 μmol/ml
   Dilute 10 ml stock solution to 50 ml with water.

D. Procedure

1. Warm the saturated substrate solution by allowing it to sit in a water bath at 37°C for at least 15 min.
2. Pipette 0.2 ml serum into a 12 X 75 mm tube. Add 1.8 ml of previously warmed substrate solution and mix on a vortex mixer. Immediately withdraw 0.5 ml of the mixture and add to 2 ml 10% acetic acid in a test tube marked "control."
3. Incubate the 12 X 75 mm tube containing the rest of the substrate-serum mixture for

exactly 10 min, withdraw 0.5 ml of the mixture and add to 2 ml 10% acetic acid in a test tube marked "test."

4. Pipette 0.5 ml working standard and 0.5 ml water in two similar test tubes marked "standard" and "blank," respectively. Add 2 ml 10% acetic acid to each.

5. Add 1 ml sodium nitrite solution to all four tubes and mix. Allow to stand for 3 min.

6. Add 1 ml ammonium sulfamate solution to all four tubes and mix. Allow to stand for at least 3 min.

7. Add 1 ml NED solution to all four tubes and mix. Measure the absorbance at 540 nm of all four tubes in a suitable spectrophotometer or colorimeter adjusted to zero with water.

E. Calculation

$$\frac{Abs_T - Abs_C}{Abs_S - Abs_B} \times 200 = \text{units GGTP/liter of serum}$$

where

Abs_T = absorbance of "test" at 540 nm
Abs_C = absorbance of "control" at 540 nm
Abs_S = absorbance of "standard" at 540 nm
Abs_B = absorbance of "blank" at 540 nm

A unit of GGTP activity is that which releases 1 μmol *p*-nitroaniline/min at 37°C.

Note: Because the standard is equivalent to 0.1 μM, the factor used in the calculation is arrived at as follows:

$$\frac{Abs_T - Abs_C}{Abs_S - Abs_B} \times 0.1 = \mu\text{mol } p\text{-nitroaniline released/10 min/0.05 ml serum}$$

$$\frac{Abs_T - Abs_C}{Abs_S - Abs_B} \times \frac{0.1}{10} = \mu\text{mol } p\text{-nitroaniline released/min/0.05 ml serum}$$

$$\frac{Abs_T - Abs_C}{Abs_S - Abs_B} \times \frac{0.1}{10} \times 20{,}000 = \mu\text{mol } p\text{-nitroaniline released/min/liter serum}$$

$$\frac{Abs_T - Abs_C}{Abs_S - Abs_B} \times 200 = \text{units/liter serum}$$

F. Accuracy and Precision

Experimental results in the author's laboratory indicate that this technique is well within the acceptable limits of accuracy and precision for enzymes. Reproducibility of this methodology is approximately 10%. Szasz described a kinetic procedure for performing gamma-glutamyl transpeptidase.[9] This method is available in kit form from the Boehringer-Manheim Corporation. The substrate is the same as the method described above except that glycine is used as an acceptor for liberated glutamate residue. The technique is based on the increased liberation of *p*-nitroaniline measured over a period of time. Since the optimal wavelength of *p*-nitroaniline is 405 nm, this methodology is not recommended unless a sensitive spectrophotometer is available, for the changes in absorbance per minute in the normal range are relatively small. In addition, body fluids and tissues other than serum in which low levels of enzyme activity are expected are more easily assayed by the method described above than by Szasz's methodology. It is advisable to use serum for the determination of gamma-glutamyl transpeptidase since a slight inhibition of the enzyme does occur when oxalate, citrate, or fluoride is present in the usual concentrations found in plasma collecting tubes.

Several investigators have detected induction of this enzyme by pharmacological agents.[10] It is well documented that alcohol, Dilantin®, and phenobarbital will, given prior to the drawing of the sample, produce increased levels of gamma-glutamyl transpeptidase. The patient should not be given such drugs prior to the blood collection for at least 48 hr.

G. Interpretation

Various normal ranges for this enzyme have been reported in the literature. In the author's experiences the normal range in both males and females is 5 to 40 units/liter of serum. Some individuals have described sex differences in levels between males and females. However, in a large series of cases this phenomenon has not been observed. Even if there are some moderate differences in the means between the two sexes, the upper limits of normal tend to overlap. The major value of this enzyme determination in neoplastic disease lies in the ability to separate liver from osseous metastases. It can be especially helpful when the alkaline phosphatase levels are elevated and the patient's serum is anicteric. In these cases the normal gamma-glutamyl transpeptidase levels would tend to indicate that the alkaline phosphatase elevation is not due to liver involvement. On the other hand, elevated alkaline phosphatase and gamma-glutamyl transpeptidase in patients with anicteric sera and primary carcinoma would confirm the impression that at least a portion of the alkaline phosphatase elevation is due to liver involvement. Lum and Gambino have stated that gamma-glutamyl activity is greatest in patients with primary carcinoma at the head of the pancreas and adenocarcinoma of the bile duct.[11]

VII. LACTIC ACID DEHYDROGENASE

A. Principle

Lactic acid dehydrogenase (E.C. 1.1.1.27.) is a glycolytic enzyme which is nicotinamide-adenine dependent. The enzyme can be measured by fluorometry, colorimetry, or spectrophotometry. The spectrophotometric method depends on the effect of oxidation or reduction on the absorption of $NADH_2$ at 340 nm. Wacker devised a procedure which measures the reduction of NAD to $NADH_2$, utilizing L-lactate as a substrate. This technique is available commercially in the Statzyme LDH and marketed by the Worthington Biochemical Company. This is the well-known "forward" reaction. There is still some controversy over the preferred technique for lactic acid dehydrogenase assay. The method to be described below is the "backward" spectrophotometric reaction as modified by Henry et al. in 1960.[12] This method depends on the oxidation of $NADH_2$ to NAD, as pyruvate is enzymatically converted to lactate by lactic acid dehydrogenase. The decreasing absorbance of $NADH_2$ is considered to be related to consumption of pyruvate and the unit of activity is described in terms of micromole pyruvate transformed per minute per milliliter of serum at 37°C.

B. Sample

Unhemolyzed serum. Hemolyzed samples cannot be used since erythrocytes contain approximately 100 times the activity of normal serum. Serum should be removed from the clot at the earliest opportunity to cancel LDH release from the formed elements of the blood. Oxalated, citrated, or heparinized samples should not be used. If the test cannot be performed within 8 hr the serum should be stored in the refrigerator until assayed. Freezing of serum results in an alteration of LDH isoenzymes, specifically LDH-5.

C. Reagents

1. Hydrochloric acid, 6 *N*

   Carefully dilute 50 ml of concentrated hydrochloric acid to 100 ml with water.

2. Tris-$NADH_2$ mixtures:

   Directions are for the preparation of 100 tubes; if less tubes are required, then the concentration of the reagents can be proportionally decreased. The reagents are stable for 10 days.

a. Dissolve 3.0 g of tris (hydroxymethyl) aminomethane (Sigma 121) in 100 ml distilled water.

b. Adjust the pH to 7.45 ± 0.5 with dropwise addition of 6 *N* HCl.

c. Add 50 mg of the highest purity dehydronicotinamide-adenine dinucleotide ($NADH_2$) and mix well. Check the absorbance on a spectrophotometer at 340 nm. It should be at least 1.30. Add more $NADH_2$ if the absorbance is less than the above figure.

d. Pipette 2.25 ml of the finally adjusted mixture in dry screw cap tubes and store at 4°C.

3. Tris-pyruvate mixture

   Add 1.2 g tris and 150 mg of sodium pyruvate to 100 ml of water. Mix well and adjust the pH with the dropwise addition of 6 *N* HCl to 7.45 ± 0.05. Store at 4°C and prepare fresh weekly.

4. Potassium dichromate, 3 mg%

   Dissolve 300 mg of potassium dichromate in 100 ml of water. Prepare an approximately 1 in 100 dilution of the above potassium dichromate solution. The absorbance of this solution should be adjusted to read 0.500 to 0.600, so that absorbances of the tests fall between 0.400 and 0.600.

TABLE 2

Temperature Correction Factors for Lactic Acid Dehydrogenase

Temperature °C	Correction factor Tf
37	1.00
32	1.47
30	1.70
28	1.81
25	2.22
23	2.64
20	3.21

LDH act. at 37.5°C X T_f 1.00 (n) = LDH observed at T_f (n) in I.U. liter.

D. Procedure

1. Remove the tris $NADH_2$ mixture tube from the refrigerator and bring to room temperature. Add 0.050 ml of nonhemolyzed serum to the tube, cap, mix by inversion several times, and incubate at 37°C for 20 min to remove endogenous serum pyruvate. If a high activity is suspected, the specimen should be diluted 1 to 10 with heat inactivated sera (56°C for 30 min).

2. Warm the tris-pyruvate mixture for 10 min at 37°C. Add 0.20 ml of this warmed reagent to the reaction tube. Cap. Mix quickly by inversion and transfer to a square glass cuvette with a light path of 1.00 cm.

3. Place in the temperature-controlled chamber of a spectrophotometer capable of measuring reactions at 340 nm.

4. Record the absorbance change at 30-sec intervals for 2 to 3 min at 340 nm.

E. Calculation

The international unit of micromoles per milliter per minute at 37.5°C is

$$\text{LDH}_{\text{I.U./liter at }37.5^\circ\text{C}} = \frac{\Delta \text{ A per minute}}{6.22} \times \frac{V}{D}$$

where

ΔA = change in absorbance
6.22 = absorbance of 1 μmol $NADH_2$ per milliliter at 340 nm for 1 cm light path
V = total volume of reaction mixture in milliliters
D = serum sample size in milliliters

If a temperature other than 37°C is used, then a conversion factor must be introduced into the formula. This fact is based on a Q_{10} (the increase in enzyme activities at 10°C intervals) of 2.1 for LDH between 30 and 40°C. Table 2 gives a partial list of the connection factors (Tf).

F. Interpretation

There is a moderate sex difference in the normal levels of this enzyme as measured by the Henry technique. Females appear to have slightly lower means. The range of normal values for females is 400 to 650 I.U. liter, while males have a range of 450 to 700 I.U. liter. The reproducibility of the method appears to be well within the acceptable 10% limits for enzyme determinations. The choice of the so-called "backward" reaction (pyruvate to lactate) for LDH is predicated on several factors. At 37.5°C, the reaction rate of pyruvate to lactate is approximately threefold that of the opposite reaction. In addition, $NADH_2$ is stable at the alkaline pH of this method, while NAD decomposes in alkaline solution and mixtures must be lyophilized for stability. However, $NADH_2$ solutions should not be frozen since there is development of an inhibitor to $NADH_2$ when the compound is stored in the frozen state.[13]

Elevation of LDH in the serum has been noted in almost all neoplastic processes, including carcinomas and malignant blood disorders.[14] It is a ubiquitous enzyme and cannot be used as a screening test for occult carcinoma. Investigation in the author's laboratory over several years confirmed by many other oncologists indicates lactic acid dehydrogenase changes will be noted as patients respond to various forms of therapy. However, these alterations in LDH levels are not as sensitive a monitor of the clinical condition of the cancer as some of the other enzymes. LDH isoenzyme separation has been used to obtain more sensitive parameters of clinical status. This technique will be described in a separate section of this chapter.

VIII. LACTIC ACID DEHYDROGENASE ISOENZYMES

A. Principle

In 1957, Vesel and Bearn first described the electrophoretic heterogeneity of serum lactic acid dehydrogenase.[15] Using starch zone electrophoresis, they originally reported three peaks of

LDH activity. The introduction of modified electrophoretic procedures and different supporting media such as cellulose acetate has confirmed the presence of at least five different electrophoretic zones. The technique described below can be used to separate LDH enzymes by electrophoresis of the serum on cellulose acetate in a barbital buffer at pH 8.6. The various fractions are visualized by incubation with a mixture of lactate, NAD, phenazine methosulfate, and nitroblue-tetrazolium (NBT). The LDH in the separated fraction reduces the NAD to $NADH_2$ which, in the presence of phenazine methosulfate, converts NBT to a violet color.[16]

B. Sample

Serum should be collected and separated rapidly from the clot. If the test cannot be performed within 2 hr, the sample should be stored at 4°C. It should not be frozen since LDH-5, one of the isoenzyme fractions, is inactivated at temperatures below freezing. If the sample must be stored for longer than 48 hr then NAD (10 mg/ml) or glutathione should be added to decrease the rate of inactivation.

Reagents

1. Phosphate buffer, pH 7.5, 0.1 *M*
   15.01 g of disodium monohydrogen phosphate ($Na_2HPO_4 \cdot 2H_2O$) and 2.17 g of sodium dihydrogen phosphate ($NaH_2PO_4 \cdot H_2O$) are placed in a 200-ml beaker and dissolved in 800 ml of water. If the pH does not equal 7.5 ± 0.05, add the appropriate salt to adjust the pH and dilute to 1,000 ml with water.
2. Sodium lactate, 1 *M*
   Transfer 8 ml of 70% sodium lactate syrup to a 100-ml volumetric flask. Add phosphate buffer to 100 ml.
3. Nicotinamide-adenine dinucleotide (NAD) 10 mg
   On an analytical balance weigh out 10 mg aliquot of NAD and store in a dry stoppered 10-ml volumetric flask in a dessicator at −10°C.
4. Phenazine methosulfate (PMS), 1 mg/ml
   Dissolve 10 mg of PMS in 10 ml of water and store at 4°C. The solution is stable for 1 week.
5. Nitroblue-tetrazolium (NBT), 1 mg/ml
   Dissolve 100 mg of NBT in 100 ml of water and store at 4°C. The solution is stable for at least 2 weeks.
6. Working solution
   To the 10-ml volumetric flask containing 10 mg NAD add the following reagents:

   1 ml of 1 *M* sodium lactate
   3 ml of NBT solution
   0.3 ml of PMS solution
   1 ml of phosphate buffer
   Make up to the mark with distilled water.

7. Electrophoresis buffer (barbital-barbituric acid, pH 8.6, ionic strength, 0.05)
   Dissolve 1.84 g of diethyl barbituric acid and 10.30 g of sodium diethyl barbituric acid in 1,000 ml of water.

D. Apparatus

1. Electrophoresis tank capable of holding cellulose acetate strips approximately 1 x 6¾ in.
2. A plastic box with lid to act as moist chamber. Moistened filter paper is placed in the bottom of the chamber.

E. Procedure

1. The manufacturer's instructions should be followed for the preparation of the strips and the setting up of the electrophoretic tank. A horizontal electrophoresis system is required.
2. The strips are usually immersed in the barbital buffer for at least 10 min, blotted to remove excess buffer, and placed in the electrophoresis chamber. Mark the cathode end of the strip with a pencil.
3. The application of the serum will depend on the nature of the electrophoretic tank. Usually a 5- to 10-μl sample is applied about 2 cm from the center of the strip on the cathode side.
4. Approximately 160 V DC is applied for 1¼ hr. This will, with most currently available electrophoresis chambers, provide a current flow of 0.5 A/cm of strip.
5. During the electrophoresis run, microscope slides are placed on the filter paper in the moist chamber.

6. Pour the working solution (Reagent 7) into a wide shallow vessel and float fresh cellulose acetate strips (cut to the size of the microscope slide) on the surface of the solution.

7. After the undersurface of the acetate strip has become completely wet, with a forceps, gently transfer the strip to a microscope slide in the moist chamber; wet the surface down.

8. At the completion of the electrophoresis, the test strip is removed from the box, *inverted*, and placed carefully, without trapping air bubbles, on top of the reagent strip.

9. Cover the moist chamber and incubate at 37°C in the dark for approximately 30 min.

10. The pairs of strips, both the reagent and the test strip, are carefully immersed in the methanol-acetic acid solvent for 10 min.

11. The test strips are then removed, blotted, and pressed between sheets of filter paper.

12. The isoenzymes will appear as violet bands.

F. Interpretation

The strips may be evaluated by a recording densitometer. However, the author's experience indicates that the semiquantitative results obtained on densitometer tracings have shown such a wide variation in the ranges of LDH isoenzyme values, especially in patients with neoplasms, that quantitation of LDH bands is virtually valueless. Since isoenzyme evaluation should only be performed on samples with gross LDH elevations, the author has found that simple visualization of the bands is adequate for his purpose. Using this technique five LDH bands can be observed in most serum samples. The predominant bands are usually cathodal and are called LDH-1, LDH-2, and LDH-3. The anodal bands LDH-4 and LDH-5 stain very lightly in sera unless there is some hepatic or skeletal muscle pathology. In malignancy, there is often a tendency for the proportion of the slower isoenzymes LDH-3, LDH-4, and LDH-5 to be increased.[17] The author has observed patients with hepatic metastasis to breast carcinoma, demonstrating heavy stained LDH-5 bands. The status of LDH isoenzyme in the diagnosis of tumors is still unsettled. Some investigators feel that malignant change may alter genetic patterns, resulting in the formation of an entirely new system of isoenzymes. Latner has detected at least one such aberrant isoenzyme in neoplastic tissue and in sera from patients suffering from cancer.[18]

IX. PHOSPHOHEXOSEISOMERASE

A. Principle

Recently a simple rapid method for the determination of phosphohexoseisomerase (phosphoglucoseisomerase) (e.c. 5.3.1.9.) has been reported by Schwartz et al.[19] The method uses a coupled reaction and was first described by Beuding in 1965.[20] It is commercially available from the Worthington Biochemical Corporation. The reaction followed in the methodology is shown below.

$$\text{1. Fructose-6-phosphate} \overset{\text{PHI}}{\rightleftharpoons} \text{Glucose-6-phosphate}$$

$$\text{2. Glucose-6-phosphate} + NAD^+ \xrightarrow{\text{GP6DH}} \text{6-phosphogluconate} + NADH + H^+$$

The reduced nucleotide (NADH) produced in a coupled reaction is directly proportional to the glucose-6-phosphate produced in the first reaction. The rate of increase in NADH measured at 340 nm is an indication of the level of PHI activity.

B. Sample

Methods will be described for the evaluation of PHI in serum, vaginal fluid, and tissue sections. Unhemolyzed serum may be kept at 25°C for at least 8 hr and up to 2 weeks at 4°C without significant loss of PHI activity. Vaginal or cervical fluid should be analyzed within 8 hr. Tissue sections should be obtained by a cold knife technique and analyzed immediately. There is loss of PHI activity even at −20°C in tissue sections if the sample is not immediately analyzed.

C. Reagents

1. Glycine buffer, 100 m*M*, pH 8.5
   A. Sodium hydroxide, 0.1 *M*
      Dissolve 400 mg of sodium hydroxide in 100 ml water.
   B. Glycine, 100 m*M*
      Dissolve 750 mg of glycine in 100 ml

water. Transfer 50 ml of glycine and 3.8 ml of sodium hydroxide to a 200-ml volumetric flask and dilute to the mark with water. Check the pH; this should be 8.5 ± 0.1. If the pH is beyond these limits, adjust to the proper pH with reagents A and B, starting with a new mixture. Add more Reagent A if pH is below 8.5 or more Reagent B if pH is above 8.5. The reagent is stable if stored at 4°C for 1 month.

2. Reaction mixture: glycine buffer, fructose-6-phosphate, nicotinamide-adenine-dinucleotide (NAD), and glucose-6-phosphate dehydrogenase. The following instruction will allow the preparation of a sufficient quantity of reagent mixture for the performance of approximately 30 tests.

A. Fructose-6-phosphate, barium salt, 2.2 m*M* (Sigma)

Place 143.5 mg of fructose-6-phosphate in 100 ml volumetric flask.

B. Nicotinamide-adenine-dinucleotide (NAD), 0.7 m*M*

Add 47 mg NAD to the same volumetric flask.

C. Glucose-6-phosphate dehydrogenase, Type XV (Sigma) lyophilized vials, each vial containing 250 units of G6PD

Add 1 ml of glycine buffer to one vial, be sure all of the lyophilized material is completely dissolved, then transfer 0.5 ml to the 100-ml volumetric flask.

D. Glycine buffer

Dilute to 100 ml with glycine buffer and dispense 2.9 ml of the buffered reaction mixture to cuvettes or test tubes. The constituted reaction mixture is stable for 24 hr at 4°C.

D. Procedure

1. To the cuvette or test tube containing 2.9 ml of reconstituted reagent add 0.1 ml serum and mix. If in a test tube, transfer to a 1-cm cuvette and place in a spectrophotometer after 1 min. Record the absorbance at 340 nm at 30-sec intervals for approximately 5 min and read the temperature.

2. The rate of change of absorbance per minute should be linear. If the rate is greater than 0.05 Δ absorbance per minute, 50 μl of serum should be used and the test repeated, reconstituting the PHI system with 2.95 ml buffer.

E. Calculation

One absorbance unit is defined as an increase in absorbance of 0.001 per min at 30°C. One international unit is the reduction of 1 μmol of NAD per min at 30°C. Therefore, the absorbance units can be calculated as follows:

1. The absorbance units/ml serum at 340 nm is equal to

$$\frac{\Delta A/\text{min} \times 1{,}000 \times \text{temp coefficient}}{1 \text{ ml serum}}$$

2. International units/ml can be determined by the

$$\frac{\Delta A/\text{min} \times 1{,}000 \times \text{temp. coefficient} \times 3 \times 1{,}000}{6.2 \times 10^3 \times \text{ml of serum used}}$$

where 6.2×10^3 = extinction coefficient of NADH at 340 nm.

A simplified calculation of I.U./liter when the temperature coefficient is equal to one and the volume of serum used is 0.1 is

I.U./liter at 340 nm = $\Delta A/\text{min} \times 4{,}840$

Temperature coefficients are shown in Table 3.

This technique can be performed on any instrument that can measure absorbance at 340 nm. The author's experience has been with the Gilford Laboratory Spectrophotometer® using a flow through cuvette which can be adjusted to 0.000 absorbance at 340 nm against the buffer solution. He has not had any experience with the Coleman Jr. II or the Bausch and Lomb Spectronic 20, Spectronic 70, or any other spectrophotometer, but Worthington's brochure on the PHI system suggests that a 3 mg/100 ml potassium

TABLE 3

Temperature Correction Factors For Phosphohexose Isomerase

Temperature of reaction mixture	T_f
25°C	1.29
30°C	1.00
32°C	0.89
37°C	0.72

dichromate solution may be used as a reference absorbance for these instruments or a reference may be prepared from serum and neutral phosphate buffer.

F. Vaginal Fluid Content

A few investigators have been intrigued with enzyme evaluations of vaginal fluid as a possible supplement to the exfoliative cytology for the diagnosis of gynecological neoplasia. Several enzymes have been investigated including beta glucuronidase, 6-phosphogluconate dehydrogenase, and phosphohexoseisomerase.[21,22] Levels of all these enzymes have been found to be elevated in carcinoma *in situ* and in invasive carcinoma, but the technique used by the original investigators produced a large amount of false positives in benign disease. Although the author has not had a great deal of experience with vaginal fluid analysis, using his current PHI methodology, it might well be a valid tool for the investigation of this body fluid.

1. Collection and Preservation of Vaginal Fluid

A glass tissue grinding tube is carefully weighed on an analytical balance. A standard Papanicolaou glass pipette, approximately 16 cm long and 0.6 cm in diameter with a polished distal end, fitted with a 2-oz rubber suction bulb is inserted in the vaginal orifice. The bulb is compressed, the pipette inserted high into the posterior vaginal vault, and the fluid collected in the pipette by releasing the bulb and slowly sweeping the pipette across the vagina as it is withdrawn.[21] The fluid is then expelled into a glass tissue grinding tube which is reweighed and 2 ml of buffer (Reagent 1 in serum PHI method) is added to the tube. A piston rod with a Teflon® tip is inserted into the homogenizer tube and the solution is ground for 30 sec. Then 0.1 ml of the fluid is used for the enzyme analysis.

2. Calculation

The results can be reported in terms of I.U./mg of fluid. The calculations described in the serum methodology are valid for the fluid analysis.

For vaginal fluid, the only additional modification would be as follows:

$$\text{I.U./mg of fluid} = \frac{\text{I.U./liter}}{\text{weight of vaginal fluid in mg}}$$

The activity of PHI in vaginal fluid is stable if stored at 4°C. Freeze drying of the fluid, a technique first suggested by Muir for enzyme levels in vaginal fluid, should not be used since PHI activity is irreversibly lost by this manipulation.

G. Tissue Analysis

1. Collection of Samples

Hilf et al. have recently studied PHI levels in breast tissue of patients with benign and malignant disease.[23] By modifying the ultraviolet method of PHI, Akhtar et al. have been able to analyze the PHI activity in frozen sections obtained at the time of breast biopsy.[24] The technique is relatively simple. A tissue homogenate is prepared as follows: Two frozen sections, immediately following the one taken for histological analysis, approximately 1 X 1 cm in size and 8 μm in thickness, are cut by a cold knife technique. The sections are individually transferred from the cold knife by means of a small glass rod to a glass tissue grinding tube containing 2 ml of PHI buffer (Reagent 1 in serum PHI method). The tissue is homogenized for approximately 15 to 30 sec and immediately centrifuged at 3,000 R.P.M. for 5 min. Then 0.1 of the clear supernatant is used for the enzyme analysis.

2. Calculation

Since the major purpose of this analysis is to confirm histological impressions, we simply report results in terms of ml of homogenate. The calculation, therefore, is extremely simple. Units are calculated by the method described in the serum section and the final value is divided by the milliliters of homogenate which is usually 2.

The tissue should not be stored, but should be assayed immediately. The author has noted loss of PHI activity in tissue stored on a frozen block. Homogenates should also be assayed immediately.

H. Interpretation

1. Serum

The normal range of PHI is 20 to 90 I.U./liter at 30°C with one international unit defined as the reduction of 1 μmol of NAD per min at 30°C. Elevated PHI activity has been reported in a high percentage of cases with active metastasizing cancer of the breast in the female.[25,26] However, this enzyme may be normal in the early stages of cancer and its primary usefulness lies in following the course of the disease since changes in enzyme

TABLE 4

Tissue Levels of PHI

Type of tissue	No. of cases	Phosphohexose-isomerase/ml	Range I.U./ml
Nontumorous	22	17.8 ± 6.5	5–28
Benign tumors	8	35.5 ± 14.9	15–59
Malignant tumors	23	124 ± 108.8	30–500

activity following regression or progression of tumor growth appear to antedate other laboratory evidence by a considerable period of time. Regression in metastatic breast disease induced by a wide variety of chemical moieties, such as steroid and androgen therapy, oophorectomy, and chemotherapy, seem to be mimicked by decreases in enzyme levels.[27]

2. Vaginal Fluid

Experience with vaginal fluid PHI is extremely limited and normal values cannot be provided. However, Muir and Valters, using a modified collection technique and different PHI methods, reported in 25 cases of cervical carcinoma that the levels of vaginal fluid ranged 10 to 1,000 times that found in normal fluid.[28] However, they also observed, even with PHI, a high false positive rate which seems to preclude the use of PHI activity in vaginal fluid as a screening test for cervical carcinoma. However, it might possibly be useful as an adjunct for histological evaluation in difficult diagnostic cases.

3. Tissue

Akhtar et al. obtained the results shown in Table 4 in a series of frozen sections.[29]

The nontumorous samples were obtained from patients who had cystic and inflammatory breast disease, fibrosis fatty tissue, and at least two samples of normal breast obtained at autopsy. The benign conditions included fibroadenomas and intraductal papillomas. The patients with histological proven malignancy had invasive ductal carcinoma, medullary carcinoma, and a malignant intraductal papilloma. Although there is some overlap in the ranges, there is obviously a significant difference between the nontumorous tissue and the neoplastic tissue.

REFERENCES

1. Karker, H., *Scand. J. Clin. Lab. Invest.*, 16, 570, 1964.
2. Koehler, L. H. and Benz, E. J., *Clin. Chem.*, 8, 133, 1962.
3. Coleman, C. M. and Stroge, R. C., *Clin. Chem. Acta*, 13, 401, 1966.
4. Dalal, F. R., Akhtar, M., Shin, K. H., and Winsten, S., *Clin. Chem.*, 17, 323, 1971.
5. Roy, A. U., *Clin. Chem.*, 16, 431, 1970.
6. Reynolds, M. N., Lemmon, H. M., and Bryner, W. W., *Can. Res.*, 16, 943, 1956.
7. Lemon, H. M. and Reynolds, M. N., *Proc. Ann. Assoc. Can. Res.*, 2, 129, 1956.
8. Naftalin, L., Sexton, M., Whitaker, J. F., and Tracy, D. F., *Clin. Chim. Acta*, 26, 293, 1969.
9. Szasz, G., *Clin. Chem.*, 15, 124, 1969.
10. Rosalki, S. B., *Lancet*, ii, 376, 1971.
11. Lum, G. and Gambino, S. R., *Clin. Chem.*, 18, 358, 1972.
12. Henry, R. J., Chiamori, N., Golub, O. J., and Berkman, S., *Am. J. Clin. Pathol.*, 34, 381, 1960.
13. Erickson, R. J. and Merales, D. R., *N. Engl. J. Med.*, 265, 478, 1961.
14. Fawcett, C. P., Ciotti, M. M., and Kaplan, N. O., *Biochim. Biophys. Acta*, 54, 210, 1961.
15. Vessel, E. S. and Bearn, A. G., *Proc. Soc. Exp. Biol. Med.*, 94, 96, 1957.
16. Preston, J. A., Brien, R. O., and Batsakis, J. G., *Am. J. Clin. Pathol.*, 43, 256, 1965.

17. Barnett, H. and Gibson, A. J., *Clin. Pathol.*, 17, 201, 1964.
18. Latner, A. C., *Proc. Assoc. Clin. Biochem.*, 3, 120, 1964.
19. Schwartz, M. K., Bethune, V. C., Bochi, B. L., and Woodbridge, J. E., *Clin. Chem.*, 17, 656, 1971.
20. Beuding, F. and MacKinnon, J. A., *J. Biol. Chem.*, 215, 507, 1955.
21. Kasdon, W. H., Dasdon, S. C., and Homburger, F., *J.A.M.A.*, 143, 350, 1953.
22. Bonham, D. J. and Gibbs, D. F., *Br. Med. J.*, 2, 823, 1962.
23. Hilf, R., Goldberg, H., Orlander, R. A., and Archer, F. L., *Proc. Soc. Exp. Biol. Med.*, 132, 613, 1969.
24. Akhtar, M., Dalal, F. R., and Winsten, S., *Clin. Chem.*, 17, 657, 1971.
25. Rose, A., West, M., and Zimmerman, H. T., *Cancer*, 14, 726, 1961.
26. Griffitt, M. M. and Beck, J. F., *Cancer*, 16, 1032, 1963.
27. Winsten, S., Levick, S., and Cohn, E., *Cancer*, 18, 1066, 1965.
28. Muir, G. G. and Valters, G. J., *J. Clin. Pathol.*, 21, 24, 1968.

THE SELECTION, PERFORMANCE, AND INTERPRETATION OF SERUM ENZYME TESTS IN PREGNANCY

D. W. Moss

TABLE OF CONTENTS

I. Introduction 237

II. Enzymes Specifically Related to Pregnancy 237
- A. Placental Alkaline Phosphatase 238
- B. Oxytocinase (Arylamidase) 241

III. Serum Enzyme Tests Not Specifically Related to Pregnancy 244
- A. Enzymes Which Reflect Cellular Damage 244
- B. Enzymes Which Reflect Biliary Function 246

IV. Effects of Oral Contraceptives on Serum Enzyme Activities 251

V. Other Enzyme Tests in Pregnancy and Possible Future Developments 252

VI. Aspects of Quality Control of Serum Enzyme Assays 254

References 255

I. INTRODUCTION

The value of measurements of enzyme activities in serum as aids to the diagnosis and treatment of a wide range of diseases is firmly established. In particular, such tests are useful in conditions that cause damage to skeletal and cardiac muscles and in hepatobiliary diseases of various kinds. The value of serum enzyme tests in these respects is maintained throughout pregnancy. However, normal pregnancy, particularly in its later stages, produces a pattern of changes in serum enzyme activities upon which pathological abnormalities are imposed, so that the interpretation of serum enzyme levels in pregnancy must take into account these physiological variations. Furthermore, pregnancy is accompanied by entry into the maternal circulation of enzymes or isoenzymes which are not normally present in it or which are present in low activities only. These "pregnancy-specific" enzymes offer means of monitoring the progress of pregnancy, in association with other measurements such as hormone levels, and potentially of detecting deviations from the normal course.

The determination and interpretation of the activities of two pregnancy-specific enzymes, *placental alkaline phosphatase* and "oxytocinase," are discussed in this chapter, together with the effects of pregnancy on the selection and interpretation of other serum enzyme estimations.

II. ENZYMES SPECIFICALLY RELATED TO PREGNANCY

The activities of several enzymes have been shown to be increased regularly in the serum or plasma of women during normal pregnancy. The presence in the plasma of pregnant women of an enzyme which oxidizes oxytocin was reported by Fekete.[1] Oxytocinase activity was measured by means of a biological assay. An increased serum alkaline phosphatase in late pregnancy was observed by Coryn,[2] but it was only after an interval of 30 years that this was shown to be due to the entry of placental alkaline phosphatase into the maternal circulation.[3] Other serum enzyme activities that increase during the later stages of pregnancy include β-glucuronidase, β-acetylglucosaminidase, hyaluronidase, renin, and diamine oxidase.[4-6] Changes in some or all of these enzyme levels have been studied in both normal and abnormal pregnancies, but most attention has been concentrated on oxytocinase and the placental isoenzyme of alkaline phosphatase.

A. Placental Alkaline Phosphatase

The alkaline phosphatases are a group of enzymes which catalyze the transfer of inorganic phosphate from a range of organic phosphate monoesters, and also from inorganic pyrophosphate and pyrophosphate esters, such as ADP and ATP, to a hydroxylic acceptor. Under the conditions usually chosen for the reaction in vitro, the acceptor is water so that the reaction is one of hydrolysis. Alkaline phosphatases are widely distributed, but the enzymes from different tissues vary to some extent in their relative activities toward various substrates. This and other variations in properties, including sensitivity to certain inhibitors and denaturating agents, electrophoretic mobility, and reaction with specific antisera, have resulted in the recognition of at least three classes of human alkaline phosphatases:[7] placental phosphatase, intestinal phosphatase, and a third category which includes the enzymes from bone and liver and which may itself be heterogeneous. The tissue-specific differences between the several alkaline phosphatases are retained when the enzymes are released into the circulation and can be exploited to determine the probable tissue of origin of a raised serum alkaline phosphatase activity.

Several criteria have been used to identify placental alkaline phosphatase in the serum of pregnant women, including resistance to denaturation by heat,[8] inhibition by L-phenylalanine, and precipitation by antiplacental phosphatase antiserum.[9] However, intestinal alkaline phosphatase, which may occur in the serum of some individuals, is also inhibited by L-phenylalanine and may cross-react to some extent with antisera raised against placental phosphatase. On the other hand, the stability of placental phosphatase to heat is considerably greater than that of intestinal phosphatase and even more marked than that of liver and bone phosphatases, so that the alkaline phosphatase activity remaining after a suitable period of heating at an elevated temperature can be used as a reliable indication of the amount of the placental isoenzyme present.

Earlier investigators exposed serum to a temperature of 56°C for 30 min to inactivate nonplacental phosphatases.[3] However, a higher temperature is preferable to ensure complete inactivation of nonplacental enzymes and 65°C is usually chosen. There have been some variations in the duration of heating at 65°C, Green et al.[10] preferring 5 min and Hunter[11] 30 min. Although the difference in heat stability between placental and nonplacental phosphatases is very marked, placental phosphatase exists in several genetically determined and electrophoretically distinct variants;[12,13] those of intermediate electrophoretic mobilities have significantly lower heat stabilities than other forms.[14] These differences are seen most clearly at temperatures above 65°C, but the possibility arises that prolonged exposure to 65°C might cause inactivation of placental phosphatase, if this is of the comparatively more labile variety. However, experiments with different placental variants have shown no significant loss of activity in 30 min at 65°C, even with intermediate forms; 30 min exposure to 65°C is adopted in the method set out here.

1. Determination of Placental Alkaline Phosphatase Activity in Serum

a. Principle

The alkaline phosphatase activity of the serum sample is determined before and after heating at 65°C for 30 min. Since nonplacental alkaline phosphatases are destroyed by this treatment, the remaining activity represents placental phosphatase and the difference between the two activities the nonplacental contribution to the serum enzyme activity.

Alkaline phosphatase activity may be determined by any of the commonly used procedures, e.g., the manual colorimetric method of Kind and

King[15] given here, the rate-of-reaction methods of Hausamen et al.[16] or Bowers and McComb,[17] or an AutoAnalyzer® method (Technicon Instruments, Ltd., Basingstoke, Hants.). However, whatever method is chosen, the same procedure should be used for both heated and unheated samples because the various phosphatase isoenzymes have different relative activities toward a range of substrates. If two different substrates are used (e.g., phenyl phosphate and *p*-nitrophenyl phosphate), the difference between total and residual activities will not be identical to that observed when a single substrate is used because of the change in isoenzyme composition brought about by heating.

b. Apparatus

A mechanically stirred, thermostatically controlled water bath of large volume is required with temperature stability to ± 0.05°C and temperature accuracy within the same limits. Suitable baths are Colora Messtechnik GMBH (Lorch, Germany) Model NB, and Haake (Berlin) Model NB22.

c. Reagents

Substrate: 2.18 g disodium phenyl phosphate (2.54 g of the dihydrate) is dissolved in a final volume of 1 liter of distilled water. The solution is brought quickly to the boil and cooled. As a preservative 4 ml of chloroform per liter is added. Store at 4°C.

Buffer (0.1 *M* sodium carbonate-bicarbonate): 6.36 g anhydrous sodium carbonate (AR) and 3.36 g sodium bicarbonate (AR) are dissolved in 1 liter CO_2-free distilled water. The pH of this mixture should be 10.14 ± 0.1 at 20°C (9.90 ± 0.1 at 37°C). Store at 4°C.

Standard: 1 ml of a stock solution containing 1 g AR phenol per liter of 0.1 *M* HCl is diluted to 100 ml with distilled water for use. (A stock phenol solution is available from BDH Ltd., Poole, Dorset.) The absorbance of a new batch of standard should be checked against that of a standard of known composition, e.g., one which has been standardized by titration.[18] The diluted standard is stable for several months at 4°C.

Sodium bicarbonate (0.5 *M*): 42 g sodium bicarbonate AR per liter in water.

Sodium hydroxide (0.5 *M*): 20 g sodium hydroxide AR per liter in water.

4-Aminoantipyrine (4-aminophenazone): 6 g 4-aminoantipyrine per liter in water. Store in a brown bottle.

Potassium ferricyanide: 24 g potassium ferricyanide AR per liter in water. Store in a brown bottle.

d. Procedure

Put approximately 0.5 ml serum in a small pointed glass test tube (Dreyer tube) which is then sealed with plastic film ("Parafilm," Gallenkamp, London). Place the tube in the water bath, already set to 65.0°C, so that the surface of the serum in the tube is below the level of the water, and time 30 min with a stopwatch. At the end of this period the tube is immediately removed and placed in ice.

The alkaline phosphatase activities of the heated sample and of an unheated portion of the same serum (kept at 4° or in ice, but not frozen) are then determined as follows. Total and heat-stable activities: mix 1.0 ml buffer and 1.0 ml substrate in each of two hard-glass test tubes and warm to 37.0 ± 0.2° in a water bath. Add 0.1 ml unheated serum to the first tube (T_1) and 0.1 ml heated serum to the second (T_2) without removing the tubes from the water bath (a constriction pipette is the most suitable type for these additions). Mix and start timing with a stopwatch after the first addition, noting the interval between the two additions. After exactly 15 min add 0.8 ml of 0.5 *M* sodium hydroxide to the first tube and also to the second tube, in this case allowing for the delay in initiating the reaction. Mix well immediately after these additions and remove the tubes from the water bath.

Controls: Mix 1.0 ml buffer, 1.0 ml substrate, and 0.8 ml sodium hydroxide in each of two tubes, then add 0.1 ml unheated serum to the first tube (C_1) and 0.1 ml heated serum to the second (C_2).

Standard: Mix 1.1 ml buffer, 1.0 ml working phenol standard (1 mg per 100 ml), and 0.8 ml 0.5 *M* sodium hydroxide.

Blank: Mix 1.1 ml buffer, 1.0 ml water, and 0.8 ml 0.5 *M* sodium hydroxide.

To all tubes, add 1.2 ml 0.5 *M* sodium bicarbonate followed by 1.0 ml aminoantipyrine solution and 1.0 ml potassium ferricyanide solution. Mix the contents of each tube thoroughly after each addition: The successive additions adjust the pH and develop the color; inadequate mixing at this stage leads to irregular results.

Measure the absorbance of each tube immediately at 510 nm in a spectrophotometer (or in a photoelectric colorimeter with an Ilford® 624 green light filter) avoiding exposure to sunlight.

e. Calculation

The standard contains 10 μg phenol per ml; thus, the amount of phenol produced in the two tests in 15 min is

$$\frac{T_1 - C_1}{\text{Std.-blank}} \times 10\ \mu g \text{ and } \frac{T_2 - C_2}{\text{Std.-blank}} \times 10\ \mu g.$$

These amounts are produced by 0.1 ml of each sample, so that 100 ml of each would produce

$$\frac{T_1 - C_1}{\text{Std.-blank}} \times 10 \text{ mg and } \frac{T_2 - C_2}{\text{Std.-blank}} \times 10 \text{ mg, respectively.}$$

Since 1 King-Armstrong unit corresponds to the production of 1 mg phenol in 15 min under the conditions of the assay,

$$\text{Total serum alkaline phosphatase (King-Armstrong units per 100 ml)} = \frac{T_1 - C_1}{\text{Std.-blank}} \times 10$$

and

$$\text{Heat-stable alkaline phosphatase (King-Armstrong units per 100 ml)} = \frac{T_2 - C_2}{\text{Std.-blank}} \times 10$$

The difference between these results represents the activity derived from nonplacental sources.

For activities between 30 and 60 King-Armstrong units per 100 ml, the final colors of tests and controls may be diluted with 6 ml of water and the result multiplied by two. For activities above 60 King-Armstrong units per 100 ml, the determination must be repeated with a 5 min incubation time, the result being multiplied by three.

f. Precision

The coefficient of variation of replicate determinations of the heat-stable fraction on the same serum should not exceed that inherent in the method of measuring alkaline phosphatase activity. For the manual colorimetric procedure described, the coefficient of variation is ±4% at the 25 King-Armstrong units per 100 ml level and ±2 to 3% for an AutoAnalyzer method.

2. Normal Values

Heat-stable alkaline phosphatase is detectable in the serum of pregnant women between the 16th and 20th weeks of pregnancy and its amount increases progressively up to the onset of labor. There may be a slight increase during the second stage of labor. After the delivery of the placenta, heat-stable alkaline phosphatase begins to decrease and disappears within 3 to 6 days. The absolute amount of heat-stable phosphatase in the serum depends on the method used to measure activity and the units in which the results are expressed. However, if the heat-stable fraction is expressed as a percentage of the total serum alkaline phosphatase activity, the following average figures are obtained: 26th week of pregnancy, 25%; 30th week, 35%; 34th week, 45%; 38th week, 55%; and 40th week, 58%. When activity is expressed in King-Armstrong units per 100 ml of serum, these percentages correspond to about 2.5, 4, 6, 8, and 8.5 King-Armstrong units per 100 ml, respectively. Thus, the total serum alkaline phosphatase activity approximately doubles between the 20th and 40th weeks of pregnancy because of the entry of placental phosphatase into the maternal circulation so that, on average, the upper limit of normal for this enzyme activity in nonpregnant subjects is reached or slightly exceeded by the end of pregnancy.

However, it must be emphasized that these are average values and published reports agree that the range of heat-stable alkaline phosphatase activity in the serum is very wide at all stages of normal pregnancy. This is illustrated by the data of different authors who have used 65°C for inactivation of nonplacental phosphatase and have expressed their results in King-Armstrong units per 100 ml (Table 1). This normal variation must be taken into account when interpreting serum heat-stable phosphatase levels in suspected pathological conditions.

3. Heat-stable Alkaline Phosphatase in Complications of Pregnancy

Earlier studies on serum heat-stable alkaline phosphatase showed that this enzyme was often increased in pregnancies which were complicated by hypertension, pre-eclampsia, or eclampsia. For example, Curzen and Morris,[22] who used a temperature of 56°C to inactivate nonplacental phosphatases, observed that about 20% of all estimations were abnormal (mostly elevated) in essential hypertension and pre-eclampsia, while in severe pre-eclampsia the percentage of elevated values rose to 54%. Thus, the probability of an abnormal heat-stable alkaline phosphatase increased with the

TABLE 1

Upper and Lower 95% Confidence Limits for Heat-stable Alkaline Phosphatase (King-Armstrong Units/100 ml) in Serum at Different Stages of Pregnancy

	Duration of pregnancy (weeks)							
Authors	25/26	30	32	34	36	38	40	1st stage labor
Hunter et al.[19]	–	0.7–4.6	1.2–5.9	1.7–7.9	2.3–10.1	3.2–12.1	–	–
Pirani et al.[20]	0.6–5.0	0.5–8.5	–	1.8–10.6	–	2.3–13.9	–	1.6–18.0
Aleem[21]	3.2–7.3	4.7–10.8	4.8–12.6	3.5–16.6	4.7–15.6	7.7–17.1	7.3–19.5	–

severity of the disorder. However, as the rather low percentage of abnormal results indicates, there was considerable overlapping of the serum heat-stable alkaline phosphatase activities of the normal and abnormal groups of patients.

These results have been confirmed by later workers who have used 65°C for heat treatment of the serum; the discrimination between normal and complicated pregnancies has been somewhat improved by this more specific inactivation procedure. Hunter et al.[19] found an incidence of 59% abnormal values in patients with pre-eclampsia at the 30th week of pregnancy, rising to 90% by the 38th week. In severe pre-eclampsia the incidence of abnormality was consistently 100%. However, a rather lower frequency of abnormality was found by Aleem,[21] who reported that almost half of 162 results on 67 patients with pre-eclampsia were above normal.

More information can be obtained from serial estimations than from single measurements of heat-stable alkaline phosphatase in hypertensive complications of pregnancy and an abnormally steep rise usually precedes the onset of clinical signs by 2 to 3 weeks.[19,21] Clinical improvement may be mirrored by a fall in serum heat-stable alkaline phosphatase activity.

High levels of heat-stable alkaline phosphatase may accompany threatened abortion; the probability of this is greater in patients with a history of previous miscarriage. A sharp rise or fall in the heat-stable enzyme in serum may precede intra-uterine death.

A tendency to low serum heat-stable alkaline phosphatase activities has been noted in pregnancies accompanied by diabetes mellitus.[21,23] However, Hunter[11] reported normal values for well-controlled pregnant diabetics.

The value of estimation of serum heat-stable alkaline phosphatase activity as an aid to fetal prognosis has been studied by Curzen.[24] Heat-stable alkaline phosphatase had little value in predicting fetal dysmaturity or fetal distress, compared with simultaneous assays of total estrogens in 24-hr urine collections which correctly predicted these conditions in about 60% of cases. These results are in accordance with the observations of Pirani et al.,[20] who found that in normal pregnancy there was no correlation between serum heat-stable alkaline phosphatase activity and crude baby weight or placental weight. No correlation existed between the results of the enzyme estimations and urinary pregnanediol output, but there was a significant correlation with urinary estriol output at 30, 34, and 38 weeks.

In summary, although frequent elevations of serum heat-stable alkaline phosphatase activity accompany hypertensive disorders of pregnancy, interpretation of single estimations in an individual patient is made difficult by the wide normal range of this activity. Serial estimations are, therefore, necessary and sharp rises or falls are probably of clinical significance. However, these estimations should be accompanied by other tests of placental function.

B. Oxytocinase (Arylamidase)

Inactivation of the peptide hormone oxytocin by the serum of pregnant women has been shown to involve the removal of the aminoterminal cystine residue from the octapeptide by hydrolytic cleavage of the cystinyl-tryosyl peptide bond.[25] Thus, the "oxytocinase" involved is an aminopeptidase. Although the enzyme present in pregnancy serum has not been purified, a parallel increase in hydrolysis of synthetic amino acid amides such as L-cystinyl-di-β-naphthylamide, L-leucyl-β-naphthylamide, and L-alanyl-β-naphthylamide suggests that the enzyme present in pregnancy serum belongs to the category of arylamidases which are widely distributed in tissues and plasma. An enzyme in serum with the ability to hydrolyze L-leucyl-β-naphthylamide was formerly identified with leucine aminopeptidase, an enzyme which is specific for the removal of aminoterminal L-leucyl residues joined by peptide bonds to other amino acids or polypeptides. However, it is now clear that these activities are functions of distinct enzymes. Therefore, enzymes with the ability to hydrolyze derivatives of β-naphthylamine and similar aromatic compounds are now generally referred to as "arylamidases," although the term "leucine aminopeptidase" (LAP) is still sometimes used to describe activity toward substrates such as L-leucyl-β-naphthylamide. Since synthetic substrates are usually employed to monitor the increasing serum enzyme activity of pregnancy and are presumed also to reflect oxytocinase activity, the results obtained with pregnancy serum are usually referred to in terms of the substrate used, e.g., as leucyl-β-naphthylamidase or cystinyl-di-β-naphthylamidase activities, etc. However, it seems probable that all these activities are functions of a single enzyme or group of closely related enzymes.

Since arylamidases apparently reach the circulation from several different tissue sources, the measurement of the particular arylamidase (or isoenzyme of arylamidase) which is believed to enter the serum from the placenta in late pregnancy is beset with the same difficulties as the measurement of placental alkaline phosphatase, that is, the need to measure the activity of the placental isoenzyme against a background of closely similar activities derived from other organs. Electrophoretic separation of arylamidase isoenzymes in serum has demonstrated the appearance during pregnancy of one or more additional enzyme zones with mobilities similar to those in extracts of placenta.[26] L-cystinyl-di-β-naphthylamide has been considered by some workers to approximate more closely to the structure of oxytocin, the presumed physiological substrate of the placental arylamidase and, therefore, has been chosen for the measurement of "oxytocinase" activity in pregnancy serum.[27] However, this substrate also is hydrolyzed to some extent by arylamidases apparently unrelated to pregnancy.

Other substrates have been chosen on grounds of analytical convenience. L-alanyl-β-naphthylamide is hydrolyzed about twice as fast as the L-leucyl derivative by arylamidases of several tissues, but probably the same enzymes act on the two substrates.[28] Use of L-leucyl-ρ-nitroanilide[29] (available in kit form from the Boehringer Corporation Ltd., London) permits the direct determination of the liberated *p*-nitroaniline by spectrophotometry at 405 nm. However, most studies have involved the use of L-leucyl-β-naphthylamide as substrate.

An early colorimetric method using this substrate is that of Goldbarg and Rutenburg.[30] A reagent kit based on a modification of this method is commercially available (Sigma London Chemical Co.). A disadvantage of the original method was that the possibility of binding of β-naphthylamine to serum proteins restricted the volume of serum used to 0.02 ml; an incubation period of 2 hr was, therefore, necessary. Measurement of the product β-naphthylamine by its fluorescence greatly increases the sensitivity of the method and reduces the incubation period required. The fluorimetric procedure described by Roth[31] is given below.

1. Determination of L-Leucyl-β-Naphthylamidase Activity in Serum

a. Principle

Serum is incubated with a solution of L-leucyl-β-naphthylamide at 37°C and pH 7.2. After 30 min β-naphthylamine is measured directly in the reaction mixture by fluorescence.

b. Apparatus

A filter-fluorimeter or spectrofluorimeter is required. The latter offers the advantage that the optimal excitation wavelength of 335 nm for the fluorescence of β-naphthylamine can be selected, whereas the nearest useful line in the spectrum of the mercury arc lamp, with which most filter-fluorimeters are equipped, is at 365 nm so that maximum activation of fluorescence is not obtained with this light source. However, usable excitation of fluorescence of β-naphthylamine by a mercury lamp is possible and filter instruments are considerably less expensive than spectrofluorimeters. Many filter-fluorimeters are now equipped with mercury lamps coated with phosphors to convert the line spectrum to a nearly continuous emission, or with zinc lamps which have an emission line at 333 nm.

In the author's laboratory the Locarte Mk 5 filter-fluorimeter (The Locarte Company, 24

Emperor's Gate, London, S.W.7) and the Perkin-Elmer Model 203® spectrofluorimeter (Perkin-Elmer Ltd., Beaconsfield, Bucks) have been found to be suitable for this determination. (A useful list of filter-fluorimeters, drawn up by S. S. Brown and H. Braunsberg, can be obtained from the Association of Clinical Biochemists, 7 Warwick Court, London W.C.1; Technical Bulletin No. 21.)

c. Reagents

Buffer (pH 7.2): 14.4 g K_2HPO_4 (AR) and 3.1 g KH_2PO_4 (AR) are dissolved in double-distilled water and the volume is made up to 1 liter.

Buffered substrate solution: 20 mg L-leucyl-β-naphthylamide hydrochloride is dissolved in 2.5 ml distilled water in a 100-ml graduated flask and the volume is adjusted to the mark with phosphate buffer.

Stock standard solution (100 μg β-naphthylamine per ml) in AR methanol: This solution is stable for 1 week if stored at 4°C in a brown bottle and should, therefore, be made in small volumes only.

Working standard solution (1 μg/ml): Made by adding 99 ml buffer to 1 ml stock standard solution immediately before use. Note: β-Naphthylamine is carcinogenic and can be absorbed by contact with the skin or by inhalation. The code of practice for handling such compounds drawn up by The Chester Beatty Research Institute, London SW3 6JB should be followed strictly. A mask and disposable gloves should be worn when weighing out β-naphthylamine and gloves should be used when handling the stock standard solution. Glassware that has contained the solution should be washed out with cold water to avoid vaporization. Several fluorimeters are now available which are equipped with permanent fluorescence standards in the form of solid solutions of fluorphors in acrylic plastic rods or blocks. These can be used as secondary standards so that the calibration of the instrument in terms of the β-naphthylamine standard need only be checked occasionally. Solutions of quinine sulfate (50 μg/ml in 0.1 N H_2SO_4, stored at 4°C in a brown bottle and diluted 1 in 50 v/v before use) form readily available alternative secondary standards.

d. Procedure

Warm 4.0 ml buffered substrate solution to 37.0°C in a thermostatted water bath. Add 10 μl of serum with a constriction pipette and mix. After exactly 30 min at 37°C, measure the fluorescence (T) of the solution, with an activation wavelength of 335 nm and emission wavelength of 415 nm if a spectrofluorimeter is used, or with activation by the 365 nm line of the mercury spectrum and an appropriate filter on the emission side to cut out scattered ultraviolet light from the source if a filter fluorimeter is employed.

A reagent blank (B) consisting of 4.0 ml buffered substrate solution + 10 μl water is also measured. The fluorescence of a serum blank (4.0 ml buffer + 10 μl of serum) is usually negligible. The fluorescence of the working standard solution (S) is measured after warming the solution to 37°C since fluorescence is temperature-dependent.

e. Calculation

Leucyl-β-naphthylamidase activity = $\frac{(T-B)}{(S-B)} \times 93$ (I.U./1. of serum).

f. Precision

The coefficient of variation of the method is ±2.5% at the 40 I.U./1. level.

2. Normal Values

The normal range of activity of leucyl-β-naphthylamidase in the serum of nonpregnant women is 15 to 35 I.U./1. A slightly higher range, 20 to 42 I.U./1., is found in men.

In normal pregnancy, increased serum leucyl-β-naphthylamidase activity may become apparent after the 16th week of pregnancy. Between the 20th and 40th weeks, activity increases exponentially, the mean value rising approximately three and a half-to fourfold. However, the range of activity at all stages of normal pregnancy is wide: about 15 to 65 I.U./1. at 20 weeks; 25 to 90 at 25 weeks; 35 to 125 at 30 weeks; 45 to 175 at 35 weeks; and 65 to 240 at 40 weeks. Thus, suspected pathological changes in the activity of this enzyme have to be interpreted against a background of considerable normal variation, as is the case for placental alkaline phosphatase.

3. Leucyl-β-naphthylamidase in Complications of Pregnancy

Lower levels of serum leucyl-β-naphthylamidase activity have been observed in pregnancies resulting in stillbirths than in those in which healthy babies were born, with intermediate enzyme levels accompanying live births of babies in poor condition.[32] Kleiner et al.[33,34] reported significantly lower serum enzyme activities in pregnancies in which placental insufficiency was deduced from the birth of dysmature babies, i.e., babies whose birthweight was more than two standard deviations below normal for the corresponding length of pregnancy.

Hensleigh and Krantz[27] correlated serum enzyme activity (with cystine-di-β-naphthylamide as substrate) with evidence of placental dysfunction in 186 pregnant women in whom the possibility of placental insufficiency was increased by factors such as hypertension, pre-eclampsia, or a bad obstetric history. Twelve patients with an abnormal enzyme level also had other evidence of inadequate placental function. However, a further seven patients had normal serum enzyme activities, but showed other evidence of placental malfunction. In some of these patients, only single enzyme assays were possible.

These and other reports of experience with "oxytocinase" measurements in pregnancy emphasize that the wide range of activity encountered in normal pregnancy makes the interpretation of results of single estimations extremely difficult, although differences between groups of patients can be established. As with serum heat-stable alkaline phosphatase, serial measurements of "oxytocinase" during pregnancy are more valuable than single determinations. A progressive rise in "oxytocinase" activity during the last 3 months of pregnancy suggests a normal outcome, while persistently low or falling levels suggest a risk of fetal dysmaturity, premature labor, or intrauterine death.

Increased "oxytocinase" levels have been reported in groups of normal twin pregnancies by some workers[35] but not by others,[33] so that also in this instance measurements in a single patient are unlikely to be decisive.

III. SERUM ENZYME TESTS NOT SPECIFICALLY RELATED TO PREGNANCY

A. Enzymes Which Reflect Cellular Damage

Serum aspartate and alanine transaminases (GOT and GPT) remain within normal limits during uncomplicated pregnancy. Lactate dehydrogenase and isocitrate dehydrogenase are similarly unaffected in most cases, although slight increases (particularly of isocitrate dehydrogenase) at the end of pregnancy have been reported in some cases. Elevations of creatine kinase (ATP: creatine phosphotransferase; formerly called creatine phosphokinase and often abbreviated as CPK) are more frequently encountered. During labor, the activities of several enzymes in serum are frequently increased, including the transaminases and lactate dehydrogenase.[36] The increases in the activities of these enzymes are usually slight, but CPK may rise to 5 to 20 times the upper limit of normal, i.e., to levels which resemble those seen following myocardial infarction. Raised serum enzyme levels return to normal within 2 to 5 days after delivery.

The transaminases and lactate dehydrogenase can, therefore, be applied to the detection of tissue damage during pregnancy as in the nonpregnant subject. However, some caution is necessary in the interpretation of isocitrate dehydrogenase or CPK activities at the end of pregnancy.

1. Determination of Serum Transaminase Activity

Numerous colorimetric, spectrophotometric, or fluorimetric methods have been described for the determination of these enzyme activities. However, spectrophotometric methods have proved to be the most reliable of manual techniques and that of Wilkinson et al.[37] is given here. Automated methods based on the Technicon AutoAnalyzer® or special purpose enzyme analyzers such as the LKB 8600 reaction-rate analyzer are also widely used.

a. Principle

Aspartate transaminase catalyzes the reaction:

$$\left.\begin{matrix}\text{L-aspartate}\\+\\\text{2-oxoglutarate}\end{matrix}\right\} \rightleftharpoons \left\{\begin{matrix}\text{glutamate}\\+\\\text{oxaloacetate}\end{matrix}\right.$$

and alanine transaminase catalyzes the reaction:

$$\left.\begin{matrix}\text{L-alanine}\\+\\\text{2-oxoglutarate}\end{matrix}\right\} \rightleftharpoons \left\{\begin{matrix}\text{glutamate}\\+\\\text{pyruvate}\end{matrix}\right.$$

Oxaloacetate producted by aspartate transaminase reacts with reduced NAD in the presence of added malate dehydrogenase to form malate and oxidized NAD:

$$\text{oxaloacetate} + \text{NADH} \rightleftharpoons \text{malate} + \text{NAD}^+$$

In the case of alanine transaminase, lactate dehydrogenase is added to catalyze the conversion of pyruvate to lactate with oxidation of reduced NAD:

pyruvate + NADH ⇌ lactate + NAD^+

In each case, the progress of oxidation of NADH is followed by observing the fall in absorbance of light at 340 nm in a spectrophotometer, the rate of this reaction being proportional to the rate of transamination and, hence, to the activity of the transaminases.

b. Apparatus

A spectrophotometer capable of giving readings at 340 nm is required. Preferably this should be equipped with a chart recorder. It is even more desirable that the instrument should be provided with a means of controlling the temperature of the cuvettes, e.g., by circulation of water through the cuvette holder from an external thermostat, since fluctuations in temperature of 1°C during measurement of the enzymic reaction rate can result in a 10% error. (A useful comparative list of spectrophotometers compiled by C. E. Wilde and P. Sewell is obtainable from the Association of Clinical Biochemists, 7 Warwick Court, London W. C. 1; Technical Bulletin No. 19.)

c. Reagents

Phosphate buffer (1.0 *M*, pH 7.4): Dissolve 136 g of KH_2PO_4(AR) and 33 g of NaOH (AR) in distilled water and adjust the volume to 1 liter.

Phosphate buffer (0.1 *M*, pH 7.4): Dilute 100 ml of 1.0 *M* phosphate buffer to 1 liter with water and adjust to pH 7.4 at 25°C.

NADH solution: Dissolve 2.5 mg of reduced β-nicotinamide adenine dinucleotide, disodium salt, per ml in 0.1 *M* phosphate buffer. (Prepare fresh each day.)

2-Oxoglutarate solution (0.1 *M*): Dissolve 0.73 g of 2-oxoglutaric acid in 35 ml of distilled water. Add 5 ml of 1.0 *M* phosphate buffer and adjust to pH 7.4 at 25°C by the addition of 1 *M* NaOH. Dilute to 50 ml with water. Store at 4°C. (Stable for 1 month.)

L-Aspartate solution (0.375 *M*): Dissolve 5.0 g of L-aspartic acid with heating in a mixture of 50 ml of distilled water and 35 ml of 1 *M* NaOH. After cooling, add 10 ml of 1.0 *M* phosphate buffer and adjust to pH 7.4 at 25°C with 1 *M* NaOH. Dilute to 100 ml with water. Store at 4°C. (Stable for 1 month.)

L-Alanine solution (0.75 *M*): Dissolve 6.67 g of L-alanine in 75 ml of water and add 10 ml of 1.0 *M* phosphate buffer. Adjust to pH 7.4 at 25°C with 1 *M* NaOH. Dilute to 100 ml with water. Store at 4°C. (Stable for 1 month.)

Malate dehydrogenase solution: Dilute a commercial preparation fresh each day with distilled water to an activity of 5 I.U. per ml. The malate dehydrogenase must be free of aspartate transaminase activity and also of apotransaminase. Commercial preparations now available are usually satisfactory in these respects. Absence of transaminase can be verified by carrying out an aspartate transaminase determination from which serum is omitted and absence of apotransaminase by determination in which serum is replaced by 0.2 ml of a solution containing 0.2 μg of phosphopyridoxal, the prosthetic group of transaminases.[38]

Lactate dehydrogenase solution: Dilute a commercial preparation of the rabbit muscle enzyme fresh each day with distilled water to an activity of 5 I.U. per ml.

d. Procedure

For the determination of aspartate and alanine transaminases, respectively, prepare the following mixtures in spectrophotometer cuvettes (1 cm path length) and incubate at 25 ± 0.2°C for 15 min (volumes are given in ml);

	Aspartate transaminase	Alanine transaminase
Phosphate buffer (0.1 *M*)	1.3	1.3
Serum	0.2	0.2
NADH solution	0.2	0.2
L-aspartate solution	1.0	–
L-alanine solution	–	1.0
Malate dehydrogenase	0.1	–
Lactate dehydrogenase	–	0.1
Total Volume	2.8	2.8

Transfer the cuvettes to the spectrophotometer, which should be already set to measure at 340 nm and with the cuvette holder at 25 ± 0.2°C if the equipment is thermostatically controlled. Start the reaction by adding 0.2 ml of 2-oxoglutarate solution to each cuvette, mixing thoroughly with a plastic paddle. Record the change in absorbance at 340 nm for a sufficient period (usually 4 to 6 min) to ensure that a linear decrease with time is registered, ignoring any short lag phase at the beginning of the reaction.

If the spectrophotometer is not fitted with a recorder, readings should be taken at intervals of 30 sec, timed with a stopwatch, over a similar period. Where only one cuvette at a time can be accommodated for reading or recording, other reaction mixtures can be left at 25°C until convenient since the length of the pre-incubation period is not critical, provided that at least 15 min is allowed for side reactions involving the oxidation of NADH to reach completion.

If temperature control of the cuvette compartment is not available, the temperature inside it should be noted at the beginning and end of the reaction and an average figure calculated. Results can then be corrected to 25°C by multiplying by the following factors for temperatures below 25°C or dividing by them for temperatures above 25°C: aspartate transaminase, 1 + 0.098 X (number of degrees difference from 25°C); alanine transaminase, 1 + 0.08 X (number of degrees difference from 25°C).

e. Calculation

The absorbance at 340 nm of 1 μmol of NADH in 1 ml in a cuvette of 1 cm path length is 6.22. The volume of the reaction mixture is 3 ml. Hence, the aspartate or alanine transaminase activity is given by

$$(\text{Absorbance change per min}) \times \frac{3}{6.22}\ \mu\text{mol per min}$$

since for each molecule of NADH oxidized one molecule of aspartate or alanine is converted to its corresponding ketoacid.

This activity is contained in 0.2 ml serum; therefore

$$\begin{array}{l}\text{Transaminase activity} = (\text{Absorbance change per min}) \\ (\text{I.U. per liter}) \quad \times \frac{3}{6.22} \times \frac{1000}{0.2}.\end{array}$$

The accuracy of the method depends on accurate calibration of the absorbance and wavelength scales of the spectrophotometer. Methods for checking these have been described by Rand.[39]

f. Precision

The coefficient of variation of the method is ±5% at the upper limit of the normal range.

2. Normal Values

The normal ranges for serum aspartate transaminase activity by this method are 5 to 15 I.U./liter (males) and 5 to 12 I.U./liter (females). For alanine transaminase the ranges are 5 to 19 I.U./liter (males) and 5 to 12 I.U./liter (females). Raised values do not occur in normal pregnancy, except during labor.

3. Serum Transaminase Activity in Complications of Pregnancy

Probably the most frequent use of transaminase estimations in pregnant women is in the detection of liver damage; in this application aspartate and alanine transaminase are of equal value. When liver damage is due to acute infective hepatitis, the pronounced increases in serum transaminase activities characteristic of this disease in non-pregnant subjects occur also in pregnant patients. In contrast, normal or only moderately increased serum transaminase activities are seen in the idiopathic jaundice which may cause slight or moderate elevations of serum bilirubin, particularly in the last trimester of pregnancy. Similarly, acute fatty liver that may occur (though rarely) in the last month of pregnancy and give rise to high serum bilirubin levels is not accompanied by marked elevations of the transaminases, unlike infective hepatitis.

Elevated serum transaminase activities are associated with toxemia of pregnancy and, although the origin of these elevations is not established beyond doubt, the most probable cause of them is necrosis of liver cells. Crisp et al.[40] found that serum aspartate transaminase was normal in essential hypertension of pregnancy, but that in almost all cases of pre-eclampsia and eclampsia there was some elevation of the activity of this enzyme. Furthermore, the degree of elevation of aspartate transaminase was proportional to the severity of the condition, a fall in the serum enzyme level paralleling clinical improvement.

These results were confirmed by Dass and Bhagwanani,[41] who found, however, a rather lower total incidence of serum enzyme abnormalities in toxemic conditions. When eclampsia with convulsions was present, peak serum transaminase activities were reached about 6 days after the onset of convulsions.

B. Enzymes Which Reflect Biliary Function

The activities of several enzymes in serum have

been shown to undergo marked increases in hepatobiliary diseases in which the flow of bile is impaired; the determination of one or more of these enzymes has become one of the most useful applications of diagnostic enzymology. The enzymes that are most frequently estimated in this category of diseases include alkaline phosphatase, arylamidase ("leucine aminopeptidase"), 5′-nucleotidase, and γ-glutamyl transpeptidase. Of these, most experience has been gained with alkaline phosphatase.

Investigation of suspected hepatobiliary disease or jaundice is frequently required in pregnancy. However, the use of alkaline phosphatase or arylamidase estimations is complicated by the entry into the maternal circulation of enzymes or isoenzymes derived from the placenta which have similar substrate specificities to those which accumulate in the blood when biliary function is impaired. Thus, it becomes difficult to assign a pathological imterpretation to activities of these two enzymes in serum, particularly in late pregnancy, because of the considerable physiological variation in the activities of the placenta-derived isoenzymes. Although it is possible to improve the organ specificity of both alkaline phosphatase and arylamidase estimations, e.g., by heat treatment of the serum or selection of an appropriate substrate, as has already been discussed, a simpler solution is to select as an index of hepatobiliary disease an enzyme the activity of which is not increased in normal pregnancy. Two enzymes fulfill these criteria, γ-glutamyl transpeptidase and 5′-nucleotidase.

γ-Glutamyl transpeptidase catalyzes the cleavage of peptide-type bonds joining the γ-carboxylic group of L-glutamic acid to amino groups, either in naturally occurring substances such as glutathione (γ-glutamylcysteinylglycine) or in synthetic substrates in which the amino group is contributed by such compounds as β-naphthylamine or *p*-nitroaniline. The glutamyl residue is transferred to an acceptor molecule which is usually another amino acid or a peptide. An increased serum γ-glutamyl transpeptidase activity is probably the most sensitive enzymic indicator of hepatobiliary disease, being superior in this respect to both arylamidase[42] and 5′-nucleotidase,[43] as well as to alkaline phosphatase.[42,43] Some elevations of the activity of the enzyme have been reported in conditions other than hepatobiliary disease, including various renal diseases, pancreatitis, and after myocardial infarction, but it is not certain that hepatic disease could be definitely excluded in all these cases. Serum γ-glutamyl transpeptidase activity remains normal in osteoblastic bone diseases, in which alkaline phosphatase is raised,[42,43] and in normal pregnancy.[42]

There is, however, one factor which indicates the need for caution in the use of serum γ-glutamyl transpeptidase measurements for the investigation of suspected hepatobiliary disease in pregnancy: that the activity of the enzyme in serum may rise in response to the administration of certain hypnotic or sedative drugs, without any apparent hepatobiliary disorder being present.[44,45] Drugs that have been shown to produce this effect include barbiturates, dichloralphenazone, and phenazone. The increase in enzyme activity was seen after about 1 week of drug administration and the changes reached a maximum (which in some cases represented a fivefold increase over the basal level) in 10 to 15 days. When drug administration ceased, the enzyme activity returned to pretreatment values in 15 to 20 days.[45] Some evidence that the plasma enzyme changes are dose-dependent was obtained by Whitfield et al.,[45] who found that 100 mg of quinalbarbitone at night caused no significant rise in plasma γ-glutamyl transpeptidase activity in one subject, but that an increase in dose to 200 mg at night significantly increased the enzyme level. Chlordiazepoxide, diazepam, nitrazepam, and methaqualone produced slight increases in plasma γ-glutamyl transpeptidase in two out of four patients.

The most probable explanation of these findings seems to be that certain drugs induce an increased synthesis of γ-glutamyl transpeptidase in the liver, as they do of enzymes which are concerned in drug metabolism, although it is not yet known whether γ-glutamyl transpeptidase itself has any role in the metabolism of the enzyme-inducing drugs. Overflow of the newly synthesized enzyme from the liver into the circulation may then account for the increased serum activity. Sedatives or hypnotics are often administered during pregnancy, and in such cases the use of serum γ-glutamyl transpeptidase activities to investigate hepatobiliary function is complicated by the possibility of iatrogenic changes.

Serum 5′-nucleotidase activity does not appear to increase after the administration of enzyme-inducing drugs; if some response in enzyme activity does take place it does not seem to be so marked as to raise the serum 5′-nucleotidase above the upper limit of normal. As an indicator of hepatobiliary disease, serum 5′-nucleotidase is rather less sensitive than γ-glutamyl transpeptidase, but perhaps slightly more sensitive than alkaline phosphatase. Its activity remains normal almost invariably in those diseases of bone in which serum alkaline phosphatase is elevated; only a few exceptions to this generalization have been reported.

Significantly higher serum 5′-nucleotidase activities have been reported in the last trimester of pregnancy than at earlier stages, reaching or even slightly exceeding the upper limit of normal for nonpregnant subjects.[46,47] However, Seitanidis and Moss[48] detected no upward trend in 5′-nucleotidase activity in 52 women at different stages of pregnancy, including 34 in the last third in whom the expected rise in serum alkaline phosphatase was apparent, and concluded that measurement of serum 5′-nucleotidase activity could be used to investigate suspected hepatobiliary dysfunction in pregnancy.

The placenta contains 5′-nucleotidase;[49] this enzyme might, therefore, be expected to enter the maternal circulation as do placental alkaline phosphatase and arylamidase. The different behavior of 5′-nucleotidase may be due to its apparently higher molecular weight, as determined by gel filtration[50] and more restricted diffusion, or perhaps to a different localization within the placenta.

5′-Nucleotidase is a phosphatase which acts only on nucleoside 5′-phosphates, releasing inorganic phosphate. The substrate usually used in assays of the enzyme is adenylic acid (adenosine-5′-phosphate, AMP). At the pH optimum of the serum enzyme, pH 7.5, this substrate is also hydrolyzed by nonspecific alkaline phosphatase: Methods for the estimation of 5′-nucleotidase in serum must, therefore, include some means of correcting for the action of alkaline phosphatase on the substrate. Several ways of making this correction have been described. 5′-Nucleotidase differs from alkaline phosphatase in being inhibited by nickel ions; this property is used to distinguish the two enzymes in the method of Campbell[51] described here. Another approach is to add a large excess of a phosphate ester, such as β-glycerophosphate or phenyl phosphate, which is a substrate for alkaline phosphatase but not for 5′-nucleotidase, thus reducing the proportion of the total alkaline phosphatase activity which is directed to the hydrolysis of adenosine-5′-phosphate. This approach cannot be used when the estimate of reaction rate is based on the determination of inorganic phosphate, since this is a product of the reactions catalyzed by both enzymes, but methods have been described in which adenosine (the second product of the 5′-nucleotidase reaction) is determined either colorimetrically or spectrophotometrically after conversion to ammonia by the action of added adenosine deaminase.[52,53]

Automated methods for the estimation of 5′-nucleotidase based on nickel inhibition or addition of a nonspecific substrate have been devised.

1. *Determination of Serum 5′-Nucleotidase Activity*

a. Principle

Serum is incubated with adenosine-5′-phosphate at pH 7.5 and 37°C, with and without added nickel ions. After 30 min, the amount of inorganic phosphate liberated is determined. Phosphate produced in the absence of nickel results from the combined activities of alkaline phosphatase and 5′-nucleotidase, whereas that produced in the presence of nickel is due to the activity of alkaline phosphatase alone. Thus, the difference between these two amounts corresponds to the activity of 5′-nucleotidase in the serum sample. Manganese ions are added as an activator of 5′-nucleotidase.

b. Apparatus

A thermostatically controlled water bath with temperature stability to ±0.2°C and a spectrophotometer or photoelectric colorimeter are required.

c. Reagents

Buffer (0.04 *M*, pH 7.5): Dissolve 8.25 g of sodium diethylbarbiturate (barbitone sodium) in 140 ml of 0.2 *M* HCl and dilute to 1 liter with water. Check the pH and adjust to pH 7.5 if necessary.

Substrate (10 m*M*): Dissolve 0.347 g of adenosine-5′-phosphoric acid in 18 ml of 0.1 *M* NaOH and dilute to 100 ml with water.

Manganese sulfate solution (20 m*M*): 0.446 g of $MnSO_4 \cdot 4H_2O$ (AR) per 100 ml in water.

Nickel chloride solution (0.1 *M*): 2.4 g of $NiCl_2 \cdot 6H_2O$ (AR) per 100 ml in water.

Trichloracetic acid (10% w/v): 100 g of trichloracetic acid per liter in water.

Acetate buffer (2 *M* pH 4.0): Dissolve 2.5 g of $CuSO_4 \cdot 5H_2O$ (AR) and 46 g of sodium acetate trihydrate in 1 liter of 2 *M* acetic acid. Check the pH and adjust to pH 4.0 if necessary.

Ammonium molybdate solution (5% w/v): 5 g of $(NH_4)_6Mo_7O_{24} \cdot 4H_2O$ (AR) per 100 ml in water.

Reducing agent: Dissolve 2 g of *p*-methylaminophenol sulfate (Rhodol® or Metol®) in 80 ml of water. Add 10 g of $Na_2SO_3 \cdot 7H_2O$, dilute to 100 ml with water, and filter. Store in a dark bottle at 4°C.

Stock phosphate standard (100 mg of P per 100 ml): 2.19 g of KH_2PO_4 (AR) per 500 ml in water. Add a few drops of chloroform to prevent bacterial contamination and store at 4°C.

Working phosphate standard (1 mg of P per 100 ml): Measure 1.0 ml of stock standard into a 100-ml graduated flask, add 50 ml of 10% w/v trichloracetic acid, and dilute to 100 ml with water. This solution is stable for 4 weeks if stored at 4°C.

d. Procedure

Conical centrifuge tubes are used, which must be of hard glass (Pyrex®). Test: Mix 1.5 ml of buffer (pH 7.5) with 0.1 ml of manganese sulfate solution. Control: Mix 1.3 ml of buffer (pH 7.5), 0.1 ml of manganese sulfate solution and 0.2 ml of nickel chloride solution.

Add 0.2 ml of serum to both test tubes and place them in the water bath at 37.0 ± 0.2°C. Allow 5 min for the mixtures to attain 37°C, then add 0.2 ml of substrate solution to each and mix, without removing the tubes from the water bath.

After exactly 30 min, stop the enzymic reaction by the addition of 2 ml of 10% trichloracetic acid to each tube. Shake well, remove from the water bath, and centrifuge at 3,000 r.p.m. for 15 min.

Pipette 2.0 ml of the clear supernatant from the test (T) and control (C) mixtures into clean hard-glass test tubes. (These tubes should be washed with dilute hydrochloric acid and thoroughly rinsed with distilled water. Some detergents contain phosphate and their use should be avoided.)

Prepare a standard (S) by pipetting 1.0 ml of working standard and 1.0 ml of 10% trichloracetic acid into an acid-washed hard-glass tube and a reagent blank (B) by mixing 1.0 ml of water with 1.0 ml of 10% trichloracetic acid.

To all tubes add 3 ml of acetate buffer followed by 0.5 ml of ammonium molybdate solution and 0.5 ml of Rhodol solution, mixing well after each addition.

Measure the intensities of the blue solutions after at least 5 min in a spectrophotometer, preferably at a wavelength of 880 nm; if this wavelength is beyond the range of the instrument, a wavelength of 700 nm should be used. A photoelectric colorimeter with an Ilford 608 red light filter or an equivalent filter may be used. The presence of copper in the acetate buffer accelerates color development. The color is stable for at least 30 min.

e. Calculation

The standard tube contains 10 μg of phosphate (as P). Therefore, the amount of phosphate produced by the action of 5′-nucleotidase is given by

$$\frac{T - C}{S - B} \times 10\ \mu g, \text{ i.e., } \frac{T - C}{S - B} \times \frac{10}{31}\ \mu mol$$

This amount of phosphate is released by the enzyme present in 0.1 ml of serum (since half of the total reaction mixture is taken for estimation of phosphate), acting for 30 min. Therefore, serum 5′-nucleotidase activity in μmoles of substrate hydrolyzed per min per liter of serum (I.U. per liter)

$$= \frac{T - C}{S - B} \times \frac{10}{31} \times \frac{1}{30} \times \frac{1000}{0.1}$$

$$= \frac{T - C}{S - B} \times 108 \text{ I.U./liter}$$

If the activity is greater than 150 I.U./liter, the test should be repeated with a shorter incubation time and a corresponding adjustment made to the calculation.

f. Precision

The coefficient of variation of the method is ±10% at the upper limit of the normal range.

2. Normal Values

The normal range of serum 5′-nucleotidase

activity in normal men and in normal nonpregnant and pregnant women is 2 to 17 I.U./liter.

3. Serum 5′-Nucleotidase Activity in Complications of Pregnancy

Serum 5′-nucleotidase activity resembles alkaline phosphatase in being elevated in hepatobiliary disease of all types in which there is interference with the elaboration and excretion of the bile, whether this is due to intrahepatic causes or extrahepatic obstruction. A correlation coefficient of 0.71 was found for measurements of these two enzymes in 240 sera from a series of patients investigated for liver disease, with approximately equal incidences of positive results and degrees of elevation above normal being recorded for the two activities in 174 samples drawn from 81 patients with proven liver disease.[43] However, the correlation between serum alkaline phosphatase and 5′-nucleotidase activities in liver disease is not absolute and elevations of one enzyme activity accompanied by normal levels of the other do occur. This is more common in chronic liver disease, in which changes in the two activities follow different courses with respect to time, and 5′-nucleotidase activity is generally more persistently elevated. In a series of simultaneous estimations of alkaline phosphatase and 5′-nucleotidase activities on 66 serum specimens from patients with chronic liver disease (chronic active hepatitis, primary biliary cirrhosis, extrahepatic obstruction due to cancer of the head of the pancreas, and Weil's disease), both enzyme levels were abnormal on 39 occasions, 5′-nucleotidase was abnormal and alkaline phosphatase normal 21 times, and on only 6 occasions was a normal 5′-nucleotidase associated with an elevated alkaline phosphatase.[54] On this evidence, therefore, an abnormality is more likely to be detected by a measurement of serum 5′-nucleotidase than alkaline phosphatase in suspected chronic liver disease.

Highest activities of serum 5′-nucleotidase are found in those conditions in which involvement of the biliary tree is most pronounced,[55] including acute cholangitis, primary and secondary biliary cirrhosis, and malignant infiltration. High levels of serum 5′-nucleotidase may also be encountered in jaundice induced by drugs which particularly affect the biliary system, such as chlorpromazine.

As is the case with alkaline phosphatase, normal levels of 5′-nucleotidase or only slight to moderate elevations are seen in acute infective hepatitis: When elevations do occur in this disease, maxima are reached on average about 5 days after the peak of activity of enzymes such as transaminases or isocitrate dehydrogenase which indicate hepatocellular destruction.[54] This pattern of serum enzyme changes is reproduced in infective hepatitis affecting both pregnant and nonpregnant subjects, and the combination of a steep rise in enzymes which reflect hepatocellular damage (e.g., the transaminases) with normal or only moderately raised levels of 5′-nucleotidase is useful in differentiating infective hepatitis from other causes of jaundice in pregnancy.

As well as reflecting hepatobiliary diseases that may affect both pregnant and nonpregnant individuals, changes in serum 5′-nucleotidase activity occur in pathological states involving the liver which are peculiar to pregnancy; in these conditions, measurements of the enzyme activity also provide useful aids to diagnosis and assessment.

Idiopathic jaundice of pregnancy, which occurs in a small proportion of women, usually in the last trimester and which is characterized by pruritus and mild jaundice (serum bilirubin usually less than 10 mg/100 ml), is typically accompanied by markedly elevated serum 5′-nucleotidase activity. However, the liver is not usually enlarged much and serum transaminases are rarely more than slightly elevated, if at all. Pruritus of pregnancy appears to be similar to idiopathic jaundice of pregnancy,[46] except for the absence of jaundice in the former, and similar serum enzyme changes are seen in the two conditions. Both resolve completely and spontaneously after parturition with an accompanying return of the serum enzyme levels to normal and a progression to chronic liver disease does not appear to take place, although recurrent episodes in pregnancy are common.[56]

The principal appearances in liver biopsy specimens taken from patients with either idiopathic jaundice or pruritus of pregnancy are of canalicular bile stasis, with no necrosis of parenchymal cells.[47] That impaired biliary function is common to both conditions is shown by the observation of a reduced transport maximum for bromsulfothalein in idiopathic jaundice and pruritus of pregnancy without jaundice.[47] A possible causative factor in these conditions is sensitivity on the part of some individuals to the increased level of hormone production in the later stages of pregnancy.

The rare condition of acute fatty liver (or obstetric fatty liver) that occurs in the last 3 months of pregnancy has the clinical appearances of a severe, fulminating infective hepatitis. Renal failure usually occurs and this, together with hepatic failure, accounts for the high fatality rate in this disease. In contrast to infective hepatitis, however, only slight or moderate elevations of serum transaminases are seen, but these are accompanied by marked increases in 5′-nucleotidase. This pattern of serum enzyme changes is consistent with the microscopic appearance of the affected liver, which shows diffuse and widespread fatty infiltration but little cellular necrosis.

The possibility that damage to the liver may occur in toxemia of pregnancy, giving rise to raised serum transaminase activities, has already been mentioned. However, interference with hepatobiliary function is not pronounced and serum 5′-nucleotidase is not markedly elevated.

In all of the hepatobiliary diseases of pregnancy in which serum 5′-nucleotidase is raised, an increased activity of an alkaline phosphatase which, by various criteria, resembles that of the liver also appears in the circulation. If measurements are made solely of the total alkaline phosphatase activity, it is difficult to demonstrate the increased phosphatase activity against the wide range of normal activity of this enzyme in pregnancy resulting from the entry of variable amounts of placental alkaline phosphatase into the circulation. Thus, if only one type of enzyme assay is to be relied upon for the investigation of suspected hepatobiliary disease in pregnancy, this and other factors discussed earlier indicate that assay of 5′-nucleotidase is the most satisfactory choice. However, if total and heat-stable alkaline phosphatase measurements are also being undertaken for the separate purpose of assessing placental function, the heat-labile fraction of alkaline phosphatase (i.e., the difference between the activities before and after heating at 65°C for 30 min) includes the liver component and is consequently increased in hepatobiliary disease. Increases in the heat-labile alkaline phosphatase activity, therefore, provide data which reinforce findings of raised 5′-nucleotidase levels in suspected hepatobiliary disease superimposed on pregnancy.

IV. EFFECTS OF ORAL CONTRACEPTIVES ON SERUM ENZYME ACTIVITIES

The possibility that sensitivity to increased circulating levels of steroid hormones accounts for some of the hepatobiliary disorders associated with pregnancy suggests that similar processes may operate in nonpregnant women taking oral contraceptives. Jaundice has been reported in a variously estimated proportion of women taking these drugs, while in other subjects changes in liver function without overt jaundice can be demonstrated. In both these situations, the activities in serum of enzymes that reflect hepatocellular integrity may be altered, as well as those of enzymes related to biliary function.

Changes in enzyme activity in serum of women taking oral contraceptives may be transient, tending to occur particularly in the early months of exposure to the drugs; this may explain the complete absence of raised transaminase levels reported by some authors[57] and an incidence of nearly 20% of elevated values found by others.[58] Raised levels also seem more frequent in the latter part of the menstrual cycle.[58] When elevations of transaminases are observed, these are usually of the order of twice the upper limit of normal[59] although activities of up to seven or eight times normal have been reported.

Increases of the order of twofold in serum alkaline phosphatase activity have been observed in 2 to 3% of nonjaundiced women taking oral contraceptives[59,60] and some elevation of serum 5′-nucleotidase has been reported.[61] Increased bromsulfothalein retention confirms the existence of impaired biliary function in these subjects; as in pruritus of pregnancy, this is due to a decreased transport maximum for the dye.[62]

When jaundice results from the administration of oral contraceptives, liver biopsy shows hepatocellular and canalicular bile stasis, consistent with the moderate elevations of the activities in serum of enzymes which reflect biliary function such as alkaline phosphatase and 5′-nucleotidase. This type of lesion would not be expected to produce greatly raised serum transaminase activities and, in general, such elevations do not occur, increases being usually of a slight or moderate order. However, transaminase activities greater than 30 times the upper limit of normal have been reported, accompanied by normal or only slightly raised alkaline phosphatase levels. In such cases, differentiation of oral contraceptive-induced jaundice from infective hepatitis solely on biochemical evidence becomes impossible, although the two conditions can be distinguished by examination of biopsy specimens. High activities of transaminases in serum may perhaps be related to the extent of hepatocellular degeneration and necrosis. This has been noted in biopsy specimens from postmenopausal patients; if present, it is limited to centrilobular areas.[64] However, very high serum transaminase levels are certainly uncommon in jaundice associated with oral contraceptives and if they occur during exposure to these drugs, one must strongly direct attention to a possible diagnosis of infective hepatitis.

V. OTHER ENZYME TESTS IN PREGNANCY AND POSSIBLE FUTURE DEVELOPMENTS

The enzyme tests for which methods are described in this chapter are those that are capable of yielding the most valuable information, both in diagnostic situations peculiar to pregnancy and in those of a more general nature which are likely to occur in pregnant women. Thus, the transaminases remain the most generally useful indicators of cellular damage, particularly in conditions which affect the liver. If a test with greater specificity for liver damage is required, the enzyme of choice is probably isocitrate dehydrogenase, elevations of which in serum are highly specific for damage to that organ. An alternative is the determination of lactate dehydrogenase and its isoenzymes, LD5 (the electrophoretically slowest, most heat-labile fraction) being the form predominant in liver; however, this is a less-sensitive test than either transaminase or isocitrate dehydrogenase determinations.

Damage to skeletal or cardiac muscle is a less common complication of pregnancy, but the most sensitive enzyme test for detection of damage to these tissues is creatine kinase. However, serum creatine kinase activity is usually somewhat depressed during normal pregnancy so that small increases in the activity of this enzyme in the sera of pregnant women may pass unnoticed unless normal ranges established for pregnant women are used as the basis of comparison. The rather small differences between normal creatine kinase levels in pregnant and nonpregnant subjects are unlikely to cause diagnostic difficulties in the case of major tissue damage, such as a myocardial infarction, but in one application of this test which is particularly relevant to pregnancy erroneous interpretations can arise. This is in the use of creatine kinase estimations to detect the carrier state of Duchenne-type, X-linked muscular dystrophy. Female carriers of this disease frequently have significantly elevated serum creatine kinase activities. However, Blyth and Hughes[65] reported that creatine kinase estimations on sera taken during pregnancy from three potential carriers of the disease and one known carrier gave results within the nonpregnant normal range. Emery and King[66] similarly noted normal values in two carriers of Duchenne muscular dystrophy during pregnancy, with elevated values when the same subjects were not pregnant. The converse possibility, that false positives rather than false negatives might arise in the investigation of suspected carriers of muscular dystrophy, was pointed out by Emery and Pascasio,[67] who observed a rapid postpartum rise in serum creatine kinase activity following normal pregnancy with a return to normal values by the tenth day after delivery. Thus, in this period the interpretation of slight or moderate changes in creatine kinase activity is also difficult.

Specific and nonspecific phosphatase (5′-nucleotidase and alkaline phosphatase) remain the best indicators of impaired hepatobiliary function in pregnant patients, although as the effects of administration of drugs on serum γ-glutamyl transpeptidase activities become more clearly defined, it is probable that this enzyme will replace 5′-nucleotidase in this application, not only because its response in hepatobiliary disorders is more sensitive than that of 5′-nucleotidase, but also because methods for the estimation of γ-glutamyl transpeptidase are simpler and the results more reproducible.

The place of estimations of enzymes in serum specific to pregnancy, such as heat-stable alkaline phosphatase or arylamidase, still remains debatable. The activities of these enzymes do reflect the course of pregnancy but they seem, at best, to be rather insensitive indicators of placental development and function, or of damage to that organ, and at present cannot be used in isolation from other clinical or laboratory assessments. Isoenzyme studies of the kind that have led to the recognition of placental alkaline phosphatase and arylamidase in the maternal circulation may also improve the possibility of discriminating between maternal and fetoplacental tissues as the sources of other enzymes. An increase in the proportion of lactate dehydrogenase fractions 4 and 5 in the maternal serum has been found to be a useful sign in some cases of antepartum hemorrhages[68] and a difference in pH optimum between the isocitrate dehydrogenase activities of the sera of pregnant and nonpregnant women has been noted[69] which may represent the appearance of a different isoenzyme in the former. However, studies of this type have not yet disclosed differences of general diagnostic utility.

Now that amniocentesis is established as a safe procedure, the possibility of a more direct approach to assessment of fetoplacental function by analysis of samples of amniotic fluid has become available. So far, analysis of amniotic fluid for nonenzymic constituents has proved most rewarding in assessing fetal maturity as, for example, in the determination of creatinine as an index of developing muscle mass and renal function[70] and lecithin and sphingomyelin derived from fetal lung tissue.[71] Recent reports have suggested that increased levels of creatine kinase in amniotic fluid may indicate fetal death in utero.[72]

The most important advance made possible by the combination of enzyme analysis with amniocentesis, however, is the opportunity that it offers of determining whether a fetus is affected by a serious genetic abnormality at a stage of gestation at which termination of the pregnancy is still possible or of allowing palliative measures to be instituted from the moment of birth in those inherited conditions in which early treatment can prevent the more serious manifestations. The basis of this approach to prenatal diagnosis is that the cells present in the amniotic fluid are mainly fetal in origin, deriving from fetal skin and amnion, so that if a genetic defect has been transmitted to the fetus it will be present also in the chromosomes of the amniotic cells. Like the other cells of the fetus, therefore, those shed into the amniotic fluid are unable to produce the enzyme of which the absence is characteristic of the inborn error of metabolism in question. Since the amniotic fluid is both ingested by the fetus and receives its urine, abnormal metabolites can in some cases be detected by analysis of the amniotic fluid itself, in a manner analogous to the detection of abnormal metabolites in the urine of the newborn infant. Prenatal diagnosis of the adrenogenital syndrome, an autosomal recessive defect in the synthesis of adrenocorticosteroids, has been achieved by the demonstration of increased amounts of 17-oxosteroids and pregnanetriol in amniotic fluid. However, the ability of the placental circulation to remove accumulating metabolites complicates the interpretation of amniotic fluid composition (e.g., with respect to amino acids) so that in most cases prenatal diagnosis will probably concentrate on the amniotic cells.

For the diagnosis of many diseases, the number of cells obtained by a single amniocentesis is too few for reliable biochemical or histochemical enzyme studies. However, human fibroblasts such as those from fetal skin can be cultured in vitro for up to 50 to 70 generations, retaining their characteristics and thus increasing the amount of material available for analysis. However, the time required for cell growth does impose some restrictions; the amniotic fluid can be sampled safely at about 16 weeks' gestation when its volume has reached about 250 ml, leaving 3 to 4 weeks for cell culture and analysis if 20 weeks is regarded as the latest date for abortion.

Prenatal diagnosis is now possible in about 40 genetic abnormalities, including galactosemia, a disease amenable to dietary control, as well as a range of lipoidoses, glycogen storage diseases, and derangements of amino acid metabolism.[73] Although the demonstration of specific enzyme defects has been emphasized in this discussion, the availability of fetal cells also allows the morphology of the chromosomes to be examined, thus disclosing gross abnormalities such as Down's syndrome (trisomy of chromosome 21, resulting in mongolism) and many others which together probably occur once in every 200 live births.

Amniocentesis is a procedure which carries a small but finite risk. Therefore, it is unlikely at present to develop into a general screening procedure, but rather to be applied in selected cases where a previous abnormal child or an unexplained death in infancy suggests an increased risk of inherited disease or when the mother is in an age group with a greater incidence of chromosome abnormalities. The specialized biochemical and especially cytological techniques needed to examine amniotic cell samples, particularly the maintenance of cell lines of known enzymic make-up, are likely to favor the establishment of regional or national reference centers for this type of diagnosis.

VI. ASPECTS OF QUALITY CONTROL OF SERUM ENZYME ASSAYS

In any series of laboratory measurements there exists an inherent experimental error which, for a particular analytical method, is usually expressed in terms of its *accuracy* (i.e., the degree to which the analytical result approaches the true or most probable value of the substance being estimated) and *precision* (i.e., the agreement between replicate determinations of the same quantity). Precision may be further subdivided into *repeatability* (the agreement between the results of replicate analyses when these are carried out without change of operator or apparatus) and *reproducibility* (the agreement between the results of replicate analyses which are carried out with different apparatus and reagents, by different operators, over a period of weeks or months).[74] Precision is usually expressed as the standard deviation or coefficient of variation of a series of replicate determinations. If the results of laboratory measurements are to be able to sustain a useful clinical function, it is essential that the accuracy and precision of the methods used should be adequate and that steps are taken to ensure that the performance of a method in these respects is maintained for each analysis or batch of analyses of patients' specimens. This is achieved by the interpolation among the unknown samples of solutions of known composition (controls), the results of which must lie within specified limits. The nature of the controls, the frequency with which they are analyzed, and the formulation of allowable limits of variation together constitute a quality control program.[74]

Certain problems arise when these principles are extended to the quality control of serum enzyme assays, due to the nature and properties of enzymes themselves.[75,76] The concept of accuracy is difficult to apply to enzyme assays since it is virtually impossible to define the amount of an enzyme except in terms of its effect on the rate of the catalyzed reaction, which in turn is a function of the method used to measure it. Enzymes of clinical interest from human sources have not been completely purified and it is not possible, therefore, to refer to the absolute amount of an enzyme in the way that it is possible to speak of an absolute amount in grams of a substance of defined composition such as urea or glucose. A further consequence of this is that units of enzyme activity are not absolute quantities but are functions of the methods to which they refer: Results of one type of enzyme assay cannot be regarded as equivalent to or interchangeable with those given by a different method of assaying the same enzyme. This is true even when the results of both assays are expressed as "international units" (micromoles of substrate transformed per minute), since differences between methods in such parameters as pH, temperature, substrate concentration, etc. will alter the significance of the respective units.

Accuracy of an enzyme assay method is, therefore, usually expressed in terms of the agreement between it and an alternative, accepted method of assaying the enzyme in question. Reference methods of enzyme assay are currently being formulated by various national and international bodies.[37,77-79] When these become widely adopted, they will provide standards against which existing methods of enzyme assay or secondary methods such as automated procedures can be evaluated and should lead to better agreement between results obtained in different clinical enzymology laboratories.

The problem of determining and monitoring precision in enzyme assays is also greater than in assays for most nonenzymic constituents because of the lack of enzyme preparations of assured stability which can be used as control samples. The susceptibility of enzymes to denaturation, with an associated loss of activity, makes it difficult to distinguish between poor analytical performance and denaturation as possible causes of a low result obtained for a control sample introduced into a batch of analyses. Lyophilized preparations containing various enzymes are available from commercial sources; these have a useful function in quality control since their day-to-day reproducibility is generally good. However, it is often difficult to reproduce the values for activity stated by the manufacturer, even when the same methods are apparently in use, so that recalibration in the user's laboratory becomes necessary.[76,80] Alternatives to lyophilized preparations are serum pools, prepared in the laboratory, assayed for enzyme activity, and stored at $-20°C$ in small portions for daily use. Care must be taken to avoid contamination with hepatitis-associated antigen. The use of either type of control material should be standardized within the laboratory with respect to such factors as reconstitution or thawing procedures. For example, the activity of alkaline phosphatase undergoes a slow, progressive increase after reconstitution of freeze-dried samples or thawing of some types of specimen; this effect must be taken into account in the use of control materials which contain this enzyme.[81]

A further useful control procedure in enzyme analysis is to carry forward one or two samples from one batch of analyses to the next, so that successive results can be compared. Some loss of activity usually takes place between the first and second analyses, but for many enzymes this is less than 10% with appropriate interim storage of the samples.

In spite of the difficulties of selecting suitable control samples, some such samples should nevertheless be included with every batch of unknowns and the results used to prepare control charts. Probably the best solution is to include as many different types of control materials as possible: lyophilized preparations (if these are obtainable containing the enzyme under test), laboratory-prepared pools, and carry-over specimens.

Where alternative procedures for estimating a particular enzyme are available, quality control surveys have shown that some methods give more acceptable levels of precision than others. In general, methods in which the progress of the reaction is monitored continuously (e.g., the method for transaminases described above) are preferable to fixed-time methods (such as the Kind-King alkaline phosphatase procedure[82]). However, continuous monitoring methods for some enzymes are still comparatively new and analytical and clinical experience with them is still being accumulated. Some well-tried fixed-time assay methods have been recommended in this chapter for this reason, but it is anticipated that these older procedures will eventually give way to ones in which the enzymic reaction is monitored continuously.

REFERENCES

1. **Fekete, K.,** *Endokrinologie,* 7, 364, 1930.
2. **Coryn, G.,** *J. Chir.,* 33, 213, 1934.
3. **McMaster, Y., Tennant, R., Clubb, J. S., and Neale, F. C.,** *J. Obstet. Gynaecol. Br. Commonw.,* 71, 735, 1964.
4. **Platt, D. and Platt, M.,** *Klin. Wochenschr.,* 46, 768, 1968.
5. **Geelhoed, G. W. and Vander, A. J.,** *J. Clin. Endocrinol.,* 28, 412, 1968.
6. **Southren, A. L., Kobayashi, Y., Carmody, N. C., and Weingold, A. B.,** *Am. J. Obstet. Gynecol.,* 95, 615, 1966.
7. **Moss, D. W.,** *Ann. N. Y. Acad. Sci.,* 166, 641, 1969.
8. **Kitchener, P. N., Neale, F. C., Posen, S., and Brudenell-Woods, J.,** *Am. J. Clin. Pathol.,* 44, 654, 1965.
9. **Sussman, H. H., Bowman, M., and Lewis, J. L.,** *Nature,* 218, 359, 1968.
10. **Green, S., Pirnik, M. P., Sharkey, L. J., Babson, A. L., and Fishman, W. H.,** *Enzymologia,* 38, 243, 1970.
11. **Hunter, R. J.,** *J. Obstet. Gynaecol. Br. Commonw.,* 76, 1057, 1969.
12. **Boyer, S. H.,** *Science,* 134, 1002, 1961.
13. **Robson, E. B. and Harris, H.,** *Nature,* 207, 1257, 1965.
14. **Thomas, D. M. and Harris, H.,** *Ann. Hum. Genet.,* 35, 221, 1971.
15. **Kind, P. R. N. and King, E. J.,** *J. Clin. Pathol.,* 7, 322, 1954.
16. **Hausamen, T. U., Helger, R., Rick, W., and Gross, W.,** *Clin. Chim. Acta,* 15, 241, 1967.
17. **Bowers, G. N., Jr. and McComb, R. B.,** *Clin. Chem.,* 12, 70, 1966.
18. **Treadwell, F. P. and Hall, W. T.,** *Analytical Chemistry,* Vol. 2, 7th ed., John Wiley & Sons, New York, 1930, 591.
19. **Hunter, R. J., Pinkerton, J. H. M., and Johnston, H.,** *Obstet. Gynecol.,* 36, 536, 1970.
20. **Pirani, B. B. K., MacGillivray, I., and Duncan, R. O.,** *J. Obstet. Gynaecol. Br. Commonw.,* 79, 127, 1972.
21. **Aleem, F. A.,** *Obstet. Gynecol.,* 40, 163, 1972.
22. **Curzen, P. and Morris, I.,** *J. Obstet. Gynacecol. Br. Commonw.,* 75, 151, 1968.
23. **Watson, D., Weston, W., and Porter, R.,** *Enzymol. Biol. Clin.,* 5, 25, 1965.
24. **Curzen, P.,** *J. Clin. Pathol.,* Suppl. 24, 4, 90, 1971.
25. **Tuppy, H. and Nesvadba, H.,** *Mh. Chem.,* 88, 977, 1957.
26. **Jones, D. D., Williams, G., and Prochazka, B.,** *Enzymologia,* 43, 325, 1972.

27. **Hensleigh, P. A. and Krantz, K. E.,** *Am. J. Obstet. Gynaecol.*, 107, 1233, 1970.
28. **Panveliwalla, D. K. and Moss, D. W.,** *Biochem. J.*, 99, 501, 1966.
29. **Nagel, W., Willig, F., and Schmidt, E. H.,** *Klin. Wochenschr.*, 42, 447, 1964.
30. **Goldbarg, J. A. and Rutenburg, A. M.,** *Cancer*, 11, 283, 1958.
31. **Roth, M.,** *Clin. Chim. Acta*, 9, 448, 1964.
32. **Klimek, R. and Bieniasz, A.,** *Am. J. Obstet. Gynecol.*, 104, 959, 1969.
33. **Kleiner, H., Brouet-Yager, M., and Graff, G.,** *J. Obstet. Gynaecol. Br. Commonw.*, 76, 127, 1969.
34. **Kleiner, H.,** in *The Foeto-Placental Unit*, Pecile, A., Ed., Excerpta Medica Foundation, Amsterdam, 1969, 363.
35. **Babuna, C. and Yenen, E.,** *Am. J. Obstet. Gynecol.*, 94, 868, 1966.
36. **Meade, B. W. and Rosalki, S. B.,** *J. Obstet. Gynaecol. Br. Commonw.*, 70, 693, 1963.
37. **Wilkinson, J. H., Baron, D. N., Moss, D. W., and Walker, P. G.,** *J. Clin. Pathol.*, 25, 940, 1972.
38. **Rosalki, S. B. and Wilkinson, J. H.,** *J. Clin. Pathol.*, 12, 138, 1959.
39. **Rand, R. N.,** *Clin. Chem.*, 15, 839, 1969.
40. **Crisp, W. E., Miesfeld, R. L., and Frajola, W. J.,** *Obstet. Gynecol.*, 13, 487, 1959.
41. **Dass, A. and Bhagwanani, S.,** *J. Obstet. Gynaecol. Br. Commonw.*, 71, 727, 1964.
42. **Lum, G. and Gambino, S. R.,** *Clin. Chem.*, 18, 358, 1972.
43. **Whitfield, J. B., Pounder, R. E., Neale, G., and Moss, D. W.,** *Gut*, 13, 702, 1972.
44. **Rosalki, S. B., Tarlow, D., and Rau, D.,** *Lancet*, 2, 376, 1971.
45. **Whitfield, J. B., Moss, D. W., Neale, G., Orme, M., and Breckenridge, A.,** *Br. Med. J.*, 1, 316, 1973.
46. **Kater, R. M. H. and Mistilis, S. P.,** *Med. J. Austral.*, 1, 638, 1967.
47. **Mistilis, S. P.,** *Austral. Ann. Med.*, 17, 248, 1968.
48. **Seitanidis, B. and Moss, D. W.,** *Clin. Chim. Acta*, 25, 183, 1969.
49. **Ahmed, Z. and Reis, J. L.,** *Biochem. J.*, 69, 386, 1958.
50. **Moss, D. W.,** *Nature*, 209, 806, 1966.
51. **Campbell, D. M.,** *Biochem. J.*, 82, 34P, 1962.
52. **Persijn, J. P. and Van der Slik, W.,** *Proc. 7th Intl. Congr. Clin. Chem.*, Vol. 2, Karger, Basel, 1970, 108.
53. **Ellis, G. and Goldberg, D. M.,** *Anal. Lett.*, 5, 65, 1972.
54. **Phelan, M. B., Neale, G., and Moss, D. W.,** *Clin. Chim. Acta*, 32, 95, 1971.
55. **Hobbs, J. R., Campbell, D. M., and Scheuer, P. J.,** *Proc. 6th. Intl. Congr. Clin. Chem.*, Vol. 2, Karger, Basel, 1968, 106.
56. **Haemmerli, U. P. and Wyss, H. I.,** *Medicine*, 46, 299, 1967.
57. **Swaab, L. I.,** *Brit. Med. J.*, 2, 755, 1964.
58. **Larsson-Cohn, U.,** *Acta Obstet. Gynecol. Scand.*, 45, 499, 1966.
59. **Eisalo, A., Jarvinen, P. A., and Luukkainen, T.,** *Brit. Med. J.*, 1, 1416, 1965.
60. **Larsson-Cohn, U.,** *Br. Med. J.*, 1, 1414, 1965.
61. **Kreek, M. J., Weser, E., Sleisenger, M. H., and Jeffries, G. H.,** *N. Engl. J. Med.*, 277, 1391, 1967.
62. **Kleiner, G. J., Kresch, L., and Arias, I. M.,** *N. Engl. J. Med.*, 273, 420, 1965.
63. **Thulin, K. E. and Nermark, J.,** *Br. Med. J.*, 1, 584, 1966.
64. **Stoll, B. A., Andrews, J. T., Motteram, R., and Upfile, J.,** *Br. Med. J.*, 1, 723, 1965.
65. **Blyth, N. and Hughes, B. P.,** *Lancet*, i, 855, 1971.
66. **Emery, A. E. H. and King, B.,** *Lancet*, i, 1013, 1971.
67. **Emery, A. E. H. and Pascasio, F. M.,** *Am. J. Obstet. Gynecol.*, 91, 18, 1965.
68. **Hawkins, D. F. and Whyley, G. A.,** *J. Obstet. Gynaecol. Br. Commonw.*, 73, 140, 1966.
69. **Pulkkinen, M. O. and Willman, K.,** *Acta Obstet. Gynecol. Scand.*, 47, 273, 1968.
70. **Donnai, P., Gordon, H., Harris, D. A., and Hughes, E. A.,** *J. Obstet. Gynaecol. Br. Commonw.*, 78, 603, 1971.
71. **Gluck, A. L., Kulovich, M. V., Borer, R. C., Jr., Brenner, P. H., Anderson, G. G., and Spellacy, W. N.,** *Am. J. Obstet. Gynecol.*, 109, 440, 1971.
72. **Sarkosi, L., Kerenyi, T., Saary, Z., and Hutterer, F.,** *Res. Commun. Chem. Pathol. Pharm.*, 3, 189, 1972.
73. **Milunsky, A., Littlefield, J. W., Kanfer, J. N., Kolodny, E. H., Shih, V. E., and Atkins, L.,** *N. Engl. J. Med.*, 283, 1370, 1441, 1498, 1970.
74. **Whitby, L. G., Mitchell, F. L., and Moss, D. W.,** *Adv. Clin. Chem.*, 10, 65, 1967.
75. **Moss, D. W.,** *Clin. Chem.*, 16, 500, 1970.
76. **Moss, D. W.,** *J. Clin. Pathol.*, suppl. 24, 4, 22, 1971.
77. **Moss, D. W., Baron, D. N., Walker, P. G., and Wilkinson, J. H.,** *J. Clin. Pathol.*, 24, 740, 1971.
78. Recommendations of the German Society for Clinical Chemistry, Standardization of methods for the estimation of enzyme activity in biological fluids, *Z. Klin. Chem. Klin. Biochem.*, 8, 659, 1970.
79. Draft proposals, I.F.C.C. Reference Methods for Enzymes. General considerations concerning the standardization of measurements of enzyme activity in human serum or plasma, *Clin. Chem.*, 19, 268, 1973.
80. **Dobrow, D. A. and Amador, E.,** *Am. J. Clin. Pathol.*, 53, 60, 1970.
81. **Brojer, B. and Moss, D. W.,** *Clin. Chim. Acta*, 35, 511, 1971.
82. **Moss, D. W.,** *Clin. Chem.*, 18, 1449, 1972.

PROBLEMS OF THYROID ANALYSES IN PREGNANCY

J. D. Acland

TABLE OF CONTENTS

I. Iodine Metabolism in the Female . . . 258
A. Introduction . . . 258
B. Extrathyroidal Iodine Metabolism . . . 258
C. Intrathyroidal Iodine Metabolism . . . 258
D. The Transport of Thyroid Hormones in Plasma . . . 260
E. Free T_4 and T_3 in Serum . . . 260
F. The Physiological Action and Peripheral Metabolism of Thyroid Hormones . . . 261
G. The Hormonal Control of Iodine Metabolism . . . 261
H. Changes in Iodine Metabolism during Pregnancy . . . 261
I. Oral Contraceptives . . . 262

II. The Assessment of Thyroid Status in Pregnancy . . . 262
A. Tests of Thyroid Iodine Uptake . . . 263
B. Measurement of Total Circulating Thyroid Hormone . . . 264
C. Comparison Between Measures of Circulating Thyroid Hormone . . . 268
D. Measurement of Serum TBG Capacity and Free T_4 and T_3 Levels . . . 270
E. Thyroid Stimulation and Suppression Tests . . . 273
F. Tests for the Presence of Circulating Thyroid Autoantibodies . . . 273
G. Tests of the Peripheral Action of Thyroid Hormones . . . 274
H. Measurement of Serum TSH and LATS Levels . . . 275
I. Choice of Thyroid Function Tests in Pregnancy . . . 276

III. Methods . . . 276
A. The Serum PBI Test . . . 276
B. Serum Total Thyroxine by Displacement Analysis . . . 281
C. The Resin T_3 Uptake Test . . . 283
D. The Free T_4 Factor . . . 284
E. Basal Metabolic Rate . . . 285

IV. The Interpretation of Thyroid Function Tests . . . 288
A. The Diagnosis of Thyroid Status . . . 288
B. Special Problems in Pregnancy . . . 289
C. The Interpretation of the Serum PBI and Total Thyroxine . . . 291
D. The Interpretation of the Resin T_3 Uptake . . . 291
E. The Interpretation of the Free T_4 Factor . . . 292
F. The Interpretation of the Basal Metabolic Rate . . . 292
G. The Clinical Assessment of Thyroid Status . . . 293

References . . . 294

I. IODINE METABOLISM IN THE FEMALE

A. Introduction

The interpretation of thyroid function tests in pregnancy requires a knowledge of iodine metabolism in the nonpregnant female and of the changes in iodine metabolism which accompany pregnancy. A schematic representation of iodine metabolism in the female is presented in Figure 1. Iodine is absorbed from the small intestine in the inorganic form and transported by the plasma to the tissues as iodide, a proportion of which is excreted by the kidneys. Iodide is taken up by the thyroid gland and incorporated into the thyroid hormones, which are then secreted into the plasma. The rate of synthesis of the thyroid hormones is controlled by the anterior pituitary under the influence of a feedback control mechanism actuated by the level of circulating thyroid hormones. The thyroid hormones are metabolized peripherally, the liver playing the major role in the degradation process. As a result of peripheral metabolism, iodine is returned to the plasma both as inorganic iodide and as an iodoalbumin. Conjugates of thyroid hormones that are formed by the liver are excreted in the bile into the small intestine and subsequently partly lost in the feces and partly reabsorbed.*

B. Extrathyroidal Iodine Metabolism

Absorption of iodide after oral administration in normal human subjects is virtually complete within 24 hr.[1] Iodide secreted into the gastric contents[2] and saliva[3] is reabsorbed by the intestine. The level of plasma inorganic iodide (PII) is largely determined by the iodide intake.[4] Iodide is cleared from the plasma by the thyroid gland and the kidney. An increase in the thyroid iodide clearance (milliliters of blood cleared of iodide per minute) is associated with a decrease in the renal excretion of iodide and, conversely, a decrease in thyroid iodide clearance is associated with an increase in renal iodide excretion.[5] This means that thyrotoxicosis is associated with a decrease and hypothyroidism with an increase in renal iodide excretion. Similarly, an acute increase or decrease in renal iodide clearance causes the proportion of the PII taken up by the thyroid gland to change in the opposite direction to the change in renal excretion. Adaptation of the normal thyroid to chronic changes in renal iodide clearance occurs so that the absolute amount of iodide taken up by the thyroid remains essentially unaltered. This is achieved through changes in the plasma inorganic iodide concentration and in the rate of TSH secretion by the pituitary. For further discussion, see Blahd.[6]

Iodide is concentrated by the ovaries,[7] the placenta,[8] and the mammary glands.[9] It is not known whether these concentrating mechanisms have a physiological function in the development of the ovum and embryo, fetus, or neonate.

C. Intrathyroidal Iodine Metabolism

Iodine is concentrated in the thyroid gland by an active transport mechanism known as the 'iodide trap.'[10] The stages in the conversion of inorganic iodide into the thyroid hormones have not yet been fully elucidated.[11] Iodide is oxidized by the thyroid peroxidase to an active form, which then iodinates tyrosine residues to form mono- and di-iodotyrosines (MIT and DIT, respectively). DIT and MIT residues are coupled in pairs by a coupling enzyme, as yet uncharacterized, to form iodothyronines, of which the most important both quantitatively and physiologically are thyroxine (T_4) and 3,5,3'-tri-iodothyronine (T_3). The formulas of some relevant compounds are given in Figure 2. The product of iodination is the protein thyroglobulin, which is stored in the thyroid follicles. The iodination and coupling reactions, which result in iodothyronines being included in the peptide chains of the thyroglobulin molecule, are thought to occur in tyrosine residues already present in the peptide chain of a protein precursor of thyroglobulin. The circulating thyroid hormone is a mixture of T_4 and T_3, which are released from thyroglobulin by pinocytosis and proteolysis at the follicular borders. The breakdown of thyroglobulin is stimulated by TSH, which causes iodide, T_4, T_3, and thyroglobulin to enter the

*The following abbreviations will be used throughout this chapter: T_4 = Thyroxine; T_3 = 3,5,3'-Tri-iodothyronine; T_2 = 3,3'-Di-iodothyronine; BEI = Butanol-extractable iodine; BMR = Basal metabolic rate; DIT = 3,5-Di-iodotyrosine; LATS = Long-acting thyroid stimulator; MIT = 3-Mono-iodotyrosine; PBI = Protein-bound iodine; PII = Plasma inorganic iodide; STP = Standard temperature and pressure; TBG = Thyroxine-binding globulin; TBPA = Thyroxine-binding prealbumin; TRH = Thyrotropin-releasing hormone; and TSH = Thyroid stimulating hormone.

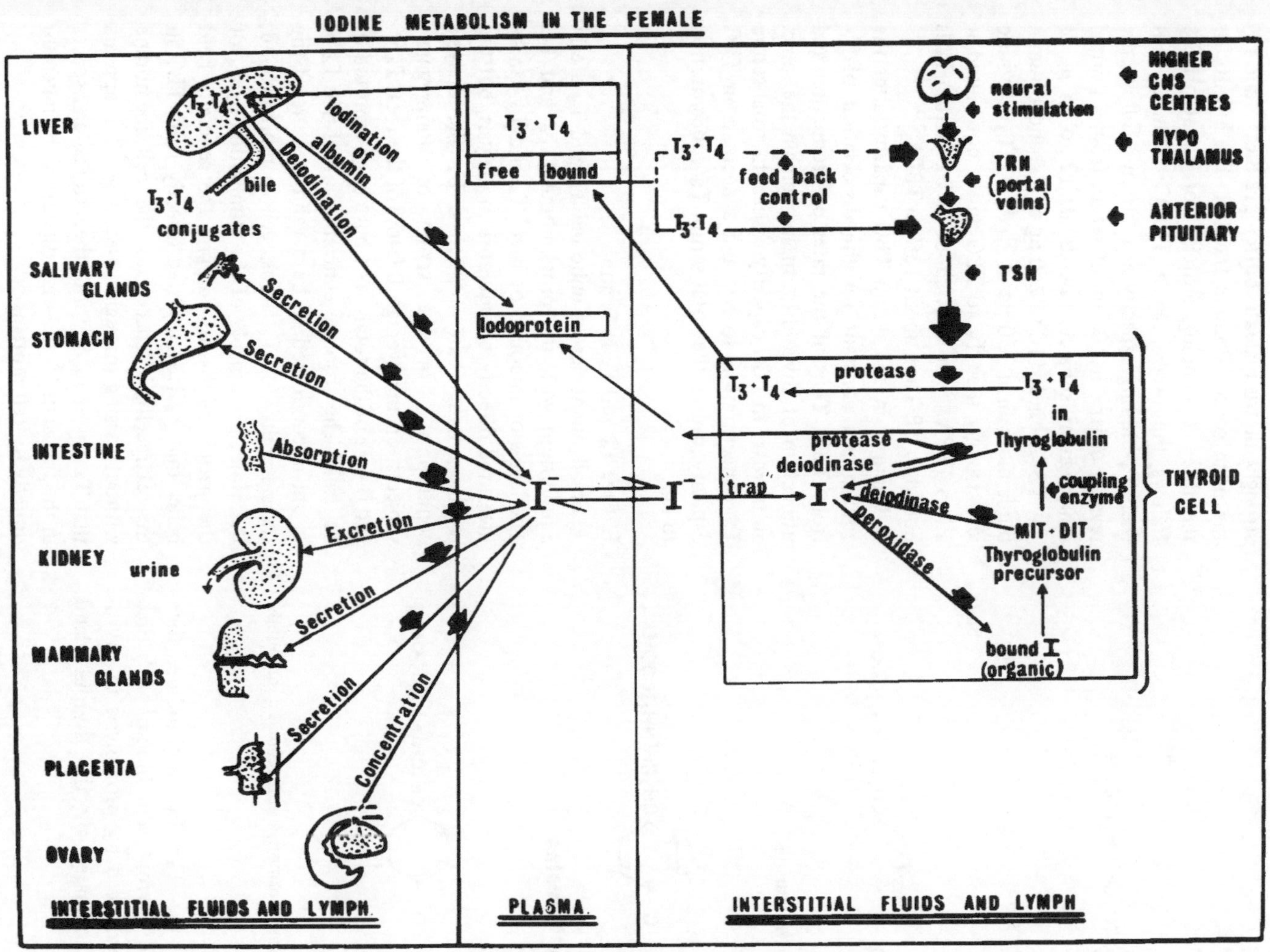

FIGURE 1. Iodine metabolism in the female.

——— Proven pathways

- - - - Hypothetical pathways

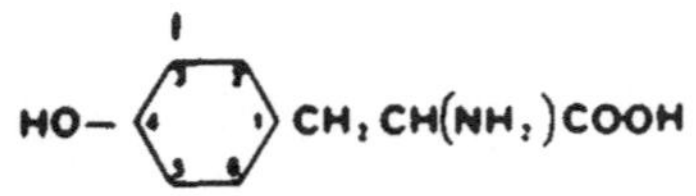

3-mono-iodotyrosine

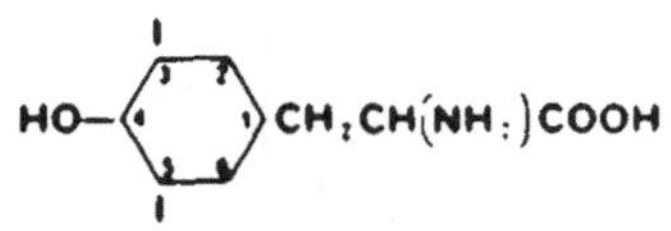

3,5-di-iodotyrosine

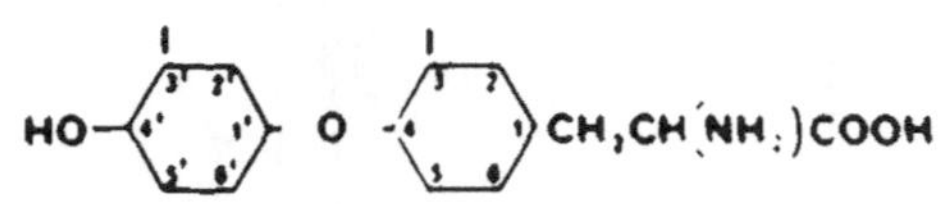

3,3'-di-iodothyronine

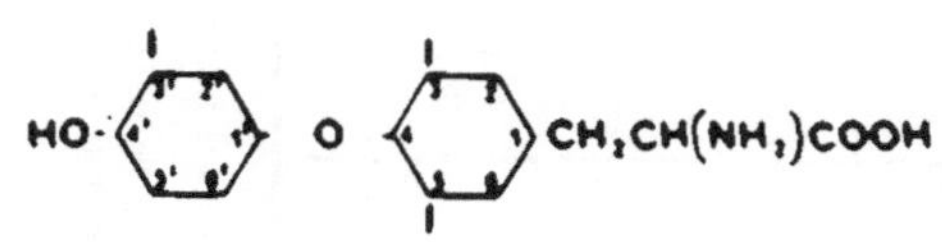

3,5,3'-tri-iodothyronine

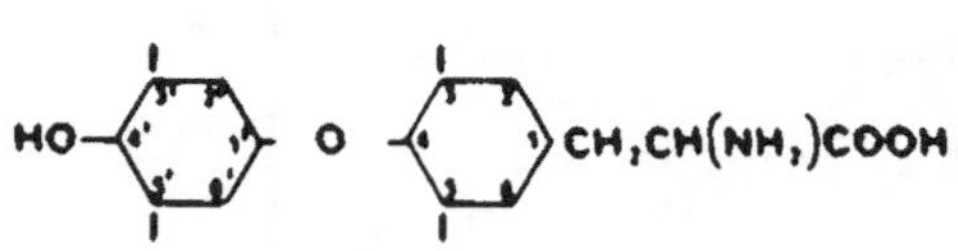

thyroxine

FIGURE 2. Thyroid hormones and related compounds.

thyroid lymph, the T_4, T_3, and iodide, then passing into the thyroid vein blood.[12,13] Free MIT and DIT produced by proteolysis of thyroglobulin are deiodinated by the thyroid gland deiodinase.[14]

D. The Transport of Thyroid Hormones in Plasma

The thyroid hormone circulating in plasma is composed of more than 95% T_4 and less than 5% T_3.[15] T_4 and T_3 in plasma are nondialyzable and precipitate with the serum proteins from which they can be extracted by organic solvents. In vitro, T_4 can be shown to combine with three protein fractions: (1) a thyroxine-binding globulin (TBG) probably a glycoprotein,[16] which migrates between the α_1 and α_2-globulins; (2) a thyroxine-binding pre-albumin (TBPA), which binds T_4 most strongly in tris-maleate buffer pH 8.6;[17] and (3) the main serum albumin fraction. The affinity of the three T_4-binding proteins for T_4 at pH 8.6 follows the order TBG> TBPA> albumin, but their T_4-binding capacity at saturation follows the reverse order. In 100 ml of serum from a normal nonpregnant subject, TBG binds 12 to 28 μg of T_4, TBPA binds 110 to 260 μg of T_4, and albumin binds more than 750 μg of T_4.[18,19] T_3 is bound less strongly than T_4 to TBG and is not bound at all to TBPA.[16] In vivo, about 85% of serum T_4 is bound to TBG and about 15% to TBPA.[20]

Changes in the serum TBG capacity alter the equilibrium constant for the dissociation of T_4 from TBG. Thus, if the rates of production and utilization of thyroid hormones remain the same, an increase in TBG capacity causes the total serum T_4 concentration to rise and a decrease in TBG capacity causes the total serum T_4 concentration to fall.

E. Free T_4 and T_3 in Serum

Small quantities of unbound T_4 and T_3 are in equilibrium with the protein-bound T_4 and T_3. The levels of unbound or 'free' T_4 and T_3 can be measured directly by equilibrium dialysis, but the results obtained are influenced by the dialysis conditions.[21] In the serum of nonpregnant subjects, on the average, 0.046% of the total T_4[22] and 0.46% of the total T_3[23] have been reported to be 'free.' The absolute concentration of free T_4 in serum is considered to be a valid measure of the transferability of T_4 across cell membranes and, therefore, presumably of the stimulation by T_4 of the peripheral tissues, despite the generally held belief that T_4 is transferred directly from TBG in the extracellular spaces to the cell membrane without passing through a free stage.[24] It appears that T_3 passes readily into the extracellular spaces in the free state[25] and may, therefore, be taken up directly by cell proteins.

Within the range of serum total T_4 concentrations seen in clinical practice, the absolute concentration of free T_4 is determined by the serum TBG capacity. Thus, at a given total serum T_4 concentration, a reduced TBG capacity is associated with an increased absolute free T_4 concentration and an increased TBG capacity with a reduced absolute free T_4 concentration.

F. The Physiological Action and Peripheral Metabolism of Thyroid Hormones

Thyroid hormones appear to increase the metabolic rate by inducing the synthesis of oxidative enzyme systems.[26] Of the iodoamino acids present in thyroid hydrolysates, 3,5,3′-T_3 is three to four times as potent as thyroxine, but MIT, DIT, 3,3′-T_2, and 3,3′,5′-T_3 are inactive, as are the nonphysiological dextro-isomers of T_4 and T_3.[4]

Peripheral deiodination is a significant metabolic pathway of the thyroid hormones and T_4 may give rise to significant quantities of T_3 after administration to athyreotic subjects.[27,29] Nevertheless, it is not known whether deiodination is necessary for thyroid hormones to exert their action or indeed whether T_3 is in fact the active component of the thyroid secretion and T_4 is thus exerting its action as a result of conversion to T_3. Deiodination of T_4 is rapid in the liver[29] and is associated with the formation of an iodoalbumin, which is secreted into the plasma,[30] where it may constitute up to 15% of the iodine precipitable with the serum proteins (PBI).

T_4 and T_3 are conjugated by the liver and the conjugates are excreted in the bile.[31,32] There is a variable amount of reabsorption of the conjugates from the gut. On a normal diet, the daily fecal excretion of iodine ranges from 4 to 35 μg, compared with a daily urinary iodine excretion of 44 to 171 μg.[4] The urinary iodine consists of iodide, T_4, and T_3. The iodide output is determined by the dietary iodide intake,[4] the T_4 and T_3 outputs by the thyroid status.[33,34]

The daily rate of peripheral utilization of T_3 in healthy nonpregnant subjects is about one quarter that of T_4, approximately half the T_3 being derived from T_4 by deiodination.[35,36] Hence, T_3 is of comparable physiological importance to T_4, although plasma concentrations of T_3 are much lower than those of T_4 because T_4 is the more strongly protein-bound. The well-known greater potency of administered T_3, compared with T_4,[4] can be explained by its relatively weak combination with plasma proteins and by its distribution largely in the unbound state within the extracellular fluid.

G. The Hormonal Control of Iodine Metabolism

The rate of release of the thyroid hormone is controlled by the anterior pituitary. TSH is secreted by the pituitary when the level of circulating thyroid hormone is low. The effect of circulating thyroid hormone on TSH secretion enables the activity of the thyroid gland to be maintained by a feedback control mechanism. The evidence for the feedback mechanism is largely circumstantial.[11] A thyrotropin-releasing hormone (TRH) has been isolated from the hypothalamus, characterized as a tripeptide, pyroglutamyl-histidinyl-proline amide, and synthesized. The synthetic hormone has been shown to cause an increase in circulating TSH and has been used in the diagnosis of thyroid disorders.[37] It is believed that TRH is carried to the pituitary along the pituitary stalk in the portal veins, but in such minute quantities that it would be difficult, if not impossible, to detect. Evidence has been presented to show that an increase in serum T_3, but not serum T_4, concentration inhibits the effect of TRH on the pituitary-thyroid axis.[38] The existence of a feedback mechanism controlled by the level of circulating T_3 would enable the level of peripheral thyroid activity to be kept constant by appropriate variation in the activity of the thyroid gland. Thus, in dietary iodine deficiency the supply of thyroid hormone to the tissues is maintained by an increase in thyroid activity under the influence of a raised level of circulating TSH. Conversely, when the level of circulating thyroid hormone is raised, as in thyrotoxicosis, TSH secretion is reduced below the detectable limit.

The hypothalamic effect on TSH secretion, which is mediated by TRH, also provides a pathway whereby the activity of the thyroid gland could be influenced by higher neural centers.

H. Changes in Iodine Metabolism during Pregnancy

Iodine metabolism is influenced in a complex manner by pregnancy. The pregnant woman is in a mild iodine deficiency state with a low plasma inorganic iodide level, probably as a result of the increased renal iodide clearance in pregnancy; the thyroidal radioiodine clearance is increased, although the thyroid absolute iodide uptake is normal.[39] Levels of circulating TSH, determined by radioimmunoassay[40] and by bioassay,[41] have been reported to be raised in pregnancy. The radioimmunoassay results might indicate that TSH secretion was increasing in response to iodine deficiency, or they might represent a cross-reaction in the assay with chorionic TSH, which has been shown to be secreted by the placenta.[41] The data are difficult to interpret, since TSH has been shown to cause an increase in renal iodide clearance in thyroidectomized rats.[42] Thus, the secretion of chorionic TSH could be responsible for the changes in iodine metabolism in pregnancy which have been detailed so far. Nevertheless, administration of T_4 or T_3 in pregnancy still suppresses the thyroid iodine uptake, indicating that the normal feedback mechanism has not been overridden, and TSH administration still stimulates uptake, showing that the response of the thyroid to TSH has not been affected.[43]

The increase in circulating estrogens during pregnancy stimulates synthesis of TBG by the liver, giving rise to a raised serum TBG capacity,[44] although the serum TBPA capacity is not affected by pregnancy.[17] An increase in TBG capacity by itself could explain the high levels of circulating thyroid hormones in pregnancy, which may be within the range characteristic of thyrotoxicosis in the nonpregnant subject. However, direct determination of serum free T_4 by equilibrium dialysis has been reported to show an increase on average in pregnancy, as compared with the nonpregnant state.[45] On the other hand, assessment of free T_4 activity by a refined indirect technique based on measurements of the serum PBI and T_3-binding capacity has shown that the estimated serum free T_4 activity in pregnancy is within the normal limits for the nonpregnant state[46,47] and urinary excretion of T_3 and T_4 is normal.[33,34] It is difficult to explain this discrepancy, but it should be noted that determinations of serum free T_4 levels by equilibrium dialysis are very sensitive to methodological variations[21] and are not necessarily more reliable than indirect estimates of free T_4 activity.

A raised serum free T_4 level in pregnancy would be consistent with a primary increase in circulating TSH, resulting from the secretion of chorionic TSH. On the other hand, if the raised level of circulating TSH in pregnancy is a response to iodine deficiency, the serum free T_4 level in pregnancy would not be expected to be higher than in the nonpregnant state.

The basal metabolic rate increases gradually during the course of pregnancy, but there is general agreement that these changes result from the metabolic requirements of the uterus and fetus and from an increase in cardiac and respiratory work, rather than from a change in thyroid status.[48]

I. Oral Contraceptives

Estrogenic oral contraceptive preparations cause an increase in TBG capacity and total serum T_4 in subjects with a normal thyroid function, but progestational contraceptive preparations are without effect on the level of circulating thyroid hormones.[49] Serum total T_4 levels in women taking estrogenic oral contraceptives are similar to those found in pregnant subjects.[50] It is generally believed that oral contraceptives do not cause fundamental changes in iodide metabolism, such as those found in pregnancy. However, a recent unconfirmed report has suggested that low or low-normal thyroid radioiodine uptakes may occur in some instances after a long-term course of high-dose oral contraceptives.[51] This finding was attributed to a direct effect of the estrogen on the thyroid gland.

Some values for parameters of circulating thyroid hormone activity in women taking oral contraceptives (estrogenic), derived from the papers of Goolden et al.[46] and Schatz and co-workers,[45] are given in Table 1. The evidence suggests that, despite the changes in total T_4, the free T_4 activity remains the same when oral contraceptives are being used.

II. THE ASSESSMENT OF THYROID STATUS IN PREGNANCY

The object of clinical assessment of thyroid status is to decide whether the physiological action of the thyroid gland in a patient is reduced (hypothyroidism), normal (euthyroid state), or increased (thyrotoxicosis). The final diagnosis is based on a combination of clinical observations and laboratory tests. Laboratory tests of thyroid function are divided into in vivo and in vitro tests. In vivo tests require administration of radioisotopes to the patient, followed by measurements of radioactivity on the whole subject, on plasma, or on urine. The use of in vivo tests in pregnancy is limited by the necessity to avoid radiation damage to the fetal thyroid.

Thyroid function tests are considered under the following headings: (1) tests of thyroid iodine

TABLE 1

Parameters of Circulating Thyroid Hormone Activity in Women Taking Estrogenic Oral Contraceptives

	Euthyroid nonpregnant	Pregnant	Oral contraceptives
Total serum T_4 (μg/100 ml)	4.5–11.5	8–16	6–14
Serum TBG capacity (μg of T_4 bound per 100 ml serum)	12–28	26–75.5	24.5–52
Resin (Triosorb) T_3 uptake ratio [% uptake (test): % uptake (mean normal)]	0.81–1.20	0.44–0.75	0.59–1.00
Free T_4 factor* (arbitrary units)	0.215–0.555	0.165–0.580	0.290–0.540

*See Methods section

uptake, (2) measurement of total circulating thyroid hormone, (3) measurement of serum TBG capacity and free T_4 and T_3 levels, (4) tests of the peripheral action of thyroid hormones, (5) measurements of serum TSH and LATS levels, (6) thyroid stimulation and suppression tests, (7) tests for the presence of circulating thyroid autoantibodies, and (8) choice of thyroid function tests in pregnancy.

A. Tests of Thyroid Iodine Uptake

The standard selection of thyroid function tests includes a simple measurement of radioiodine uptake as a percentage of the administered radioactive dose. The maximum uptake is normally reached in 24 to 48 hr and it is usual to make measurements of thyroid uptake at either 24 or 48 hr. However, thyroid radioiodine uptake may be disproportionately accelerated in the early period (up to 6 hr after a dose) in thyrotoxicosis, in autoimmune thyroiditis, and in iodine-deficiency goiter.[4] Hence, a measurement of early uptake at 2½ or 4 hr is usually also included to increase the diagnostic accuracy in these conditions.

Since the thyroid and the kidney compete for the plasma inorganic iodide, measurements of the thyroid radioiodine uptake will be influenced by changes in renal iodide clearance. When variations in renal clearance interfere with the interpretation of the thyroid radioiodine uptake, more reliable results are obtained by measuring the thyroid radioiodine clearance. This is expressed in the usual way as the number of milliliters of plasma cleared of radioiodine per minute. It is calculated from two measurements of the thyroid uptake, at 1 hr and at 2½ hr after a dose of radioiodine, and one measurement of plasma radioactivity at the midpoint.

If the plasma inorganic iodide level is low, due to iodine deficiency, or high, due to administration of iodine drugs, the thyroid radioiodine uptake and clearance do not accurately reflect the iodine-concentrating power of the thyroid. In the first instance, the radioiodine uptake may be above the normal range and in the second instance the radioiodine uptake may be below the normal range, although normal amounts of thyroid hormone are being secreted. In these cases, measurement of the absolute thyroid iodine uptake is required to give an accurate quantitative estimate of thyroid iodine uptake. The absolute iodine uptake is obtained by multiplying together the thyroid radioiodine clearance and the plasma inorganic iodide concentration, with the results expressed in μg/hr. The plasma inorganic iodide concentration is too low to be estimated conveniently by chemical means and although a method has been published,[52] it is usually determined by an indirect radiochemical method.[53] After a dose of radioiodine, during determination of the thyroid radioiodine clearance, specimens of plasma and either saliva or urine are collected. The radioactivity of the plasma and saliva (or urine) specimens and the iodide concentration in the saliva (or urine) are measured. Iodide concentrations are considerably higher in saliva and in urine than in plasma; iodide estimations can conveniently be performed by chloric acid digestion, using the radioactivity of the specimens to correct for incomplete recovery in the chemical analysis. The plasma inorganic iodide (PII) is then obtained by the formula:

$$\underset{(\mu g/100\ ml)}{PII} = \underset{(\mu g/100\ ml)}{\text{saliva (or urine) } I^-} \times \frac{\text{plasma radioactivity (\% dose/ml)}}{\text{saliva radioactivity (\% dose/ml)}}$$

It is usually better to use saliva than urine to determine the PII since saliva, unlike urine, contains insignificant amounts of organic iodine,[54] unless the salivary glands have been damaged by radiation.[55]

The measurement of the thyroid absolute iodine uptake may be of value in doubtful cases, especially in the light of an overall reduction in the percentage radioiodine uptake in normal subjects which has occurred in the U.S. in recent years, possibly as a result of an increase in the dietary intake of iodide, principally in the form of additions to bread.[56]

Thyroid function tests, such as the T index[5] based on the 48-hr urinary excretion of radioiodide after a radioactive dose, are mainly of value in the diagnosis of hypothyroidism, but are no longer widely employed. They are not suitable for pregnant subjects, since the long-lived isotope ^{131}I has to be used.

The reader is referred to Wayne and coworkers[4] for a full discussion of thyroid uptake tests.

Thyroid scanning is a specialized technique requiring an automatic gamma scanning apparatus, which uses a highly collimated scintillation head to obtain a graphic picture of the pattern of radioiodine uptake by the thyroid gland.[6] Thyroid scanning is of particular value in screening the thyroid for 'hot' or 'cold' nodules. Gamma scanning can also be used quantitatively for thyroid uptake measurements after an intravenous dose of pertechnetate ($^{99m}TcO_4{}^-$) produced locally by elution from a sterile generator (e. g., that supplied by the Radiochemical Centre, Amersham, England). Pertechnetate is concentrated by the thyroid iodide trap, but is not metabolized further. Since ^{99m}Tc has a short half-life and pertechnetate is rapidly discharged from the thyroid, relatively high doses of radioactivity can be used safely. Measurements of pertechnetate uptake can be made less than half an hour after a radioactive dose. However, the background extrathyroid neck radioactivity is high and a correction for it must be included when the thyroid uptake is calculated. The pertechnetate uptake is a good discriminator of thyrotoxic from euthyroid subjects, although, as might be expected, raised uptakes are found in some cases of simple goiter (for an account of routine $^{99m}TcO_4{}^-$ uptake measurement, see Goolden et al.[57]).

The iodine isotopes used routinely in thyroid uptake measurements are ^{131}I, with a half-life of 8.1 days, and ^{132}I, with half-life 2.3 hr. Because of their respective physical properties, ^{131}I can be used for both early and late uptake measurements, whereas ^{132}I can only be used for early uptake measurements.

The long-lived isotope ^{131}I should not be used for diagnostic purposes in pregnancy, as it may damage the fetus. The short-lived isotope ^{132}I may be used diagnostically in the first half of pregnancy, when the fetal thyroid is inactive,[58] if the patient's clinical state necessitates a thyroid uptake test. The preferred ^{132}I test is the thyroid radioiodine clearance because of the increased renal clearance of iodide in pregnancy.[39] Each center should establish its own normal ranges in pregnant subjects. The pertechnetate uptake may also be measured in early pregnancy (Goolden, A. W. G., personal communication). In both instances in the U.K., specific permission for the administration of a radioactive isotope to the patient in question must be obtained from the Medical Research Council (20 Park Crescent, London W1N 4AL).

B. Measurement of Total Circulating Thyroid Hormone

1. Serum PBI

The first method to be developed that enabled circulating levels of thyroid hormone to be estimated with a satisfactory degree of accuracy was the serum protein-bound iodine (PBI) determination, in which the serum hormonal iodine was digested to iodide or iodate, and the inorganic iodine determined by its catalytic effect on the reaction between ceric and arsenite ions. The analytical technique was gradually improved over the years and at the present time the test has an efficient diagnostic performance.[59] Nevertheless, it has some analytical drawbacks. The serum PBI is not a completely specific measure of circulating thyroid hormones. Up to 15% of the iodine determined as PBI may be derived from an iodoprotein, which is synthesized in the liver following deiodination of the thyroid hormones.[30,60] In addition, variable amounts of iodotyrosines, which may be present, for instance, in the serum of familial goitrous cretins,[61] are estimated with the PBI. Most seriously, iodinated radiographic contrast media and many iodine drugs are persistent in the body and may cause elevation of the serum PBI lasting for weeks, months, or even years.[59]

The manual techniques originally used to determine serum PBI were time-consuming and exacting, and were subject to interference by iodine contamination from the environment. A considerable advance was made when the PBI analysis by acid digestion was completely automated and adapted to the Technicon® Continuous Digestor.[62] This apparatus enables a complete analysis to be performed with minimal handling of the specimen. Strictly speaking, the Technicon Continuous Digestor measures total serum organic iodine rather than PBI because an ion-exchange resin is used instead of protein precipitation to remove inorganic iodide from the serum.[63] The apparatus requires careful handling because of the corrosive nature of the reagents. Nevertheless, analyses can be performed at 20 to 30 per hr, relatively cheaply, after the initial capital outlay.

2. Serum BEI

Considerable effort has been put into devising possible ways of improving the specificity of serum PBI determinations. The first technique to be tried was solvent extraction. Serum is extracted with butan-1-ol and the extract washed with 4 *N* sodium hydroxide in 5% (w/v) sodium carbonate[64] to remove iodide and iodotyrosines. This procedure, used, for example, by Klein and Chernaik,[65] followed by chemical determination of the iodine in the extract, i.e., the butanol-extractable iodine (BEI), was supposed to give a more accurate measure of circulating thyroid hormones than the PBI. Unfortunately, the BEI procedure is subject to variable (10 to 30%) losses of T_4[66] and extracts some contaminating iodine drugs from serum.[67]

3. Serum T_4 by Column

Column chromatography on an ion-exchange resin was introduced by Pileggi et al.[68] to separate T_3 and T_4 from iodotyrosines and iodide prior to chemical determination of the $(T_3 + T_4)$I fraction. The technique was further improved by Pileggi and Kessler,[69] who introduced a bromination step, which enabled them to estimate iodine liberated from iodothyronines in the column eluate without the preliminary digestion step required by the technique of Pileggi et al.[68] However, the chromatography and bromination did not eliminate contamination by all iodine drugs.[69]

The bromination technique has been adapted to the AutoAnalyzer II® (Methodology II-28, Technicon Instruments, Tarrytown, N.Y.); a manual ion-exchange chromatography step is followed by automated estimation of iodine in the eluate after bromination.

Treatment with ion-exchange resin has to be done manually the day before the automated iodine estimation, thereby increasing the handling of specimens and necessitating precautions against contamination of the column eluates before the automated analysis. Since the column eluate from each serum is analyzed in two portions to detect contamination, the analysis of 60 eluate samples per hour does not confer any advantage in speed over the Technicon Continuous Digestor running at 30 samples per hour on serum samples. However, the AutoAnalyzer II equipment might be considered as an alternative to the Continuous Digestor if iodine drugs were commonly found in the serum in a population being screened for thyroid status. The technical problems of partially automated determinations of T_4 by column are discussed by Hanok et al.[70] The column chromatographic technique for determining T_3 and T_4 is, in any case, not so specific for T_4 and T_3 as the displacement methods to be described next.

4. Serum T_4 by Displacement

With the increasing availability of radioactively labeled T_3 and T_4, more specific techniques for determining serum T_4 and serum T_3 were developed using a method variously described as competitive protein-binding analysis, saturation analysis, or displacement analysis.[71] The term 'displacement analysis' is preferred.[72] In the original technique,[73] T_4 is extracted from serum by ethanol. The ethanol is evaporated off and the T_4 in the residue taken up in a preparation of TBG (diluted human serum or globulin fraction) dissolved in barbiturate buffer pH 8.6, which also contains radioactively labeled T_4. The nonradioactive T_4 competes with the labeled T_4 for the binding sites on TBG. The higher the concentration of nonradioactive T_4, the more radioactive T_4 will be displaced from TBG. The free and bound radioactivity are separated by passing the mixture through a column of Sephadex® G25, which retains the free labeled T_4. The percentage of labeled T_4 bound to TBG is then found by measuring the activity in the eluate and comparing it with the amount of radioactivity put onto the column. A calibration curve is constructed from standards, which allows the amount of T_4 in the residue from the ethanol extraction to be calculated from the percentage of bound radioactivity. The serum T_4 concentration is then calculated, allowance being made for the incomplete extraction of T_4 from serum by ethanol. An average extraction of 77% is usually assumed to have been achieved,[73] although differences occur in the extraction of T_4 from different sera.[71] In the Thyopac-4 kit (Radiochemical Centre, Amersham, Bucks, England) for determining serum T_4 by displacement, the reciprocal of the unbound radioactivity is plotted against the T_4 concentration to give a calibration curve which is linear between 0 and 20 μg T_4/100 ml, according to the manufacturers, justifying two-point standardization for routine purposes. The displacement technique is exacting, but can give good results in routine practice.[74] The advantage of the technique is that drugs, other than D-thyroxine and phenytoin which give falsely elevated results, do not cause interference.[73,75] The disadvantages of the technique derive from the ethanol extraction step which is difficult to control.

Mougey and Mason[76] found that T_4 was separated from TBG in 0.015 *N* sodium hydroxide and that free T_4 could be completely separated from TBG by passage down a column of Sephadex G25 ('T_4 stripping'[77]). This principle was applied to displacement analysis of T_4 in the Tetralute® kit (Ames Laboratories, Elkhart, Ind.). Serum and $[^{125}I]$ - T_4 in 0.1 *N* sodium hydroxide are passed down a column of Sephadex G25. The T_4 adsorbed on the column is then equilibrated with a solution containing TBG at pH 8.6 and the T_4 bound to TBG washed out with barbiturate buffer at pH 8.6. The percentage of T_4 retained on the column is dependent on the amount of T_4 present initially on the column, and the T_4 concentration of an unknown serum is read off a calibration curve prepared from standards run through the whole procedure. T_4 determinations by the Tetralute kit gave results which did not differ significantly from T_4 determinations by the method of Murphy and Pattee[73] on the same sera.[77] The manufacturers recommend that only the region of the calibration curve between 2.5 and 10 μg of T_4-I/100 ml should be used, owing to nonlinearity and deterioration in accuracy beyond these limits.

In the manufacturers' recommended procedure for the Tetralute kit, standardization is carried out by means of different volumes of a standard serum whose T_4 concentration has been given an assigned value by the manufacturers. This is not an acceptable procedure for an analytical laboratory, which should be in control of its own standards. Seligson and Seligson[78] used solutions of pure T_4 in human serum albumin to standardize their own version of the T_4 stripping technique. This is a more satisfactory procedure on general principles than the use for standardization of lyophilized sera of varying composition, which are provided by Ames Laboratories and other manufacturers of kits for determining serum T_4 (e.g., Abbott Laboratories, Chicago, Ill. and Radiochemical Centre, Amersham, Bucks, England). Such reference sera do not give results identical with their assigned values when analyzed by other manufacturers' kits, probably due to differences in the recovery of T_4 from dried standards by ethanol extraction.[79]

5. Serum T_3

Measurements of the serum T_3 concentration are becoming increasingly important clinically since it has been recognized that the serum T_3 level is a better discriminator than the serum T_4 level in the diagnosis of thyroid status.[15] This is particularly true in certain patients who have previously been treated by radioiodine for thyrotoxicosis, in iodine deficiency, and in 'T_3 toxicosis,' a condition in which the clinical signs of thyrotoxicosis are associated with normal values of the serum T_4 level, the serum TBG, and the thyroid iodine uptake. In each of these groups, the ratio of T_3 to T_4 in the serum is increased and the serum T_3 level correlates the best with the clinical condition.[15]

Early methods of determining serum T_3 levels[23,80] employed chromatographic separation of T_3 from T_4, followed by determination of the eluted T_3 by displacement analysis. Labeled T_4 was added to the serum to enable correction to be made for any T_4 which might appear in the T_3 fraction. The techniques were exacting, the results difficult to reproduce, and considerable technical controversy was generated.[81-85]

Gas chromatographic techniques for estimating T_3 and T_4 require the preparation of suitable derivatives before putting them on the column.[86,87] Preliminary purification by extraction with an ion-exchange resin and paper chromatography is required in the technique of Nihei et al.[87]

Both types of chromatographic method are subject to artifacts as a result of paper blanks, the presence of some T_4 in the T_3 fraction, deiodination or esterification of T_4, resulting in an enlargement of the T_3 fraction, and introduction of T_3 into the system as a contaminant of the labeled T_4 preparations added to estimate the amount of T_4 in the T_3 fraction. Radioimmunoassays have now replaced the chromatographic techniques.

6. Serum T_3 and T_4 by Radioimmunoassay

The chromatographic techniques for determining T_3 were so exacting that other methods were sought which would enable determinations to be made on unextracted serum. Several groups of workers turned to radioimmunoassay. The method of radioimmunoassay is similar in principle to the displacement analysis for T_4, with the difference that the labeled and unlabeled molecules compete for binding sites on an antibody instead of on a specific binding protein. The use of an antibody in the reaction makes it possible to increase the specificity of the reaction above that achieved by the use of binding proteins.

It is first necessary to raise an antibody to T_3. This is achieved by injecting doses of a protein, to which T_3 has been attached chemically, into a rabbit according to a regime which has been found to stimulate the production of circulating antibodies in the rabbit's serum. After a suitable time lapse to allow for the production of a high titer of antibodies, the rabbit is bled and dilutions of the serum used as a source of antibodies in the assay. Most workers prepare the T_3-protein antigen by modifications of the method of Brown et al.,[88] which couples T_3 methyl ester with synthetic succinylated polylysine by a reaction involving a substituted carbodiimide. Some groups use the carbodiimide reaction to couple T_3 in the free acid form with bovine, human, or rabbit serum albumin.[89] Other groups raise the antibody by injecting purified thyroglobulin into the rabbit.[90] This last procedure introduces difficulties because the antiserum contains antibodies to T_4 as well as T_3, since both T_4 and T_3 residues are present in thyroglobulin, and special precautions have to be taken to avoid cross-reaction with T_4 in the assay. For this reason, the use of thyroglobulin to raise antibodies to T_3 does not appear to be desirable.

In the assay, labeled T_3 and serum containing unlabeled T_3 are incubated with a dilution of the antibody and the ratio of free T_3 to bound T_3 measured. A calibration curve is constructed relating the value of this ratio to the serum T_3 concentration using T_3 standards and the serum T_3 concentrations of the unknown specimens are then read from this graph.

Antibodies vary in their affinity for T_3. The final concentration of antibody is chosen by trial and error so that the region of maximum slope on the calibration curve coincides with the range of T_3 concentrations which it is wished to determine.

Different groups use different methods of separating free T_3 from T_3 bound to antibody. Some use a 'double antibody' technique in which T_3 combined with its antibody (the 'first antibody') is precipitated by an anti-rabbit serum, raised in a goat, which contains an anti-rabbit globulin antibody, the 'second antibody.'[90] Other groups use charcoal-coated dextran[91] or a mixture of activated charcoal and methylcellulose[88] to adsorb free T_3. If a constant amount of labeled T_3 is added to each tube, as can be achieved by automatic pipetting, it is only necessary to measure the radioactivity in the final precipitate after it has been separated and washed by centrifugation. These measurements are conveniently carried out in an automatic gamma-counter. The precipitate contains the bound T_3 in the double antibody technique and the free T_3 in the charcoal adsorption technique.

T_3 standards are prepared by adding pure T_3 to pooled human serum which has been freed from T_4 and T_3 by treating 100 ml of the serum with

20 g of Norit A charcoal (Sigma Chemical Co., St. Louis, Mo.) for 24 hr at 4°C, followed by three centrifugations at 20,000 X g.[91] Triosorb resin (Abbott Laboratories, Chicago, Ill.) in the presence of tetrachlorothyronine can also be used to render serum free from T_3.[92]

Other methods have been published, e.g., by Lieblich and Utiger.[28] The main difficulties in the radioimmunoassay of T_3 in unextracted serum are (1) TBG in the test sera interferes with the combination of T_3 and the anti-T_3 antibody, (2) T_4 may cross-react with the anti-T_3 antibody, and (3) the test sera may precipitate the anti-T_3 antibody, giving spuriously high values for T_3 if the charcoal adsorption techniques are used.

Interference by TBG is reduced or abolished by adding to the reaction medium tetrachlorothyronine,[92] phenytoin,[28] 8-anilino-1-naphthalene sulfonic acid,[91] salicylate, or merthiolate.[93] All of these compounds block the combination of T_3 with binding sites on TBG.

The degree to which T_4 cross-reacts with the anti-T_3 serum varies in different systems. Relative potencies of T_4 with references to T_3 ranging from <0.1%[92] to 1.3%[90] have been reported. It is likely that these differences result from different amounts of T_3 contaminating manufacturers' samples of T_4.[28] This would make the addition of T_4 to all sera, as recommended by Chopra et al.,[90] an unreliable method of compensating for cross-reaction, since changes in the amount of T_3 added to the system as a contaminant in different samples of T_4 might cause the sensitivity of the assay to be inconsistent.

Correction for nonspecific precipitation of the antibody by the test serum is made by including suitable control tubes in the assay.[93]

Hufner and Hesch[93] found that the TBG antagonists salicylate and aminonaphthalene sulfonic acid caused a decrease in the binding of T_3 to their antibody, while the TBG antagonist tetrachlorothyronine cross-reacted with their antibody. They recommend the use of merthiolate (0.1% w/v) as a TBG antagonist in 0.05 *M* tris-HC1 buffer pH 8.9 containing 0.25% (w/v) human serum albumin.

There seems to be no agreed standard technique for the radioimmunoassay of T_3 at the time of writing. The success of the assay depends on the production of a satisfactory antibody and on the use of a reliable method of inhibiting TBG, which also depends in part on the properties of the antibody.

The pressure for developing a radioimmunoassay technique for serum T_4 is not great because the standard displacement analysis techniques give satisfactory results. Furthermore, TBG interferes more seriously with the radioimmunoassay of T_4 than of T_3, because T_4 is bound the more strongly to TBG. Nevertheless, a radioimmunoassay of T_4, after its extraction from serum with butan-1-ol:ethanol (50:50 by vol), has been published in which thyroglobulin is used to raise the antibody.[94] This technique does not eliminate the troublesome solvent extraction step, but the assay is stated not to be affected by phenytoin, which interferes with T_4 analyses by displacement (see above). A combined radioimmunoassay of T_3 and T_4 in unextracted serum has been described[91] using antibodies raised against either T_3 or T_4 coupled to serum albumin. This is a convenient procedure since samples can be prepared for both assays in the same manner. However, anilinonaphthalene sulfonic acid is used to inhibit TBG, and may not give satisfactory results with all antibodies.[93]

C. Comparison Between Measures of Circulating Thyroid Hormone

Besides T_3 and T_4, the serum PBI contains an iodoprotein synthesized by the liver, which may comprise 10 to 15% of the total PBI.[30] In some pathological conditions, iodotyrosines may be estimated as PBI. Thus, the serum PBI may give spuriously high estimates of the level of circulating thyroid hormones. In addition, the serum PBI is affected by iodine drugs. In T_3 toxicosis, the serum PBI or T_4 is not the appropriate measure since the serum T_3 level is raised with a normal serum level of total PBI and T_4.[15] Nevertheless, the results of the serum PBI determination correlate well with the more specific serum T_4 analyses by displacement.[95] These authors found that the serum PBI concentration on the average was 0.35 μg of I/100 ml higher than the T_4-I concentration in euthyroid subjects; there was no significant difference between the results of the two methods in hypothyroid patients, but the (PBI minus T_4I) difference was greater in thyrotoxicosis, in some cases of goiter, and in patients on thyroxine replacement therapy. Differences in the concentration of iodoproteins, and possibly iodotyrosines, in the sera of the different groups could account for these findings. Lee et al.[96] report larger mean differences between the serum PBI and the serum T_4-I by displacement: 0.7 μg/100 ml in hypothyroid patients, 1.1 μg/100 ml in euthyroid patients, and 1.8 μg/100 ml in patients of thyroxine maintenance; in thyrotoxic patients the mean difference between the serum PBI and the T_4-I by column[69] was 1.5 μg/100 ml.

Serum T_4 levels determined by column do not differ significantly from the levels determined by displacement.[15,96]

The original chromatographic techniques for estimating T_3 gave mean values for euthyroid subjects of 3.3 ng/ml[23] and 2.2 ng/ml.[80] Fisher and Dussault[81] introduced corrections into the method of Sterling et al.[80] to take into account a blank effect associated with the chromatography steps. They added [^{131}I]-T_4 initially to allow for deiodination of T_4 and [^{125}I]-T_3 as a recovery check, and used [^{131}I]-T_3 in the displacement analysis. The corrections reduced the average serum T_3 level in euthyroid subjects from 2.4 to 0.98 ng/ml. Radioimmunoassay techniques showing the least cross-reaction with T_4[92] gave results between 1.2 and 1.4 ng/ml for the mean euthyroid serum T_3 level, so these values may be presumed to be of the right order.

Serum T_4 levels, like serum PBI levels, are raised in pregnancy in euthyroid subjects.[50] Serum T_3 levels, estimated by the method of Sterling et al.,[80] increase in pregnancy,[97] but some of the increase in T_3 found by this technique may have arisen by artifactual deiodination of T_4 since the serum total T_4 concentration increases during pregnancy. Levels of serum T_3, determined by radioimmunoassay, are elevated in maternal sera at term.[191]

The choice of a method of determining circulating thyroid hormone depends on circumstances. In a referral center that has to carry out several hundred screening tests per week, determination of serum PBI by the Technicon Continuous Digestor is the most economical solution in terms of time and money. Although the serum PBI test is the least specific of the tests of circulating thyroid hormone, it is sufficiently precise and accurate to have clinical value.[59] The semiautomated determination of T_4 by column with bromination on the AutoAnalyzer II does not convey any advantages in speed over the Continuous Digestor (see above) and requires more manipulation of the specimen, but it might be preferable if many specimens contaminated with iodine drugs were being screened. In a reference laboratory, T_4 determination by displacement analysis should be available to deal with the problem of iodine contamination detected in specimens sent for PBI analysis. The capital cost of automatic gamma-counting equipment and the cost of the labeled T_4 would usually preclude the displacement method from use as a screening technique of first resort in a reference laboratory, although it is more specific than the semiautomated determination of T_4 by column in the analysis of contaminated specimens. In an unspecialized laboratory that might perform up to 50 thyroid screens per month, the displacement technique for T_4 is appropriate. Suitable manual gamma-counting equipment is not expensive. For research purposes and in a clinical unit specializing in thyroid disorders, serum T_3 determinations are essential, and a radioimmunoassay method would be the method of choice. The amount of additional information provided by estimation of serum T_3 levels does not justify introducing the method into a general or routine screening laboratory at the present time in view of the technical difficulties involved.

In the section on methods, details are given of the determination of the serum PBI by the Technicon Continuous Digestor, and of the determination of the total serum T_4 by displacement, using the T_4 stripping technique with Sephadex G25 and standards of pure T_4 in 0.25% human serum albumin. These techniques are suitable for general analytical laboratories operating a thyroid screening service. A technique for determining serum T_3 levels by radioimmunoassay is not described. Radioimmunoassay methods that require the analyst to raise his own antibody are not suitable for introduction into a general analytical laboratory and are best developed in a unit specializing in radioimmunoassays.

In screening pregnant patients for thyroid disorders, it is not enough to provide an estimate of circulating thyroid hormone level. It is also necessary to include in the screen a measure of TBG capacity and free thyroid hormone. These tests are discussed in the next section.

D. Measurement of Serum TBG Capacity and Free T_4 and T_3 Levels

1. Direct Measurement of TBG Capacity

The T_4 binding capacity of serum is measured by electrophoresis on paper[98] or cellulose acetate.[99] Portions of the test serum are incubated with different amounts of added T_4 so that a series of final T_4 concentrations up to 150 μg/100 ml is obtained in the enriched serum samples. [^{131}I]-T_4 is added to the enriched sera, which are then subjected to electrophoresis on paper or cellulose acetate. The strip is stained for protein and then either scanned for radioactivity in a radiochromatogram scanner, using a proportional counter (e.g., the model supplied by Atomic Accessories, Valley Stream, N.Y.) or cut into sections which are then put into plastic tubes and counted in a standard well-type scintillation counter. The radioactivity in the inter-alpha globulin area of each strip, which is due to the T_4 associated with TBG, is measured and expressed as a percentage of the total radioactivity on the strip. From the total concentration of thyroxine in each enriched specimen (the sum of the concentration of added T_4 and the T_4 concentration in the original serum, which must be measured), the total concentration in the serum of T_4 bound to TBG is calculated. This concentration tends toward a maximum value as the concentration of added T_4 increases; this maximum concentration is taken to represent the TBG capacity of the serum. Values for serum TBG capacity in nonpregnant subjects range from 12 to 28 μg of T_4/100 ml of serum and in pregnant subjects from 24 to 74 μg of T_4/100 ml of serum.[45,100]

The method is too complicated and time-consuming for general routine use, and, except for research purposes, has been replaced by measurements of the resin T_3 uptake.

2. The Resin T_3 Uptake Test[101]

This test is an inverse measure of free T_4 binding sites on TBG.[46] Radioactive T_3 is added to serum and the mixture incubated for a fixed time with an anion-exchange resin. The resin is separated from the serum and washed. The percentage of the added radioactivity remaining in the resin (the T_3 uptake) is then measured. When the number of free binding sites is increased, the resin uptake is decreased, and vice versa. Since the number of free sites depends on both the TBG capacity and the total circulating T_4, the resin T_3 uptake is influenced by both these factors and is not solely a measure of T_4 binding capacity. However, low values of the T_3 uptake are usually found when the TBG capacity is increased, as in pregnancy, and high values of the T_3 uptake when the TBG capacity is reduced, as in the nephrotic syndrome.[19] On the whole, high values for T_3 uptake are found in thyrotoxicosis and low values in hypothyroidism,[19] but the differences in total thyroxine may not be the full explanation, as TBG capacity is often low in thyrotoxicosis and high in hypothyroidism.[103] The resin T_3 uptake is reduced in pregnancy in keeping with the increase in TBG capacity.[45,100]

Radioactive iodide, which may be present in varying proportions in different samples of T_3, causes interference with the resin T_3 uptake test because iodide is taken up completely from serum by the ion-exchange resin. In view of this, it is preferable to report the results of the resin uptake test as the T_3 uptake ratio or the ratio of the percentage uptake from the test serum to the percentage uptake from a sample of pooled normal human serum measured at the same time.[102] The percentage uptake may be influenced by the time and temperature of incubation, since the resin may detach some combined T_3 from combination with TBG, as well as adsorbing free T_3.[104] Heparin interferes with the resin binding of T_3 and it is preferable to use serum (or plasma) consistently.[19]

Other materials have been used besides ion-exchange resin to measure T_3 uptakes, e.g., erythrocytes,[105] Sephadex,[106,107] and charcoal.[47,108] The percentage uptake of T_3 from serum varies between the different materials and it is necessary to determine the normal ranges for the percentage T_3 uptake and T_3 uptake ratio appropriate to each adsorbent material. Resin uptake tests using T_4 have also been described,[109] but have not been widely used because the specific activity of the T_4 preparations must be higher than that of the T_3 preparations to compensate for the greater affinity of the TBG sites for T_4 and allow sufficient radioactivity to enter the resin for accurate counting.

3. Direct Measurement of Serum Free T_4 and Free T_3 Concentrations

Serum free T_4 concentrations are usually determined by equilibrium dialysis of serum to which has been added labeled T_4, followed by radiochemical estimation of the T_4 in the diffusate.[21,22] The passage of serum down columns of Sephadex G25 to separate free from bound T_4 did not prove satisfactory because deiodination of T_4 on the column caused losses of free T_4.[110]

In the equilibrium dialysis methods, labeled T_4 in the diffusate is most simply separated from iodide by precipitation as the magnesium salt in the presence of excess nonradioactive T_4.[22] The methods are sensitive to radioactive impurities in the labeled T_4 preparation, principally triiodothyronines, which are not so strongly bound to TBG as T_4 and, consequently, cause spurious elevation of free T_4 levels.[111,112] Other technical difficulties arise from contamination of the diffusate with concentrated T_4 leaking out of the dialysis cell, from adsorption of free T_4 on the walls of reaction vessels, and from variations in the measured percentage of free T_4 caused by differences in pH, temperature, concentration of buffer, concentration of the serum albumin used to stabilize the labeled T_4, and dilution of serum in the dialysis vessel.[21]

In keeping with expectation, the percentage of the serum T_4 in the free state was found to be lower in pregnancy than in euthyroid nonpregnant subjects, but the absolute free T_4 concentrations in the pregnant and nonpregnant groups did not show a significant difference.[22] A statistical difference between absolute serum free T_4 levels in pregnant and nonpregnant subjects was later reported, with values ranging from 1.6 to 4.4 ng/100 ml in nonpregnant subjects and from 2.2 to 7.0 ng/100 ml in pregnant subjects.[45] The conflicting results may have a methodological explanation, in view of the sensitivity of the method to changes in technique,[21] or they may reflect differences in the selection of pregnant subjects in the two investigations.

The serum free T_3 concentration has been measured by equilibrium dialysis in a similar manner to the serum free T_4 concentration.[23] This estimation is mainly of importance in studies on the peripheral metabolism of thyroid hormones.

Measurement of serum free T_4 and free T_3 concentrations by equilibrium dialysis is exacting and in any case gives results which are arbitrary, since the values obtained vary according to the conditions of the analysis and a standard procedure has not been agreed upon. For these reasons, determination of the serum free T_4 level is not practicable in the routine laboratory and indirect estimates of free T_4 activity are used instead.

4. Indirect Measures of Serum Free T_4 and T_3 Activity

Since the resin T_3 uptake is an inverse measure of the number of unoccupied binding sites on TBG and since by the Law of Mass Action the concentration of unoccupied sites is inversely proportional to the ratio of free to total serum T_4,[46] it is possible to derive a measure of free T_4 activity by combining measurements of total serum T_4 with measurements of the resin T_3 uptake.

Initially, the resin T_3 uptake was assumed to have a simple inverse relation with the concentration of unoccupied TBG sites and a 'free T_4 index' was constructed by multiplying the total serum T_4 (or the serum PBI) concentration by the resin T_3 uptake, expressed as a percentage or as a ratio.[113] Theoretically, the calculation of the free T_4 index should, by its derivation, compensate exactly for the effects on total T_4 of variations in TBG capacity, given a constant value for the serum free T_4 concentration. In confirmation, Clark and Horn[113] found that the free T_4 index in pregnant patients fell within the normal range for euthyroid nonpregnant subjects. However, other workers have found that in some pregnant subjects, values for the free T_4 index, calculated as the product of the total serum T_4 and the resin (or charcoal) T_3 uptake, may be above the normal range.[46,47] This finding was attributed by Goolden et al.[46] to the fact that the relation between the resin T_3 uptake and the number of unoccupied TBG sites was not a simple inverse proportionality, but included a constant term, which could be related to irreversible uptake of T_4 by the resin.

These workers also defined another measure of free T_4 activity, which they termed the 'free T_4 factor.' This is the ratio of the serum total T_4 (in μg/100 ml) to the concentration of unoccupied TBG sites, estimated from the resin T_3 uptake by a regression equation relating T_3 uptake to the free TBG capacity measured by electrophoresis.

Thus, inserting the values they determined for the constants in the regression equation:

$$\text{Free } T_4 \text{ factor} = \frac{\text{total } T_4}{\text{unoccupied TBG sites}}$$
$$= \text{total } T_4 \div \left(\frac{1090}{T_3 \text{ uptake (\%)} + 3.5} - 15.75 \right)$$

Using the free T_4 factor so derived, it was found that sera from pregnant women and from women taking oral contraceptives, which gave values above the normal range for the uncorrected free T_4 index, gave values not significantly different from the normal range for the corrected free T_4 factor. Standeven[47] reported similar findings based on measurements of the charcoal T_3 uptake, using the appropriate values of the regression constants to calculate a free T_4 factor for this test.

Use has been made of the renal excretion of T_4 and T_3 to provide indirect measures of the serum free T_4 and T_3 activity.[33,34] The total 24-hr urinary outputs of T_4 and T_3 are determined, and the results termed the 'urinary free T_4' and the 'urinary free T_3,' respectively. Interpretation is based on the assumption that the level of serum free T_4 and free T_3 determines the 24-hr output of T_4 and T_3 in the urine. This assumption is valid, provided that the glomerular filtration rate is constant. The tests showed good discrimination between euthyroid patients and patients with proven thyrotoxicosis or hypothyroidism, although in some thyrotoxic patients either the urinary free T_4 or the urinary free T_3, but not both, were within the range for euthyroid subjects.[34] Urinary free T_4 and free T_3 excretion in neonates was found to be considerably less than the adult levels. In neonates, on the average, compared with normal values for adults, the serum PBI is raised, with slightly raised TBG capacity, slightly lowered erythrocyte T_3 uptake, and unchanged serum free T_4 concentration.[114] Thus, it would seem that the low urinary T_4 and T_3 outputs in neonates are related to the glomerular filtration rate, which is low in neonates as compared with adults. In confirmation of this conclusion, Burke et al.[115] have reported that the urinary free T_4 and T_3 in neonates, expressed per gram of creatinine excreted, are similar to adult normal values. Urinary free T_4 and T_3 levels in pregnancy are within the normal euthyroid range.[33,34]

To carry out measurements of urinary free T_4 and free T_3, 3 ml of a 24-hr sample of urine is acidified to pH 4 with hydrochloric acid, extracted with 7.5 ml of ethyl acetate by vigorous shaking for 20 min, and 6.5 ml of the resulting extract evaporated to dryness at 70°C under a gentle stream of air; the residue is then dissolved in 650 μl of 0.06 *M* barbiturate buffer, 500 μl being added to reaction tubes containing TBG with $[^{125}I]$-T_4 and an ion-exchange resin (Thyopac-4 kit, Radiochemical Centre, Amersham, England) for displacement analysis of T_4, and 100 μl being taken for radioimmunoassay of T_3.[34] Normal ranges are 4 to 12.5 μg/24 hr for T_4 and 2 to 4 μg/24 hr for T_3.

In routine practice, the measurement of urinary free T_4 and free T_3 has the disadvantage that a full 24-hr collection of urine is required, which may be difficult to achieve, especially in outpatients. Burke et al.[115] reported that both the urinary free T_4 and T_3 tests were invalidated by proteinuria, which would tend to reduce their value in pregnancy when proteinuria is a common finding. The methods require further clinical assessment before introduction on a routine basis.

A test kit (Res-O-Mat® ETR, Mallinckrodt Chemical, St. Louis, Mo.) has been marketed for determining an indirect measure of the serum free T_4, called the 'Effective Thyroxine Ratio,' in which a modification of the displacement technique for estimating serum T_4 concentration enables the TBG capacity of the test serum to be taken into account (Mallinckrodt Chemical, St. Louis, Mo.). T_4 is extracted in the usual manner, but a 5-μl portion of the test serum is added to the reaction vial containing the T_4 extract, buffered TBG, and $[^{125}I]$-T_4 before treatment with an ion-exchange strip to remove free T_4. A reconstituted pooled normal serum is taken through the whole procedure and the result in the test serum is expressed as the ratio of the bound T_4 in the standard serum to the bound T_4 in the test serum (Effective Thyroxine Ratio). An increase in this ratio is caused either by the serum T_4 concentration in the test serum being greater than that of the standard serum, or by the TBG capacity of the test serum being lower than that of the standard. Conversely, a low Effective Thyroxine Ratio results if either the T_4 concentration in the test serum is reduced or the TBG concentration in the test serum is raised. In this way, the test tends to compensate for the effect of TBG capacity on the T_4 concentration in the test serum. Good discrimination between groups of euthyroid, thyrotoxic, and hypothyroid patients, in whom the diagnosis of thyroid status was unequivocal, has been reported by Thorson et al.[116] but a greater overlap between patients in the thyrotoxic or hypothyroid groups with the euthyroid group was reported by Wellby et al.[117] The Effective Thyroxine Ratio shows a significant correlation with the serum free T_4 level directly determined. The normal range for the Effective Thyroxine Ratio quoted by the manufacturers (0.68 to 1.13) appears to be clinically satisfactory.[117]

The Effective Thyroxine Ratio method has the disadvantage that it includes the extraction of T_4 from serum by ethanol which may give inconsistent recoveries, as discussed under the determination of total serum T_4. It is worth mentioning that the use of a vortex mixer in the alcohol extraction is recommended by the manufacturers of the Effective Thyroxine Ratio kit (Mallinckrodt Chemical), whereas the manufacturers of the Thyopac-4 test kit for total serum T_4 (Radiochemical Centre) state that a vortex mixer is undesirable for the extraction of T_4 from serum by ethanol, and recommend a rotating mixer of the Matburn type. Further work is necessary to determine whether the Effective T_4 Ratio technique is an advance on the calculation of the free T_4 factor from the separately determined total serum T_4 and resin T_3 uptake.

E. Thyroid Stimulation and Suppression Tests

Stimulation and suppression tests of thyroid function are based on the normal physiological control of the thyroid gland by the pituitary and hypothalamus.[118] The T_3 suppression test depends on the suppression of thyroid activity in the euthyroid subject by administering thyroid extract, T_4, or T_3. In practice, T_3 is used for the test because it has the highest potency. The thyroid ^{132}I uptake is measured before and after giving 120 μg of T_3 orally per day for 6 days. Euthyroid subjects show a reduction in thyroid ^{132}I uptake after T_3 administration; the thyroid ^{132}I uptake is not suppressed in thyrotoxicosis. The ^{132}I uptake is also used to measure thyroid reserve in the TSH stimulation test, in which the uptake is measured before and 18 to 24 hr after an intramuscular dose of 2.5 units of TSH. Euthyroid subjects show an increase in uptake; patients with primary hypothyroidism (i.e., not due to hypopituitarism) do not respond.

The pituitary TSH reserve can be assessed by a thyrotropin-releasing hormone (TRH) stimulation test, in which the response of the pituitary to a dose of TRH is measured.[119] In most instances, TRH causes an increase in circulating TSH levels in euthyroid subjects, and an exaggerated increase in patients with primary myxedema who have high initial TSH concentrations, but has little or no effect in patients with hypothyroidism of pituitary origin or in thyrotoxic patients.[119]

Recently, the inhibition by pharmacological doses of glucocorticoids of TRH release from the hypothalamus and the rebound in TRH secretion which follows withdrawal of the steroid have been used to devise a test of TRH reserve.[120] Special techniques are required. The thyroid gland is labeled by a dose of 125-iodide. The release of radioiodine from the gland before and after a dose of steroids is followed by measuring the ratio in the plasma of ^{125}I to ^{131}I after an intravenous dose of [^{131}I]-thyroxine. An increase in the ratio of ^{125}I to ^{131}I occurs if radioiodine release from the gland is stimulated by the secretion of TSH under the influence of TRH. This test is of research interest only at present.

It is not justifiable to submit pregnant patients to stimulation or suppression tests of thyroid function. Such tests are employed in investigating borderline cases, in whom a full thyroid investigation can be deferred until the postpartum period.[121]

F. Tests for the Presence of Circulating Thyroid Autoantibodies

In thyroid diseases, autoantibodies may be produced either to the thyroid colloid or to the thyroid microsomes.[122] Autoantibodies to thyroglobulin are detected by Latex® precipitation or tanned red cell agglutination reactions, antibodies to a second thyroid colloid antigen by immunofluorescence, and microsomal antibodies by complement fixation. A high titer of autoantibodies to thyroglobulin is characteristic of Hashimoto's Thyroiditis. Many patients with thyrotoxicosis and primary myxedema show positive tests for at least one of the thyroid autoantibodies. These tests may be of value in patients under the age of 50 if thyroid disease is suspected. Over the age of 50, the proportion of euthyroid subjects showing positive test results for thyroid antibodies increases.[121] However, pregnant patients are usually younger than 50, and the tests may be of value in such cases to investigate a suspected thyroid disorder.

Although test reagents and kits for the precipitin, tanned red cell agglutination, and complement fixation tests are available (e.g., Wellcome Research Laboratories, Beckenham, Kent, England), these investigations are best carried out in a laboratory which has experience in immunology or serological testing.

G. Tests of the Peripheral Action of Thyroid Hormones

Logically, it would be desirable to base assessments of thyroid status on metabolic tests of the tissue response to thyroid hormones. Unfortunately, many other influences besides the thyroid hormones affect the metabolism of tissues. As a result, metabolic tests of thyroid function require careful interpretation, and there may be considerable overlap between the normal range for a particular investigation and the ranges in thyrotoxic or hypothyroid patients. Nevertheless, such tests may prove valuable in cases of diagnostic difficulty or in pregnancy.[123]

1. Serum Cholesterol and Triglyceride Levels

The serum cholesterol level tends to be raised in hypothyroidism and in the lower part of the normal range in thyrotoxicosis, but has little diagnostic value because it is affected by many other conditions besides thyroid disease and because of overlap with the normal range.[123] The fasting serum triglyceride level shows similar changes to the serum cholesterol in thyroid disorders.[124] The main value of serum cholesterol determinations in thyroid disorders is in the follow-up of hypothyroid patients on replacement therapy with thyroxine.[125]

In pregnancy, there is a statistical increase in the level of serum cholesterol, becoming most marked in the third trimester,[126] but this increase is not universal – for instance, it is not found in Masai Africans[127] – and may not be demonstrable in an individual patient followed during pregnancy.[100] During pregnancy, there is also an increase in serum triglycerides which is more marked than the rise in serum cholesterol, but there is no close correlation between the levels of the two serum lipid fractions.[128] The cause of hyperlipidemia in pregnancy is not fully established, but has been attributed to an increase in serum β-lipoproteins[129] and to accelerated hepatic lipogenesis.[130]

High-dose estrogenic oral contraceptives cause the serum levels of triglycerides and of cholesterol to rise; low-dose estrogenic oral contraceptives cause a rise in serum triglycerides and a fall in serum cholesterol followed by a delayed rise over a period of a year; progestational oral contraceptives have no effect on serum lipid levels.[131]

2. Erythrocyte Sodium Concentration

The concentration of sodium in red cells is increased in thyrotoxicosis because the activity of the sodium pump in the erythrocyte membrane is decreased.[132] Goolden et al.[133] have employed the erythrocyte sodium concentration as a test of peripheral thyroid hormone activity in suspected thyrotoxicosis, but found it of no value in hypothyroidism, which was not differentiated from the euthyroid state. Smith and Samuel[132] reported a normal range of 4.6 to 9.5 m*M*, and found that 65% of their thyrotoxic patients had erythrocyte sodium concentrations above their normal limits. Goolden and co-workers[133] reported a normal range of 6.1 to 8.4 m*M*, and found that 90% of their thyrotoxic patients had erythrocyte sodium concentrations above their normal limits. Erythrocyte sodium concentrations in nonthyroid diseases have not been well documented, but high levels might be expected in patients with muscle disorders, renal failure, or in patients on digoxin therapy. Further assessment of the erythrocyte sodium concentration as a possible test of the peripheral action of thyroid hormone is required.

3. Tyrosine Tolerance

Fasting plasma tyrosine levels are raised in thyrotoxicosis, and the increase in plasma tyrosine level after a loading dose of 50 mg/kg of tyrosine is also often greater than in euthyroid subjects; in hypothyroidism, there is considerable overlap with the normal range.[134] Although the oral tyrosine tolerance test has been proposed as a supplementary test of thyroid status in pregnancy, it has not yet been sufficiently validated clinically for it to be recommended as a routine test.[123]

4. Basal Metabolic Rate (BMR)

The heat production of the body under resting conditions (BMR) is increased in thyrotoxicosis and reduced in hypothyroidism. Clinical measurement of the BMR is carried out by the method of indirect calorimetry, in which the rate of oxygen consumption is measured and the heat output is calculated by assuming a value of 0.82 for the respiratory quotient ($= \frac{\text{volume of } CO_2 \text{ produced}}{\text{volume of } O_2 \text{ absorbed}}$) and a factor of 4.825 for the number of kilocalories liberated as a result of the consumption of 1 liter of oxygen (measured at STP). To get reliable results from the determination, it is necessary to pay particular attention to consistency of technique. This question was thoroughly investigated by Robertson,[135] who recommended a technique which was sufficiently consistent to be used on outpatients.

The result of a BMR determination is first expressed in kcal/m^2 body surface/hr. This result is then converted to the percentage deviation from a standard value appropriate to the age and sex of the patient. The normal range is taken to lie between ±15% on either side of the standard values. There is some controversy about the appropriate standard values to use. Crooks et al.[136] found that the classical standards of Aub and Dubois were too high by comparison with clinical criteria and tended to give a preponderance of negative deviations in euthyroid subjects. On the other hand, when the standards of Robertson and Reid[137] were used, there was a symmetrical distribution of the results in euthyroid subjects around the 0% deviation. Crooks et al.[136] recommended the use of the Robertson and Reid standards and they have also been generally used by workers in the U.K.

Although there is no significant difference between the BMR of obese and normal women when calculated per unit of surface area, the BMR of obese women has been found to be lower than that of normal women when calculated with reference to a measure of the cell mass derived arithmetically from the total body water.[138] However, there is evidence to suggest that the total body water may be disproportionately high in obesity[139] which would explain the apparently anomalous results. Thus, this evidence does not support a modification of the accepted standards for the BMR when studying obese subjects, as was recommended by Wayne[125] and Hall.[123] However, in extreme obesity, the BMR could be low because of the relatively low oxygen consumption in adipose tissue.[11]

Because of the difficulties that may be experienced in performing the test on a nervous patient, it has been proposed that the test be performed under light hypnosis with barbiturates. Although good discrimination can be obtained in this way between thyrotoxic, euthyroid, and hypothyroid groups,[140] it was later found that phenobarbitone increased the metabolic rate in some patients with psychiatric disturbances;[141] hypnotics or sedatives have not been widely used in the performance of the test.

If carefully performed, the BMR is a valuable supplementary test in thyrotoxicosis and hypothyroidism.[11,123] Interpretation of the BMR must take account of the increase that occurs during the course of pregnancy. On the average, taking the initial value as 0%, the BMR rises to +5% by the end of the first half of pregnancy and to +20% at term, as a result of the growth of the uterus and fetus and of circulatory and respiratory adaptations.[48]

H. Measurement of Serum TSH and LATS Levels

Estimation of the serum TSH concentration is mainly of value for assessing pituitary function in disorders affecting the thyroid.[119,142] Determinations of serum TSH are carried out either by bioassay or by immunoassay. Bioassay techniques[143] are not suitable for routine analyses on serum, and radioimmunoassay of TSH in serum gives different results according to the antiserum used[144] because antisera to TSH may cross-react with chorionic and other pituitary hormones.[145] Raised levels of circulating TSH occur in hypothyroidism; TSH is not detectable in the serum in thyrotoxicosis.[119,142]

In pregnancy, the serum contains relatively high concentrations of chorionic TSH which cross-reacts in radioimmunoassays with antisera to pituitary TSH.[41] However, chorionic tumors may produce a bioassayable chorionic TSH which does not cross-react in radioimmunoassays with antisera to pituitary TSH.[146] Determination of the serum TSH concentration is of little value for diagnosing thyroid disorders in pregnancy and could not be used to monitor the treatment of hypothyroidism with thyroxine, as can be done in nonpregnant patients.[147]

The long-acting thyroid stimulator (LATS), present in the serum of some thyrotoxic patients, is an immunoglobulin which has a thyroid-stimulating action in bioassays differing from that of TSH in that the response to LATS continues to increase between 3 and 12 hr after its addition to the system, at a time when the response to TSH is diminishing.[148,149] LATS is not demonstrable in the serum of all thyrotoxic patients, even after preliminary concentration of the immunoglobulins; the bioassay is exacting and false positive results may occur.[150] High serum levels of LATS are more consistently found in pretibial myxedema, a complication both of treated and of untreated thyrotoxicosis, than in uncomplicated thyrotoxicosis.[151] Thus, determination of LATS in serum in not of much value in the diagnosis of thyrotoxicosis.

I. Choice of Thyroid Function Tests in Pregnancy

The initial requirement in pregnancy is for a rapid screening test to confirm the clinical diagnosis of thyroid status. In the event that the clinical and laboratory findings show a discrepancy, further tests can be used to establish a provisional diagnosis.

The initial screening of a pregnant patient for thyroid disorders should consist of a measure of total circulating thyroid hormone (e.g., serum PBI or T_4) and a measure of serum TBG capacity (e.g., resin T_3 uptake), from which an indirect estimate of the serum free T_4 (e.g., the free T_4 factor) should be calculated. If there is still doubt, tests of the peripheral action of the thyroid hormone should be tried in the first instance. The BMR test, if properly carried out, is the most satisfactory of this group because it has discriminatory power both in thyrotoxicosis and in hypothyroidism.

If tests of the peripheral action of thyroid hormones have not enabled a firm diagnosis to be made, measurement of the thyroid uptake might be considered in a limited number of cases. The administration of radioactive preparations to pregnant patients is to be discouraged, in view of the possibility of harmful effects on the fetus. Thus, in the first instance, in vitro thyroid function tests that do not require the administration of radioactive doses should be used in the pregnant subject. However, in borderline cases during the first half of pregnancy, where there is clinical doubt whether the patient is thyrotoxic and where it is important to establish the diagnosis, it would be permissible to administer the short-lived isotope ^{132}I,[39] or pertechnetate for an in vivo test of thyroid uptake (Goolden, A. W. G., personal communication). In the U.K., permission to administer any radioisotope to patients has to be obtained from the Medical Research Council (20 Park Crescent, London, W1N 4AL) and such permission for thyroid uptake tests is normally given for nonpregnant patients only. Before carrying out a thyroid uptake test with ^{132}I or pertechnetate in a pregnant subject, in the U.K. it is necessary to submit a reasoned request to the Medical Research Council for authority to administer the test to the particular patient.

During the second half of pregnancy, thyroid uptake tests should not be performed because the fetal thyroid starts to take up radioiodine at about the 14th week.[58]

Methods for determining the thyroid radioiodine or pertechnetate uptake are outside the scope of this chapter. The reader interested in techniques for measuring thyroid uptakes is referred to Aboul-Khair et al.,[39] Blahd,[6] Gillespie and Keyes,[152] Goolden et al.,[57] International Atomic Energy Agency,[153] Wagner,[154] and Wayne et al.[4]

In considering the adoption in a routine laboratory of a thyroid function test which is not well established in clinical practice, it should be remembered that in many instances the initial published assessments of a new test have been made on groups of patients in whom the diagnosis of thyroid status was unequivocal. It is well known that in cases where the diagnosis is definite, both clinical findings and any of the usual routine laboratory tests of thyroid status have diagnostic efficiencies of around 95% or better, but their respective diagnostic performances fall off in borderline cases.[125] Thus, a new test of thyroid status should not be accepted into routine practice until it has been shown to provide an advance over existing tests in cases where there is diagnostic difficulty.

III. METHODS

A. The Serum PBI Test[62]

1. Principle

Serum samples are treated manually with an ion-exchange resin to remove inorganic iodide. The treated samples are aspirated in sequence into the Digestor, where organic material is destroyed by heating with a mixture of sulfuric, nitric, and perchloric acids and organic iodine is converted to iodate. The digestion mixture diluted with water is sampled and mixed with arsenious acid and ceric ammonium sulfate at 55°C. Iodate in the digest catalyzes the reduction of ceric ion by arsenite ion; the loss of ceric ion color, which is proportional to iodate concentration, is measured at 410 nm. Serum PBI levels are determined by comparison with iodate standards.

The same system is used to screen for contaminated specimens by sampling at 120 per hr.

2. Apparatus

a. Technicon Continuous Digestor, comprising Sampler, Input Manifold and Proportioning Pump, Digestor with heaters and rotating glass helix, Output Manifold and Proportioning Pump, Heating Bath at 55°C, Colorimeter and Recorder (Technicon Instruments, Tarrytown, N.Y.).

b. Plain and conical AutoAnalyzer sample cups (Technicon Instruments, Tarrytown, N.Y.), stored in closed containers to prevent contamination.

c. Standard plastic transmission tubing, Tygon® tubing, polyethylene tubing, clear pump tubing, and Acidflex® pump tubing, as specified in the manufacturers' instruction sheet.

3. Reagents

Arsenic trioxide and ceric ammonium sulfate should be graded 'for protein-bound iodine determination' (British Drug Houses, Poole). Each new batch of a reagent should be tested against the old batch before use. Water should be deionized by passage through an ion-exchange resin.

a. Sulfuric acid: concentrated analytical reagent.

b. Perchloric acid (70 to 72%): General Purpose Reagent grade (Hopkin and Williams, Romford, Essex, England) has been found satisfactory.

c. Nitric acid: concentrated analytical reagent.

d. Digestion mixture: to 50 ml of concentrated nitric acid in a 500-ml conical flask add 200 ml of perchloric acid and mix well. Transfer to an amber glass reagent bottle. Make up freshly the amount of reagent which will be required for each run.

e. Sodium hydroxide pellets: analytical reagent grade.

f. Arsenious acid solution: dissolve 19.6 g of arsenic trioxide, As_2O_3, and 14 g of sodium hydroxide in about 200 ml of deionized water in a 2-liter volumetric flask. When the solids have dissolved, dilute to about 800 ml with deionized water, add a few drops of phenolphthalein indicator solution, and add concentrated sulfuric acid carefully until the pink color is just discharged. To the neutralized solution carefully add a further 56 ml of concentrated sulfuric acid and make up to 2 liters with deionized water. Dissolve 50 g of sodium chloride (analytical reagent grade) in the solution and store in a polyethylene bottle.

g. Ceric ammonium sulfate solution: carefully add 52 ml of concentrated sulfuric acid to 700 ml of deionized water. Dissolve 6.3 g of ceric ammonium sulfate, $Ce(NH_4)_4(SO_4)_4 2H_2O$, in the solution, allow to stand overnight, and dilute to 1 liter with deionized water. Filter through glass wool and store in a polyethylene bottle.

h. Resin: Iobeads® (Technicon Instruments, Tarrytown, N.Y.) used in the moist state.

i. Standards: potassium iodate is used for the standards as iodide is adsorbed by the system.

1. Stock iodate (100 μg of I/ml): dissolve 168.5 mg of potassium iodate, KIO_3 (analytical reagent grade), in deionized water and dilute to 1 liter with deionized water.

2. Dilute stock iodate (1 μg of I/ml): dilute 10 ml of stock iodate to 1 liter with deionized water.

3. Working iodate standards

Dilute stock iodate (ml)	Dilute to (ml)	μg of I/100 ml
5	200	2.5
5	100	5
10	100	10
15	100	15
20	100	20

4. Screening standard (50 μg of I/100 ml): dilute 50 ml of dilute stock iodate to 100 ml with deionized water. Keep all standards in polyethylene bottles.

5. Serum standards for quality control: pool sera from hypothyroid patients to obtain a low PBI standard (approximately 2.5 μg I/100 ml). Use pooled normal serum to obtain a standard of approximately 5 μg I/100 ml. Prepare standards containing approximately 10, 15, and 20 μg I/100 ml by addition of T_4 to the pooled sample.

Prepare an aqueous T_4 standard. Dissolve 17.51 mg of L-thyroxine Na salt pentahydrate (Koch-

Light, Colnbrook, Bucks, England) in 2 ml of 0.1 *N* sodium hydroxide, and dilute to 50 ml with deionized water. Store frozen in glass vials (stable 3 months). Freshly dilute 1 ml of stock standard to 200 ml with deionized water to give a working standard containing 100 μg I/100 ml. Add the appropriate volumes of working standard to three portions of pooled normal serum to bring the serum PBI to about 10, 15, and 20 μg I/100 ml, respectively. (Approximately 5, 10, and 15 ml of working standard, respectively, made up to 100 ml with pooled serum, will be required.) Distribute the standards into AutoAnalyzer cups and store frozen at −20°C.

Analyze about 30 portions of each standard to obtain a mean value, including quality control sera in the run to check on the accuracy of the analyses. Most manufacturers of quality control sera (e.g., Dade, Miami, Fla., Hyland, Costa Mesa, Calif., Warner-Lambert, Morris Plains, N.J.) include serum PBI determinations in their batch analyses.

Include a full range of serum standards in each run.

4. Procedures

The flow diagrams are shown in Figures 3 and 4 and it is assumed that the operator will have a copy of Technicon Method Sheet N56, which gives the method recommended by the manufacturers. The following account amplifies these instructions.

a. Screening for Contaminated Specimens

Sera containing iodine drugs or high concentrations of inorganic iodide will give spuriously high PBI results and may contaminate the whole apparatus, rendering it useless for several hours until decontamination can be achieved. Place a cup containing the 50 μg/100 ml screening standard in position 1 and then in every tenth position on an AutoAnalyzer plate. Place the sera on the plate in series, leaving a water cup on either side of each screening standard. Sample the cups for 1 to 2 sec at 125 per hr using the special plastic sampling cam. Any contaminated samples can be picked out from the record by their position relative to the screening standards. In case of doubtful identification, repeat the screening on the suspect sample and the samples on either side of it, separated from each other by water cups. Repeat the screen of the contaminated specimens after treatment with the Iobeads. Contamination with inorganic iodide will be removed by the resin treatment, and the specimen can be run in the usual manner, the result being given with the comment 'high inorganic iodide.' Contamination

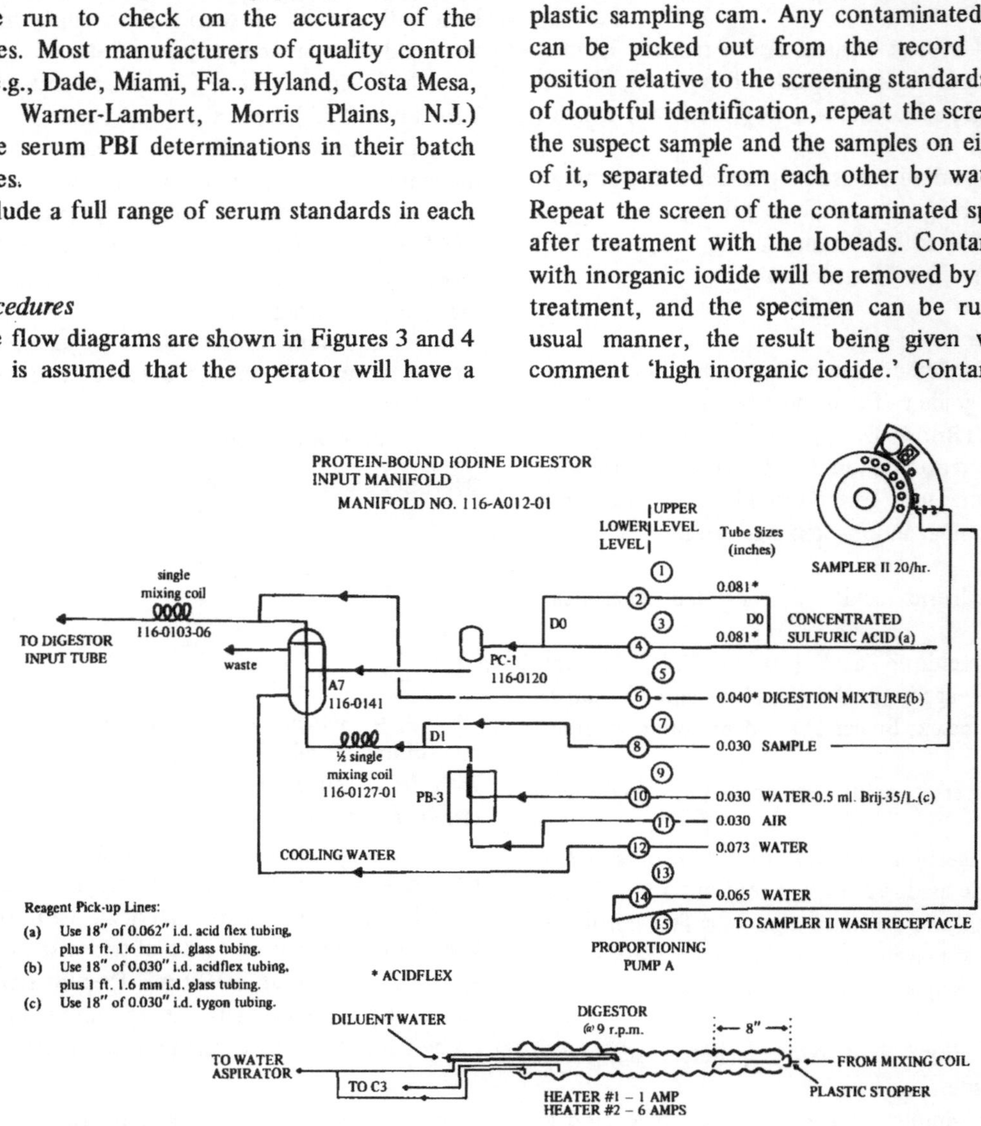

FIGURE 3. Protein-bound iodine digestor input manifold. (Courtesy Technicon Instruments Corp.)

OUTPUT MANIFOLD

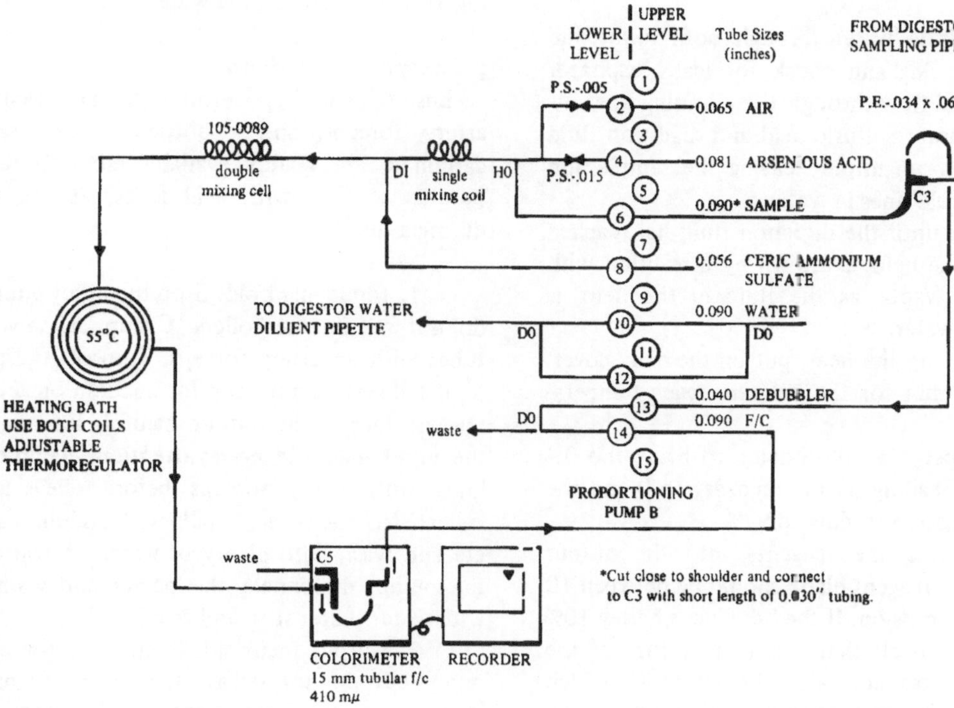

FIGURE 4. Protein-bound iodine digestor output manifold. (Courtesy Technicon Instruments Corp.)

by organic iodine drugs is not removed by the resin and specimens contaminated in this way must not be run. If a contaminated specimen is introduced into the apparatus by mistake, pooled serum must be aspirated continuously through the apparatus until a chart reading shows that the contamination has been removed. This process may take several hours.

b. Resin Treatment of Serum Samples

Transfer about 1.5 ml of serum with an acid-washed Pasteur pipette into a plain AutoAnalyzer cup, add about 0.3 g of Iobead resin to the cup with the scoop provided, and cap. Place the capped cups on a rotary mixer (e.g., Matburn, London, W.C.1) and mix for 5 min at about 60 rev./min. Allow the resin to settle and decant the serum into a conical AutoAnalyzer cup, taking care not to allow the resin beads to enter the conical cup. Conical cups are used to prevent resin beads introduced by accident from being sucked into the sampler, where they may cause blockage. The resin must not be allowed to dry out, as the dry resin will concentrate the serum proteins with the PBI by taking up water from the serum and will cause results to be spuriously high.

c. Sampling

Run at 30 determinations per hour, using a cam with a variable sample-to-wash ratio set at a value (approximately 1·2:1) to minimize carryover without serious reduction in peak height. Include a sample of pooled normal serum every ten samples to act as a drift and run standards at the beginning and end of each run. Analyze each specimen in duplicate on separate runs. Make sure the sampler does not suck air, as this causes a spurious peak to appear on the chart.

d. Manifolds

Change the Acidflex pumping tubes before each run. The pumping rates of sulfuric acid (3.5 ml/min) and of perchloric/nitric acid mixture (0.5 ml/min) are critical. Aspirate distilled water through the systems and check that the flow rates are correct before starting a run. Make sure the Acidflex lines do not 'snake.'

e. Operating Sequence

1. Start the aspirator (three water pumps connected to a reservoir) and switch on the colorimeter and recorder.

2. Start the helix rotating without turning on the heat.

3. Aspirate water through both input and output manifolds and check for leaks, aspirator efficiency, and flow through the pumping lines.

4. Aspirate sulfuric acid and digestion fluid into the input manifold, leaving the sample and output manifold lines in water.

5. Wait until the digestion fluid has reached the end of the helix as shown by a reaction with evolution of vapor as the fluid in the helix is diluted with water.

6. Turn on the heat, put on the helix cover, and wait 30 min for the helix to reach temperature.

7. Adjust the colorimeter to bring the 0% transmission reading on the recorder to 0 and the 100% transmission reading to 97%.

8. Aspirate the reagents into the output manifold. The reagent blank should be between 10 and 20% transmission. If the baseline is below 10% transmission, check that the temperature of the heating bath has not fallen below 55°C, which would cause the percent transmission to be low. Alternatively, a new batch of ceric ammonium sulfate may have given a solution with too high an optical density which must be diluted with 1.86 *N* sulfuric acid to bring the optical density within range. If the percent transmission is above 20, the temperature of the heating bath may be too high. Alternatively, one of the reagents may have become contaminated. A batch of sodium chloride is sometimes unsatisfactory for this reason.

9. When the baseline is stabilized, check that the temperature of the first stage of the digestor is 55°C and that of the second stage 280°C, then start the screening run and analyses.

f. Closedown

1. After the last peak appears on the chart, turn off the heat, remove the helix cover, and place the sample and output manifold lines in distilled water. Leave the input manifold lines pumping sulfuric acid and perchloric/nitric acid.

2. When the helix temperature has fallen to about 80°C, put the acid lines into distilled water and aspirate until the pen returns to the 100% transmission setting.

3. Switch off the pumps and release the manifolds from the end blocks.

4. Allow the helix to rotate until the wash water has been sucked off, then switch off the digestor, and turn off the water pumps.

g. Emergency Shutdown

Ensure that hypodermic syringes, plenty of artery forceps, and a bottle of solid sodium carbonate are readily available close to the apparatus to deal with acid leaks, etc. occurring during a run.

1. Input manifold. Stop the input pump but do not release the rollers. Clamp off any loose tubes with an artery forceps. Turn off the power to the digestor, but keep the vacuum on. Put the reagent lines of the output manifold into water. If the input manifold needs attention, clamp off all lines with artery forceps before releasing the rollers. Neutralize acid spills with sodium carbonate and wash with plenty of water. If acid enters the pump, disconnect the power and wash well with water. Later strip and re-oil.

2. Output manifold. Switch off the output pump but do not release the rollers. Clamp off loose lines with artery forceps. Disconnect the sampling line from the output end of the digestor. Clamp off all lines with artery forceps before releasing the rollers of the output pump to attend to the manifold. Put the input serum sampling line into distilled water, but keep the input manifold, digestor, and helix running. The vacuum will empty the reagents from the helix. After a long shutdown the system must be restandardized.

h. Loss of Sensitivity

This may arise because of deterioration on the input or output side. A protein coat may form on the input side, either in the mixing coil where the serum samples are diluted with water, or in the jacketed T-fitting (Type A7) in which the diluted sample is mixed with sulfuric acid. To remove the protein coat, disconnect the input line from the digestor and aspirate 2.5 *N* sodium hydroxide through the sample and sample diluent lines for 15 min. Then wash well by aspirating distilled water and reconnect the input to the digestor. Repeat twice weekly as part of routine maintenance. On the output side, the glass coil in the heating bath may become coated with precipitated material, thereby interfering with the progress of the reaction between ceric and arsenite ions, which is probably glass-dependent. To remove encrusted precipitate, fill the coil with 10% alcoholic potassium hydroxide and leave overnight at 55°C. Wash well with deionized water before the next run. Precipitation in the reagent stream in the output manifold is usually due to the sulfuric acid concentration being too high. This occurs if the contents of the helix are not sufficiently diluted by the water stream before entering the output manifold.

5. Accuracy and Precision

Recovery of added T_4 from serum is better than 95% and the coefficient of variation of a single analysis varies from 2 to 6.5%, according to the serum PBI content.[155] Up to 10% of the normal serum PBI consists of iodoprotein, the rest mainly of T_4 with small amounts of T_3. Many iodine drugs, if present in the serum, precipitate with the PBI. In some thyroid disorders, a larger percentage of iodoproteins may be found in the serum PBI; in some cases of familial goiter, significant quantities of iodotyrosines may be present in the serum and determined as PBI. Less than 2% of the serum inorganic iodide remains in the serum after treatment with Iobeads and heavy iodide contamination can be removed by passing serum down a column of Iobeads.[156]

Quality control sera should be included in each run as a check on accuracy. Samples of human origin are preferable.

Loss of accuracy is associated with incorrect proportions of sulfuric acid and nitric/perchloric acid mixture and also with the causes of loss of sensitivity noted above. A peculiar feature of these faults is that inorganic iodate may continue to be estimated accurately although the organic iodine in serum samples is not, presumably because iodine is caught up in a layer of protein precipitated on the walls of the tubing.

B. Serum Total Thyroxine by Displacement Analysis[77,78]

1. Principle

Serum mixed with $[^{125}I]$-T_4 is passed down a column of Sephadex G25 at pH 11, in which conditions T_4 is split from TBG and adsorbed on the Sephadex. The column is washed with barbiturate buffer pH 8.6 to remove $[^{125}I]$-iodide and serum proteins and the radioactivity remaining on the column is counted. The column is then equilibrated with a preparation of human globulin which contains TBG. At this pH the T_4 binding sites on the TBG compete with the Sephadex for T_4. As the ratio of T_4 to $[^{125}I]$-T_4 on the column increases, more unlabeled T_4 is taken up by TBG, and consequently a higher percentage of $[^{125}I]$-T_4 is retained by the Sephadex. The TBG with its bound T_4 is washed out of the column by barbiturate buffer pH 8.6 and the radioactivity retained by the column is measured as a percentage of the initial count. Using aqueous T_4 standards stabilized with human serum albumin, a calibration curve is constructed relating the percentage retention of T_4 by the column to the serum T_4 concentration. The unknown sera are taken through the procedure and their total T_4 concentrations read from the calibration curve.

2. Apparatus

Tetralute kit (Ames Laboratories, Elkhart, Ind.) comprising Sephadex G25 columns, $[^{125}I]$-thyroxine, eluting reagent (dried lyophilized human globulin), dried lyophilized standard serum, and barbiturate buffer preparation.

A well-type scintillation crystal head with power supplies amplifier, pulse height analyzer, and scaler suitable for counting ^{125}I.

3. Reagents

a. Barbiturate buffer (0.075 *M*), pH 8.6: Dissolve 4.15 g of sodium barbiturate and 0.74 g of barbituric acid in 270 ml of copper-free deionized water.

b. $[^{125}I]$- Thyroxine solution: 0.02 μg/100 ml and approximately 12.5 μCi/100 ml in 0.1 *N* NaOH containing 1 ml of propane-1,1,2-diol per 100 ml.

c. Eluting reagent: Dissolve 30 mg of lyophilized human globulin in 15 ml of barbiturate buffer, pH 8.6.

d. Sephadex G25 columns (19 mm X 12 mm diameter) equilibrated with 0.1 *N* sodium hydroxide, ready-prepared, and protected from spillage by a porous plastic disc.

e. Water is glass-distilled or deionized and should not come into contact with copper.

f. Serum standard: Dissolve the sample of dried lyophilized serum in 1 ml of deionized water to obtain a serum sample containing the stated concentration of T_4.

All the above reagents are provided in the Tetralute kit except water. They are stable at room temperature (below 30°C) until the expiration date of the kit.

g. Stock thyroxine standard (100 μg/100 ml): Dissolve 28.6 mg of L-thyroxine Na salt pentahydrate (Koch-Light, Colnbrook, Bucks, England) in about 1 ml of 0.1 *N* sodium hydroxide and dilute to 25 ml with deionized water. Store in glass vials at -20°C. Stable at least 3 months.

h. Working thyroxine standards: Prepare a 20-μg/ml stock solution of T_4 by diluting 1 ml of stock thyroxine to 50 ml with 0.5% albumin. This solution is stable for 1 week at 4°C. Prepare freshly for each run a 20 μg/100 ml working thyroxine standard by diluting 1 ml of the stock 20 μg/ml solution to 100 ml with 0.5% albumin. Make up working thyroxine standards of 2, 5, 10, and 15 μg/100 ml by diluting, respectively, 1, 2.5, 5, and 7.5 ml of the 20 μg/100 ml working standard to 10 ml with 0.5% albumin.

i. Albumin solution (0.5 g/100 ml): Dilute 1 ml of 25% normal human serum albumin solution (Squibb, New Brunswick, N.J.) to 50 ml with deionized water.

4. Procedure

Bring all reagents and serum samples to room temperature before starting the analysis.

Remove and discard the top caps of the columns and pour off the excess alkali. Add 0.5 ml of the $[^{125}I]$-T_4 solution from an automatic dispenser (or seven drops from the dropper bottle in the kit) to the column. Do not spatter the sides of the column, as this may result in adsorption of radioactivity onto the column walls.

Test: Add 0.1 ml of serum to the column and mix with the $[^{125}I]$-T_4 by gentle swirling.

Blank: 0.1 ml of water treated in the same way as the test.

Standards: 0.1 ml of each dilute working standard treated in the same way as the test. Include fresh standards with each batch of tests.

Remove the bottom cap from the columns and allow to drain into a waste receptacle. Wait until no further fluid drains off, then add 4 ml of barbiturate buffer, and allow to drain. After drainage, dry the tip of the column with a paper towel, cap the tip, and place the column in the well of the crystal. Ensure that the column fits closely into the well or that a plastic liner is inserted into the well to maintain a constant counting geometry throughout. Time at least 20,000 counts and record the count for each column. This is the *initial count*. Remove the cap from the tip of the column and place over the waste receptacle. Add 1.0 ml of eluting reagent, allow to drain, and stand for 5 min. Then add 4 ml of barbiturate buffer and allow to drain. Dry the tip of the column with a paper towel, cap the tip, replace the column in the scintillation crystal well, and record the count again. This is the *final count*.

5. Calculation

Calculate the T_4 retention (%) for each column:

$$T_4 \text{ retention } (\%) = \frac{\text{final count}}{\text{initial count}} \times 100$$

Construct a calibration curve relating T_4 retention (%) to serum T_4 concentration from observations on the blank and standards. The form of the curve is sigmoid, with the maximum slope corresponding to sera containing between 4 and 15 μg of T_4 per/100 ml. Read the T_4 concentrations of the unknown sera from the calibration curve. More accurate analyses on sera whose T_4 concentrations are outside this range may be achieved by repeating the analyses on smaller or larger volumes of sample, as necessary, to bring the observations within the range of maximum sensitivity for the method. The appropriate correction for the sample volume is made to the T_4 concentration read from the calibration curve.

The T_4 concentration of the standard serum in the kit is given in μg of T_4-I/100 ml, whereas the concentrations of the T_4 standards are in μg of T_4/100 ml. To convert from μg of T_4-I to μg of T_4, multiply by 1.53.

6. Accuracy and Precision

Recoveries of T_4 from serum range from 95 to 99% and the coefficient of variation of replicates is 5.1%.[77] Determinations of serum T_4 by this method do not differ significantly from determinations by the reference method of Murphy and Pattee[73] on the same sera and iodine drugs do not interfere with the analysis.[77]

Use the standard serum provided by the manufacturers as a quality control specimen, setting the 95% limits at ±10% of the value assigned by the manufacturers. Also include a sample of pooled normal human serum in each run as an additional quality control check.

C. The Resin T_3 Uptake Test[101]

1. Principle

Serum and $[^{131}I]$- or $[^{125}I]$-T_3 are incubated with an ion-exchange resin at room temperature. After a fixed time, the resin is washed with water and the proportion of the radioactivity that remains in the resin is measured as a percentage of the original radioactivity. The proportion of radioactive T_3 adsorbed by the resin is inversely related to the number of free T_4 binding sites in the serum TBG. The results of the test are best expressed as the ratio of the T_3 uptake of the test serum to the T_3 uptake of a sample of pooled normal human (nonpregnant) serum run in the same batch.[102] This helps to eliminate between-batch variation in uptake due to differences between the properties of different lots of ion-exchange resin.

2. Reagents and Equipment

Kits for determining the T_3 resin uptake are marketed by a number of manufacturers, e.g., Abbott Laboratories, Chicago, Ill. (Triosorb), Ames Laboratories, Elkhart, Ind. (Trilute), and Radiochemical Centre, Amersham, England (Thyopac-3). Kits should be stored at 4°C and should not be used after the expiration date given by the manufacturers. Old samples of labeled T_3 may contain radioiodide, which is adsorbed by the resin and gives falsely elevated values for T_3 uptake. All the methods require a standard well-type crystal scintillation head, with suitable power supplies, amplifier, single-channel pulse height analyzer, and scaler.

3. Procedure

The use of the Triosorb kit is described here. In this kit, the ion-exchange resin is incorporated into a plastic sponge and the labeled T_3 is supplied in a disposable syringe. Plastic plungers, aspirators, and capped plastic test tubes are also supplied.

Separate the serum from clotted blood. Store at 4°C if determinations are to be done within 24 hr; otherwise, store frozen at -20°C.

Allow kits and sera to reach room temperature. Transfer 1 ml of serum to a plastic test tube. Dispense into each serum the $[^{131}I]$- or $[^{125}I]$-T_3 contained in one of the syringes provided in the kit. Mix well and add a resin sponge to the tube. Compress and release the sponge several times with the plunger provided in the kit to allow the serum to penetrate the sponge. Stand for exactly 60 min at room temperature.

During the 60-min incubation period, place the test tube in the well counter and count the total activity.

After 60 min, aspirate the contents of the test tube, compressing the sponge with the aspirator attachment provided in the kit to ensure that the liquid in the interstices of the sponge is completely removed. Wash three times with about 5 ml of distilled water, compressing and releasing the sponge several times with the aspirator before aspirating off the wash water.

After the last wash, replace the tube in the well counter and count the activity remaining in the resin sponge.

Include a sample of pooled normal human (nonpregnant) serum in each series of analyses. Check the T_3 uptake value for each new batch of pooled serum against the value for the old batch. About 30 determinations on each are sufficient for a comparison.

4. Counting Technique

Set the E.H.T. voltage and the pulse height and 'window' voltages in the pulse height analyzer to give maximum counting rates with a standard preparation of the isotope being used, contained in a plastic tube placed in the well counter. The contents of a T_3 syringe can be used as a standard. Start by counting the background. In each sample, time 10,000 counts or count for 5 min, whichever gives the shorter counting period. Count the standard preparations of the isotope after every tenth sample and adjust the pulse height analyzer if necessary to maintain a constant counting rate, thereby compensating for drifts in the response of the crystal.

5. Calculation

Correct the counting rates if necessary for background by subtracting the background counting rate, and for radioactive decay by multiplying by the appropriate decay factor.

$$T_3 \text{ uptake } (\%) = \frac{\text{corrected counts/min remaining in resin}}{\text{corrected initial counts/min}} \times 100$$

Correct the percentage uptake for any discrepancy in temperature from 25°C or any difference from a 60-min incubation time by means of correction factors read from a graph. Correction graphs are provided by the manufacturers or can be constructed from observations.

Express the results as a ratio:

$$T_3 \text{ uptake ratio} = \frac{T_3 \text{ uptake \% (test)}}{T_3 \text{ uptake \% (normal pooled serum)}}$$

6. Accuracy and Precision

T_3 is bound to TBG but not to TBPA.[16] Thus, the T_3 resin uptake test is an inverse measure of free TBG capacity and is, consequently, influenced both by the total serum TBG capacity and by the total serum T_4 concentration. A high T_3 uptake is associated with a low free TBG capacity and vice versa. The serum total T_3 concentration is too small for the variations in T_3 concentration encountered clinically to affect the serum T_3 uptake directly.

The inverse relation between the resin T_3 uptake and the free TBG capacity can be expressed as a linear regression equation.[46] In practice, the results of the analysis are expressed simply as the resin T_3 uptake and the free TBG capacity is not calculated. The values of the constants in the regression equation may differ when different techniques are used to determine the resin T_3 uptake.[46,47] The coefficient of variation of a T_3 uptake determination is about 2%.[157] The Thyopac-3 test gives results which have an inverse numerical relation to the results of other tests of resin T_3 uptake since in the Thyopac-3 test; the radioactivity measurements are made on the supernatant fluid, not on the resin as in the other tests.

D. The Free T_4 Factor[46]

1. Principle

The free T_4 factor is a measure of the serum free T_4 level, calculated from the serum total T_4 and resin T_3 uptake. By the Law of Mass Action,

$$\text{Free } T_4 = k \times \frac{\text{total } T_4}{\text{free TBG capacity}} \qquad (1)$$

where k is the dissociation constant for the reaction between T_4 and TBG. Although the value of k has not been accurately determined, a free T_4 factor ($= \frac{1}{k} \times$ free T_4) defined as the total serum T_4 divided by the free TBG capacity can be used in practice as a test of thyroid status. The free TBG capacity is inversely related to the resin T_3 uptake from which it can be derived by means of a regression equation. The total serum T_4 is either measured directly by displacement analysis or calculated from the serum PBI.

2. Procedure

Prepare a specimen of serum from clotted blood. Measure the total serum T_4 (or PBI) and the resin T_3 uptake as described above. Express the resin T_3 uptake of test and pool sera as percentages. A reference value for the resin T_3 uptake of the pooled normal serum is also required. This is obtained by taking the mean of at least 30 determinations of the resin T_3 uptake of the freshly prepared pool.

3. Calculation

Multiply the sample resin T_3 uptake (%) by the ratio of the reference value for the pool T_3 uptake (%) to the pool T_3 uptake (%) observed in the batch. Calculate the free T_4 factor.

$$\text{Corrected resin } T_3 \text{ uptake (\%)} = \text{sample } T_3 \text{ uptake (\%)} \times \frac{\text{reference pool } T_3 \text{ uptake (\%)}}{\text{batch pool } T_3 \text{ uptake (\%)}} \qquad (2)$$

$$\text{Free } T_4 \text{ factor} = \frac{\text{total serum } T_4\ (\mu g/100\ ml)}{\left(\frac{1090}{(\text{Corrected } T_3 \text{ resin} + 3.5)} - 15.75\right)} \qquad (3)$$

If the serum PBI has been measured, multiply it by 1.53 to obtain the total serum T_4. The calculation of the free T_4 factor is facilitated by the use of a programmable electronic calculator, such as the Wang® 600 (Wang Electronics Inc., Tewksbury, Mass.) used in the present author's laboratory. Alternatively, the result can be read from a nomogram.[158]

The constants in Equation 3 refer to the resin T_3 uptake measured by the resin sponge method. The values of the constants must be redetermined if another technique is adopted. To do this, the resin T_3 uptake is determined in a series of sera both by the resin sponge method and by the alternative technique. A regression equation relating the T_3 uptake by the alternative method to the T_3 uptake by the resin sponge method is then calculated and substituted into Equation 3. This derivation cannot be used with the Thyopac-3 kit because the regression of Thyopac-3 results on Triosorb results is nonlinear.[157]

4. Accuracy and Precision

There is no satisfactory test of the accuracy of determinations of the free T_4 factor. Although the serum free T_4 concentration can be measured directly by equilibrium dialysis, the values so obtained are highly sensitive to the conditions in which they are determined[21] and cannot be satisfactorily used as primary reference values. Thus, the free T_4 factor can only be assessed by its performance in clinical diagnosis, which is satisfactory.[46] Values of the free T_4 factor calculated from the serum PBI are about 10% higher on the average than those calculated from the serum total T_4 measured by displacement analysis as a result of the differences between the accuracy of the two methods of estimating the total T_4. The precision of estimates of the free T_4 factor corresponds to a coefficient of variation for a single estimate of about 10%. The determination of serum total T_4 has the larger method error of the two analyses used in deriving the free T_4 factor and, consequently, has the larger influence in determining the precision of calculations of the free T_4 factor.

E. Basal Metabolic Rate[135,137]

1. Principle

The patient is in 'basal' conditions (i.e., at rest in the postabsorptive state, recumbent in bed, and kept comfortably warm covered with one blanket). The rate of body heat production (metabolic rate) measured in these conditions is the 'basal metabolic rate.' The subject breathes pure oxygen from a closed circuit, which contains a soda-lime trap to absorb carbon dioxide. The volume of gas in the closed circuit is recorded continuously on a spirometer and the consumption of oxygen calculated from the spirometer tracing. The volume of oxygen consumed per minute is corrected to STP and from it is derived the heat production in kcal/m^2 body surface/hour. The surface area is calculated from the subject's height and weight using tables. It is assumed that the respiratory quotient (the ratio by volume $\frac{CO_2 \text{ produced}}{O_2 \text{ absorbed}}$) has the average value of 0.82 in the postabsorptive state and that at this value 1 liter of oxygen consumed is equivalent to 4.825 kcal of heat produced. The heat production is expressed as the percentage deviation from a standard value, which is the mean heat production for the subject's sex and age group.

2. Equipment and Materials

Benedict-Roth closed circuit metabolism apparatus with kymograph, e.g., the model supplied by J. E. Kendrick, Ltd. (5 and 6 Lodge Villas, Woodford Green, Essex, England):

Spirometer charts with vertical graduations in minutes ('minute lines') and horizontal graduations in 100 and 500 ml.

Barometer.

Soda-lime 4–8 mesh.

Cylinder of oxygen with pressure gauge and reducing valve.

3. Procedure

It is essential that this test is performed by an experienced member of the laboratory staff who is used to putting nervous patients at their ease. Only in this way will consistent results be obtained from the test. The investigation can be carried out successfully on outpatients. The patient attends on two successive mornings after fasting for 12 hr. The patient's weight and height are measured, she voids urine, and is rested for half an hour on a couch covered with one blanket in a room at 20 to 23°C. The barometric pressure is recorded. The plug on the soda-lime CO_2 absorber is greased and fitted onto the entry tube in the spirometer with precautions to ensure an air-tight fitting, and the sealing jacket of the spirometer is filled with water to 7 or 8 cm from the top. A chart is wrapped round the recorder. The apparatus is tested for leaks by partly filling the bell with O_2, closing the oxygen inlet tap and the three-way tap in the mouthpiece, and placing a weight on top of the bell. The pen should rise a little and then stay stationary, but will continue to rise if there is a leak. The bell of the spirometer is filled with oxygen until the pen on the recorder falls to the base-line. The temperature of the instrument is recorded from the thermometer on the spirometer bell. The metal mouthpiece attachment and the rubber mouthpiece are washed and sterilized by boiling before and after each test.

Allow the patient to test the feel of the mouthpiece and nose clip. Then connect the mouthpiece to the apparatus with the three-way stop-cock open to the air. Close the patient's nostrils with the spring nose clip. When the patient's respirations have settled down, turn the stop-cock to connect the patient to the spirometer, switch on the spirometer drum, and record a spirometer trace for 10 min. The bell rises at expiration and the pen forms a trace which slopes upwards at a rate corresponding to the oxygen consumption. Disconnect the patient after 10 min. Record the spirometer temperature again. Discard the run if the respiratory movements are irregular or the slope is inconsistent. Remove the chart and draw a straight line through consecutive bottom (expiratory) peaks (the 'slope line') during a 5-min period in which the slope is constant and respirations regular. Rest the patient for half an hour, then carry out another run. Repeat the whole procedure on the following day. On the second day, if the results are lower than the previous day, and if the two results on the second agree to within 5%, the lower result is reported. Otherwise, the patient is asked to attend on the subsequent day until these criteria are fulfilled.

4. Calculation

Make a volume reading where the slope line crosses a minute line at the beginning and end of the 5-min period. The difference between the two readings gives the volume of oxygen consumed. Calculate the oxygen consumption per minute and reduce this volume to the volume of dry gas at STP. The temperature is taken to be the average of the two readings on the spirometer thermometer:

$$\underset{\text{(ml/min at STP)}}{O_2 \text{ consumption}} = \underset{\text{(ml/min)}}{O_2 \text{ consumption (obs.)}} \times \frac{B - V}{760} \times \frac{273}{273 + t}$$

where B = barometric pressure (mm Hg), t = temperature of spirometer (°C), V = vapor pressure of water at t°C (mm Hg) from Table 2.

Calculate the patient's body surface area in m^2 from the DuBois' formula:

$$S = W^{0.425} \times H^{0.725} \times 71.84 \times 10^{-4}$$

where S is the surface area in m^2, W is the patient's weight in kg, H is the patient's height in cm.

A nomogram for this calculation is given in Figure 5. Calculate the heat production in kcal/m^2/hour, using a factor of 4.825 to convert liters of O_2 at STP to kcal. This factor assumes the average figure of 0.82 for the respiratory quotient in the post-absorptive state.

TABLE 2

Vapor Pressure of Water (mm Hg)

Temp. °C	Pressure	Temp. °C	Pressure	Temp. °C	Pressure
7	7.49	15	12.70	23	20.89
8	8.02	16	13.54	24	22.18
9	8.57	17	14.42	25	23.55
10	9.17	18	15.36	26	24.99
11	9.79	19	16.35	27	26.51
12	10.46	20	17.39	28	28.10
13	11.16	21	18.50	29	29.78
14	11.91	22	19.66	30	31.55

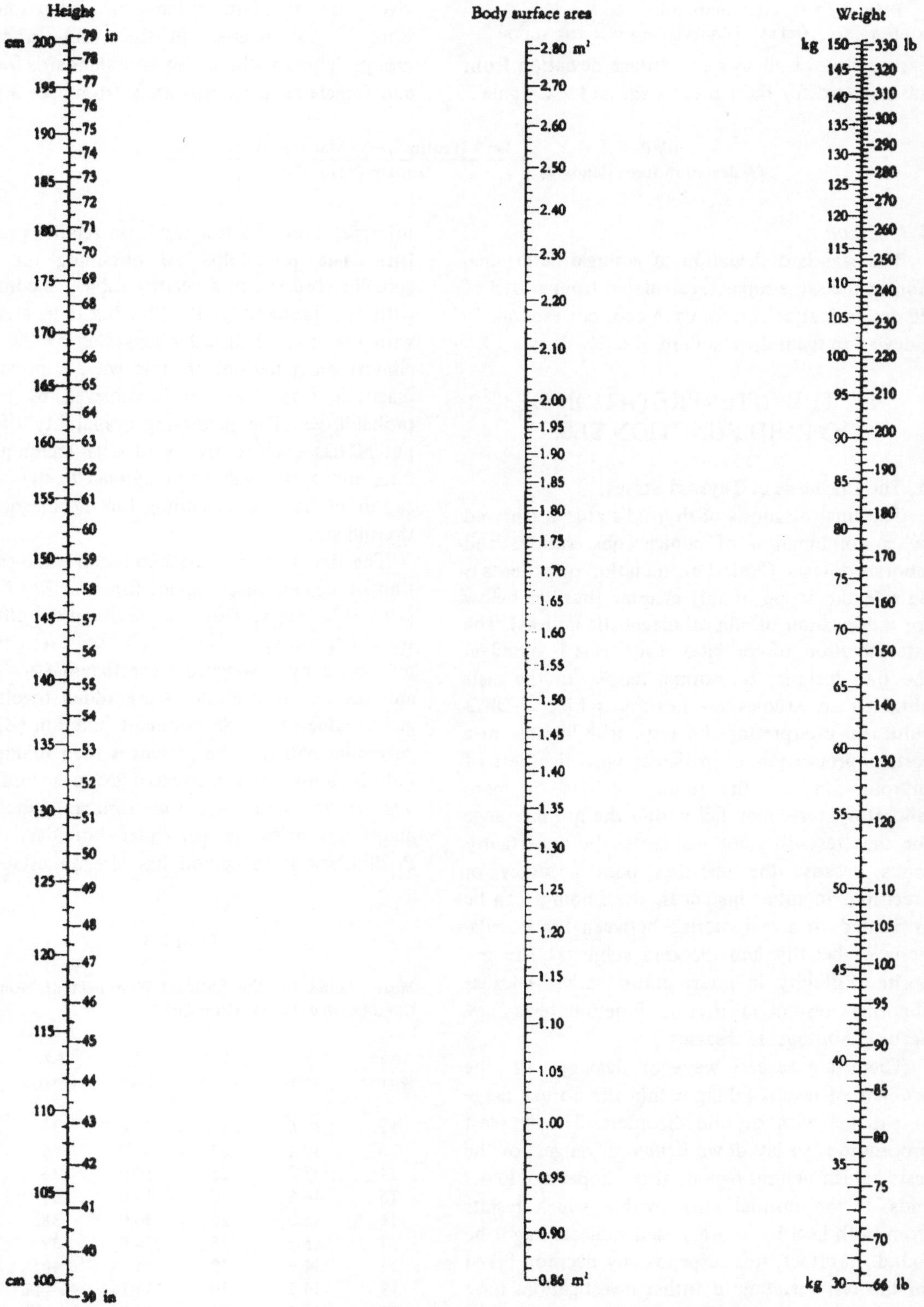

FIGURE 5. Nomogram for determining the body surface area from the height and body weight. (From *Documenta Geigy Scientific Tables,* 7th ed., Basle, 1970, 537. Courtesy CIBA-GEIGY Ltd., Basle, Switzerland.)

$$\underset{(\text{kcal/m}^2\text{/hr})}{\text{BMR}} = \underset{(\text{ml at STP/min})}{O_2 \text{ consumption}} \times \frac{60 \times 4.825}{\text{surface area (m}^2)}$$

Express the result as a percentage deviation from the standard for the patient's age and sex. Table 3 gives the standard values of Robertson and Reid[137] for women in the child-bearing age groups. These authors also give standards for male and female subjects separately from ages 3 to 75.

$$\underset{(\%\text{ deviation from standard})}{\text{BMR}} = \frac{\text{BMR (kcal/m}^2\text{/hr)} - \text{standard (kcal/m}^2\text{/hr)}}{\text{standard (kcal/m}^2\text{/hr)}} \times 100$$

5. Precision

The standard deviation of a single determination in the same subject, calculated from a total of 20 determinations on four subjects, corresponds to a deviation from the standard of ±2%.[135]

IV. THE INTERPRETATION OF THYROID FUNCTION TESTS

A. The Diagnosis of Thyroid Status

The final diagnosis of thyroid status is achieved by a combination of clinical observations and laboratory tests. Clinical examination of patients is outside the scope of this chapter (but see below for a discussion of clinical diagnostic indices). The interpretation of the laboratory tests is based in the first instance on normal ranges for the tests obtained in studies on healthy subjects. Difficulties in interpreting the tests arise because in a certain proportion of patients with disorders of thyroid function the results of one or more laboratory tests may fall within the normal range for that test. In some instances, the uncertainty arises because the test has poor accuracy or precision. In other instances, the findings can be attributed to a real overlap between the populations of healthy and diseased subjects. The opposite difficulty in interpretation arises because abnormal results in thyroid function tests may occur in nonthyroid diseases.

There are several ways of dealing with the problem of results falling within the normal range in patients with thyroid disorders. The simplest procedure is to lay down equivocal ranges for the tests, which delimit regions at the upper and lower ends of the normal range within which results from both healthy and diseased subjects might be found. In effect, this suspends any decision based on the test result until futher investigations have been carried out. Very often the results of all or most of the tests are borderline in these circumstances, in which case it may become extremely difficult to reach a final diagnosis.

A more rational procedure is to base the interpretation of a test result on relative probabilities. The probability of obtaining the result actually observed in a healthy subject is compared with the probability of obtaining it in a subject with the thyroid disorder suggested by the initial clinical examination. In this way, a provisional diagnosis from the test is achieved by relative probabilities. The increasing availability of computers has enabled two statistical techniques in discriminant analysis to be applied to the classification of patients attending for the diagnosis of thyroid status.

The first statistical method requires the calculation of a linear discriminant function $Z = b_1x_1 + b_2x_2 + \ldots + b_px_p$ from the results of the different tests $(x_1 \ldots x_p)$. The result of each test is multiplied by a weighting coefficient $(b_1 \ldots b_p)$ and the resulting products are added together to give a value of the discriminant function (Z) for a particular patient. The patient is then adjudged to fall into a normal or a diseased group, according to whether the value of the discriminant function is above or below a particular boundary value. Preliminary investigation has already established

TABLE 3

Mean Values for the BMR in Nonpregnant Women of Child-bearing Age (kcal/m² /hr)

Age (years)	BMR	Age (years)	BMR	Age (years)	BMR
12	40.6	23	34.0	34	33.7
13	39.1	24	33.9	35	33.5
14	37.8	25	34.0	36	33.3
15	36.8	26	34.0	37	33.3
16	36.0	27	34.0	38	32.9
17	35.3	28	34.0	39	32.8
18	34.9	29	34.1	40	32.6
19	34.5	30	34.1	41–44	32.5
20	34.3	31	34.0	45–49	32.2
21	34.1	32	33.9	50–54	31.9
22	34.0	33	33.8	55–59	31.6

From Robertson, J. D. and Reid, D. D., *Lancet*, 1, 940, 1952. With permission.

the values of the weighting coefficients and the boundary as those which give the minimum percentage of misclassifications of healthy subjects and patients with the disorder in question. The technique is discussed by Armitage[159] and has been used in studies on the laboratory diagnosis of thyroid disorders.[160,161] Linear discriminant functions were also used to construct the clinical indices for diagnosing thyroid disorders discussed below.

The second statistical technique is based on a mathematical approach in probability theory known as Bayes' Theorem.[159,162,163] The respective probabilities of finding in each of the diagnostic groups a patient with the test results actually observed are first calculated. The probability that the patient belongs to any one group is then given by the ratio of the probability calculated for that group to the sum of the probabilities for all the diagnostic groups.

$$P_{i|y} = \frac{P_{y|i} \cdot P_i}{P_y} = \frac{P_{y|i} \cdot P_i}{P_{y|1} \cdot P_1 + P_{y|2} \cdot P_2 + \cdots + P_{y|i} \cdot P_i + \cdots + P_{y|k} \cdot P_k}$$

where

y consists of the results of p investigations $(x_1 \ldots x_p)$,

$P_{i|y}$ is the ('posterior') probability that a patient with the clinical features y is in group i,

$P_{y|i}$ is the probability that a patient in group i has the clinical features y, and

P_i is the ('prior') probability of observing a member of the ith group in the hospital or clinic.

A recent study[164] suggests that the Bayes' technique is most efficiently used in a sequential manner, the probabilities for each test being examined in sequence and the next test selected as the one most likely to lead to a diagnosis given the findings of the previous tests.

Both linear discriminants and Bayesian analysis were reported by the workers quoted to achieve at least as good a diagnostic efficiency as a clinician experienced in the particular field.

The method of classical linear discriminant functions has the disadvantage that the values of the discriminant function coefficients apply only to the actual population for which they were determined and they are likely to prove less efficient in diagnosis in populations of patients which differ widely from the original population. The Bayesian method has the disadvantage that the probabilities of observing patients in each of the relevant clinical groups at the clinic must be known. Also, the statistical theory requires that the results of each of the tests should be independent of all others. This latter condition is not always fulfilled, for instance, the resin T_3 uptake is influenced by the total serum T_4 and the free T_4 factor is calculated from the total serum T_4. Thus, care has to be taken in selecting the tests to include in a Bayesian diagnostic scheme.

It is assumed that those workers with access to a computer will develop their own program and method of interpretation. Here, interpretation of thyroid function tests is restricted to a consideration of the normal values in pregnant and nonpregnant patients, for the tests described in the Methods section, to an indication of the percentage of hypothyroid and of thyrotoxic patients in whom the results may be expected to fall within the normal range and to references to nonthyroid disorders affecting the test results. When assessing thyroid status in pregnancy, it is recommended that laboratories should use diagnostic ranges appropriate to their own population of euthyroid pregnant subjects.

Normal ranges for the thyroid function tests described in the Methods section are given in Table 4. They should be regarded as guides to be modified where necessary in the light of local experience. Interpretation of the tests in individual patients should take account of results outside the normal range occurring in nonthyroid disease and of the overlap between the results in thyroid disease and the range in euthyroid subjects.

B. Special Problems in Pregnancy

The diagnosis of mild thyrotoxicosis in a pregnant patient is often difficult.[121] In pregnancy, the thyroid gland is often visible and palpable and many of the symptoms of thyrotoxicosis, such as heat intolerance and increases in heart and respiratory rate, are also found in pregnant subjects. However, thyrotoxicosis tends to improve symptomatically in pregnancy and does not result in abortion. Hence, definitive diagnosis and treatment of thyrotoxicosis in the

TABLE 4

Normal (95%) Ranges for Selected Thyroid Function Tests in Nonpregnant and Pregnant Women

	Normal range		
Test	Nonpregnant women and 2 weeks postpartum	Pregnant women	Reference
Serum PBI (μg of I/100 ml)	4–8	1st and 2nd trimester 4.5–11.5; 3rd trimester 7.5–13.5	Acland[59]
Serum total T_4 (μg of T_4/100 ml)*	4.5–11.5	6–16	Ekins et al.[50]
Resin T_3 uptake as %	23–33	15–25	Mestman et al.[100]
as ratio of test to normal serum	0.81–1.22	0.44–0.75	Schatz et al.[45]
Free T_4 factor (arbitrary units)	0.215–0.555	0.165–0.580	Goolden et al.[46]
Basal metabolic rate (% deviation from† standard)	L.L. –15 U.L. +15	At 18 weeks L.L. –10 U.L. +20	Javert[165]
		At term L.L. +5 U.L. +35	Freedberg et al.[48]

*To convert μg of T_4 to μg of T_4-I, divide by 1.53.
†The mean values of the Robertson and Reid[136] standards are used.
L.L. = Lower limit
U.L. = Upper limit

pregnant subject may well be left until the postpartum period. It must, nevertheless, be remembered that if the mother is thyrotoxic, the neonate may suffer from a self-limiting form of thyrotoxicosis,[166] possibly as a result of placental transfer of LATS.[167]

Authenticated cases of pregnancy occurring in an already hypothyroid patient are rare because hypothyroidism does not usually develop until later in life, often after the end of the child-bearing period, and is associated with menstrual disturbances leading to impaired fertility or (in hypothyroidism of pituitary origin) with failure of ovulation. Hypothyroidism may, however, occur in pregnancy due to the overenthusiastic treatment of thyrotoxicosis with antithyroid drugs. This is a serious complication as it is likely to lead to spontaneous abortion or hypothyroidism in the fetus with goiter and irreversible mental retardation.[168]

There is no conclusive evidence that thyroid deficiency plays a part in the etiology of habitual abortion.[11] The serum PBI and TBG may be reduced in habitual abortion, which could be a secondary effect of a reduction in fetoplacental estrogen synthesis.[169] However, the serum PBI and TBG levels are not closely correlated with the urinary estrogen output in habitual abortion and although the serum TBG level is low on average in patients with placental insufficiency as compared with patients whose placental function is normal, it is not of diagnostic value in the individual patient.[170]

Hyperthyroidism occurs in a proportion of cases of hydatidiform mole[171] and choriocarcinoma[172] as a result of secretion of chorionic TSH by the tumor.

The serum PBI is within the normal range for pregnancy in treated pregnant diabetics.[173] In panhypopituitarism due to postpartum pituitary necrosis (Simmonds' disease or Sheehan's syndrome), hypothyroidism is often a prominent feature.[4] In antepartum pituitary necrosis, which is a rare complication of diabetes in pregnancy, TSH secretion is commonly spared, although pituitary TSH reserve may be reduced.[174]

C. The Interpretation of the Serum PBI and Total Thyroxine

Raised serum levels of PBI and total T_4 are found in thyrotoxicosis or in diseases where the serum TBG capacity is increased. Lowered serum levels of PBI and total T_4 are found in hypothyroidism or in diseases where the serum TBG capacity is reduced. Normal ranges for the serum PBI and total T_4 are given in Table 4 for both pregnant and nonpregnant subjects. It should be noted that the serum PBI concentrations are given in μg of I/100 ml, whereas the serum T_4 concentrations are given in μg of T_4/100 ml. This is in keeping with general practice, as the serum T_4 determination is specific for T_4, whereas the serum PBI is not. However, numerical confusion in interpreting serum PBI and T_4 determinations may arise if this difference in units is not appreciated. T_4 concentrations can be converted to T_4-I concentrations by dividing by 1.53.

The interpretation of the serum PBI in thyroid and nonthyroid diseases has been reviewed.[59] In most instances, the interpretation is the same whether the serum PBI or the serum T_4 estimation is used. However, the serum T_4 concentration is a reliable measure of the level of circulating thyroid hormone in certain conditions when the serum PBI concentration is not. This occurs notably when iodine drugs or radiographic contrast media have been administered and also when the serum contains abnormal iodine metabolites, e.g., iodotyrosines, as in some types of familial goitrous cretinism, or thyroglobulin as in Hashimoto's thyroiditis and in some cases of thyroid carcinoma. In these circumstances, the serum PBI is inappropriately elevated. The average difference between the serum PBI and T_4-I concentrations in euthyroid subjects is 0.35 μg/100 ml, due to the presence of non-T_4 iodine, probably iodoprotein, in the serum; average differences are, respectively, negligible in hypothyroidism and 3.25 μg/100 ml in thyrotoxicosis.[95] Differences greater than 1 μg/100 ml indicate that increased amounts of non-T_4 iodine, i.e., iodine drugs or contaminants, iodotyrosines, or iodoproteins, are present. Non-T_4 iodine in the serum may be increased for instance in diffuse simple goiter,[95] Hashimoto's thyroiditis,[175] and carcinoma of the thyroid.[176] The only drugs extractable from serum that are known to affect the determination of serum T_4 by displacement are phenytoin, which may cause low results, and D-thyroxine, which is estimated as L-thyroxine.[73,75]

The percentage of patients whose thyroid status is misdiagnosed by the serum PBI or T_4 determination varies from clinic to clinic, according to the accuracy and precision of the analytical method, the criteria used in making the final diagnosis, and the composition of the population of patients attending for diagnosis. In practice, among patients finally diagnosed as euthyroid, the serum PBI or T_4 level would be expected to be above the normal range in about 5% of cases and below the normal range in 5 to 10% of cases; in about 5% of thyrotoxic patients and 20 to 25% of hypothyroid patients, the serum PBI or T_4 level would be expected to be within the normal range (see, for example, Clark and Horn[113]).

The serum PBI and total T_4 levels are affected in a similar manner by changes in the serum TBG capacity. Thus, in pregnancy and after the administration of estrogens (including estrogenic oral contraceptives), when the serum TBG capacity is increased, both the serum PBI and the T_4 concentrations may be elevated above the normal range. Conversely, in the nephrotic syndrome, when the serum TBG capacity is reduced, both the serum PBI and T_4 levels may be reduced below the normal range. In neither instance is there any evidence of disordered function of the thyroid gland. These discrepancies can be resolved by interpreting the serum PBI and T_4 levels in relation to a measure of serum TBG capacity, of which the simplest for routine purposes is the resin T_3 uptake.

It has been reported that the gestagen clomiphene blocks the effects of estrogens on the serum T_4 level,[177] whereas the gestagen dydrogesterone has little or no influence on the increase in serum T_4 caused by estrogens.[178] In patients receiving only clomiphene, the serum PBI is normal[179] or occasionally low.[180]

D. The Interpretation of the Resin T_3 Uptake

The resin T_3 uptake measures the proportion of labeled T_3 present in serum which is taken up by the resin added to the serum and is an inverse measure of the number of free T_4 binding sites on the serum TBG. There is, consequently, a reduced resin T_3 uptake when either the serum TBG capacity is raised or the total serum T_4 concentration is lowered and an increased uptake when either the serum TBG capacity is lowered or the total serum T_4 concentration is raised. The value obtained for the resin T_3 uptake depends on the resin preparation used. The test is fully discussed by Robbins and Rall.[181]

The main difficulty in determining the resin T_3 uptake is caused by the presence of variable amounts of radioiodide in the labeled T_3 preparation. This affects the measurements of resin uptake because iodide is completely adsorbed by the resin. It is, therefore, advisable to correct for this error by including in each batch a sample of pooled normal human serum and adjusting the results on the test sera with reference to the T_3 uptake of the sample of pooled normal serum. A test result may be reported as the ratio of the T_3 uptake of the unknown serum to the T_3 uptake of the pooled serum. Alternatively, a test result may be reported as a percentage uptake, but adjusted by multiplying this percentage by the T_3 uptake assigned initially to the pooled serum, and dividing the result by the T_3 uptake of the pooled serum as measured in the batch. Normal ranges for the resin T_3 uptake in pregnant and nonpregnant subjects, both as a percentage and as a ratio, are given in Table 4.

The resin T_3 uptake discriminates between patients of different thyroid status, uptakes being raised in thyrotoxicosis and low or low normal in hypothyroidism; 3 to 4% of thyrotoxic patients and 10 to 25% of hypothyroid patients have resin T_3 uptakes within the normal range.[113,162,182] Variable findings are obtained in different series because some, but not all, thyrotoxic patients have a reduced serum TBG capacity and some, but not all, hypothyroid patients have an increased serum TBG capacity.[103] In pregnancy and in patients taking estrogens (including estrogenic oral contraceptives), the serum TBG capacity increases and the resin T_3 uptake is reduced.[46,102] In conditions such as the nephrotic syndrome and in patients taking androgens, the serum TBG capacity decreases and the resin T_3 uptake is raised.[113,182] Drugs such as salicylate and phenytoin, which interfere with T_4 binding by TBG, cause elevation of the resin T_3 uptake.[182]

Although the resin T_3 uptake has discriminatory power as a single test in the diagnosis of thyroid status, it is most efficiently used to adjust measurements of the serum PBI or T_4 level to take into account variations in TBG capacity. Since a change in serum TBG capacity has the opposite effect on the serum PBI (or T_4) level to its effect on the resin T_3 uptake, a combination of the two analyses enables allowance to be made for the change in TBG capacity, for instance, by calculation of the free T_4 factor, as described in the Methods section.

When screening pregnant patients for thyroid disorder, it is recommended that the resin T_3 uptake should be determined routinely as well as the serum PBI (or T_4) level, and the free T_4 factor calculated.

E. The Interpretation of the Free T_4 Factor

The free T_4 factor is defined as the total serum T_4 concentration divided by the concentration of free T_4 binding sites and is proportional to the serum free T_4 concentration. Its estimation from the serum T_4 (or PBI) concentration and the resin T_3 uptake is described in the Methods section.

The use of the free T_4 factor was proposed by Osorio et al.[183] and by Goolden et al.[46] The ranges for the test in pregnant and nonpregnant subjects are shown in Table 4. Changes in TBG capacity are compensated. The increase in serum TBG capacity in pregnancy and in women taking estrogenic oral contraceptives does not influence the average value of the free T_4 factor, although the scatter of results is somewhat greater than in euthyroid nonpregnant patients.[46,47] About 5% of thyrotoxic patients and 10% of hypothyroid patients give results within the normal range.[183]

If the serum T_4, estimated by displacement, is used to calculate the free T_4 factor, iodine drugs (other than D-thyroxine) and iodine contamination in the serum do not interfere, but low results may be obtained when phenytoin is present in the serum due to interference with the displacement analysis for T_4. If the serum PBI is used to calculate the free T_4 factor, non-T_4 iodine in the serum can cause spuriously high results, but phenytoin does not interfere.[183]

F. The Interpretation of the Basal Metabolic Rate

The normal limits of the Basal Metabolic Rate (BMR) are taken to be within ±15% of the standard value for the patient's height and weight.[136] In thyrotoxicosis, 95% of clinically unequivocal cases and 75 to 80% of clinically doubtful cases have a BMR above the normal range; in hypothyroidism, between 75 and 80% of cases have a BMR below the normal range.[125] The BMR increases during the course of pregnancy, the increase being more rapid in the second half.[48] In

Table 4, the normal range for nonpregnant subjects is given, together with ranges in pregnancy at 18 weeks and at term. Normal ranges at intermediate times can be obtained by interpolation.

The interpretation of the BMR has been discussed by Wolff.[11] It is essential in the first place that the technique and the apparatus should be reliable and the patient properly prepared in basal conditions. Spuriously high BMR's occur if there is an oxygen leak. Other diseases beside thyroid disorders may give rise to changes in oxygen consumption and may cause diagnostic difficulty. Possible causes of a low BMR relevant to pregnancy are sleep or sedation (e.g., with barbiturates), extreme obesity, and an acute depressive illness. Possible causes of a high BMR relevant to pregnancy include anxiety, fever, heavy smoking, a large intake of caffeine, and excess cardiac work due to hypertension, anemia, or cardiac valvular disease. Many nonthyroid conditions that can cause changes in the BMR are either rare in or not compatible with pregnancy, e.g., malnutrition, Addison's disease (low BMR); acromegaly, Cushing's syndrome, and diabetes insipidus (high BMR). Other nonthyroid conditions, which may be incidentally present in pregnant as well as in nonpregnant subjects and may cause the BMR to be raised, include myeloproliferative diseases (leukemias, polycythaemia). Postpartum pituitary necrosis (Simmonds' disease or Sheehan's syndrome) is usually associated with some degree of hypothyroidism, including a low BMR, but the diagnosis is usually clinically obvious without the necessity of carrying out a BMR.

Since the BMR is likely to be determined in pregnancy only if there is clinical doubt about the diagnosis of thyroid status, it is advisable when interpreting the BMR in a pregnant patient to bear in mind those nonthyroid diseases which are not clinically obvious and which may cause changes in oxygen consumption.

G. The Clinical Assessment of Thyroid Status

The clinical assessment of thyroid status is outside the scope of this chapter. Mention must, however, be made of attempts to increase the objective accuracy of a clinical examination by the use of weighted diagnostic indices. In the construction of these indices, the presence or absence of selected symptoms and signs characteristic of thyroid disorders are allotted scores corresponding to their discriminatory power in differentiating between a euthyroid patient and one suffering from thyroid disease. After the clinical examination has established whether or not each of the symptoms and signs is present, the scores are totaled to give a diagnostic index which is interpreted in relation to euthyroid and equivocal ranges for the value of the index. Diagnostic indices have been constructed for differentiating thyrotoxic from euthyroid patients by Crooks et al.[184] and by Gurney et al.,[185] the index of the last-named workers being particularly designed to differentiate thyrotoxicosis from an anxiety state in women. Diagnostic indices for differentiating hypothyroid from euthyroid patients have been published by Murray,[186] Billewicz et al.,[187] and Philp et al.,[188] the third group of workers being concerned with the detection of hypothyroidism following treatment of thyrotoxicosis with radioiodine.

The use of the diagnostic index technique has been criticized on the grounds that it gives an illusion of objectivity and provides reproducible results merely because everyone using the system looks for the same clinical signs.[189] The various diagnostic indices would be expected to perform well on populations of patients resembling those who were studied initially to derive the respective indices, but might not be so discriminating when applied to patients of a different type. In particular, pregnant patients might be difficult to distinguish from thyrotoxic patients by means of a diagnostic index which gives weight to signs such as a palpable thyroid gland or an increased heart rate which are commonly found in both states. The practical value of the diagnostic indices is that they tend to prevent the clinician either from overlooking an important clinical sign or from concentrating on the presence of one clinical sign which happens to be prominent to the exclusion of other signs of diagnostic importance.

REFERENCES

1. Perlman, I., Chaikoff, I. L., and Morton, M. E., *J. Biol. Chem.*, 139, 433, 1941.
2. Davenport, H. W., *Gastroenterology*, 1, 1055, 1943.
3. Fletcher, K., Honour, A. J., and Rowlands, E. N., *Biochem. J.*, 63, 194, 1956.
4. Wayne, E. J., Koutras, D. A., and Alexander, W. D., *Clinical Aspects of Iodine Metabolism*, Blackwell, Oxford, 1964.
5. Fraser, R., Hobson, Q. J. G., Arnott, D. G., and Emery, E. W., *Q. J. Med.*, 22, 99, 1953.
6. Blahd, W. H., *Nuclear Medicine*, 2nd ed., McGraw Hill, New York, 1971.
7. Roche, J. and Desruisseaux, G., *C. R. Seanc. Soc. Biol.*, 144, 1179, 1950.
8. Logothetopoulos, J. and Scott, R. F., *J. Physiol.*, 132, 365, 1956.
9. Honour, A. J., Myant, N. B., and Rowlands, E. N., *Clin. Sci.*, 11, 447, 1952.
10. Wolff, J., *Physiol. Rev.*, 44, 45, 1964.
11. Wolff, J., in *Biochemical Disorders in Human Disease*, 3rd ed., Thompson, R. H. S. and Wootton, I. D. P., Eds., Churchill, London, 1970, 379.
12. Daniel, P. M., Pratt, O. E., Roitt, I. M., and Torrigiani, G., *Immunology*, 12, 489, 1967.
13. Daniel, P. M., Plaskett, L. G., and Pratt, O. E., *J. Physiol.*, 188, 25, 1967.
14. Roche, J., Michel, R., Michel, O., and Lissitzky, S., *Biochim. Biophys. Acta*, 9, 161, 1952.
15. Hollander, C. S., Shenkman, L., Mitsuma, T., Blum, M., Kastrin, A. J., and Anderson, D. G., *Lancet*, 2, 731, 1971.
16. Robbins, J. and Rall, J. E., *Physiol. Rev.*, 40, 415, 1960.
17. Ingbar, S. H., *Ann. N.Y. Acad. Sci.*, 86, 440, 1960.
18. Robbins, J. and Rall, J. E., *J. Clin. Invest.*, 34, 1324, 1955.
19. Clark, F., *J. Clin. Pathol.*, 20, 344, 1967.
20. Woeber, K. A., Ingbar, S. H., and Traenkle, U. I., *J. Clin. Invest.*, 47, 1710, 1968.
21. Lee, N. D. and Pileggi, V. J., *Clin. Chem.*, 17, 166, 1971.
22. Sterling, K. and Brenner, M. A., *J. Clin. Invest.*, 45, 153, 1966.
23. Nauman, J. A., Nauman, A., and Werner, S. C., *J. Clin. Invest.*, 46, 1346, 1967.
24. Oppenheimer, J. H., Bernstein, G., and Hasen, J., *J. Clin. Invest.*, 46, 762, 1967.
25. Zaninovitch, A. A., Farach, H., Ezrin, C., and Volpe, R., *J. Clin. Invest.*, 45, 1290, 1966.
26. Tata, J. R., *J. Clin. Pathol.*, 20, 323, 1967.
27. Braverman, L. E., Ingbar, S. H., and Sterling, K., *J. Clin. Invest.*, 49, 855, 1970.
28. Lieblich, J. and Utiger, R. D., *J. Clin. Invest.*, 51, 157, 1972.
29. Lissitzky, S., Benevent, M. T., Roques, M., and Roche, J., *C. R. Seanc. Soc. Biol.*, 152, 1490, 1958.
30. Surks, M. I. and Oppenheimer, J. H., *J. Clin. Invest.*, 48, 658, 1969.
31. Myant, N. B., *Clin. Sci.*, 15, 227, 1956.
32. Myant, N. B., *Lectures on the Scientific Basis of Medicine*, 6, 313, 1958.
33. Chan, V. and Landon, J., *Lancet*, 1, 4, 1972.
34. Chan, V., Besser, G. M., Landon, J., and Ekins, R. P., *Lancet*, 2, 253, 1972.
35. Fisher, D. A., Oddie, T. H., and Thompson, C. S., *J. Clin. Endocrinol. Metab.*, 33, 647, 1971.
36. Pittman, C. S., Chambers, J. B., and Read, V. H., *J. Clin. Invest.*, 50, 1187, 1971.
37. Hall, R., Amos, J., Garry, R., and Buxton, R. L., *Br. Med. J.*, 2, 274, 1970.
38. Shenkman, L., Mitsuma, T., and Hollander, C. S., *J. Clin. Invest.*, 52, 205, 1973.
39. Aboul-Khair, S. A., Crooks, J., Turnbull, A. C., and Hytten, F. E., *Clin. Sci.*, 27, 195, 1964.
40. Lemarchand-Beraud, T. and Vannotti, A., *Acta Endocrinol.*, 60, 315, 1969.
41. Hennen, G., Pierce, J. G., and Freychet, P., *J. Clin. Endocrinol. Metab.*, 29, 581, 1969.
42. Rosenberg, I. N., Ahn, C. S., and Chalfen, M. H., *J. Clin. Endocrinol. Metab.*, 21, 554, 1961.
43. Halnan, K. E., *Clin. Sci.*, 17, 281, 1958.
44. Dowling, J. T., Freinkel, N., and Ingbar, S. H., *J. Clin. Endocrinol. Metab.*, 16, 1491, 1956.
45. Schatz, D. L., Palter, H. C., and Russell, C. S., *Can. Med. Assoc. J.*, 99, 882, 1968.
46. Goolden, A. W. G., Gartside, J. M., and Sanderson, C., *Lancet*, 1, 12, 1967.
47. Standeven, R. M., *J. Endocrinol.*, 43, 217, 1969.
48. Freedberg, I. M., Hamolsky, M. W., and Freedberg, A. S., *N. Engl. J. Med.*, 256, 505, 551, 1957.
49. Winikoff, D., *Med. J. Aust.*, 2, 13, 1968.
50. Ekins, R. P., Williams, E. S., and Ellis, S., *Clin. Biochem.*, 2, 253, 1969.
51. Vega de Rodriguez, G., Fuertes de la Haba, A., and Pelegrina, I., *Obstet. Gynecol.*, 39, 779, 1972.
52. Mantzos, J. D. and Malamos, B., *Clin. Chim. Acta*, 21, 501, 1968.
53. Harden, R. McG., Mason, D. K., and Buchanan, W. W., *J. Lab. Clin. Med.*, 65, 500, 1965.
54. MacFarlane, S., Papadopoulos, S., Harden, R. McG., and Alexander, W. D., unpublished work, 1967. Quoted by Wayne.[190]
55. Fitting, W., *Klin. Wochenschr.*, 42, 744, 1964.
56. Pittman, J. A., Jr., Dailey, G. E., and Beschi, R. J., *N. Engl. J. Med.*, 280, 1431, 1969.
57. Goolden, A. W. G., Glass, H. I., and Williams, E. D., *Br. Med. J.*, 4, 396, 1971.

58. Aboul-Khair, S. A., Buchanan, T. J., Crooks, J., and Turnbull, A. C., *Clin. Sci.*, 31, 415, 1966.
59. Acland, J. D., *J. Clin. Pathol.*, 24, 187, 1971.
60. Surks, M. I., Schwartz, H. L., and Oppenheimer, J. H., *J. Clin. Invest.*, 48, 2168, 1969.
61. Rhodes, B. A., *Acta Endocrinol.*, Suppl. 127, 1, 1968.
62. Riley, M. and Gochman, N., *Clin. Chem.*, 10, 649, 1964.
63. Austin, E. and Koepke, J. A., *Am. J. Clin. Pathol.*, 45, 244, 1966.
64. Blau, N. F., *J. Biol. Chem.*, 102, 269, 1933.
65. Klein, D. and Chernaik, J. M., *Clin. Chem.*, 6, 476, 1960.
66. Taurog, A., Chaikoff, I. L., and Tong, W., *J. Biol. Chem.*, 184, 99, 1950.
67. Levin, K., Josephson, B., and Grunewald, G., *Acta Endocrinol.*, 52, 627, 1966.
68. Pileggi, V. J., Lee, N. D., Golub, O. J., and Henry, R. J., *J. Clin. Endocrinol. Metab.*, 21, 1272, 1961.
69. Pileggi, V. J. and Kessler, G., *Clin. Chem.*, 14, 339, 1968.
70. Hanok, A., Kuo, J., Kuhn, J., and Trivelli, L., *Am. J. Clin. Pathol.*, 54, 542, 1970.
71. Murphy, B. E. P., *Recent Progr. Horm. Res.*, 25, 563, 1969.
72. Solomon, D. H., Benotti, J., DeGroot, L. J., Greer, M. A., Pileggi, V. J., Pittman, J. A., Robbins, J., Selenkow, H. A., Sterling, K., and Volpe, R., *J. Clin. Endocrinol. Metab.*, 34, 884, 1972.
73. Murphy, B. E. P. and Pattee, C. J., *J. Clin. Endocrinol. Metab.*, 24. 187, 1964.
74. Cassidy, C. E., Benotti, J., and Peno, S., *J. Clin. Endocrinol. Metab.*, 28, 420, 1968.
75. Murphy, B. E. P., *Can. Med. Assoc. J.*, 106, 360, 1972.
76. Mougey, E. H. and Mason, J. W., *Anal. Biochem.*, 6, 223, 1963.
77. Braverman, L. E., Vagenakis, A. G., Foster, A. E., and Ingbar, S. H., *J. Clin. Endocrinol. Metab.*, 32, 497, 1971.
78. Seligson, H. and Seligson, D., *Clin. Chim. Acta*, 38, 199, 1972.
79. Watson, D. and Lees, S., *Ann. Clin. Biochem.*, 10, 14, 1973.
80. Sterling, K., Bellabarba, D., Newman, E. S., and Brenner, M. A., *J. Clin. Invest.*, 48, 1150, 1969.
81. Fisher, D. A. and Dussault, J. H., *J. Clin. Endocrinol. Metab.*, 70, 675, 1971.
82. Larsen, P. R., *Metabolism*, 20, 609, 1971.
83. Radichevich, I. and Werner, S. C., *J. Clin. Endocrinol. Metab.*, 32, 350, 1971.
84. Row, V. V., McConnon, J., and Volpe, R., *Clin. Chim. Acta*, 31, 473, 1971.
85. Wahner, H. W. and Gorman, C. A., *N. Engl. J. Med.*, 284, 225, 1971.
86. Jaakonmaki, P. I. and Struffer, J. E., *J. Gas Chromatogr.*, 5, 303, 1967.
87. Nihei, N. N., Gershengorn, M. C., Mitsuma, T., Stringham, L. R., Cordy, A., Kuchmy, B., and Hollander, C. S., *Anal. Biochem.*, 43, 433, 1971.
88. Brown, B. L., Elkins, R. P., Ellis, S. M., and Reith, W. S., *Nature*, 226, 359, 1972.
89. Gharib, H., Ryan, R. J., Mayberry, W. E., and Hockert, T., *J. Clin. Endocrinol. Metab.*, 33, 509, 1971.
90. Chopra, I. J., Solomon, D. H., and Beall, G. N., *J. Clin. Invest.*, 50. 2033, 1971.
91. Mitsuma, T., Colucci, J., Shenkman, L., and Hollander, C. S., *Biochem. Biophys. Res. Commun.*, 46, 2107, 1972.
92. Mitsuma, T., Nihei, N., Gershengorn, C., and Hollander, C. S., *J. Clin. Invest.*, 50, 2679, 1971.
93. Hufner, M. and Hesch, R. D., *Clin. Chim. Acta*, 44, 101, 1973.
94. Chopra, I. J., Solomon, D. H., and Ho, R. S., *J. Clin. Endocrinol. Metab.*, 33, 865, 1971.
95. Farran, H. E. A., Haiste, C., and Hoffenberg, R., *Acta Endocrinol.*, 68, 451, 1971.
96. Lee, N. D., Catz, B., Margolese, M. S., and Pileggi, V. J., *Clin. Chem.*, 17, 174, 1971.
97. Hotelling, D. R. and Sherwood, L. M., *J. Clin. Endocrinol. Metab.*, 33, 783, 1971.
98. Elzinga, K. E., Carr, E. A., and Beierwaltes, W. H., *Am. J. Clin. Pathol.*, 36, 125, 1961.
99. Burke, G., Metzger, B. E., and Goldstein, M. S., *J. Lab. Clin. Med.*, 63, 708, 1964.
100. Mestman, J. H., Niswonger, J. W. H., Anderson, G. V., and Manning, P. R., *Am. J. Obstet. Gynecol.*, 103, 322, 1969.
101. Mitchell, M. L., Harden, A. B., and O'Rourke, M. E., *J. Clin. Endocrinol. Metab.*, 20, 1474, 1960.
102. Clark, F., *Lancet*, 2, 167, 1963.
103. Inada, M. and Sterling, K., *J. Clin. Invest.*, 46, 1442, 1967.
104. Goolden, A. W. G., Gartside, J. M., and Osorio, C., *J. Clin. Endocrinol. Metab.*, 25, 127, 1965.
105. Hamolsky, M. W., Golodetz, A., and Freedberg, A. S., *J. Clin. Endocrinol. Metab.*, 19, 103, 1959.
106. Hansen, H. H., *Scand. J. Clin. Lab. Invest.*, 18, 240, 1966.
107. Gimlette, T. M. D., *J. Clin. Pathol.*, 20, 170, 1967.
108. Braverman, L. E., Foster, A. E., and Mead, L. W., *J.A.M.A.*, 199, 469, 1967.
109. Levy, R. P., Marshall, J. S., and McGuire, W. L., *Metabolism*, 13, 557, 1964.
110. Lee, N. D., Henry, R. J., and Golub, O. J., *J. Clin. Endocrinol. Metab.*, 24, 486, 1964.
111. Schussler, G. C. and Plager, J. E., *J. Clin. Endocrinol. Metab.*, 27, 242, 1967.
112. Volpert, E. M., Martinez, M., and Oppenheimer, J. H., *J. Clin. Endocrinol. Metab.*, 27. 421, 1967.
113. Clark, F. and Horn, D. B., *J. Clin. Endocrinol. Metab.*, 25, 39, 1965.
114. Russell, K. P., Rose, H., and Starr, P., *Am. J. Obstet. Gynecol.*, 90, 682, 1964.
115. Burke, C. W., Shakespeare, R. A., and Fraser, R., *Clin. Sci.*, 44, 5P, 1973.
116. Thorson, S. C., Mincey, E. K., McIntosh, H. W., and Morrison, R. T., *Br. Med. J.*, 2, 67, 1972.
117. Wellby, M. L., O'Halloran, M. W., and Marshall, J., *Clin. Chim. Acta*, 45, 225, 1973.

118. Bayliss, R. I. S., *J. Clin. Pathol.*, 20, 360, 1967.
119. Ormston, B. J., Garry, R., Cryer, R. J., Besser, G. M., and Hall, R., *Lancet,* 2, 10, 1971.
120. Singer, P. A. and Nicoloff, J. T., *J. Clin. Invest.*, 52, 1099, 1973.
121. Crooks, J., *J. Clin. Pathol.*, 20, 373, 1967.
122. Doniach, D., *J. Clin. Pathol.*, 20, 385, 1967.
123. Hall, R., *J. Clin. Pathol.*, 20, 365, 1967.
124. Tulloch, B. R., Lewis, B., and Fraser, T. R., *Lancet,* 1, 391, 1973.
125. Wayne, E. J., *Br. Med. J.*, 1, 1, 78, 1960.
126. De Alvarez, R. R., Gaiser, D. F., Simkins, D. M., Smith, E. K., and Bratvold, G. E., *Am. J. Obstet. Gynecol.*, 77, 743, 1959.
127. Mann, G. V. and Shaffer, R. D., *J.A.M.A.*, 197, 1071, 1966.
128. Peters, J. P., Heinemann, M., and Man, E. B., *J. Clin. Invest.*, 30, 388, 1951.
129. Watson, W. C., *Clin. Sci.*, 16, 475, 1957.
130. Richardson, D. P. and Naismith, D. J., *Proc. Nutr. Soc.*, 31, 7A, 1972.
131. Barton, G. M. G., Freeman, P. R., and Lawson, J. P., *J. Obstet. Gynaecol. Br. Commonw.*, 77, 551, 1970.
132. Smith, E. K. M. and Samuel, P. D., *Clin. Sci.*, 38, 49, 1970.
133. Goolden, A. W. G., Bateman, D., and Torr, S., *Br. Med. J.*, 2, 552, 1971.
134. Williams, T. and Besser, G. M., *Clin. Chem.*, 17, 148, 1971.
135. Robertson, J. D., *Br. Med. J.*, 1, 617, 1944.
136. Crooks, J., Murray, I. P. C., and Wayne, E. J., *Lancet,* 1, 604, 1958.
137. Robertson, J. D. and Reid, D. D., *Lancet,* 1, 940, 1952.
138. Bernstein, L. M., Johnston, L. C., Ryan, R., Inouye, T., and Hick, F. K., *J. Appl. Physiol.*, 9, 241, 1956.
139. Passmore, R. and Draper, M. H., in *Biochemical Disorders in Human Diseases,* 3rd ed., Thompson, R. H. S., and Wootton, I. D. P., Eds., Churchill, London, 1970, 15.
140. Fraser, R. and Nordin, B. E. C., *Lancet,* 1, 532, 1955.
141. Hortling, H. and Hiisi-Brummer, L., *Scand. J. Clin. Lab. Invest.*, 9, 1, 1957.
142. Gaitskell, R. E., Kendall-Taylor, P., Munro, D. S., Amos, J., and Hall, R., in *Radioimmunoassay Methods,* Kirkham, K. E. and Hunter, W. M., Eds., Churchill Livingstone, London, 1971, 563.
143. Brown, J. and Munro, D. S., *J. Endocrinol.*, 38, 439, 1967.
144. Hall, R. and Amos, J., in *Radioimmunoassay Methods,* Kirkham, K. E. and Hunter, W. M., Eds., Churchill Livingstone, London, 1971, 143.
145. Kirkham, D. E. and Hunter, W. M., Eds., *Radioimmunoassay Methods,* Churchill Livingstone, London, 1971.
146. Galton, V. A., Ingbar, S. H., Jimenez-Fonseca, J., and Hershman, J. M., *J. Clin. Invest.*, 50, 1345, 1971.
147. Evered, D., Young, E. T., Ormston, B. J., Menzies, R., Smith, P. A., and Hall, R., *Br. Med. J.*, 3, 131, 1973.
148. Adams, D. D., *J. Clin. Endocrinol. Metab.*, 18, 699, 1958.
149. McKenzie, J. M., *Endocrinology,* 62, 865, 1958.
150. McKenzie, J. M., *Metabolism,* 21, 883, 1972.
151. Kriss, J. P., Pleshakov, V., and Chien, J. R., *J. Clin. Endocrinol. Metab.*, 24, 1005, 1964.
152. Gillespie, P. J. and Keyes, W. I., *Br. J. Radiol.*, 44, 319, 1971.
153. International Atomic Energy Agency, *Br. J. Radiol.*, 35, 205, 1962.
154. Wagner, H. N., *Principles of Nuclear Medicine,* W. B. Saunders, Philadelphia, 1968.
155. Simpson, D., *Clin. Chem.*, 13, 890, 1967.
156. Ahuja, J. N., Kaplan, A., and Van Dreal, P., *Clin. Chem.*, 13, 708, 1967.
157. Chan, V., McAlister, J., and Landon, J., *J. Clin. Pathol.*, 25, 30, 1972.
158. Kintner, E. P., *Am. J. Clin. Pathol.*, 49, 599, 1968.
159. Armitage, P., *Statistical Methods in Medical Research,* Blackwell, Oxford, 1971.
160. Fragu, P., Thouin, A., Bazin, J. P., and Tubiana, M., *Ann. Endocrinol.*, 33, 5, 1972.
161. Barnett, D. B., Greenfield, A. A., Howlett, P. J., Hudson, J. C., and Smith, R. N., *Br. Med. J.*, 2, 144, 1973.
162. Bender, C. E., Fitzgerald, L. T., and Williams, C. M., *Am. J. Roentgenol. Radium Ther. Nucl. Med.*, 103, 886, 1968.
163. Oddie, T. H., *J. Clin. Endocrinol. Metab.*, 32, 167, 1971.
164. Knill-Jones, R. P., Stern, R. B., Girmes, D. H., Maxwell, J. D., Thompson, R. P. H., and Williams, R., *Br. Med. J.*, 1, 530, 1973.
165. Javert, C. T., *Am. J. Obstet. Gynecol.*, 39, 954, 1940.
166. Lewis, I. C. and Macgregor, A. G., *Lancet,* 1, 14, 1957.
167. Gautier, E., Juillard, E., and Lemarchand-Beraud, T., *Schweiz. Med. Wochenschr.*, 98, 8, 1968.
168. Barnes, C. G., *Medical Disorders in Obstetric Practice,* 3rd ed., Blackwell, Oxford, 1970.
169. Greenman, G. W., Gabrielson, M. O., Howard-Flanders, J., and Wessel, M. A., *N. Engl. J. Med.*, 267, 426, 1962.
170. Nicoloff, J. T., Gross, H. A., Warren, D. W., Mestman, J. H., and Anderson, G. V., *Obstet. Gynecol.*, 35, 191, 1970.
171. Kock, H., Kessel, H. V., Stolte, L., and Leusden, H. V., *J. Clin. Endocrinol. Metab.*, 26, 1128, 1966.
172. Odell, W. D., Bates, R. W., Rivlin, R. S., Lipsett, M. B., and Hertz, R., *J. Clin. Endocrinol. Metab.*, 23, 658, 1963.
173. Newhouse, S., Zarowitz, H., and Weisenfeld, S., *Obstet. Gynecol.*, 36, 892, 1970.
174. Schalch, D. S. and Burday, S. Z., *Ann. Intern. Med.*, 74, 357, 1971.

175. McConahey, W. M., Keating, F. R., Jr., Butt, H. R., and Owen, C. A., *J. Clin. Endocrinol. Metab.*, 21, 879, 1961.
176. Robbins, J., Rall, J. E., and Rawson, R. W., *J. Clin. Endocrinol. Metab.*, 15, 1315, 1955.
177. Barbosa, J., Seal, U. S., and Doe, R. P., *J. Clin. Endocrinol. Metab.*, 36, 666, 1973a.
178. Barbosa, J., Seal, U. S., and Doe, R. P., *J. Clin. Endocrinol. Metab.*, 36, 706, 1973b.
179. Cushman, P., Jr., Alter, S., and Hilton, J. G., *Acta Endocrinol.*, 50, 329, 1965.
180. Ruiz-Velasco, V., Arceo, J. R., and Gasca, A. V., *Int. J. Fertil.*, 15, 214, 1970.
181. Robbins, J. and Rall, J. E., in *Hormones in Blood*, Vol. 1, 2nd ed., Gray, C. H. and Bacharach, A. L., Eds., Academic Press, London, 1967, 383.
182. Quimby, E. H. and Hiza, E., *J. Nucl. Med.*, 5, 489, 1964.
183. Osorio, C., Jackson, D. J., Gartside, J. M., and Goolden, A. W. G., *Clin. Sci.*, 23, 525, 1962.
184. Crooks, J., Murray, I. P. C., and Wayne, E. J., *Q. J. Med.*, 28, 211, 1959.
185. Gurney, C., Owen, S. G., Hall, R., Roth, M., Harper, M., and Smart, G. A., *Lancet*, 2, 1275, 1970.
186. Murray, I. P. C., *Med. J. Aust.*, 1, 827, 1964.
187. Billewicz, W. Z., Chapman, R. S., Crooks, J., Day, M. E., Gossage, J., Wayne, Sir E., and Young, J. A., *Q. J. Med.*, 38, 255, 1969.
188. Philp, J. R., Duthie, M. B., and Crooks, J., *Lancet*, 2, 1336, 1968.
189. Trotter, W. R., *Diseases of the Thyroid*, Blackwell, Oxford, 1962.
190. Wayne, E. J., *J. Clin. Pathol.*, 20, 353, 1967.
191. Eastman, C. J., Corcoran, J. M., Jequier, A., Ekins, R. P., and Williams, E. S., *Clin. Sci. Mol. Med.*, 45, 251, 1973.

FETAL MONITORING – CURRENT CONCEPTS

D. Fahmy, J. F. Pearson, and A. C. Turnbull

TABLE OF CONTENTS

I. Introduction 299
II. Determination of Amniotic Phospholipids for the Antenatal Prediction of Respiratory Distress Syndrome 300
A. **Introduction** 300
B. **Prediction of Neonatal R.D.S. by Determination of Amniotic Fluid Lecithin Concentration** 300
III. **Prediction of Neonatal R.D.S. by Determination of L/S Ratio** 304
A. **Method** 304
B. **Clinical Usefulness of the Lecithin Assay in Amniotic Fluid and the L/S Ratio Determination** 308
IV. Prediction of Neonatal R.D.S. by a Rapid New Test for Surfactant in Amniotic Fluid 308
A. Method 308
V. Fetal Acid-base Balance 310
A. Introduction 310
References 314

I. INTRODUCTION

Present biochemical procedures are largely confined to investigations of fetal well-being in utero. These include the determination of estrogen levels,[1,2] particularly estriol, in maternal urine[3-5] and plasma,[6,7] tests of placental function, such as the urinary excretion of pregnanediol[8] and chorionic gonadotrophin,[9] and plasma levels of progesterone,[10,11] chorionic gonadotrophin,[12] and human placental lactogen,[13,14] which also reflect fetal well-being to a certain extent. In cases of Rh isoimmunization, bilirubin concentrations in amniotic fluid are particularly helpful in assessing the degree of hemolytic disease.[15]

Recent reports suggest that this spectrum of analyses should be extended to include procedures for monitoring the fetus during labor and those which give an indication of fetal lung maturity and the likelihood of respiratory distress syndrome in the newborn. Both of these procedures are described in this chapter on current concepts; the hormonal assays, which undoubtedly are of value, are excluded having been well documented.

II. DETERMINATION OF AMNIOTIC FLUID PHOSPHOLIPIDS FOR THE ANTENATAL PREDICTION OF RESPIRATORY DISTRESS SYNDROME

A. Introduction

The respiratory distress syndrome (R.D.S.) remains a serious disorder of newborn babies, affecting 10 to 15% of infants with a birthweight of 2.5 kg or less.[16] There is no specific treatment and between 20 and 40% of afflicted infants succumb.[17]

Obstetricians faced with indications of fetal jeopardy, such as subnormal maternal urinary estrogen excretion or signs of "placental insufficiency," are forced to decide whether to allow the pregnancy to continue with the risk that the fetus may die in utero or to deliver a live fetus which may then die of R.D.S. In such situations, a precise indication of the infant's risk of developing R.D.S., if delivered at this time, is clearly of value.

A crucial factor in the development of R.D.S. is deficiency of surfactant,[18,19] generally thought to be a lipoprotein,[20] that reduces surface tension within the lung, facilitating expansion during inspiration and preventing pulmonary collapse during expiration. Its main surfactive component is dipalmitoyl lecithin.[21,22]

Increasing concentrations of surfactant in the lungs toward term are mirrored by sharply rising levels in amniotic fluid from about the 33rd week of pregnancy.[23] A definite relationship between low amniotic fluid lecithin levels and the subsequent development of R.D.S. has been established by Bhagwanani and her co-workers[24,25] using a procedure that depends on the chemical determination of the lecithin content of amniotic fluid. A somewhat simpler method developed by Gluck et al.,[23,26] based on the disproportionate increase in synthesis of lecithin compared with sphingomyelin that occurs as term approaches, has shown that the lecithin:sphingomyelin ratio (L/S ratio) is low in cases associated with R.D.S. Both of these methods require skilled technical assistance and fairly expensive equipment; this caused Clements et al.[27] to develop a simple test for surfactant, utilizing bubble stability in ethanol.

These three methods will be described in this chapter since they appear to have found general acceptance. However, it should be borne in mind that other procedures that give an indirect assessment of lecithin levels, based on the lability in alkali of phosphoglycerolipids[28,29] or on palmitic acid concentration[30] in amniotic fluid, may be found more suitable for batch analysis. A critical evaluation by other workers of the value of these methods for prediction of neonatal R.D.S. is not yet available and they have, therefore, not been included.

B. Prediction of Neonatal R.D.S. by Determination of Amniotic Fluid Lecithin Concentration[24,25]

1. Principle

Lecithin concentration in amniotic fluid appears to be a sensitive index of fetal lung maturity and of the potential risk of R.D.S. developing in the neonate. Amniotic fluid lecithin levels of around 3.5 mg/100 ml would seem critical for predelivery levels above this value are invariably associated with normal neonatal respiration, while lower levels are associated with R.D.S.

2. Materials	*Grade and/or distributor*
Silica gel H (E. Merck Darmstadt, Germany)	Anderman and Co. Ltd., East Molsey, Surrey
Lecithin (Cat. No. L8878)	Sigma, London, Kingston-upon-Thames
Sodium carbonate	Analytical reagent grade
Iodine	Analytical reagent grade
Potassium dihydrogen orthophosphate	Analytical reagent grade
Glacial acetic acid	Analytical reagent grade
Perchloric acid	Analytical reagent grade
Sulfuric acid	Analytical reagent grade
880 Ammonia	Analytical reagent grade
Ammonium molybdate	Reagent grade

1-amino-2 naphthol-4 sulfonic acid	Reagent grade
Sodium metabisulfite	Reagent grade
Sodium sulfite	Reagent grade
Filter paper	Whatman No. 1
Extraction thimbles	Baird and Tatlock 231/2301
Cotton wool (absorbent)	General chemical supplier (GCS)

a. Solvents	*Grade*
Water	Bidistilled in all-glass still
Chloroform	Analytical reagent grade
Methanol	Analytical reagent grade
Petroleum ether (40° to 60°C)	Analytical reagent grade

b. Glassware	*Distributor and catalog number*
Conical tubes	E-mil, Stone, Staffordshire – G 7723
Test tubes, 15 ml, 18 mm X 125 mm	Jenkons, Hemel Hempstead, Herts – MF 24/1/5
Soxhlet extractor	Jenkons, Hemel Hempstead, Herts – 2LRCSX
Ehrlenmeyer flasks, 250 ml	Jenkons, Hemel Hempstead, Herts – FE250/3
Hamilton syringe, 100 μl	Shandon Southern Instruments Ltd., Camberley
Glass still	Quickfit Instrumentation, Stone, Staffordshire
Filter funnels (4 cm)	GCS
Measuring cylinders	GCS
Graduated pipettes	

(Note: Glassware must all be kept scrupulously clean and given two final rinses in bidistilled water before drying.)

3. Equipment	*Distributor and catalog number*
Thin-layer chromatoplates (20 X 20 cm)	Camlab, Cambridge – 12-01-40
Chromatoplate rack	Camlab, Cambridge – 12-01-30
Spreading device	Camlab, Cambridge – 12-01-01
Spreading template	Camlab, Cambridge – 12-01-30
Simultan chamber	Camlab, Cambridge – 12-01-67
Spray gun	Camlab, Cambridge – 12-40-10
Desiccator cabinet	Shandon Southern Instruments Ltd., Camberley SAB 2855
Mechanical shaker	Gallenkamp SD-130
Refrigerated centrifuge (Mistral 6L)	Measuring Scientific Equipment
Nitrogen (high purity) cylinder and reduction valve	Air Products
Water bath	Grant Instruments Ltd., Cambridge
Rotamixer	Hook and Tucker Ltd.
Dry block	Techne
Oven	Gallencamp, Christopher Street, London EC2B 2NE – OV-010
Mechanical shaker	Gallencamp, Christopher Street, London EC2B 2NE – SD-130
Refrigerated centrifuge (Mistral 5L)	Measuring Scientific Equipment, Crawley, Sussex
Nitrogen (high purity) cylinder and reduction valve	Air Products
Water bath	Grant Instruments Ltd., Cambridge
Rotamixer	Hook and Tucker Ltd., Brixton Road, London

Dry block	Jenkons, Hemel Hempstead, Herts – H32/3
Spectrophotometer	Pye Unicam Ltd., Cambridge SP 600
Syringe (2 ml)	Rocket all glass luer fitting
Needle	Luer 15G + 5

4. Reagents

a. Developing Solvent

(Quantity Sufficient for One Tank)

Chloroform	75 ml
Methanol	30 ml
Ammonia	4 ml (dispense with safety pipette)
Water	0.5 ml

b. Eluting Solvent

Chloroform	50 ml
Methanol	39 ml
Acetic acid	1 ml
Water	10 ml

c. Iodine Solution (for Spray)

Saturated solution of iodine in petroleum ether (40 to 60°C).

d. 2.5N Sulfuric Acid

Add 6.65 ml sulfuric acid (conc.) to 100 ml water.

e. Ammonium Molybdate Solution

Dissolve 2.5 g ammonium molybdate in 100 ml 2.5 *M* sulfuric acid.

f. Reducing Agent

Mix 0.2 g 1-amino-2 naphthol-4 sulfonic acid with 1.2 g each of finely powdered sodium metabisulfite and sodium sulfite and store in a dry state. Prepare an aqueous solution containing 0.25 g per 10 ml just prior to use.

g. Phosphate Stock Solution (100 mg/1 ml Phosphorus)

Dissolve 0.439 g potassium dihydrogen orthophosphate in 100 ml water. Store at 4°C.

h. Standard Phosphate Solution (4 μg/1 ml Phosphorus)

Dilute 1 ml phosphate stock solution to 250 ml with water.

i. Standard Lecithin Solution

Dissolve 3 mg dipalmitoyl lecithin in 1 ml chloroform. Store at 4°C.

j. Washed Cotton Wool

Pack cotton wool into thimbles and place in the Soxhlet extractor. Extract with chloroform:methanol (2:1 v/v) for about 24 hr. Remove thimbles, drain, and dry cotton wool in the oven.

k. Preparation of Thin-layer Plates

Wash the plates with an abrasive cleanser, rinse well, and dry in a paper towel. Place 40 g silica gel H in an Erlenmeyer flask, add about 90 ml water containing 0.2 ml 10% aqueous sodium carbonate, stopper flask, and shake well. Use this slurry to coat five chromatoplates (20 X 20 cm). When the gel sets to a matt finish, put the plates in a rack and activate by heating in an oven at 110°C for 2 hr. Store the plates in a desiccator cabinet.

l. Amniotic Fluid

Discard all samples contaminated with antiseptics, vaginal secretions, blood, or meconium. Invert the container of amniotic fluid a few times to ensure complete mixing. Do not centrifuge or filter before assay. Assay samples within 2 hr of collection or add methanol and chloroform as in Step a of the procedure and store as a one-phase system at 4°C.

5. Procedure

a. Extraction of Lipids from Amniotic Fluid

Pipette 0.8 ml amniotic fluid into a test tube,

add 2 ml methanol and 1 ml chloroform, and shake briefly. Add another 1 ml chloroform and mechanically shake the resultant two-phase system for 10 min. Centrifuge at 2,500 r.p.m. for 5 min at 0°C to facilitate phase separation.

After centrifugation, three layers are apparent: an upper aqueous methanolic layer, an intermediate white fluffy layer (denatured protein), and a lower chloroform layer. Remove the chloroform layer completely, taking care to avoid interfacial fluff, with a 2-ml syringe fitted with a long needle and filter through Whatman no. 1 paper into a conical tube. Reextract the aqueous methanolic layer twice using 1-ml portions of chloroform and pool the chloroform extracts. Place the tube in a water bath at 37°C and dry in a current of nitrogen. Concentrate the extract in the tip by washing the walls with small volumes of chloroform.

b. Chromatographic Separation of Lecithin from Contaminating Phospholipids

To the dry lipid residue add 120 μl of chloroform:methanol (2:1 v/v) and agitate for 1 min on a rotamixer to dissolve. Apply as a band (2 cm long and not more than 3 mm wide) about 2.5 cm from the base of a silica gel chromatoplate. A current of nitrogen directed over the band facilitates application. To ensure quantitative lipid transfer add further successive portions (60 and 20 μl) of solvent to the tube, rotamix, and transfer to the chromatoplate. Apply about 10 drops of the standard lecithin solution as spots on either side of this band.

Transfer the plate to a tank containing the developing solvent (about 100 ml) and allow the solvent to rise to within 3 cm from the top of the plate. Remove, allow to dry in air, and spray the plate with iodine solution. Outline the band parallel to that of the reference lecithin spots.

c. Elution of Lecithin from Silica Gel

Scrape the lecithin band from the plate after the iodine color has faded and transfer to a 2-ml syringe plugged at the base with a 3-mm layer of tightly packed washed cotton wool. Elute the lecithin from the silica gel by adding three consecutive 2.5-ml portions of eluting solvent. Make up to 10 ml.

d. Quantitative Determination of Lecithin

Transfer an aliquot of the eluate (1/4 to 1/2) to a test tube and dry in a current of nitrogen at 60°C. Digest the dry lecithin residue by heating with 0.3 ml perchloric acid (70%) for 40 min at 180° in a thermoblock (or silicone oil bath), then cool in ice. Prepare blanks and phosphate standards (triplicate) and add the indicated volumes of reagents for color development so that the final composition in each tube is as indicated.

Reagent (ml)	Test	Blank	Standard
Perchloric acid (70%)	(0.3)	0.3	0.3
Water	2.5	2.5	2.0
Phosphate standard	–	–	0.5 (= 2 μg)
Ammonium molybdate	1.0	1.0	1.0
Reducing agent	0.1	0.1	0.1
( ) acid added prior to digestion.			

Mix on a rotamixer, place the tubes in rack, and heat in boiling water bath for 7 min to develop the characteristic blue color. Cool in ice and read at 830 nm in a spectrophotometer. Subtract the mean blank reading from tests and standards.

6. Calculation

Calculate the lecithin concentration in mg per 100 ml amniotic fluid using the formula:

$$\frac{\text{O.D. test}}{\tfrac{1}{2}(\text{O.D. standard})} \times \frac{100}{0.8} \times \frac{10}{\text{vol. aliquot taken for phosphorus determination}} \times \frac{25^{a}}{1{,}000^{b}}$$

where

O.D. = optical density

a = factor converting μg phosphorus to μg lecithin based on the average molecular weight for lecithin of 775

b = converts μg lecithin to mg lecithin

7. Results

Recovery of dipalmitoyl lecithin added to a sample of amniotic fluid with known lecithin concentration (5.01 mg/100 ml) varied from 86.7 to 95.9% over the range of concentrations investigated (50 to 150 μg per 0.8 ml). The method reported is reproducible since replicate determinations (12) made on four amniotic fluid samples by two different workers assayed at 20.36 ± 0.31, 13.94 ± 0.7, 6.34 ± 0.29, and 4.01 ± 0.07 mg per 100 ml (mean ± 2 S.D.).

Storage of amniotic fluid appears to influence the lecithin content since the lecithin content of ten samples of amniotic fluid fell on average by 20.67% (range 12.1 to 28.1) after a week at -10°C. More prolonged storage for periods of up to 6 months caused no further reduction in the lecithin content. A reduction of from 7 to 10% in the lecithin content of amniotic fluid was observed when samples were allowed to stand at room temperature for periods of up to 3 hr.

Figure 1 shows the lecithin levels in 110 samples of liquor obtained 24 hr or less before delivery. In 97 infants, respiration was normal, and in each case the predelivery lecithin level had been 3.5 mg or more (range 3.5 to 37.5 mg) per 100 ml. In 13 infants, however, respiratory distress developed after birth; in 12 of these, the lecithin level ranged between 0.6 and 3.4 mg/100 ml. The infant who developed mild R.D.S. despite a lecithin concentration of 6.9 mg/100 ml had Pierre Robin's syndrome.

Table 1 shows the lecithin levels in 13 amniotic fluid samples collected 8 to 24 hr before delivery (except in case 2, where it was obtained 10 min before delivery) where infants subsequently developed respiratory difficulty. Predelivery lecithin concentration, therefore, appears to be closely related to the severity of R.D.S. in the infant. That 11 out of 13 cases are male is in keeping with the greater risk of R.D.S. developing in male infants.

Figure 2 compares the lecithin levels found in two or more serial samples of amniotic fluid obtained from ten patients taken 1 to 7 weeks apart. There was little change over the week in cases tested before 30 weeks. Rapid increases are seen in five cases tested after 32 weeks where the infant breathed normally. However, two cases where the infant subsequently developed R.D.S. were associated with a relatively slow increase.

The value of quantitative assay of lecithin in amniotic fluid has been confirmed by Nelson.[31,32]

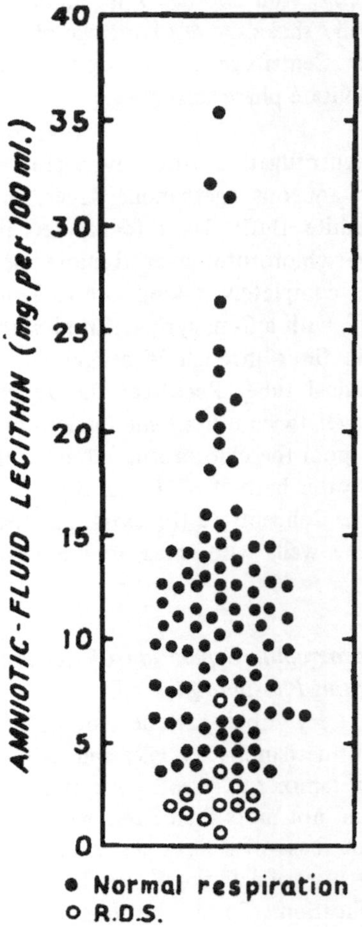

FIGURE 1. Amniotic fluid lecithin levels in 110 cases measured within 24 hr preceding delivery. (From Fahmy, D., *Lancet*, i, 159, 1972. With permission.)

III. PREDICTION OF NEONATAL R.D.S. BY DETERMINATION OF L/S RATIO

A. Method

1. Principle

The sphingomyelin content of amniotic fluid remains approximately constant throughout pregnancy, but from about 33 weeks gestation the concentration of lecithin rises sharply and continues to rise as term approaches. At 33 weeks L/S ratios of 1.5 or less are observed, while at term the ratio is 2 or more. Ratios of 1.5 or less, therefore, indicate immaturity of the fetopulmonary system and a high risk of R.D.S. if the fetus is delivered at this time.

TABLE 1

Pre-delivery Amniotic Fluid Lecithin Levels in 13 Cases Where the Infant Subsequently Developed R.D.S.

Case	Sex	Gestation (wk)	Clinical severity of R.D.S.	Lecithin (mg/100 ml)	Remarks
1	F	36	Doubtful, mild if at all	6.9	Pierre Robin's syndrome (cleft palate and micrognathia)
2	M	38	Transient and mild	3.4	Liquor collected at elective L.S.C.S.
3	M	38	Moderate	2.97	Mother had diabetes mellitus
4	M	38	Mild	2.60	
5	M	38	Mild	2.50	
6	M	37	Mild	2.45	
7	M	37	Mild	2.00	
8	M	38	Severe	1.90	Second twin-severe intrauterine anoxia; died after 82 hr
9	M	33	Mild	1.80	
10	M	35	Mild	1.64	
11	F	37	Severe	1.6	Intrauterine anoxia, failed Kielland's forceps followed by L.S.C.S.; died after 31 hr
12	M	35	Mild	1.4	
13	M	40	Severe	0.6	Mother prediabetic

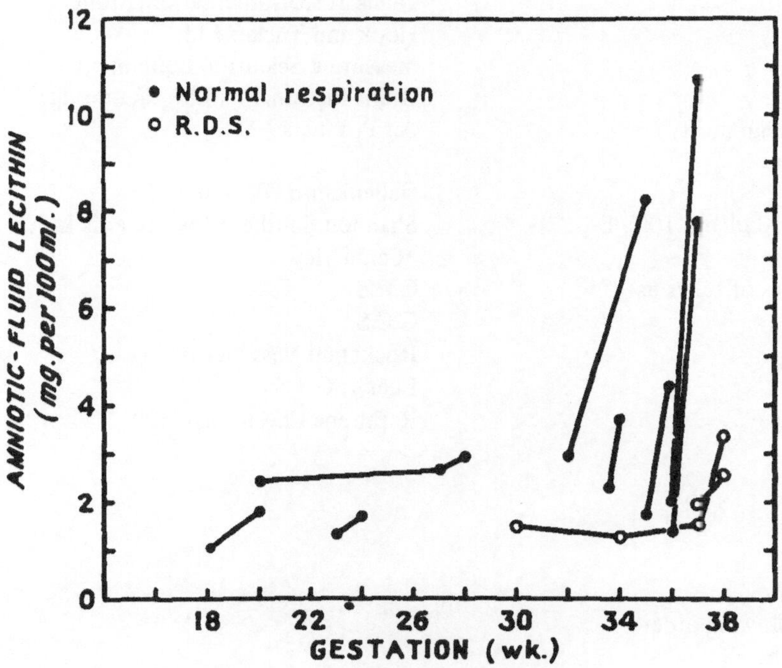

FIGURE 2. Serial estimations of amniotic fluid lecithin concentrations in ten cases including two cases where infants developed R.D.S. (From Fahmy, D., *Lancet*, i, 159, 1972. With permission.)

2. Materials

Materials	*Distributor and/or source*
Eastman chromatogram sheet 6061	Eastman Kodak Ltd., Liverpool
Silica gel without fluorescent indicator	
Bromothymol blue Na^+ salt	Eastman Kodak Ltd., Liverpool
Sodium hydroxide	Analytical reagent grade
Boric acid	Analytical reagent grade
Silica gel desiccant	G.C.S.
Acetone	Analytical reagent grade (redistilled)
Chloroform	Analytical reagent grade (redistilled)
Methanol	Analytical reagent grade (redistilled)
Ammonia	Analytical reagent grade
Water	Glass distilled

(Note: The water used throughout this procedure is all distilled and all glassware is rinsed thoroughly in distilled water before drying.)

3. Equipment

Equipment	*Source and catalog number*
Conical centrifuge tubes and stoppers	E-mil G 7723
Pasteur pipettes	E-mil DP 1009
Conical tubes (3 ml capacity)	Gallenkamp CJ 005
Large boiling tubes with stoppers	Jenkons MF 24/3/8
Dessicator	Gallenkamp DK 430
Glass still	Quickfit Instrumentation, Stone
Rotamixer	Hook and Tucker Ltd.
Centrifuge	Measuring Scientific Equipment
Water bath	Grant Instruments Ltd., Cambridge
Nitrogen cylinder (high purity) with reduction valve	Air Products
Oven	Gallenkamp OV-010
Hamilton syringes, 10 μl and 100 μl.	Shandon Southern Instruments Ltd., Camberley
Racks for three types of tubes used	G.C.S.
Small forceps	G.C.S.
Syringe (2 ml)	Rocket all-glass luer fitting
Needle	Luer 15G + 5
Vernier calliper gauge	Rathbone Chesterman 599

4. Reagents

a. Developing Solvent

Pipette in the following order into a boiling tube:

Methanol	5.0 ml
Water	0.8 ml – mix on rotamixer
Chloroform	13.0 ml – mix on rotamixer

Prepare fresh daily

b. Normal Sodium Hydroxide

Dissolve 4.0 g sodium hydroxide in 100 ml water.

c. Bromothymol Blue Detector

Dissolve 50 mg bromothymol blue Na^+ salt in 50 ml water

Add 8 ml *N* sodium hydroxide and dilute to 120 ml

Dissolve 1.25 g boric acid in this solution

(Note: This detector is stable for 3 months.)

d. Thin-layer chromatogram (TLC) Strips

Carefully cut 20 strips, 1 X 13 cm, from a sheet of silica gel with scissors and trim any uneven edges. Activate the strips by storage in a desiccator over silica gel desiccant for a minimum of 24 hr before use.

e. Amniotic Fluid

Discard all samples contaminated with antiseptics, vaginal secretions, blood, or meconium. Assay immediately if possible; alterations in the L/S ratio can occur after standing 1 hr at room temperature. If necessary store at -20°C. Centrifuge 3 ml amniotic fluid at 3,000 r.p.m. for 10 min. Assay all samples in duplicate, using 1-ml portions of this supernatant.

5. Procedure

a. Extraction of Lipids from Amniotic Fluid

Pipette 1 ml amniotic fluid into a conical tube, add 1 ml methanol, and rotamix vigorously for 15 sec. Add 2 ml chloroform and rotamix for 30 sec. Cool in ice for 2 min and then centrifuge at 3,000 r.p.m. for 10 min. After centrifugation, three layers are apparent, an upper aqueous methanolic layer, an intermediate white layer (denatured protein), and a lower chloroform layer. Transfer the chloroform layer with a 2-ml syringe fitted with a long needle to a smaller conical tube. Take care to avoid contamination with water or protein. Place in a water bath at 50 to 60°C and dry in a current of nitrogen. Concentrate the extract in the tip by washing the sides of the tube with 50 μl chloroform and redry.

b. Isolation of Surface Active Phospholipids

Cool the dry lipid residue in ice for 1 min, add 2 drops of ice cold acetone, and swirl the tube in ice. Add a further 8 drops cold acetone and stand in ice for 1 min. Decant the acetone and wash the cooled residue with another 10 drops of cold acetone. Decant the acetone and dry the residue in a current of nitrogen.

c. Chromatography of Phospholipids

Dissolve the dry residue in 6 μl chloroform. Apply as a spot (not exceeding 2 mm diameter) 2 cm from one end of a TLC strip. Lower the strip into the developing solvent (1.5 ml) in a boiling tube and stopper the tube. Remove the strip when the solvent front is 2 cm from the top of strip. Use forceps to facilitate handling.

d. Detection of Phospholipids

Air dry the strip then pass steadily through a dish of detector solution. Blot to remove the excess solvent and dry the strip in air or an oven at 70°C. Hold the strip above a beaker of ammonium hydroxide; phospholipids appear as dark blue spots on a pale blue background.

e. Interpretation of Test

The spot nearest the origin is sphingomyelin and the next is lecithin. (Note: Amniotic fluid contaminated with blood gives rise to a lysolecithin spot between sphingomyelin and the origin.) Measure the maximum length and width of the lecithin (L) and sphingomyelin (S) spots to the nearest 0.25 mm using calipers and a ruler. Calculate the ratio from formula:

$$\text{L/S Ratio} = \frac{\text{area of L spot}}{\text{area of S spot}} = \frac{\text{max. height X max. width of L spot}}{\text{max. height X max. width of S spot}}$$

Take the average of duplicate determinations and interpret results.

Average ratio	Interpretation	Risk of R.D.S.
1.5 or less	Immature lung	Very high – postpone delivery
1.51 – ≤2.0	Transitional lung	High – postpone delivery
2.0 or more	Mature lung	Negligible

TABLE 2

Comparative Value of L/S Ratio and Lecithin Concentration in Prediction of Neonatal R.D.S. in Cases Where the Amniotic Fluid was Collected Within 48 hr of Delivery

Type of assay	Number of samples	Number of cases with R.D.S.*	Accuracy of prediction (per cent)
L/S ratio			
< 1.5	13	10	76.9
1.5 ≤ 2.0	13	3	27.3
≥ 2.0	145	1	99.3
Lecithin concentration (mg/100 ml)			
< 3.5	11	11	100.0
> 3.5	109	–	100.0

B. Clinical Usefulness of the Lecithin Assay in Amniotic Fluid and the L/S Ratio Determination[33]

The L/S ratio and lecithin concentration in amniotic fluid collected within 48 hr of delivery are compared with the quality of neonatal respiration in 171 infants in Table 2. An L/S ratio of 1.5≤2 was observed in 13 cases. Only three of these infants developed R.D.S., indicating that the accuracy of L/S ratio for predicting the risk of R.D.S. was low (23.0%). Thirteen cases had a predelivery L/S ratio of <1.5; of these ten developed R.D.S. or had atelectatic hypoplastic lungs at necropsy. Thus, the accuracy of prediction in this group was good. When the ratio was 2 or more, the accuracy of prediction was even better, for of 145 cases only 1 developed R.D.S. These observations are similar to those of most other workers.[34,35]

Problems arise when the L/S ratio is between 1.5 and ≤2.0; with this ratio delivery would be postponed in all cases, as the risk of R.D.S. developing in the infant is too high to be ignored. This delay would have been unnecessary in the ten who breathed normally and could well have been dangerous if the fetus was in jeopardy. By contrast, the lecithin concentration was above 3.5 mg/100 ml in all cases where the infant breathed normally and below 3.5 mg/100 ml in those who developed R.D.S.

An L/S ratio of <1.5 observed near term may be associated with normal neonatal respiration due to unusually high sphingomyelin concentrations in the amniotic fluid.[31,33]

Gluck et al.[23] claim that acetone precipitation is an important step in determining the L/S ratio in human amniotic fluid. However, later studies indicate that there is no significant difference in the ratio when acetone precipitation is omitted.[33,36]

IV. PREDICTION OF NEONATAL R. D. S. BY A RAPID NEW TEST FOR SURFACTANT IN AMNIOTIC FLUID

A. Method[27]

1. Principle

Pulmonary surfactant, present in amniotic fluid, forms stable films capable of supporting the structure of a foam for long periods. Interference by other substances present in amniotic fluid such as proteins, bile salts, or salts of free fatty acids, which can also form stable foams, is excluded by presence of the nonfoaming competitive surfactant, ethanol.

The presence of a complete ring of bubbles at a 1:2 dilution of amniotic fluid denotes a positive result and indicates fetopulmonary maturity. Absence of such a ring at a 1:1 dilution indicates a negative result and immaturity of the fetal lungs. An intermediate result is obtained when the test is positive at 1:1 and negative at a 1:2 dilution of the amniotic fluid. The risk of R.D.S. developing in this group is too high to be ignored.

2. Materials	*Source and/or grade*
Absolute ethanol	James Burroughs Ltd.
Sodium chloride	Reagent grade
Dipalmitoyl lecithin (Cat. no. L8878)	Sigma, London, Kingston-upon-Thames
8 X 100-mm glass tubes with glass stoppers	Jenkons (Cat. no. MF 24/0/4).
Anglepoise lamp	
Stop clock	Gallenkamp TR100
Graduated pipettes	G.C.S.
Measuring cylinder	G.C.S.
Rack for tubes	G.C.S.
Glass still	Quickfit Instrumentation, Stone
Water	Glass distilled

(Note: Glassware must be kept scrupulously clean and rinsed thoroughly in distilled water before drying.)

3. Reagents

a. 95% Ethanol

Add 10 ml water to 190 ml absolute ethanol. Keep bottle tightly closed except during pipetting.

b. Normal Saline

Dissolve 0.9 g sodium chloride in 100 ml water. Store in a glass-stoppered bottle.

c. Standard Lecithin Solution

Dissolve 1 mg dipalmitoyl lecithin in 10 ml 95% ethanol. Store at 4°C and discard after 1 week. (Note: Care must be taken to ensure the lecithin is completely dissolved.)

d. Amniotic Fluid

Discard all samples contaminated with antiseptics, vaginal secretions, blood, or meconium. Invert the container of amniotic fluid several times to ensure complete mixing. Do not centrifuge or filter before assay. Assay the samples as soon as possible or store at 4°C for a maximum of 5 hr or at -20°C for longer periods.

4. Procedure

Pipette varying amounts of amniotic fluid (1.00, 0.75, 0.50, 0.25, and 0.2 ml) into 8 X 100 mm glass tubes and make up to a fixed volume of 1 ml with normal saline. Add 1 ml 95% ethanol to all tubes, stopper tightly, and shake as vigorously as possible by hand for an accurately timed interval of 15 sec. Place the tubes vertically in a rack in the order in which they were pipetted. To avoid disturbing the foam, do not move or touch the tubes until the test is complete. Exactly 15 min later, examine the air-liquid interface in each tube for the presence of small stable bubbles. To facilitate examination, use an Anglepoise lamp to give bright overhead illumination and view the tubes against a matt black screen placed just behind the rack.

Record a positive result for each tube showing a complete ring of bubbles in the meniscus. If the ring is absent or incomplete record a negative result for that tube. (Note: If the readings are missed, do not attempt to retrieve the results by shaking the tubes a second time; the procedure should be repeated.)

Tube no.	Amniotic fluid dilution	Record of bubbles in each tube					
1	1:1	+	+	+	+	+	–
2	1:1.3	+	+	+	+	–	–
3	1:2	+	+	+	–	–	–
4	1:4	+	+	–	–	–	–
5	1:5	+	–	–	–	–	–
	Result of test	Positive at 1:5 dilution	Positive at 1:4 dilution	Positive at 1:2 dilution	Intermediate	Negative	

5. Results

The usual laboratory studies of recovery, accuracy, and precision are excluded in this method. The only possible check is that the result obtained with 1 ml of a standard lecithin solution diluted with 1 ml normal saline should be comparable with the 1:2 dilution of full-term amniotic fluid.

6. Interpretation of Results

The reliability of the procedure in clinical situations has been assessed by Clements et al.[27] in 93 patients who delivered liveborn infants within 24 hr after the test. All of the 12 patients with a negative result delivered infants that developed severe R.D.S. or transient respiratory distress. Of 68 patients with clearly positive results, all delivered infants who breathed normally. In the 13 cases having an intermediate result, 8 of the infants showed mild to severe respiratory difficulty.

Table 3 compares the results obtained with the bubble stability test and the lecithin concentration in the same sample of amniotic fluid with the quality of neonatal respiration in 80 cases delivered within 48 hr of collection of the sample.[37] It would appear that the bubble stability test gives far too high a proportion of negative and intermediate results suggesting a potential risk of R.D.S. which did not exist.

Although all infants delivered within 48 hr of a positive bubble test had normal respiration, it is noteworthy that no positive test was ever obtained before the end of 36 weeks of gestation. By comparison, a lecithin level above 3.5 mg/100 ml, found within 48 hr of delivery, at any stage of gestation was associated with normal breathing in the neonate.[24]

V. FETAL ACID-BASE BALANCE

A. Introduction

In England and Wales the most common cause of perinatal mortality is intrapartum anoxia, which accounts for 22.9% of fatalities.[38] The fetus responds to intrapartum hypoxia by deriving its energy from anaerobic glycolysis, resulting in metabolic acidosis and a fall in blood pH.

Babies born in poor clinical condition due to hypoxia have long been known to exhibit metabolic acidosis at birth[39] and the degree of neonatal asphyxia correlates well with the severity of the acidosis.[40] However, some acidotic babies are born in good clinical condition and the concept that maternal infusion of the fetus with acidic metabolites (such as lactate) has been advanced.[41-43] It is currently accepted that under normal circumstances, the acid-base status of the infant at delivery is largely determined by that of the mother.[44] Fetal scalp blood sampling during labor was devised by Saling[45] to investigate the acid-base status of the fetus and, hence, its degree of oxygenation. The changes in acid-base values in fetal scalp blood have been shown by means of well-designed animal experiments to be reflective of the acid-base status of the fetus as a

TABLE 3

Comparison Between Amniotic Fluid Lecithin Concentration and Bubble Stability in Ethanol in the Antenatal Prediction of Potential R.D.S. in 80 cases

Test	No. of cases	No. of cases with normal respiration	No. of cases with R.D.S.*	Accuracy of prediction (%)
Lecithin concentration (mg/100 ml)				
<3.5	72	72	–	100.0
< 3.5	8	–	–	100.0
Bubble stability in ethanol				
Negative	25	19	6	24.0
Intermediate	18	16	2	11.1
Positive	37	37	–	100.0

whole.[46,47] It is now established that a fetal scalp blood pH of less than 7.20 is usually associated with fetal hypoxia.[48] For routine clinical work a properly taken scalp sample for pH estimation is often all that is required.[49] However, a more precise measurement of the state of the fetus needed for some research techniques requires in addition estimations of PCO_2 and base-excess on both fetal scalp and maternal arterial or "arterialized" capillary blood to be made[44] (maternal venous blood is unsuitable material for acid-base measurements).[50]

The Astrup technique[51] for measurement of acid-base balance is the one most widely used in the U.K. and the most frequently used apparatus is the Radiometer Astrup microequipment (model A.M.E.I.). For these reasons, the following methods are described preferentially.

The technique of scalp sampling briefly is as follows. The parturient patient is placed in the lithotomy position and a conical open ended endoscope is passed through the cervix to rest evenly against the fetal head; if the membranes are intact, these are ruptured. The scalp is dried, sprayed with ethyl chloride to provoke local hyperemia, and smeared with silicone grease. Small stab incisions (usually two), are made by means of a 2-mm plastic-mounted blade and the droplet of blood is directly collected into a preheparinized capillary tube.

In normal labor, the maternal arterial/fetal scalp difference in PCO_2 is about 20 mm Hg ±5 mm Hg, the maternal values vary from about 30 mm Hg at the onset of labor to about 25 mm Hg at full cervical dilatation; the fetal scalp PCO_2 values vary from 50 to 55 mm Hg during labor. The upper limit of normal for fetal scalp PCO_2 is about 60 mm Hg.

Maternal arterial base-excess values vary from about -1.5 mEq/l. ±1.0 mEq/l. at the onset of labor to -6 mEq/l. ±3 mEq/l. at delivery. Fetal scalp base-excess closely approximates that of maternal arterial blood during labor (±2 mEq/l.). Maternal-fetal differences greater than 2 mEq/l. are generally indicative of some degree of fetal acidosis which is considered severe when greater than 10 mEq/l.

The maternal-fetal pH difference is usually of the order of 0.1 pH; values greater than this are indicative of fetal acidosis.

1. The Determination of Whole Blood pH

a. Apparatus

Radiometer Type A.M.E.I., Astrup microequipment

Preheparinized glass capillary tubes (Rocket Ltd.) for fetal scalp blood (0.2-ml capacity)

Steel wire (0.5 cm lengths)

Magnet

Mercury barometer

b. Reagents

Precision buffer solutions (Radiometer) I pH 7.383, II pH 6.841

Distilled water

1 cylinder of 4% ±0.5% CO_2/96% O_2 (which has been previously accurately analyzed using the Haldane gas analysis apparatus)

1 cylinder of 8% ±0.5% CO_2/92% O_2 (which has been previously accurately analyzed using the Haldane gas analysis apparatus)

Note: All specimens of blood are heparinized.

c. Blood Samples

The pH must be read immediately. If delay is anticipated, the samples are then resealed and may be stored in ice water or in a refrigerator at 4°C for up to 40 min.

d. Procedure – Calibration of the pH Meter

1. Check that the water temperature in the circuit is 38 ± 0.3°C.
2. Check that the pool of KCl in the calomel electrode is full — top up if necessary.
3. Fill the electrode with distilled water.
4. Dry the electrode by air aspiration.
5. Break open a NEW ampoule of Radiometer precision buffer (pH 7.383).
6. Fill the electrode with the buffer.
7. Insert electrode tip into the KCl pool of the calomel electrode, place the electrode on its stand in the 'READ' position.
8. Wait until the needle on the pH meter has settled, readjust if necessary to the pH of the buffer using the fine adjustment dial.

9. Repeat procedures 3 to 8 using buffer (pH 6.841).

10. The span of the meter, if inappropriate to record accurately at the pH of both buffers, can be increased or decreased using the span adjusting dial. (Once the machine has been calibrated as above, a *full* calibration is not necessary for several days, but the meter must be recalibrated using the 7.380 buffer for each series of determinations.)

11. Rinse the electrode through with distilled water and dry with air.

12. Unseal the specimen tube and agitate the blood by inserting the steel rod and running it up and down the column gently using the magnet on the outside of the tube.

13. Remove the rod using the magnet.

14. Fill the electrode with blood, taking care that no bubbles are admitted.

15. Read and record the pH.

16. Rinse electrode with buffer.

17. Repeat procedures 11 to 16 to obtain a duplicate reading.

18. Fill electrode with distilled water and place the electrode on its stand in the 'REST' position with its tip immersed in distilled water.

2. Determination of Whole Blood Base-excess

a. Method

1. Determine the pH of the sample as in d (1 to 18).

2. Check that the four glass tonometer tubes are clean.

3. Turn on both cylinders of O_2/CO_2 mixture.

4. Adjust until the gases are bubbling steadily through each humidifier (avoid turning the gases on too suddenly or the tonometers might flood).

5. Put one drop (0.1 ml) of the specimen into each of the four tonometer tubes (two for each mixture).

6. Turn the vibrator on and equilibrate for 3 min.

7. Switch the vibrator off.

8. Determine the pH of the equilibrated samples in duplicate and record the two mean values, i.e., mean pH with 4% CO_2 (pH 4%) and mean pH with 8% CO_2 (pH 8%).

9. Record the ambient barometric pressure.

b. Calculation

1. The partial pressure of CO_2 within the tonometers is given by the formulas

Ambient barometric pressure – saturated vapor pressure of water = y (at the ambient temperature of the apparatus).

$$\frac{\text{Exact percentage of } CO_2 \text{ in "4\%" cylinders} \times y}{100} = \text{partial pressure of } CO_2 \text{ in "4\%" tonometers } (PCO_2 \text{ "4\%") in mm Hg} \qquad (a)$$

$$\frac{\text{Exact percentage of } CO_2 \text{ in "8\%" cylinders} \times y}{100} = \text{partial pressure of } CO_2 \text{ in "8\%" tonometers } (PCO_2 \text{ "8\%") in mm Hg} \qquad (b)$$

In Equations a and b "4%" and "8%" CO_2 values are approximate. It is stressed that in practice one uses the exact CO_2 percentage actually in the gas cylinders which is known from previous accurate gas analysis.

2. The values PCO_2 "4%" and PCO_2 "8%" are drawn on the Siggaard-Andersen nomogram as two horizontal lines at their respective PCO_2 values (lines A–A and B–B in Figure 1).
Figure 1).

The pH values of the samples equilibrated at "4%" and "8%" CO_2 are entered on their respective PCO_2 lines (points α and β in Figure 1) at the recorded pH value. The values of buffer base, base-excess, and standard bicarbonate are read off from the nomogram at the points where the lines passing through α and β intersect the appropriate scale (see Figure 3).

Note: These values are "fully O_2 saturated" values as recommended by Crawford and Holaday[53] and it is important to specify this in any report.

3. Determination of Whole Blood PCO_2

a. Apparatus

Oxygen saturation meter (Radiometer O.S.M.I.)

Microhaematocrit Centrifuge (Hawksley and Sons Ltd., Lancing, Sussex)

Microhaematocrit Reader (Hawksley and Sons Ltd., Lancing, Sussex)

Microhaematocrit Tubes (Hawksley and Sons Ltd., Lancing, Sussex)

Sealing compound ("Crystaseal") (Hawksley and Sons Ltd., Lancing, Sussex)

Glass ampoule file

Refrigerator with freezing compartment

b. Procedure for Arterial Blood (95% O_2 Saturated)

1. Determine the whole blood pH as in 1.
2. Determine the whole blood base-excess as in 2.
3. Draw a vertical line on the Siggaard-Andersen nomogram from the actual pH value to intercept the line already drawn to calculate base-excess, read off the PCO_2 (fully saturated) value at the side of the chart.

c. Procedure for Blood Less than 95% Oxygen Saturated

1. Determine whole blood pH as in 1.

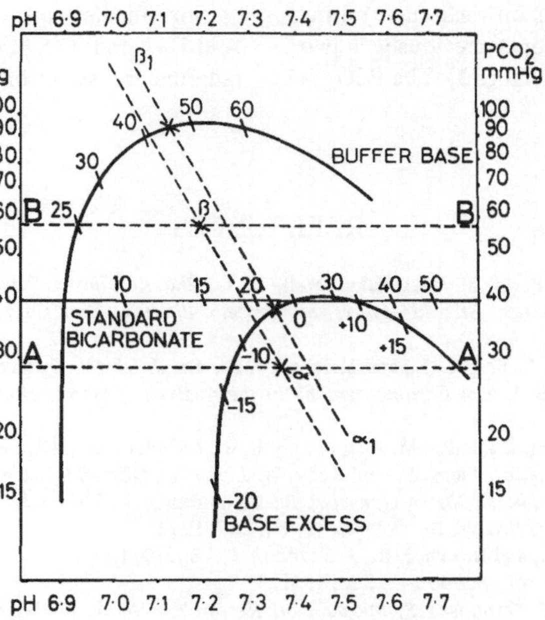

FIGURE 3. The Siggard-Andersen curve nomogram showing the derivation of acid base values. In the illustration for a specimen with a whole blood pH of 7.300, the values would be

Buffer base = 41 mEq/l.
Base excess = −7 mEq/l.
Standard bicarbonate = 18 mEq/l.
PCO_2 = 48 mEq/l.

2. Determine base-excess as in 2.

3. Fill two microhematocrit tubes with well-mixed heparinized blood.

4. Seal one end of each tube with Crystaseal.

5. Centrifuge for 2 min.

6. Determine the hematocrit using the hematocrit reader.

7. Lyse the red cells by freezing the tubes in the freezing compartment of the refrigerator. (This may fracture the tubes unless they are placed on a flat surface.)

8. Calibrate the oxygen saturation meter by turning the red and green filter adjusting knobs until both read zero without the cuvette holder in the light path.

9. Take the tube from the refrigerator and cut off the segment containing the lysed packed cells.

10. Place the cuvette in the cuvette holder and cover with coverslip (use the 0.1 mm cuvette).

11. Fill the space between the coverslip and the cuvette with lysed packed red cells.

12. Place cuvette holder into O.S.M.I.

13. Read the optical density red (O.D. red) and read the optical density green (O.D. green).

14. Place a ruler between $O.D._{red}$ and $O.D._{green}$ on the nomogram for O.S.M.I. and read percentage oxygen saturation where ruler intersects the oxygen saturation percentage line.

d. Calculation

1. Calculate the correction factor for PCO_2 using the formula:

$$\frac{\text{Hematocrit (\%)} \times \text{percentage } O_2 \text{ desaturation } (100 - \text{\% oxygen saturation})}{100}$$

2. The correction factor is added to the buffer base and base-excess curves of the Siggaard-Andersen nomogram in the alkaline direction (the same number of scale marks on each curve) using as starting point the intercepts previously drawn for base-excess (line α–β in Figure 3). The PCO_2 is read from the interconnecting line (line α_1–β_1 in Figure 3) at the actual pH of the sample.

For further information about Radiometer A.M.E.I. and O.S.M.I., consult the appropriate radiometer instruction manuals.

REFERENCES

1. **Mikhail, G., Wu, C. H., Ferin, M., and Vande Weil, R. L.,** *2nd Karolinska Symposium on Research Methods in Reproductive Endocrinology, Steroid Assays by Protein Binding,* Diczfalusy, E., Ed., Karolinska Institute, Stockholm, 1970, 347.
2. **Tulchinsky, D., Hobel, C. J., Yeager, E., and Marchall, J. R.,** *Am. J. Obstet. Gynecol.,* 112, 1095, 1972.
3. **Wotiz, H. H. and Clark, S. J.,** *Gas Chromatography in the Analysis of Steroid Hormones,* Plenum Press, New York, 1966, 258.
4. **Brown, J. B., Macnaughton, C., Smith, M. A., and Smyth, B.,** *J. Endocrinol.,* 40, 175, 1968.
5. **Gurpide, E., Giebenhain, M. E., Tseng, L., and Kelly, W. J.,** *Am. J., Obstet. Gynecol.,* 109, 897, 1971.
6. **Corker, C. S. and Naftolin, F.,** *J. Obstet. Gynaecol. Br. Commonw.,* 78, 330. 1971.
7. **Masson, G. M.,** *J. Obstet. Gynaecol. Br. Commonw.,* 80, 201, 1973.
8. **Klopper, A., Michie, E. A., and Brown, J. B.,** *J. Endocrinol.,* 18, 319, 1959.
9. **Wilde, C. E.,** *Acta Endocrinol.,* Suppl. 142, 360, 1969.
10. **Johansson, E. D. B.,** *2nd Karolinska Symposium on Research Methods in Reproductive Endocrinology, Steroid Assays by Protein Binding,* Diczfalusy, E., Ed., Karolinska Institute, Stockholm, 1970, 188.
11. **Abraham, G. E., Swerdloff, R., Tulchinsky, D., and Odell, W. D.,** *J. Clin. Endocrinol.,* 32, 619, 1971.
12. **Varma, K., Larraga, L., and Selenkow, H. A.,** *Obstet. Gynecol.,* 37, 10, 1971.
13. **Letchworth, A. T., Boardman, R. J., Bristow, C., Landon, J., and Chard, T.,** *J. Obstet. Gynaecol. Br. Commonw.,* 78, 542, 1971.
14. **Samaan, N. A., Gallager, H. S., McRoberts, W. A., and Faris, A. M.,** *Am. J. Obstet. Gynecol.,* 109, 63, 1971.
15. **Cassady, G. and Barnett, R.,** *Am. J. Obstet. Gynecol.,* 108, 1010, 1970.
16. **Scopes, J.,** *Recent Advances in Paediatrics,* 4th ed., Gairdner, D. and Hull, D., Eds., Churchill, London, 1971.
17. **Scopes, J.,** *Br. J. Hosp. Med.,* 3, 579, 1970.
18. **Avery, M. E. and Mead, J.,** *Am. J. Dis. Child.,* 97, 517, 1959.

19. Reynold, E. O. R., Robertson, N. R. C., and Wigglesworth, J. S., *Pediatrics,* 42, 758, 1968.
20. Pattle, R. E. and Thomas, L. C., *Nature,* 189, 844, 1961.
21. Buckingham, S., McNary, W. F., and Sommers, S. C., *Science,* 145, 1192, 1964.
22. Gluck, L., Landowne, R. A., and Kulovich, M. V., *Pediatr. Res.,* 4, 352, 1970.
23. Gluck, L., Kulovich, M. V., Borer, R. C., Brenner, P. H., Anderson, G. G., and Spellacy, W. N., *Am. J. Obstet. Gynecol.,* 109, 440, 1971.
24. Bhagwanani, S. G., Fahmy, D., and Turnbull, A. C., *Lancet,* i, 159, 1972.
25. Bhagwanani, S. G., Fahmy, D., and Turnbull, A. C., *Lancet,* ii, 66, 1972.
26. Borer, R. C., Gluck, L., Freeman, R. K., and Kulovich, M. V., *Pediatr. Res.,* 5, 655, 1971.
27. Clements, J. A., Platzker, A. C. G., Tierney, D. F., Hobel, C. J., Creasy, R. K., Margolis, A. J., Thibeault, D. W., Tooley, W. H., and Oh, W., *N. Engl. J. Med.,* 286, 1077, 1972.
28. Gusdon, J. P. and Waite, B. M., *Am. J. Obstet. Gynecol.,* 112, 62, 1972.
29. Bayer, H., Bonnar, J., Phizackerley, P. J. R., Moore, R. A., and Wylie, F. J., *J. Obstet. Gynaecol. Br. Commonw.,* 80, 333, 1973.
30. Warren, C. H., Allen, J. T., and Holton, J. B., *Clin. Chim. Acta,* 44, 457, 1973.
31. Nelson, G. H., *Am. J. Obstet. Gynecol.,* 112, 827, 1972.
32. Nelson, G. H. and Lawson, S. W., *Am. J. Obstet. Gynecol.,* 115, 933, 1973.
33. Bhagwanani, S. G., Fahmy, D., and Turnbull, A. C., *Br. Med. J.,* in press.
34. Bryson, M. J., Gabert, H. A., and Stenchever, M. A., *Am. J. Obstet. Gynecol.,* 114, 208, 1972.
35. Whitfield, C. R., Chan, W. H., Sproule, W. B., and Stewart, A. D., *Br. Med. J.,* 2, 85, 1972.
36. Sarkozi, L., Kovacs, H. N., Fox, H. A., and Kerenyi, T., *Clin. Chem.,* 18, 956, 1972.
37. Bhagwanani, S. G., Fahmy, D., and Turnbull, A. C., *Br. Med. J.,* i, 697, 1973.
38. Perinatal Problems, *The Second Report of the British Perinatal Mortality Survey,* Butler, N. R. and Alberman, E. D., Eds., Edinburgh and London, Livingstone, 1969.
39. Eastman, N. J., *Bull. Johns Hopkins Hosp.,* 50, 1932.
40. James, L. S., Weisbrot, I. M., Prince, C. E., Holaday, D. A., and Apgar, V., *J. Pediatr.,* 52, 379, 1958.
41. Rooth, G. and Wilson, I., *Clin. Sci.,* 26, 121, 1964.
42. Rooth, G., *Lancet,* i, 290, 1964.
43. Vedra, B., *Am. J. Obstet. Gynecol.,* 88, 802, 1963.
44. Jacobson, L. and Rooth, G., *J. Obstet. Gynaecol. Br. Commonw.,* 78, 97b, 1971.
45. Saling, E., *Arch. Gynaekol.,* 197, 108, 1962.
46. Adamsons, K., Beard, R. W., Cosmi, E. V., and Myers, R. E., in *Diagnosis and Treatment of Fetal Disorders,* Adamsons, K., Ed., Springer-Verlag, New York, 1968, 175.
47. Gare, D. J., Whetham, J. C. G., and Henry, J. D., *Am. J. Obstet. Gynecol.,* 99, 722, 1967.
48. Beard, R. W. and Clayton, S. J., *J. Obstet. Gynecol. Br. Commonw.,* 74, 812, 1967.
49. Beard, R. W., Brudenell, J. M., Feroge, R. M., and Clayton, S. G., *J. Obstet. Gynecol. Br. Commonw.,* 78, 882, 1971.
50. Report of ad hoc cimmittee on acid-base terminology, *N.Y. Acad. Sci.,* 133, 251, 1966.
51. Astrup, P., Jørgensen, K., Siggaard-Andersen, O., and Engel, K., *Lancet,* 1, 1035, 1960.
52. Siggaard-Andersen, O., *Scand. J. Clin. Lab. Invest.,* 14, 598, 1962.
53. Crawford, J. S. and Holaday, D. A., *Lancet,* 1, 834, 1964.

ASSESSMENT OF TRYPTOPHAN METABOLISM AND VITAMIN B$_6$ NUTRITION IN PREGNANCY AND ORAL CONTRACEPTIVE USERS

D. R. Rose

TABLE OF CONTENTS

I. Introduction . . . 318
  A. The Tryptophan-nicotinic Acid Ribonucleotide Pathway . . . 318
  B. Biochemical Assessment of Vitamin B$_6$ Nutrition . . . 320
  C. Tryptophan Load Test . . . 320

II. General Considerations of Methods Available for the Determination of Urinary Tryptophan Metabolites . . . 322

III. Xanthurenic Acid Determination . . . 324
  A. Basis for Selection of Method . . . 324
  B. Determination of Xanthurenic Acid by Thin-layer Chromatography and Colorimetry . . . 324
  C. Other Tryptophan Metabolites Separated by Thin-layer Chromatography . . . 327
  D. Determination of Xanthurenic Acid by Ion-exchange Chromatography . . . 327

IV. 3-Hydroxykynurenine and 3-Hydroxyanthranilic Acid Determinations . . . 327
  A. Basis for Selection of Method . . . 327
  B. Determination of 3-Hydroxykynurenine and 3-Hydroxyanthranilic Acid by Ion-exchange Chromatography and Colorimetry . . . 328

V. Normal Range of Metabolite Excretions in Urine after Tryptophan Loading . . . 330

VI. Urinary 4-Pyridoxic Acid Excretion . . . 332
  A. Determination of 4-Pyridoxic Acid by Ion-exchange Chromatography and Fluorimetry . . . 332

VII. Plasma Pyridoxal Phosphate Assay . . . 334
  A. Determination by a Radiochemical Enzyme Assay . . . 334

VIII. Erythrocyte Alanine Aminotransferase . . . 338
  A. The Method . . . 338

IX. The Clinical Significance of Tryptophan Metabolic Studies and Tests of Vitamin B$_6$ Nutritional Status in Pregnancy and Oral Contraceptive Users . . . 341
  A. Pregnancy . . . 341
  B. Oral Contraceptive Users . . . 342

Acknowledgments . . . 346

References . . . 346

I. INTRODUCTION

The administration of estrogen-progestogen preparations for contraceptive purposes is accompanied by a multitude of metabolic side effects.[1] Although there is general agreement that these biochemical changes take place, their clinical significance is less certain. For example, it remains to be established whether or not oral contraceptive-treated women who show abnormal glucose tolerance may, in later life, develop the complications of diabetes mellitus; similarly, the risks of contraceptive steroid-induced changes in lipid metabolism are still unclear.

In addition to the much studied alterations in carbohydrate, lipid, and protein metabolism, recent investigations have shown that oral contraceptives affect vitamin nutrition. The changes observed include reduced levels of folic acid, vitamin B_{12} and ascorbic acid in plasma, and increased plasma vitamin A levels.

The first clue that vitamin metabolism may be disturbed in oral contraceptive-treated women was the observation that these steroids produce an abnormality of tryptophan metabolism, which is similar to that seen in vitamin B_6 deficiency, and is reversed by treatment with large doses of pyridoxine. Subsequent investigations have extended the initial findings and provided direct evidence that true vitamin B_6 deficiency does develop in a minority of oral contraceptive users. The situation is very like that which occurs during the last trimester of pregnancy, when tryptophan metabolism is abnormal and vitamin B_6 levels reduced.

The clinical importance of this work is now emerging from data which suggest that lack of available vitamin B_6 may be at least one factor in the development of some side effects of the estrogen-containing oral contraceptives.

Because only a minority of oral contraceptive users become vitamin B_6 deficient, it is necessary to apply suitable methods to the detection of this group. An alternative would be to administer large doses of pyridoxine to all women who take these steroids, but the multiple effects that this may have on enzyme systems and metabolic pathways have not yet been assessed and could conceivably produce problems of their own.

In this chapter, most of the methods that have been published for the determination of tryptophan metabolites in urine will be reviewed and selected procedures will be described in detail. The assessment of vitamin B_6 nutritional status will also be considered, with particular reference to urinary 4-pyridoxic acid and plasma pyridoxal phosphate determinations, and the use of erythrocyte alanine aminotransferase assays. Finally, the interpretation of the data obtained by these methods in pregnant and oral contraceptive-treated women will be discussed and an appraisal made of their clinical significance.

A. The Tryptophan-nicotinic Acid Ribonucleotide Pathway

The metabolic pathway by which L-tryptophan is converted to nicotinic acid ribonucleotide is shown in Figure 1. The first step, rupture of the indole ring to yield formylkynurenine, is catalyzed by tryptophan oxygenase (tryptophan pyrrolase, E.C. 1.13.1.12). Elevated levels of this enzyme are induced by adrenocorticosteroids; treatment with hydrocortisone increases both hepatic tryptophan oxygenase activity and urinary kynurenine excretion in man.[2]

Several of the other enzymatic reactions require pyridoxal-5-phosphate as a coenzyme. Of these, the kynureninase-mediated step concerned with the further metabolism of 3-hydroxykynurenine appears to be more sensitive to a lack of the coenzyme than the transamination reaction which yields xanthurenic acid; consequently, vitamin B_6 deficiency is accompanied by very high levels of xanthurenic acid, 3-hydroxykynurenine, and kynurenine in urine collected after an oral dose of L-tryptophan.[3] In addition, the excretions of 3-hydroxyanthranilic acid[4] and quinolinic acid[5,6] are increased in experimental human vitamin B_6 deficiency, suggesting that there is an unidentified role for pyridoxal-5-phosphate beyond the kynureninase reaction.

The principal urinary derivatives of nicotinic acid are N^1-methylnicotinamide and N^1-methyl-2-pyridone-5-carboxamide (Figure 1). A proportion of these metabolites is derived from dietary nicotinyl compounds, but the extent to which their excretion is increased after a dose of L-tryptophan provides an index of the capacity for conversion of the amino acid to nicotinic acid ribonucleotide.

The tryptophan-nicotinic acid ribonucleotide metabolic pathway is of interest to the clinician and clinical biochemist for a number of diverse reasons. Several of the intermediates, including

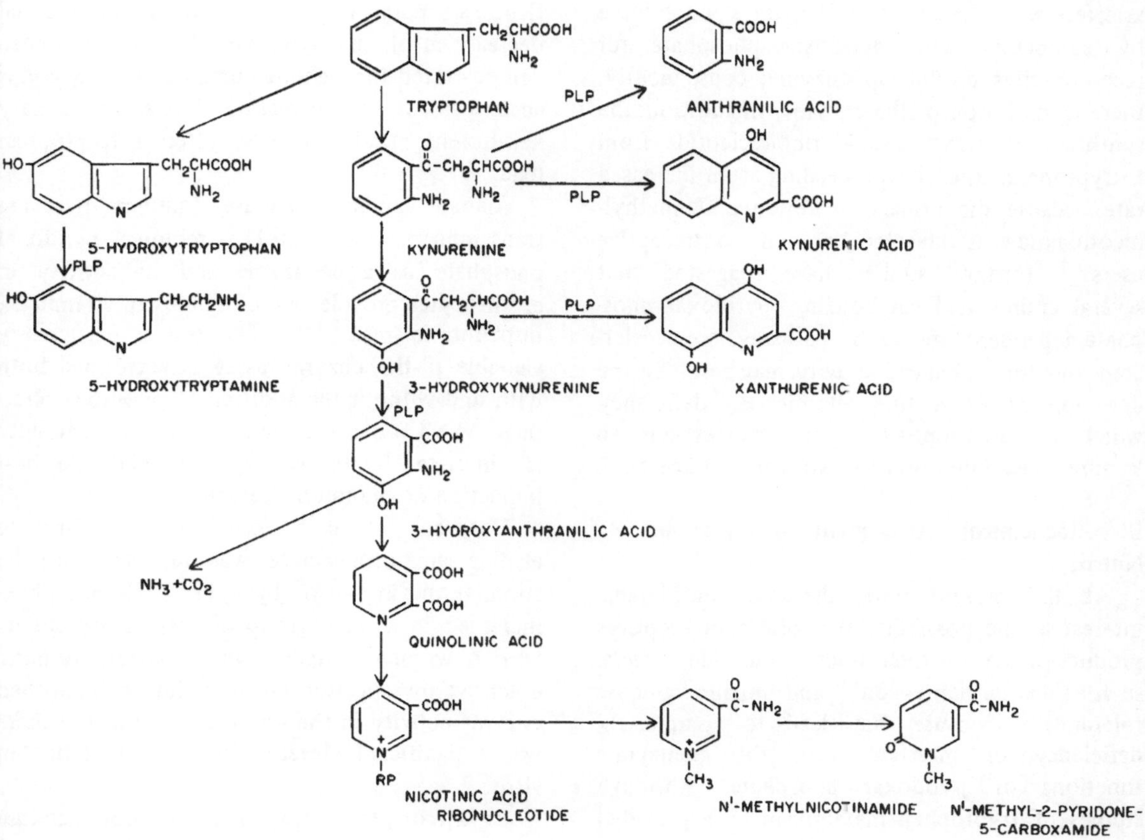

FIGURE 1. Metabolic pathways for the biosynthesis of nicotinic acid ribonucleotide and 5-hydroxytryptamine from L-tryptophan (PLP = pyridoxal phosphate-dependent reaction).

3-hydroxykynurenine and 3-hydroxyanthranilic acid, induce tumors in mice and are suspected of having a role in human bladder cancer.[7] An elevated urinary excretion of tryptophan metabolites occurs in some women with breast cancer, a hormone-responsive tumor, and there is an inverse correlation between the level of these compounds and urinary androgens in such patients.[8] The determination of tryptophan metabolites, particularly xanthurenic acid, after an oral load of the amino acid, has been used extensively for the assessment of vitamin B_6 nutrition in man,[9] and normal pregnancy is accompanied by changes in tryptophan metabolism which are similar to those of vitamin B_6 deficiency and are eliminated by pyridoxine administration.[10]

In 1966, it was reported that women taking oral contraceptives had an elevated xanthurenic acid excretion after the ingestion of a tryptophan load and that this metabolic defect was corrected by vitamin B_6 administration.[11] This observation has been confirmed and it is now known that the urinary excretions of kynurenine, 3-hydroxykynurenine, 3-hydroxyanthranilic acid,[12,13] and quinolinic acid[6] are all increased by estrogen-progestogen preparations and become normal after pyridoxine administration. An identical abnormality is produced by treatment with an estrogen alone.[12]

Recent studies indicate that although all of the available estrogen-containing oral contraceptives may cause abnormal tryptophan metabolism, some do so less frequently than others and when they do have an effect, the abnormality is relatively mild.[14] This appears to be because certain progestogens can reduce the action of the estrogen, rather than being due to differences in the nature or dose of the estrogen itself.

The possible mechanisms by which estrogens alter tryptophan metabolism have been discussed in recent publications;[15,16] the reader is referred to these for a detailed consideration of this problem. In brief, there is experimental evidence that estrogen conjugates formed in the liver interfere with the activity of hepatic kynureninase by competing with pyridoxal phosphate for receptor sites on the apoenzyme; consequently, there is inhibition of the enzyme. In addition, the syntheses of nicotinic acid ribonucleotide from L-tryptophan must be proceeding at an increased rate because the urinary excretion of *N*-methylnicotinamide is elevated in oral contraceptive users.[17] Further studies have suggested that several amino acid catabolizing, pyridoxal phosphate-dependent, metabolic pathways are accelerated and this enhanced activity may result in the development of a true vitamin B_6 deficiency which is superimposed upon changes due to kynureninase inhibition by estrogen conjugates.[16]

B. Biochemical Assessment of Vitamin B_6 Nutrition

At the present time, there is considerable interest in the possibility that oral contraceptives produce some of their undesirable side effects, such as mental depression[18] and impaired glucose tolerance[19] because they lead to vitamin B_6 deficiency or interfere with the coenzymic functions of pyridoxal phosphate. Although studies of tryptophan metabolism have provided pointers toward this possibility, assessment of vitamin B_6 nutrition in women receiving these preparations also requires the application of methods which are not affected by enzyme induction or other hormonal effects. Three of these will be considered in detail.

The major urinary metabolite of vitamin B_6 in man is 4-pyridoxic acid (2-methyl-3-hydroxy-4-carboxy-methylpyridine). Its determination provides an index of dietary intake of the vitamin relative to the body's requirements and excretion falls rapidly when a vitamin B_6-deficient diet is fed to human subjects.[20] In pregnancy the urinary 4-pyridoxic acid excretion after a test dose of pyridoxine hydrochloride is lower than in normal, nonpregnant women;[21] some oral contraceptive users also excrete reduced levels of the metabolite and usually have other biochemical abnormalities that are consistent with vitamin B_6 deficiency.[16]

The level of pyridoxal phosphate in plasma falls rapidly when a normal subject is placed on a vitamin B_6-deficient diet and returns to predepletion levels sooner than most other indices of deficiency. In the past, direct assay of the coenzyme has not been used to any great extent for the detection of deficiency because of technical difficulties. With improved methods, a gradual decrease in plasma pyridoxal phosphate has been demonstrated throughout human pregnancy and is accompanied by increasing levels of urinary xanthurenic acid excretion after a tryptophan load.[22]

Alanine aminotransferase (glutamic-pyruvate transaminase, E.C. 2.6.12) requires pyridoxal phosphate as a coenzyme and its activity in erythrocytes provides a useful index of vitamin B_6 nutrition in man.[23,24] The test is particularly valuable if the enzyme assay is performed both with and without the addition of pyridoxal phosphate in vitro because the extent to which activity is stimulated by the coenzyme provides the best indication of vitamin B_6 deficiency in vivo.[23]

Doberenz et al.[25] found that erythrocyte alanine aminotransferase was reduced and the stimulation in vitro by pyridoxal phosphate increased in a small group of oral contraceptive-treated women compared with controls. A more extensive investigation failed to detect diminished enzyme activity in the group as a whole, but there was a significant elevation in the stimulation in vitro.[26]

Erythrocyte aspartate aminotransferase (glutamic oxaloacetic transaminase, E.C. 2.6.11) activity has also been used to assess vitamin B_6 nutrition,[27,28] but this enzyme is less sensitive to mild degrees of deficiency than alanine aminotransferase.[16,23] A further disadvantage of employing this assay is that oral contraceptives actually *increase* the enzyme level,[26,29] probably because the steroids interact with the apoenzyme protein to stabilize it and so reduce the rate of enzyme degradation. A similar response has been described in the pregnant rat.[30] Because of this complicating hormonal effect, the erythrocyte aspartate aminotransferase assay is not recommended for the investigation of vitamin B_6 nutritional status in pregnant or oral contraceptive-treated women.

C. Tryptophan Load Test

Most of the studies of urinary tryptophan metabolite excretions in man have been made after administering an oral dose of the amino acid. Without prior tryptophan loading ("spontaneous excretion"), available methods fail to demonstrate elevated excretions of xanthurenic acid, or most other metabolites, either in human vitamin B_6 deficiency,[3] or oral contraceptive-treated women.[13] Brown et al.[10] have reported increased spontaneous excretions of kynurenine, 3-hydroxykynurenine, and xanthurenic acid in the urine of pregnant women, but the abnormalities were accentuated by the administration of a 2-g L-tryptophan load. Price et al.[4] applied a highly sensitive thin-layer electrophoresis method[31] to the determination of 3-hydroxyanthranilic acid in urine collected without tryptophan loading from a vitamin B_6-deficient subject, and from a group of oral contraceptive-treated women. They did find elevated levels of this metabolite compared with control values, suggesting, perhaps, that the existing methods for determining the other metabolites lack the necessary sensitivity to demonstrate small increases in excretion.

1. The Tryptophan Load

Before L-tryptophan became readily available in large quantity, the racemic mixture was used for tryptophan load tests. Although DL-tryptophan is less expensive, the D-form is not metabolized through the tryptophan-nicotinic acid ribonucleotide pathway, but only as far as D-kynurenine. This accumulates and is excreted in the urine, giving very high levels of urinary kynurenine.[32] Although D-kynurenine does not appear to interfere with the further metabolism of L-kynurenine, it seems advisable to avoid the use of tryptophan containing the nonphysiological isomer and it is recommended that L-tryptophan be administered for all metabolic studies.

2. The Dose of L-Tryptophan

A plea for the use of a standard 2-g L-tryptophan oral dose has been made by Coursin[33] and Price et al.;[34] the adoption of their recommendation would certainly aid the comparison of data between laboratories. The 2-g dose was selected after careful investigation because it produced a consistent, but not excessive, increase in urinary metabolites above their spontaneous excretion when given to normal subjects.[32,34] It also distinguishes between normal and vitamin B_6-deficient subjects;[3] control women and those who are pregnant[10] or using estrogen-containing oral contraceptives,[13] and between healthy subjects and patients with a variety of diseases.[32]

Although a 2-g L-tryptophan dose is satisfactory for most purposes, it does not demonstrate a sex difference in the excretion of metabolites; for studies concerned with the influence of endogenous estrogens on tryptophan metabolism a 5-g oral load is recommended.[35]

The tryptophan may be administered in tablet form, 500 mg each being a convenient size,[32] or, alternatively, the powder may be given as a suspension in milk.

3. Side Effects of the Tryptophan Load

Occasional subjects develop a sensation of somnolence 2 to 3 hr after taking a 2-g L-tryptophan load. These symptoms can usually be avoided by taking the amino acid at the same time as breakfast. The cause of these effects is unclear; they do not appear to result from hypoglycemia. Slight nausea is a fairly frequent side effect of a 5-g dose; occasionally this is more severe and may result in vomiting, but probably with no more frequency than occurs during oral glucose tolerance tests. The author has carried out several hundred 5-g tryptophan load tests without untoward effects, but their use is not recommended in very ill patients.

4. The Urine Collection

a. Preservatives

Urine collections should be acidified to pH 3 to 4, both to reduce the likelihood of bacterial growth and to stabilize metabolites, particularly 3-hydroxyanthranilic acid, which are labile in an alkaline medium. Some workers collect the urine into toluene,[34,36] but this is not essential and the urine still needs to be acidified. When the collection is completed, samples may be refrigerated at 0 to 4°C if they are to be analyzed within a few days; if not they should be frozen at −15°C. When stored in a frozen state, the tryptophan metabolites are stable for several weeks, but after 8 to 12 weeks, significant losses of 3-hydroxykynurenine and 3-hydroxyanthranilic acid do occur.

b. Period of Collection

There is a circadian rhythm in the levels of tryptophan metabolites excreted after an oral dose of the amino acid, the output being approximately three times higher when the load is given at 0900 hr than at 2100 hr.[37] Therefore, the tryptophan should always be given at about the same time of day; usually with breakfast at 0700 to 0800 hr is convenient.

Urine collections are most frequently made for a complete 24-hr period; this is essential if the levels of N^1-methylnicotinamide or N^1-methyl-2-

pyridone-5-carboxamide excreted in response to a tryptophan load are to be determined, as these metabolites are excreted maximally in the second 8-hr period after the dose (Rose, D. P. and Strong, R., unpublished observations). However, if interest lies in the determination of xanthurenic acid, 3-hydroxykynurenine, and 3-hydroxyanthranilic acid, the urine may be collected during the first 7 or 8 hr after tryptophan administration[38] because the bulk of the postloading rise in the excretion of these metabolites occurs in this period (Figure 2).

5. *Interference with Tryptophan Metabolic Studies by Drugs*

Drugs may cause erroneous results because either they modify tryptophan metabolism or interfere with the analytical procedures. Multiple vitamin preparations frequently contain nicotinamide and pyridoxine; and their use will then cause abnormally high urinary excretions of N^1-methylnicotinamide, *N*-methyl-2-pyridone-5-carboxamide, and 4-pyridoxic acid. Large doses of pyridoxine may also correct an abnormality of tryptophan metabolism which would otherwise be demonstrable. The effect of oral contraceptives on tryptophan metabolite excretions should always be kept in mind; unless specifically required, patients using these preparations must be excluded from studies of tryptophan metabolism. Injections of ACTH increase the urinary tryptophan metabolite excretions by induction of tryptophan oxygenase, as do large doses of hydrocortisone,[39] but in the usual therapeutic doses prednisolone is without any appreciable effect (Rose, D. P., unpublished observations). Other drugs cause abnormal tryptophan metabolism by interference with vitamin B_6 metabolism; they include isoniazid, cycloserine, and penicillamine.[40]

Interference with tryptophan metabolite analyses by drugs is of particular concern when colorimetric procedures are used in conjunction with ion-exchange chromatography. This problem has been discussed by Price et al.[34] Aromatic amines such as sulfonamides and *p*-aminosalicylic acid interfere with the determination of aromatic amines derived from tryptophan, as may acetophenetidine (APC tablets), procaine, and procaine-penicillin.

The effect of many compounds on tryptophan metabolism is unknown and the only safe course, when possible, is to exclude the administration of all drugs for several days before carrying out tryptophan metabolite studies.

II. GENERAL CONSIDERATION OF METHODS AVAILABLE FOR THE DETERMINATION OF URINARY TRYPTOPHAN METABOLITES

A variety of analytical techniques have been applied to the determination of tryptophan metabolites in urine, each new procedure attempting to overcome the difficulties associated with those which preceded it. Among the earlier methods was solvent extraction of metabolites at selective pH followed by colorimetry,[41] but this suffers from both a lack of specificity and sensitivity, and is no longer used alone although it provides a useful separation step as a preliminary to other techniques.

Coppini et al.[42] use two-dimensional paper chromatography for the separation of metabolites, followed by elution of the spots and quantitation by colorimetry or absorbance in ultraviolet light. Although this method is satisfactory for the determination of metabolites in urine collected after a tryptophan load, lack of sensitivity necessitates the application of relatively large volumes of urine to the paper making it rather laborious and time-consuming.

The advantages of thin-layer chromatography over paper chromatography have been applied to the problem.[43-46] This technique allows smaller volumes of urine to be used, reduces the length of time required for the solvent run, and yields compact spots which are readily visible under ultraviolet light and elute freely from the cellulose or silica gel to give excellent recoveries of the metabolites. After elution, quantitation may be by colorimetry,[43] fluorimetry,[44,45] or measurement of absorbance in the ultraviolet.[46] Because kynurenine and 3-hydroxykynurenine give only weak fluorescence, which is at an optimum pH (11.0) which is difficult to maintain, Bell and Mainwaring[44] use the kynureninase reaction to hydrolyze these metabolites to anthranilic acid and 3-hydroxyanthranilic acid before fluorimetry.

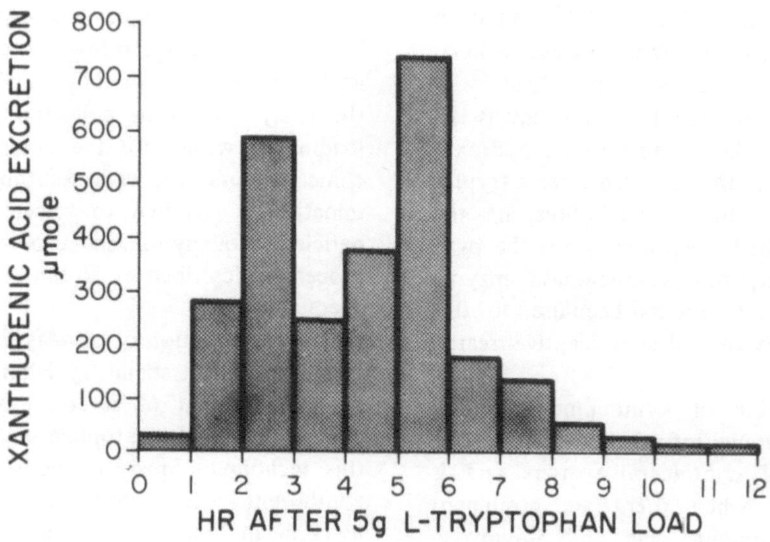

FIGURE 2. The period of elevated urinary xanthurenic acid excretion in response to a 5-g oral dose of L-tryptophan.

Several column chromatographic methods have been devised in which initial separation of metabolites is followed by their measurement by colorimetric, fluorimetric or spectrophotometric means. Probably the most widely used are the Dowex® ion-exchange procedures which were developed by Price et al.[34] and permit the determination of nine tryptophan metabolites, 4-pyridoxic acid, N^1-methylnicotinamide, and N^1-methyl-2-pyridone-5-carboxamide. Chen and Gholson[47] have recently published a procedure using DEAE-cellulose columns which gave excellent separations of 10 tryptophan metabolites and recoveries in excess of 95%.

Gas-liquid chromatography has been employed for the determination of 3-hydroxyanthranilic acids[48,49] and quinolinic acid.[50] These methods have high levels of resolution and sensitivity, but they do require careful preliminary separation of the metabolites from other urinary constituents.

Thin-layer electrophoresis gives good separations of tryptophan metabolites and has been used in conjunction with fluorimetry for the determination of 3-hydroxyanthranilic acid in urine collected without prior tryptophan loading.[31]

All of the methods mentioned so far involve standard manual techniques; they are time-consuming and unsuitable for handling large numbers of samples. Fortunately, in most situations when abnormal tryptophan metabolism is present there are elevated excretions of the diazotizable aromatic amines and assay of these compounds can be readily adapted for an automated procedure with the Technicon AutoAnalyzer® system.[51] This makes it feasible to screen urines and carry out complete manual analyses only on those samples with high levels of aromatic amines. A screening program of this type would be of interest in relation to breast cancer because it is possible that abnormal tryptophan metabolism may be demonstrable before the clinical diagnosis is apparent.[52]

Although reliable methods have now been published for the determination of many urinary tryptophan metabolites, it is usually possible to select a few for analysis, the choice depending to some extent on the purpose of the study. For the investigation of tryptophan metabolism and vitamin B_6 nutrition in pregnancy and oral contraceptive users, the key metabolites are xanthurenic acid, 3-hydroxykynurenine, and 3-hydroxyanthranilic acid.

Xanthurenic acid excretion after a tryptophan load is widely used as a test of vitamin B_6 deficiency; it is increased in pregnancy and high urinary levels are always present when there is a disturbance of tryptophan metabolism due to estrogen-containing oral contraceptives.

3-Hydroxykynurenine and 3-hydroxyanthranilic acid excretions are frequently increased by the use of oral contraceptives;[12] they are elevated in some breast cancer patients,[53,54] although after oophorectomy the levels may be subnormal.[54] The excretion of 3-hydroxykynurenine also fluctuates with the menstrual cycle[10,53] and both metabolites are higher in premenopausal women than men after a 5-g L-tryptophan load.[53] A useful indicator of vitamin B_6 deficiency is the ratio of 3-hydroxykynurenine to 3-hydroxyanthranilic acid (HK/HA) excreted after a tryptophan load because these metabolites are the substrate and product, respectively, of the pyridoxal phosphate-requiring kynureninase enzyme system.[36] The HK/HA ratio has been used for this purpose in the study of oral contraceptive-treated women.[16]

The determination of kynurenine excretion provides little additional information in studies of pregnancy or oral contraceptive users and is frequently normal when there are prominent elevations in xanthurenic acid and 3-hydroxykynurenine excretions.

III. XANTHURENIC ACID DETERMINATION

A. Basis for Selection of Method

Xanthurenic acid has been the most frequently determined of the urinary metabolites of the tryptophan-nicotinic acid ribonucleotide pathway and many different methods have been described for the assay of this compound.

Ferric salts form a complex with xanthurenic acid to give an intense green color; this reaction provided the basis for several simple colorimetric methods which do not require any special apparatus. Various modifications have been introduced,[55-57] aimed at avoiding discoloration of the green pigment when testing urine samples and removing interfering urinary chromogens. Rosen et al.[57] found that their modification was satisfactorily reproducible and that Beer-Lambert's law held over a useful working range of 10 to 250 μg of xanthurenic acid. Satoh and Price[58] compared the results of xanthurenic acid determinations made by the ferric salt method with those obtained by their ion-exchange chromatography-fluorimetric procedure, but found little similarity between the two; the values obtained by colorimetry were 1.5 to 20 times higher than those obtained by fluorimetry. Recoveries of xanthurenic acid added to urine were good by both methods, suggesting that the high values produced with the ferric salt method were due to chromogens and that the lower (fluorimetric) results were nearer to the true levels of xanthurenic acid.

The ferric complex colorimetric methods become least reliable at lower levels of xanthurenic acid excretion and so they can be used only for the analysis of urine collected after tryptophan loading. However, for the occasional needs of a clinical laboratory, when xanthurenic acid determination is required to screen for vitamin B_6 deficiency or pyridoxine-responsive anemia, the procedure described by Rosen et al.[57] is probably adequate.

The introduction of thin-layer chromatography into the clinical chemistry laboratory has led to the development of several procedures for the determination of tryptophan metabolites utilizing this technique. Most of the studies of urinary xanthurenic acid excretion by oral contraceptive users or in breast cancer have used the method developed by Walsh;[43] this will be described in detail. It is suitable for the hospital laboratory, allows the detection of several other tryptophan metabolites, is rapid in performance, and is satisfactory for all determinations of xanthurenic acid in urine collected after 2- or 5-g L-tryptophan loads. An additional bonus is that inspection of the developed chromatogram under ultraviolet light provides a visual check on the quantity of metabolites present, and so helps avoid errors due to the presence of foreign compounds which interfere with the colorimetric procedures for the aromatic amines.

The ion-exchange chromatography method of Satoh and Price[58] allows for the fluorimetric determination of both xanthurenic acid and kynurenic acid after an initial separation on a cation-exchange resin column. The procedure gives a high degree of sensitivity so that reliable determinations can be made on urine samples collected without a tryptophan load; the thin-layer chromatography method lacks this sensitivity. Its disadvantages are that it is relatively time-consuming and expensive and can only be used efficiently on a regular basis.

B. Determination of Xanthurenic Acid by Thin-layer Chromatography and Colorimetry

1. Principle

Xanthurenic acid is separated from other urinary tryptophan metabolites by one-dimensional thin-layer chromatography on cellulose with acetate-acetic acid buffer as solvent. The metabolites are identified by their fluorescence in ultraviolet light, the cellulose bearing the xanthurenic acid is removed from the plate, eluted into distilled water, and determined by its color reaction with diazotized sulfanilic acid.

2. Apparatus

1. Chromatography tanks suitable for 20 X 20-cm plates (Brinkmann Instruments Inc. or equivalent).
2. Precoated thin-layer chromatography cellulose plates, on glass; 20 X 20 cm, 0.1-mm-thick layers without fluorescent indicator (Machery, Nagel MN300).
3. Hamilton syringes, 10 μl and 100 μl.
4. Hair dryer.
5. Ultraviolet lamp with emission wavelength of approximately 360 nm.
6. Bench centrifuge.
7. Centrifuge tubes, conical, 15 ml.
8. Spectrophotometer suitable for reading at 510 nm (e.g., Beckman DU).

3. Reagents

1. Sodium acetate-acetic acid buffer; 0.2 *M*, pH 5.4 ($CH_3COONa\ 3H_2O$ 27.3 g, glacial acetic acid 1.7 ml diluted to a final volume of 1 liter with distilled water).
2. Diazotized sulfanilic acid: 1% sulfanilic acid in 10% HCl (prepared in bulk; stable at room temperature) and freshly prepared 5% sodium nitrite solution in water are cooled to 4°C and 1 vol of each are mixed together. After 15 to 20 min at 4°C and immediately before use, 2 vol of 10% anhydrous sodium carbonate solution, also cooled to 4°C, are added. It is most important that the reagent is used at once, as darkening to an orange color rapidly takes place.
3. Standard xanthurenic acid solution: xanthurenic acid (Calbiochem) is dissolved in 5 m*M* phosphate buffer, pH 7.4, to a concentration of 10 μg in 10 μl. The standard solution is stable for 3 days if kept in the dark at 4°C.

4. Procedure

The untreated urine is applied with a 100-μl Hamilton syringe, to the cellulose layer as a 1 cm line parallel with, and 2 cm from, the lower edge of the plate. The sample is dried in the warm air current from a hair dryer. The volume of urine used depends upon the circumstances; 100 μl is satisfactory for collections made from normal subjects after a 2- or 5-g L-tryptophan load, but for estrogen-treated subjects or oral contraceptive users given a 5-g dose, 25 μl or even 10 μl is frequently sufficient and larger volumes may "overload" the chromatogram. Duplicate samples are chromatographed together with a 10 μg (10 μl vol) xanthurenic acid standard.

The chromatograms are developed in sodium acetate-acetic acid buffer, the solvent front taking approximately 2 hr to rise within 2 cm of the top of the plate. Better separation of the various tryptophan metabolites is obtained if this step is carried out at 4°C, but it is not necessary for the routine quantitation of xanthurenic acid. When the separation is completed the plate is dried with a hair dryer, the spots of xanthurenic acid are located under ultraviolet light, and their positions marked with a pencil. The portions of cellulose containing the xanthurenic acid spots are scraped from the glass plate with a razor blade and the metabolite is eluted into 1.5 ml distilled water in 15-ml conical centrifuge tubes. A strip of cellulose to one side of the spots, but over which the solvent has run, is handled in an identical manner to provide a blank. Elution is completed, with occasional shaking, after 1 hr at room temperature. The cellulose is then deposited by gentle centrifugation.

Then 1 ml of the xanthurenic acid solutions and 1 ml of the "blank" solution are transferred to test tubes. To the blank 2 ml of the diazotized sulfanilic acid at 4°C is added and the absorbance is read immediately with a spectrophotometer at a wavelength of 510 nm and a 1-cm light path. The xanthurenic acid standard and the unknown solutions are then treated in the same way, a freshly mixed diazotized sulfanilic acid-sodium carbonate solution being used for each tube.

5. Calculation

$$\frac{\text{Mean value for test absorbance}}{\text{Standard absorbance reading}} \times 10$$

= urinary xanthurenic acid in μg contained in volume (usually 100 μl) applied to thin-layer plate

Mol wt of xanthurenic acid is 205.2

So:

$$\frac{\text{xanthurenic acid in } \mu g \text{ for } 100\ \mu l \times 10 \times 24\text{-hr urine volume}}{205.2}$$

= Urinary xanthurenic acid excretion in µmol/24 hr.

6. Accuracy

The accuracy of the method was assessed in the author's laboratory by recovery experiments in which pure standards were added to aliquots of a single urine sample to give a range from 2.5 to 20.0 µg/100 µl. Amounts of 100 µl each, together with an equal volume of urine without added metabolite, were chromatographed in the usual way. The absorbances after reaction with diazotized sulfanilic acid were measured, the blank was subtracted from the reading for each urine plus standard, and the absorbance of the sample without added xanthurenic acid was then subtracted from that for each of the other samples. The final values, therefore, represented the absorbance contributed by the added xanthurenic acid. The same range of xanthurenic acid standards in phosphate buffer was also reacted with diazotized sulfanilic acid without prior chromatography, so as to provide absorbance readings for 100% recovery from the cellulose plates. The results (Figure 3) show that the standard curves were linear over the entire range of 2.5 to 20.0 µg xanthurenic acid and that recoveries ranged from 80 to 95%.

7. Comparison with Two-dimensional Paper Chromatography Method

For the early work on the excretion of xanthurenic acid by women taking estrogen-containing oral contraceptives the metabolite was separated by two-dimensional ascending paper chromatography.[11] The first solvent was *n*-butanol-acetic-acid-water (12:3:5), followed by *n*-butanol-pyridine-water (1:1:1). Each urine was run in duplicate and a solution of xanthurenic acid was applied to a separate paper to give a 10-µg standard. Elution and colorimetry were performed according to the procedure described by Coppini et al.[42]

The results of these xanthurenic acid determinations made on urines collected for 8 hr after a 5-g L-tryptophan load were in good agreement with the same determinations carried out by the thin-layer chromatography method (Figure 4). In a few instances the paper chromatography method was not sufficiently sensitive to give a reliable result for a normal subject, whereas a satisfactory analysis was possible by thin-layer chromatography.

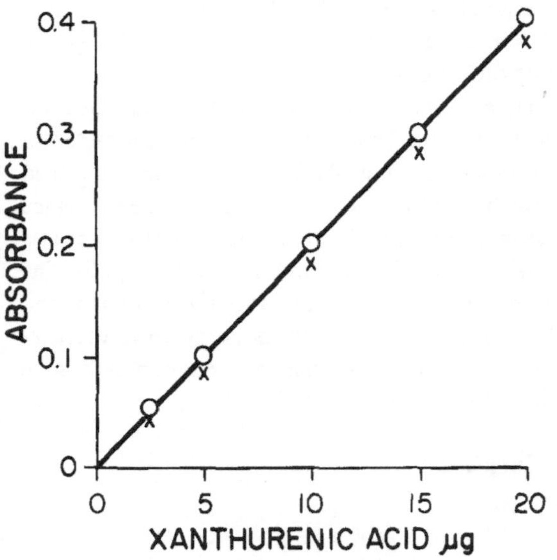

FIGURE 3. Standard curve and recoveries of xanthurenic acid from urine (x = absorbance readings of samples subjected to thin-layer chromatography).

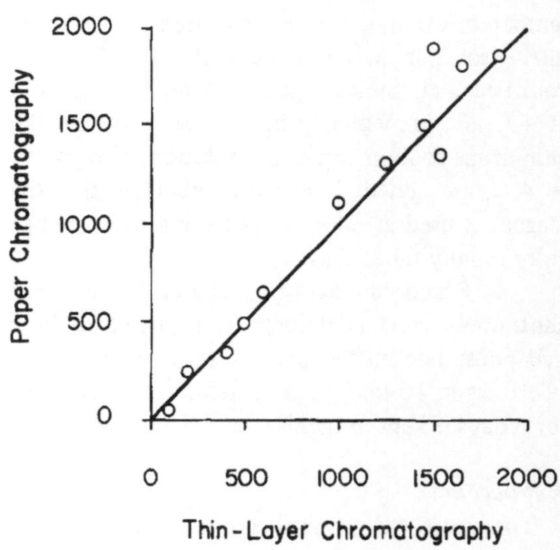

FIGURE 4. Comparison of results (in µmol/8 hr) of urinary xanthurenic acid determinations by paper chromatography and thin-layer chromatography; complete correspondence would be indicated by points falling on the 45° line.

TABLE 1

Thin-layer Chromatography of Tryptophan Metabolites on 0.1-mm-Thick Cellulose Layers with Acetic Acid-acetate Buffer: R_f Values and Color Reactions

Metabolite	R_f* (range in parentheses)	Color in ultraviolet	Color with DSA†
Acetyl-L-kynurenine	0.83 (0.80–0.86)	Bright blue	Nil
Acetyl-3-hydroxy-L-kynurenine	0.76 (0.73–0.78)	Green	Pink-purple
L-kynurenine	0.64 (0.62–0.67)	Bright blue	Nil
3-Hydroxyanthranilic acid	0.63 (0.61–0.64)	Violet	Pale pink
3-Hydroxy-L-kynurenine	0.49 (0.47–0.52)	Green	Pink-purple
Kynurenic acid	0.40 (0.34–0.44)	Grey to green	Nil
Xanthurenic acid 8-methyl ether	0.35 (0.34–0.36)	Blue	Nil
Xanthurenic acid	0.27 (0.24–0.28)	Orange-grey	Red

*Values obtained from 6 separate runs for each metabolite at 4°C.
†Diazotized sulfanilic acid.

C. Other Tryptophan Metabolites Separated by Thin-layer Chromatography

In addition to xanthurenic acid, a number of other tryptophan metabolites are detectable by thin-layer chromatography on cellulose. They are particularly prominent in urine collected after a tryptophan load from women using estrogen-containing oral contraceptives and provide a qualitative check on the ion-exchange procedures. The following metabolites have been positively identified: kynurenine, acetylkynurenine, 3-hydroxykynurenine, acetyl-3-hydroxykynurenine, kynurenic acid, and the 8-methyl ether of xanthurenic acid. Kynurenic acid is not completely separated from xanthurenic acid, but it does not react with diazotized sulfanilic acid. Kynurenine occupies almost the same position on the chromatogram as 3-hydroxyanthranilic acid and obscures the small amounts of this metabolite in urines from normal subjects. Xanthurenic acid is well separated from histidine and other natural substances in urine which give a red color reaction with diazotized sulfanilic acid.

The R_f values for tryptophan metabolites separated by acetic acid-acetate buffer on cellulose layers, their fluorescence in ultraviolet light, and color reactions after spraying with diazotized sulfanilic acid are listed in Table 1. The fluorescence of kynurenic acid takes several minutes to develop; it is a light grey color initially and then changes to an intense green-yellow on prolonged exposure to ultraviolet light.

D. Determination of Xanthurenic Acid by Ion-exchange Chromatography

Xanthurenic acid and kynurenic acid are retained when urine is passed through a column of Dowex 50 (H^+) cation-exchange resin, but they can be washed from the resin with large volumes of water, whereas other fluorescent, potentially interfering substances remain on the column. Kynurenic acid has maximal fluorescence in sulfuric acid; xanthurenic acid has a very low level of fluorescence under these conditions, but fluoresces strongly in alkali.

The characteristic fluorescence spectra of xanthurenic acid and kynurenic acid under selected alkaline or acid conditions provide a highly sensitive method for the determination of these metabolites in aqueous eluates from Dowex 50 columns.[58]

It is not proposed to describe the method here because this has been done by Price and coworkers[34] in an article in which they detail all of the methods they developed at the University of Wisconsin.

IV. 3-HYDROXYKYNURENINE AND 3-HYDROXYANTHRANILIC ACID DETERMINATIONS

A. Basis for Selection of Method

Walsh[43] used a cellulose thin-layer chromatography method for the determination of urinary kynurenine and 3-hydroxykynurenine. The metabolites were identified under ultraviolet light, eluted into distilled water, and quantitated by colorimetry. 3-Hydroxykynurenine was determined by its color reaction with diazotized sulfanilic acid, the absorbance being read at a wavelength of 450 nm 10 min after addition of the reagent. When this procedure was attempted in the author's laboratory, the deepening color of the diazotized sulfanilic acid on standing resulted in very high blank readings, even when the reagents were kept at 4°C. A solution to this problem was not found, and the method was abandoned in favor of an ion-exchange chromatography technique which was devised by Heeley[59] and is a modification of the method of Price et al.[34] This allows the determination of kynurenine, 3-hydroxykynurenine, and 3-hydroxyanthranilic acid to be made from one column run.

B. Determination of 3-Hydroxykynurenine and 3-Hydroxyanthranilic Acid by Ion-exchange Chromatography and Colorimetry

1. Principle

When urine is passed through a Dowex 50 (H^+) cation-exchange resin column, a number of tryptophan metabolites, mostly aromatic amines, are retained by the resin. They may be selectively eluted from the column by passing successive volumes of hydrochloric acid, each of increasing normality, through the resin bed and collecting the appropriate fractions. The metabolite in each hydrochloric acid fraction is then determined by colorimetry.

2. Apparatus

1. Chromatography columns: These are made by sealing a 25-cm length of glass tubing (1.2 cm external diameter) to the bottom of a 250-ml Erlenmeyer flask. The bottom of the flask is distorted slightly so that it will drain completely into the attached glass tubing. The end of the glass tube column is bevelled and a constriction is made 2 cm proximally to retain a plug of glass wool. A set of 12 flasks is supported on racks made from Dexion® 0075 strip or equivalent.
2. Measuring cylinders, 50 ml.
3. Graduated pipettes, 25 ml.
4. Erlenmeyer flasks, 100 ml.
5. Automatic pipettes to deliver 0.2 ml (Biopettes, Schwarz/Mann, Orangeburg, New York or equivalent).
6. Vortex mixer (Scientific Products, McGaw Park, Ill. or equivalent).
7. Spectrophotometer suitable for reading at 367 nm (e.g., Beckman DU).

3. Reagents

1. Dowex 50 (12% cross-linkage, 200 to 400 mesh). This is a strongly acidic cation-exchange resin and is available from Bio-Rad Laboratories, Richmond, Calif. and Sigma Chemical Co.
2. Hydrochloric acid solutions are prepared in distilled or deionized water from analytical grade HCl. Solutions of the following normalities are required: 0.1 *N*; 0.5 *N*; 1.0 *N*; 2.0 *N*; 4.0 *N*; and 8.0 *N*.
3. NaOH, 2.0 *N*.
4. Standard solutions: 3-hydroxy-DL-kynurenine and 3-hydroxyanthranilic acid (Sigma Chemical Co.) are dissolved in 0.1 *N* HCl to give standard solutions with final concentrations of 500 μg per ml. They are stable for at least 4 weeks if stored at 4°C.
5. Reagents for colorimetry: 0.25% sodium nitrite (prepare fresh) and 10% ammonium sulfamate.

4. Preparation of Dowex 50 Resin

Before a new batch of resin is used, it is prepared by removing the very fine particles and then washing with sodium hydroxide and hydrochloric acid solutions. The resin is placed in a glass cylinder, twice the volume of distilled water is added, and the resin is suspended by vigorous stirring. The suspension is allowed to settle until the supernatant contains only the very fine particles, which are then gently decanted off. This procedure is repeated four times.

The resin is washed in a large column fitted with a fritted glass disc at the point of constriction and a stopcock. If 500 ml of packed resin is to be prepared, it is washed sequentially with the following: 2 l. of 2 *N* NaOH, 5 l. of 8 *N* HCl, 4 l. of 4 *N* HCl, and 2 l. of distilled water.

5. Procedure

Preparation of the resin columns – These are prepared on the day before they are due to be used. The resin is best handled as a slurry (1:1 resin and water) and is added to the glass column, which is half filled with water, from a 10-ml graduated pipette with the tip widened to an 0.25-cm nozzle. By pipetting the resin into the tube in this manner and allowing it to settle onto the glass wool plug as the water drains away, air bubbles do not become trapped in the resin bed. The final length of the resin column is 3.5 cm.

Three columns are required for each analysis; 12 urine samples (36 columns) can be handled comfortably at each assay run. The resin is equilibrated with 0.1 *N* HCl immediately before the analysis is performed by passing 30 ml of the acid through the columns.

Preparation of the urine samples – Each urine sample is run in duplicate, together with an internal standard. For the analysis of the urine collected from normal subjects after a 5-g L-tryptophan load, pipette 1% of the total 24-hr volume into each of three 50-ml graduated measuring cylinders. If inspection of the thin-layer chromatograms prepared for the xanthurenic acid determinations indicates high excretion levels, as occurs frequently in women taking estrogen-containing oral contraceptives, 0.5% or even 0.25% of the 24-hr collection is used. After 2-g L-tryptophan loads, some urine samples are too dilute to give satisfactory absorbance readings for the 3-hydroxyanthranilic acid colorimetry if only 1% of the 24-hr volume is taken, and it is better to use 2% routinely for normal subjects.

For the internal standard, add to the third cylinder 1 ml (500 μg) of the 3-hydroxykynurenine and 3-hydroxyanthranilic acid standards.

Pipette 2 ml of 2 *N* HCl into each of the three measuring cylinders, add water to a total volume of 40 ml, and mix carefully with stirring rods.

The column run – The samples are poured onto the columns and allowed to pass through the resin under gravity. The measuring cylinders are rinsed with 30-ml volumes of 0.1 *N* HCl which are added to the columns when the urines have passed through the resin. The resin is then washed with 40 ml of 0.5 *N* HCl and the washings discarded. This is followed by 30-ml volumes of 2.0 *N* HCl, which elute the 3-hydroxyanthranilic acid from the resins and are collected into Erlenmeyer flasks. Further washes with 40 ml of 4.0 *N* HCl elute the 3-hydroxykynurenine.

Colorimetry – The two metabolites are determined by forming the yellow benzoxadiazole compounds with sodium nitrite.

First 3.0 ml of both the 2 *N* and 4 *N* HCl eluates are pipetted into two sets of three test tubes. To one tube is added 0.2 ml distilled water (blank) and into the two (duplicate) tubes is pipetted 0.2 ml of 0.25% sodium nitrite. These are mixed well (vortex mixer) and allowed to stand for 3 min (use a timer for this step) and then 0.2 ml of ammonium sulfamate is added to remove any excess sodium nitrite. After 30 min, the absorbances of the yellow-colored solutions are read at a wavelength of 367 nm against their respective blanks.

Then 1 ml of the 3-hydroxyanthranilic acid and 3-hydroxykynurenine standard solutions are added to 29 ml of 2.0 *N* HCl and 39 ml of 4.0 *N* HCl, respectively, and treated in the same way as the eluates so that the percent recovery may be calculated.

6. Calculation

$$\frac{\frac{a}{b-a} \times 500 \times f \times y}{z} = \text{Urinary 3-hydroxyanthranilic acid or 3-hydroxykynurenine excretion in } \mu\text{mol/24 hr}$$

where

a = mean value for the two test absorbances; b = absorbance of sample containing added metabolite (internal standard); f = factor to convert to μg excreted in 24 hr; y = factor to correct to 100% recovery; and z = mol wt of metabolite (3-hydroxyanthranilic acid = 153.1, 3-hydroxykynurenine = 224.2).

7. Specificity

Heeley[59] examined the 2 *N* HCl fractions by two-dimensional paper chromatography. When the chromatograms were sprayed with dilute HCl followed by sodium nitrite solution, the only yellow benzoxadiazole compound detected was 3-hydroxyanthranilic acid. This has been confirmed in the author's laboratory, using eluates from the urine samples of oral contraceptive-treated women.

Rose and Toseland[48] have determined 3-hydroxyanthranilic acid in the 2 *N* fractions by both colorimetry and gas-liquid chromatography and obtained good correspondence between the two methods of quantitation.

A number of 4 *N* HCl fractions have been evaporated to dryness under reduced pressure, the residues dissolved in 2 ml 50% isopropanol in water and subjected to two-dimensional cellulose thin-layer chromatography using acetate-acetic acid buffer, pH 5.4, followed by *n*-butanol-pyridine-water (1:1:1). Large spots of 3-hydroxykynurenine, kynurenine, and acetylkynurenine were seen on examination under ultraviolet light, together with small amounts of acetyl-3-hydroxykynurenine and xanthurenic acid. 3-Hydroxyanthranilic acid was never present, having been eluted completely from the resin by 2 *N* HCl. The chromatograms were sprayed with a solution of 2 *N* HCl, acetone, and 5% sodium nitrite (5:45:1 v/v). 3-Hydroxykynurenine gave a yellow-brown color and the small acetyl-3-hydroxykynurenine spot was a faint yellow, but the other metabolites did not react with the nitrite reagent.

8. *Accuracy and Linearity*

The accuracy of the method and the applicability of Beer-Lambert's law have been assessed by recovery experiments in which known amounts of metabolite were added to a urine sample collected without tryptophan loading and submitted to the ion-exchange procedure. A blank was prepared from urine without added metabolites and this absorbance was subtracted from those of the other samples. Corresponding amounts of the metabolites were added to 30 ml of 2.0 *N* HCl (3-hydroxyanthranilic acid) or 40 ml of 4.0 *N* HCl (3-hydroxykynurenine) and quantitated by colorimetry without prior passage through the resin. Figures 5 and 6 compare the absorbances of the samples passed through the resin columns with those of the external standards and also show that the curves were linear over the range of concentrations studied. The recoveries for 3-hydroxyanthranilic acid over the range 200 to 1,000 μ were 80 to 100%; for 3-hydroxykynurenine levels of between 400 and 2,000 μg they were 97 to 104%.

V. NORMAL RANGE OF METABOLITE EXCRETIONS IN URINE AFTER TRYPTOPHAN LOADING

The excretion of tryptophan metabolites after an oral load of the amino acid is influenced by age, sex, and the menstrual cycle.[53,60]

Table 2 gives the normal values which were obtained for 3-hydroxykynurenine, 3-hydroxyanthranilic acid, and xanthurenic acid excretions after a 5-g dose of L-tryptophan by groups of premenopausal women, paramenopausal or postmenopausal women, and men. These subjects were either healthy members of the hospital staff or patients with diseases which were not considered to have an influence on tryptophan metab-

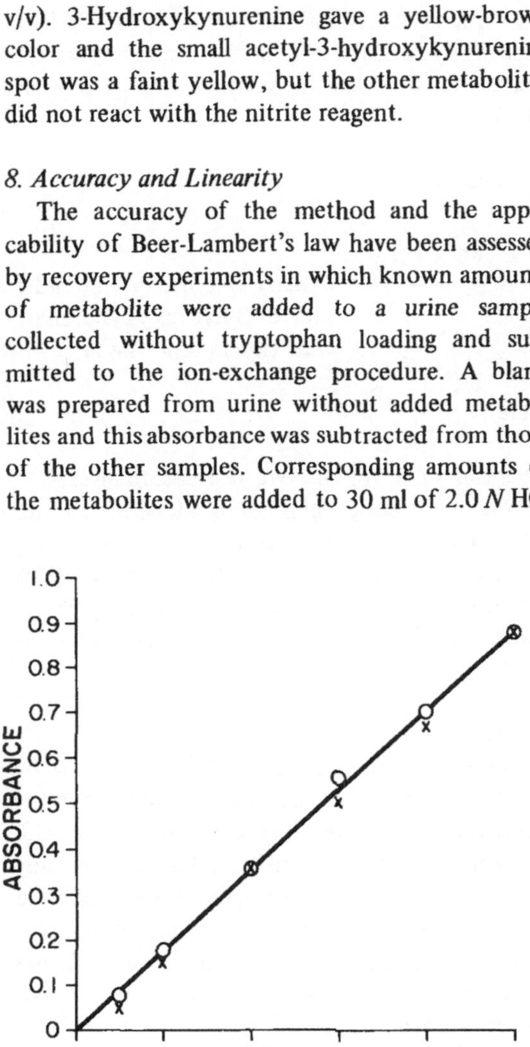

FIGURE 5. Standard curve and recoveries of 3-hydroxyanthranilic acid from urine (o = absorbance readings of external standard solutions; x = absorbance readings of samples passed through resin columns).

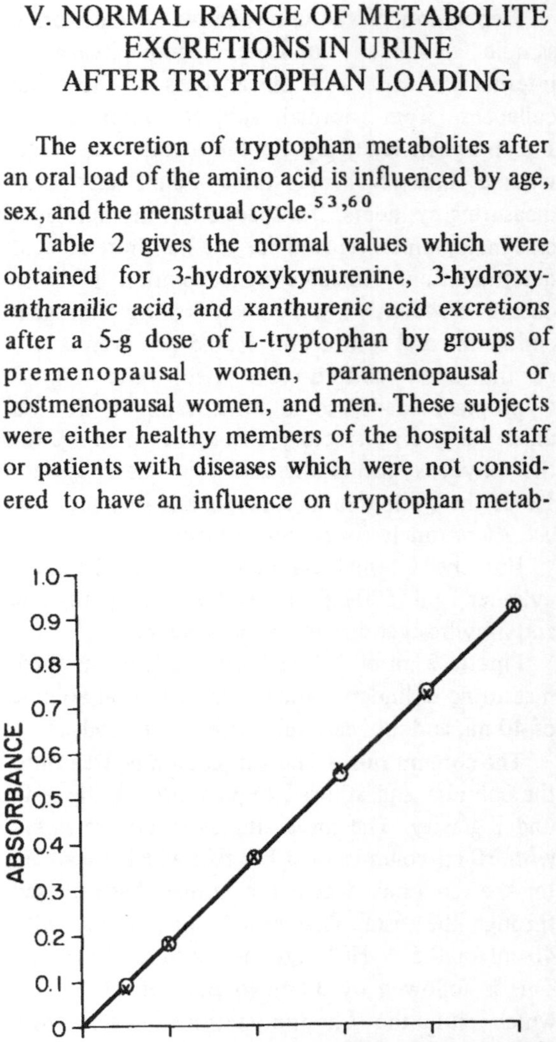

FIGURE 6. Standard curve and recoveries of 3-hydroxykynurenine from urine (o = absorbance readings of external standard solutions; x = absorbance readings of samples passed through resin columns).

TABLE 2

Excretion of Tryptophan Metabolites (Mean ± S.D.) in Urine Collected from Control Subjects for 8 hr (a, b, and c) or 24 hr (d) after 5-g Oral L-tryptophan Loads

Control subjects	Excretion in μmol		
	HK	HA	XA
a. Premenopausal females, 21–35 yr (11)	387 ± 112	189 ± 79	186 ± 82
b. Para- and postmenopausal females, 42–82 yr (10)	209 ± 105*	129 ± 77	141 ± 50
c. Males, 24–74 yr (13)	189 ± 94†	114 ± 29*	114 ± 50‡
d. Females, various menstrual status, 36–80 yr (29)	285 ± 132	175 ± 13	168 ± 68

Number of subjects in each group is given in parentheses. Abbreviations: HK, 3-hydroxykynurenine; HA, 3-hydroxyanthranilic acid; XA, xanthurenic acid.

Results significantly lower than those of premenopausal women.
*$0.01 > p > 0.001$
†$p < 0.001$
‡$0.02 > p > 0.01$

olism. They had no evidence of hepatic or renal disease and were not anemic or in congestive cardiac failure. Patients with a malignant disease were excluded, as were those who had been taking oral contraceptives or any form of hormone therapy. Xanthurenic acid was determined by the thin-layer chromatography method and 3-hydroxykynurenine and 3-hydroxyanthranilic acid by ion-exchange chromatography.

The younger, premenopausal women excreted all three metabolites in significantly greater quantity than the men and 3-hydroxykynurenine excretion was higher than in the older women. Table 2 shows also that collecting the urine for a full 24-hr period, instead of for the first 8 hr after a 5-g L-tryptophan load, produces a small, but not statistically significant, increase in the yield of these three metabolites.

Mainardi and Tenconi[61] have also reported a sex difference in the excretion of tryptophan metabolites after a 5-g load; after a 3-g dose women excrete higher levels of total urinary diazo-reacting compounds than men.[37] This difference between the sexes is not demonstrated by a 2-g L-tryptophan load; indeed after this dose, Price et al.[34] found that men excrete larger amounts of anthranilic acid glucuronide and *o*-aminophippuric acid – two metabolites derived from kynurenine.

The likelihood that the sex difference and effect of age on tryptophan metabolite excretions obtained after 5-g loads is related to differences in hormonal activity is supported by the lower excretions which occur in postmenopausal women and by the finding that tryptophan metabolism is influenced by the menstrual cycle. Thus, the excretions of 3-hydroxykynurenine, 3-hydroxyanthranilic acid and xanthurenic acid are all higher at ovulation than immediately after a menstrual period.[53] In addition, some previously premenopausal women who have been treated by prophylactic oophorectomy for early breast cancer appear to excrete subnormal levels of tryptophan metabolites.[54]

Table 3 gives the normal values obtained for the excretion of the three metabolites after 2-g L-tryptophan loading, together with the ratio of 3-hydroxykynurenine to 3-hydroxyanthranilic acid excreted. Twenty-two young adult females, aged 17 to 28 years, were studied by the methods detailed in this chapter. For comparison, the results are shown for 31 women of similar age who had been taking one of several different estrogen-containing oral contraceptives for periods ranging from 6 to 36 months. The marked variation in the effect of oral contraceptives is evident from the large standard deviations. In part, this is because the various progestogens used in these preparations modify the estrogen's effect.[14]

The influence of the menopause on tryptophan metabolism after a 2-g L-tryptophan load does not appear to have been examined. But, since this dose fails to demonstrate the sex difference in metabolite excretions which is evident with a 5-g load,

TABLE 3

Excretion of Tryptophan Metabolites (Mean ± S.D.) in Urine from Control Female Subjects or Oral Contraceptive Users for 24 hr After 2-g Oral Tryptophan Loads

	Excretion in μmol			
	HK	HA	XA	HK/HA
Control females, 17–28 yr (22)	33 ± 17	28 ± 15	63 ± 40	1.25 ± 0.50
Oral contraceptive users 18–35 yr (31)	147 ± 164	62 ± 39	292 ± 175	2.15 ± 1.20

menopausal status may well be without any significant effect.

VI. URINARY 4-PYRIDOXIC ACID EXCRETION

A. Determination of 4-Pyridoxic Acid by Ion-exchange Chromatography and Fluorimetry

1. Principle

4-Pyridoxic acid is separated from other fluorescent, potentially interfering compounds by ion-exchange chromatography using a sequence of Dowex 1 (Cl^-) and Dowex 50 (H^+) resin columns.[34,62] Quantitation is by forming the lactone of pyridoxic acid and measuring the fluorescence of this derivative. Because some 4-pyridoxic acid may already be present in urine as the lactone, or be lactonized during the ion-exchange procedure, all samples from the columns are delactonized initially, and then the test and recovery samples are fully lactonized as a preliminary to the fluorimetric step. This maneuver eliminates the possible presence of pyridoxic acid lactone in the blanks, which then give lower fluorescence values.

2. Apparatus

1. Chromatography columns: These are as described for the determination of 3-hydroxykynurenine and 3-hydroxyanthranilic acid.
2. pH meter.
3. Glass beakers, 150 ml.
4. Erlenmeyer flasks, 125 ml.
5. Measuring cylinders, 50 ml.
6. Graduated centrifuge tubes, 50 ml.
7. Aminco® Fluoro/Colorimeter with the following filters: primary Corning® 7-39; secondary Corning® 3-73; and Wratten® A, 4-70.

Alternatively, a spectrophotofluorimeter such as the Aminco-Bowman® may be used.

3. Reagents

1. Dowex 50 H^+ (12% cross-linkage, 200–400 mesh) prepared as already described.
2. Dowex 1-chloride anion-exchange resin (10% cross-linkage, 200–400 mesh): This resin requires only removal of fine particles by decantation prior to use.
3. Hydrochloric acid solutions: 0.05 *N*; 0.1 *N*; 1.0 *N*; 2.0 *N*; and 5.0 *N*.
4. NH_4OH, 1.5 *N*.
5. NaOH, 1.0 *N* and 5.0 *N*.
6. 1% sodium borate.
7. Quinine sulfate standard: 15 mg quinine sulfate is dissolved in 1 liter 0.1 *N* H_2SO_4. The working solution is prepared by taking 0.25 ml of this solution and diluting it to 100 ml with 0.1 *N* H_2SO_4. The stock standard may be stored at 4°C for 3 months, but the working standard is freshly diluted for each analysis.
8. Standard 4-pyridoxic acid solution: A stock standard is prepared by dissolving 10 mg of 4-pyridoxic acid in 100 ml of 0.1 *N* HCl. For the working standard, 10 ml of the concentrated solution is diluted to 100 ml with 0.1 *N* HCl. If stored at 4°C the stock standard is stable for 3 months and the working standard for 4 weeks.

4. Procedure

Preparation of the resin columns – These are best prepared on the day before the analyses are to be performed. Each urine sample is run with and without an added internal standard of 4-pyridoxic acid, so that two Dowex 1 (Cl′) and two Dowex 50 (H^+) columns are required for each urine analysis. Eight urinçs can be handled comfortably in one run. The resin columns are formed using the technique described for the aromatic amines. The Dowex 1 columns, which are 3.0 cm in length, are washed with 50 ml of 2.0 *N* HCl and then with 50 ml of distilled water. The Dowex 50 columns, also 3.0 cm long, are washed with 50 ml of 5.0 *N* HCl and 50 ml of water. The resins are discarded after use.

Preparation of the urine samples – Approximately 3% of a 24-hr urine collection is poured into a 150 ml beaker, the pH is adjusted to 10.6 with NaOH, and the sample filtered into a 125-ml Erlenmeyer flask using Whatman No. 1 filter paper. From this filtrate, 1% of the 24-hr volume is pipetted into each of two 50-ml measuring cylinders (a smaller percentage, 0.1 or 0.5%, is taken if the subject has been receiving pyridoxine supplements). To one of the measuring cylinders is added 2 ml (20 μg) of the 4-pyridoxic acid working standard, so as to provide an internal standard for the determination of recovery, and to the other is added 2 ml of 0.1 *N* HCl. Finally, 1.5 ml of 1.5 *N* NH_4OH is added to each measuring cylinder and the volumes are made up to 40 ml with water.

The column run – The contents of the measuring cylinders are mixed well, poured onto the Dowex 1 columns, and allowed to pass through the resins under gravity. The measuring cylinders are rinsed with 25-ml volumes of water which are added to the columns when the passage of the urine samples is completed. After this, the tips of the columns are washed with distilled water from a wash bottle, and placed in a position over the Dowex 50 columns so that the effluent from the Dowex 1 will drip onto the Dowex 50 resins. Next, the Dowex 1 columns are washed with 50 ml of 0.05 *N* HCl, which elutes the 4-pyridoxic acid from this resin and onto the Dowex 50. When elution is complete the Dowex 1 columns are removed, and the Dowex 50 columns are washed with 10 ml water. The column tips are washed with water, dried, and 125-ml Erlenmeyer flasks are placed in position under each column. Then 50 ml of 2.0 *N* HCl is added to the columns and the effluent containing the 4-pyridoxic acid is collected in the flasks.

Fluorimetry – The steps involved will be considered under five headings:

a. Delactonization. This applies to the blanks, the urine samples, and the urine samples plus internal standard (recovery samples).

b. Lactonization of the samples and recovery samples.

c. "Nonlactonization" of the blanks preparatory to fluorimetry, i.e., avoidance of the lactone formation at an acid pH.

d. Preparation, delactonization, and lactonization of an external standard so that the percentage recovery may be calculated.

e. Fluorimetric procedure itself.

a. First, 2 ml of the 2.0 *N* HCl effluent is pipetted into each of two 50-ml graduated centrifuge tubes. One of these tubes will constitute the sample tube and the other will be the blank. Then 1 ml of 5.0 *N* NaOH is added, the tubes are mixed, capped with squares of glassine paper held in place with rubber bands, and heated in a boiling water bath for 5 min to complete the delactonization process. The tubes are then removed, cooled to room temperature in a cold water bath, and the contents of each are made up to 7 ml with distilled water.

b. The contents of one of the two "delactonized" tubes, that which is to be the sample tube, are neutralized with 1 ml of 1.0 *N* HCl. Then 2.0 ml of 5.0 *N* HCl is added and the tube is heated in the boiling water bath for 15 min. After lactonization, the tube is cooled, made slightly alkaline by the addition of 11 ml of 1.0 *N* NaOH, and diluted to 30 ml with water. Next 2 ml of the 4-pyridoxic acid lactone-containing solution is immediately transferred to a fluorimeter tube containing 8 ml of 1% sodium borate solution.

c. Because lactonization occurs in an acid medium at room temperature, the order of the various additions to the blank tube is different – 11 ml of 1.0 *N* NaOH is added first, followed quickly by 1 ml of 1.0 *N* HCl and 2.0 ml of 5.0 *N* HCl. The blank tube volume is then made up to 30 ml with water and 2 ml taken into a fluorimeter tube containing 8 ml of 1% sodium borate solution.

The fluorimetric readings of samples and blanks did not change for several hours in borate buffer.

d. External standard – 1 ml of the working (diluted) standard and 9 ml of 0.1 *N* HCl are pipetted into each of four 50-ml graduated centrifuge tubes. To all four tubes 0.65 ml of 1.0 *N* NaOH is added; these are mixed and carried through the delactonization procedure. Two tubes are blanks and two are lactonized.

e. The optimum excitation wavelength for the lactone is 350 nm; the maximal fluorescence is 450 nm. If available, a spectrophotofluorimeter such as the Aminco-Bowman may be used or, alternatively, one of the filter models of fluorimeter. The quinine sulfate reference standard is used to set the instrument at 50% transmission by adjusting the sensitivity. With the Aminco Fluoro/Colorimeter, the photomultiplier setting is usually at 0.03 with a sensitivity reading of approximately 38 when the quinine sulfate reads 50. The samples are read by checking with the quinine sulfate standard every 4 to 6 tubes.

5. Calculation

The standard (nonlactonized) blanks are subtracted from the readings of the two lactonized standard solutions and the average of the duplicate standard readings is taken (equivalent to 1 μg 4-pyridoxic acid).

Average standard reading X $\frac{4}{5}$ = standard value equal to 100% recovery.

Subtract blank reading (nonlactonized) from the lactonized sample reading, and also from the lactonized recovery reading.

Subtract the two differences obtained to give the recovery value.

Then:

$\frac{\text{recovery value}}{\text{standard value equal to 100\% recovery}}$ = % recovery (the recovery should be in the range 85 to 115%)

Then

$\frac{\text{Sample} - \text{blank reading}}{\text{Recovery} - \text{blank reading}} \times \frac{50}{2} \times 100 = \mu g$ 4-pyridoxic acid/24 hr for 1% aliquot taken

Mol wt is 183 $\frac{\mu g}{183} = \mu mol$ 4-pyridoxic acid/24 hr.

Calculation example:

	Lactonized	Nonlactonized	Difference
Sample	39.5	32.0	7.5
Recovery standard	65.7	32.5	33.2
External standard	42.0	8.3	33.5
	41.5	8.3	

Recovery: 33.2 − 7.5 = 25.7

Standard value equal to 100% recovery: 33.5 X$\frac{4}{5}$= 26.8

% recovery $\frac{25.7}{26.8}$ X 100 = 95.9%

$\frac{7.5}{33.2} \times \frac{50}{2} \times 100 = 565\ \mu g$

$\frac{565}{183}$ = 3.1 μmol of 4-pyridoxic acid/24 hr.

6. Specificity and Accuracy

A fluorimetric method for the determination of 4-pyridoxic acid was developed originally by Huff and Perlzweig,[63] but it was of limited value because other fluorescent substances in the urine interfered with the assay and produced falsely elevated results.

Reddy et al.[62] compared this original method with the modified procedure described in this chapter. They found that the ion-exchange chromatography steps removed 40 to 75% of the fluorescence from urine without adversely affecting the recoveries of added 4-pyridoxic acid. Figure 7 illustrates the data obtained from one of their studies in which a normal subject, eating a self-selected diet, collected urine for 24 hr before the ingestion of 10 mg (48.6 μmol) pyridoxine hydrochloride and then made further collections on 3 successive days. The samples were analyzed by the two methods. Although the revised method gave considerably lower results, the recoveries of added pyridoxic acid were 97 to 101%. On the other hand, recoveries using the method of Huff and Perlzweig fluctuated considerably, presumably because of variation in the amounts of interfering substances consumed in the diet.

As an additional check it was shown that artificial vanilla flavoring, the ingestion of which invalidates the original method,[64] is removed during the ion-exchange chromatography and does not interfere with the revised procedure.

7. Normal Range of Urinary 4-Pyridoxic Acid Excretion

The results obtained in the author's laboratory for the excretion of 4-pyridoxic acid by 28 young adult females, none of whom had been receiving any form of vitamin preparation, ranged from 2.7 to 6.4 μmol/24 hr (mean 3.9, S.D. 0.7).

VII. PLASMA PYRIDOXAL PHOSPHATE ASSAY

A. Determination by a Radiochemical Enzyme Assay

1. Principle

L-Tyrosine is decarboxylated to yield tyramine; under appropriate assay conditions, the rate at which CO_2 is evolved gives a measure of the enzyme reaction. *Streptococcus faecalis* provides a source of the pyridoxal phosphate-dependent tyrosine decarboxylase (E.C. 4.1.1.25), from which

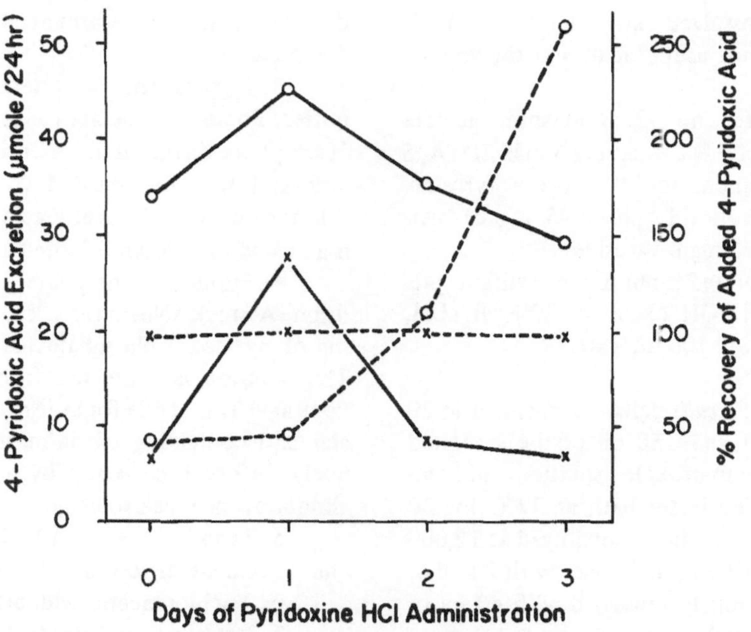

FIGURE 7. Urinary 4-pyridoxic acid excretion (–) and % recovery of added metabolite (- - - -) before and after a single 10-mg oral dose of pyridoxine HCl. The assays were performed by the methods of Huff and Perlzweig (o) and Reddy et al. (x). Note the lower values (x–x) and the consistent recoveries (x - - - x) obtained by the latter procedure. (Figure prepared from the data of Reddy et al.[62])

the coenzyme-free apodecarboxylase is prepared. The decarboxylation reaction will then be dependent on the provision of an external source of pyridoxal phosphate, which may be either a standard solution or an unknown sample such as plasma. Over a certain range, the reaction rate, and so the amount of CO_2 produced, will depend on the coenzyme concentration and a standard curve can, therefore, be prepared.

The method to be described is based on the measurement of $^{14}CO_2$ evolved during the decarboxylation of L-tryosine-1-^{14}C by trapping the gas in solution and counting the radioactivity in a liquid scintillation counter.[65,66] The detailed procedure is one developed by Dr. R. R. Brown, University of Wisconsin Medical School.

2. Apparatus

1. Incubation flasks. These are 25-ml Erlenmeyer flasks each fitted with a rubber septum cap bearing a polyethylene center well (Kontes Glass Co., Vineland, N. J., part numbers K-882310 and K-882320). To ensure ease of insertion and to provide a reliable seal, the septum caps are coated before use with a thin film of glycerol.

2. Hypodermic needles and 1-ml syringes.

3. Shaking waterbath, 37°C.

4. Refrigerated centrifuge. Beckman J-21 centrifuge with JA-20 rotor.

5. Sonicator. Branson Sonifier (Branson Sonic Power, Danbury, Conn.).

6. Liquid scintillation counter, e.g., Nuclear Chicago model 6860.

3. Reagents

Glass distilled water is used throughout for preparing aqueous solutions.

1. Tyrosine apodecarboxylase. This is prepared from an acetone powder of *S. faecalis* grown on a Vitamin B_6-deficient medium (Worthington Biochemical Co., Freehold, N. J.), as described by Sundaresan and Coursin.[65]

The following solutions are required:

i. Buffer no. 1. Potassium acetate buffer, 10 m*M* pH 5.5 containing 5 m*M* EDTA; 490.8 mg of potassium acetate and 930 mg

Na_2EDTA are dissolved in water, the pH is adjusted to 5.5 with acetic acid, and the volume brought to 500 ml.

ii. Buffer no. 2. Potassium acetate buffer, 10 m*M*, pH 5.5 containing 5 m*M* EDTA, 5 m*M* mercaptoethanol, and 0.5 m*M* L-tyrosine. Prepared as buffer no. 1 plus 195 mg of mercaptoethanol and 25 mg L-tyrosine.

iii. Saturated ammonium sulfate prepared in dilute NH_4OH (3 ml of 30% NH_4OH, specific gravity 0.9, in 100 ml water).

First, 1 g of *S. faecalis* cells is suspended in 20 ml of buffer no. 1 and 150 ml of the saturated ammonium sulfate-hydroxide solution and incubated in a shaking water bath at 37°C for 30 min. The suspension is then centrifuged at 12,000 X g for 10 min at 4°C, washed twice with 20 ml of buffer no. 1, and finally rewashed with 40 ml of buffer no. 2. The washed preparation and approximately 4 g of chromatography grade alumina (previously washed carefully with buffer no. 2) are suspended in 20 ml of buffer no. 2 and sonicated in a 125-ml stainless steel beaker using a Branson Sonifier®. Sonication is carried out in 30-sec bursts, with 30-sec cooling periods in between, for a total of 4 min. The power setting is at no. 4 and tuning is at the maximum. During this procedure the stainless steel beaker is immersed in a salt-ice bath to aid cooling. When sonication is completed the suspension is centrifuged at 27,000 X g for 10 min, the supernatant is removed and saved, and the deposit is washed with 20 ml of buffer no. 2 and recentrifuged. The two supernatant solutions are combined, the pH is adjusted to 5.5, and the purified apoenzyme is dispensed into aliquots and stored at −20°C (stable for at least 2 months).

2. L-Tyrosine-1-^{14}C substrate solution: 50 μCi of L-tyrosine-carboxyl-^{14}C, specific activity 10 mCi/mmol (Calatomic, Los Angeles, California) is dissolved in small portions of 0.1 *N* HCl totaling about 1.0 ml and transferred to a 125-ml flask. This is diluted with 30 ml of a stock tyrosine solution which is 22 m*M* L-tyrosine (Sigma Chemical Co.) in 50 m*M* HCl. The radioactivity of an aliquot of this solution is counted and further dilution is made with the 22 m*M* L-tyrosine so that the final L-tyrosine-1-^{14}C solution contains approximately 120,000 cpm/0.05 ml. This final solution is dispensed into several screw-capped plastic bottles and stored at −20°C. On refrigeration, tyrosine may crystallize out but can be dissolved readily by warming at 40 to 50°C for a few minutes.

3. Potassium acetate-EDTA incubation buffer. Potassium acetate (9.82 g) and Na_2EDTA (1.86 g) are dissolved in water, diluted almost to 1 liter and the pH adjusted to 5.5 using strong solutions of KOH or acetic acid. The final volume is adjusted to 1,000 ml. Stored at 4°C.

4. Pyridoxal phosphate stock standard solution. A stock solution is prepared containing 3.0 mg of pyridoxal phosphate per 100 ml in water. This solution is protected from light and when kept at 4°C is stable for at least 2 days. A working standard containing 10 ng/ml is prepared immediately before each assay by serially diluting an aliquot of the stock solution.

5. Perchloric acid, 1.0 *M*; 83.3 g of perchloric acid are dissolved in 500 ml of water.

6. Trichloroacetic acid, 50%.

7. Potassium hydroxide, 5.0 *M*.

8. NCS – Nuclear Chicago Solubilizer® for trapping $^{14}CO_2$ (Nuclear Chicago, Des Plains, Ill.).

9. Scintillant. 0.03% POPOP, 0.7% PPO, and 10% naphthalene in 1,4-dioxane.

4. Procedure

Deproteinization of plasma – For each sample of plasma to be assayed (collected after an overnight fast with heparin or EDTA anticoagulant), a recovery sample with added pyridoxal phosphate is also prepared. Duplicate samples of 0.5 ml of plasma are pipetted into two centrifuge tubes. To one tube ("sample") is added 0.5 ml of water and 0.6 ml of 1.0 *M* perchloric acid. To the other ("recovery") tube is added 0.5 ml of the pyridoxal phosphate working standard (10 ng/ml) and 0.6 ml of 1.0 *M* perchloric acid. The tubes are mixed well, shielded from light, and allowed to stand at room temperature for 1 hr (recoveries are consistently low if samples are stood for a shorter period of time). The samples are then centrifuged for 10 min at 23,000 X g and 4°C and the clear supernatant solutions are carefully decanted into clean tubes. These solutions are kept on ice while the pH is adjusted to 5.5 ± 0.1 with 5.0 *M* KOH dispensed from a microburette. Approximately 0.1 ml is required. The resultant potassium perchlorate settles quickly and the clear supernatant is used.

Incubation procedure – All incubations are performed in duplicate in the specially stoppered 25-ml Erlenmeyer flasks. Each flask contains 1.0 ml of 0.1 *M* potassium acetate-EDTA buffer,

0.025 ml (0.75 units) of tyrosine apodecarboxylase, and up to 0.2 ml of pyridoxal phosphate-containing sample or pyridoxal phosphate standard, plus water to bring the final volume in all flasks to 1.225 ml. Usually a water blank and 0.5, 1.0, and 2.0 ng levels of pyridoxal phosphate are run in duplicate to give a standard curve. The reaction in each flask is started by the addition of the tyrosine-1-^{14}C substrate solution. The septum cap is immediately inserted and incubation commenced in a shaking water bath at 37°C. Pairs of flasks are started at intervals of 1 min. After incubation for 9.5 min, pairs of flasks are removed from the water bath and 0.2 ml of NCS is injected through the septum into the center well, being careful that the last drop of NCS on the needle is touched onto the center well and so is not deposited on the septum when the needle is withdrawn. At exactly 10 min, 0.5 ml of 50% trichloroacetic acid is injected through the septum to stop the reaction and release $^{14}CO_2$. With this schedule it is possible to incubate and stop 20 flasks in 20 min.

After addition of the trichloroacetic acid, the flasks are allowed to stand for at least 3 hr, or preferably overnight, to allow complete trapping of CO_2 in the NCS. The flasks are then opened, the stem of the center well is cut with scissors, and the center well deposited into a counting vial containing 10 ml of scintillant. The counting vials are sealed, thoroughly mixed, and the activity of the $^{14}CO_2$ is counted.

5. Calculation

From the standard flasks a curve is plotted of cpm against ng pyridoxal phosphate (PLP) per flask.

Then

$$\text{ng PLP/ml plasma} = \text{ng} \frac{\text{PLP/flask}}{y} \times \frac{v}{p}$$

where

y = volume of neutralized deproteinized sample incubated.

v = total volume of neutralized deproteinized sample (usually 1.7 ml)

p = plasma volume used (usually 0.5 ml)

Recovery of added pyridoxal phosphate is calculated and the appropriate correction is made to the result in ng/ml of plasma.

TABLE 4

Plasma Pyridoxal Phosphate (PLP) Levels in Normal Subjects Expressed as ng/ml (Mean ± S.D.)

Subjects	No.	Plasma PLP
[a]Males aged 20–34 yrs	17	18.5 ± 5.5
35–49 yrs	7	15.8 ± 3.3
[a]Females aged 20–34 yrs	12	16.8 ± 3.6
35–49 yrs	7	11.4 ± 3.4[†*]
[b]Females aged 18–24 yrs	9	11.6 ± 4.6

[a]Chabner and Livingston[66]

[b]Brown, unpublished data

[†]Significantly different from younger women, $0.01 > p > 0.001$

*Significantly different from males of same age group, $0.05 > p > 0.02$

6. Accuracy and Precision

Although as a routine 0.5, 1.0, and 2.0 ng amounts of pyridoxal phosphate are used, the standard curve is linear up to a level of 10.0 ng per flask. Recoveries obtained from a series of 42 consecutive normal plasma samples ranged from 79.8 to 116% (mean ± S.D. 92.8 ± 8.5).

Chabner and Livingston[66] have reported that the standard error of the mean for 76 consecutive duplicate determinations was 5.8%.

7. Normal Range of Plasma Pyridoxal Phosphate Concentrations

The plasma pyridoxal phosphate levels obtained for normal subjects by Chabner and Livingston[66] are summarized in Table 4. Adult females showed a significant reduction in coenzyme levels with increasing age, which was apparent even though elderly subjects were excluded. The mechanism for this variation with age is unclear and merits further investigation. A sex difference is also evident from their data, but only in the older age groups where males had higher plasma pyridoxal phosphate levels than females.

Brown (unpublished data) is finding somewhat lower values for young women, none of whom are using oral contraceptives, than those reported by Chabner and Livingston (Table 4). His results are in agreement with those of Lumeng et al.,[67] however, who obtained a range of 4.0 to 22.5 ng/ml for 92 control women aged-matched for a group of oral contraceptive users.

No doubt, now that a satisfactory method is available for the assay of plasma pyridoxal phosphate, the influence of various physiological factors and disease states will become more clearly defined.

VIII. ERYTHROCYTE ALANINE AMINOTRANSFERASE

A. The Method

1. Principle

Alanine aminotransferase in erythrocytes is determined by incubating a hemolysate with DL-alanine and α-ketoglutaric acid to yield pyruvate. This transamination product is quantitated by forming the hydrazone with 2,4-dinitrophenylhydrazine, extracting it into toluene, and developing a color with strong alkali. The amount of pyruvate formed by the enzyme reaction is proportional to the color intensity and is obtained from a calibration curve prepared with a series of standard sodium pyruvate solutions. Giusti et al.[68] have made an appraisal of this method for the determination of aminotransferase activity in serum and found it to provide a satisfactory assay over a wide range of sensitivity.

The enzyme assay is also carried out with an optimal amount of pyridoxal phosphate added to the assay medium and the percent increase in activity due to the coenzyme supplement is calculated ("stimulation in vitro"). In vitamin B_6 deficiency the enzyme activity without added pyridoxal phosphate is reduced, but there is an elevation in the percent stimulation in vitro.

2. Apparatus

1. Shaking waterbath suitable for incubating test tubes at 37°C.
2. Bench centrifuge.
3. Spectrophotometer suitable for reading at 490 nm.
4. Vortex mixer.
5. Bench timer.
6. Centrifuge tubes, conical, 15 ml.

3. Reagents

Aqueous solutions are prepared in glass-distilled water.

1. Phosphate buffer, 0.1 *M*, pH 7.4: 1.36 g KH_2PO_4 is dissolved in 90 ml water. The pH is adjusted to 7.4 with freshly prepared 20% KOH and the volume made up to 100 ml.

2. Enzyme-substrate mixture:

DL-alanine (Sigma Chemical Co.)	0.89 g
KH_2PO_4	1.00 g
α-ketoglutaric acid (Sigma Chemical Co.)	0.30 g

The reagents are dissolved in 40 ml of water and the pH is adjusted to 7.4 with 20% KOH. The volume is made up to 50 ml. Store at 4°C. Discard after 4 days or if turbidity develops.

3. Pyridoxal phosphate solution: 26.8 mg of pyridoxal 5-phosphate $2H_2O$ (Sigma Chemical Co.) is dissolved in 50 ml of water (1 ml = 500 μg of pyridoxal phosphate). Store in the dark at 4°C and discard after 2 days.

4. Trichloroacetic acid.

5. 2,4-Dinitrophenylhydrazine reagent (Aldrich Chemical Co): 100 g is dissolved in 100 ml of 20% HCl by gentle heating and constant stirring. Stable for at least 2 weeks at room temperature.

6. Alcoholic KOH: 2.5 g of KOH is dissolved in 5 ml of water and the volume is made up to 100 ml with redistilled ethanol. Store at room temperature. Avoid disturbing the sediment.

7. Drabkin's reagent (for hemoglobinometry by a cyanmethemoglobin method[69]): 200 mg of potassium ferricyanide and 50 mg of potassium cyanide are dissolved in water and made up to a final volume of 1 liter. Store in the dark at 4°C.

8. Saline, 0.85%.

4. Procedure

Blood sampling – It is not necessary to fast the subject before taking blood for erythrocyte alanine aminotransferase assays, but if the plasmas are to be used for pyridoxal phosphate determinations, sampling should be made after an overnight fast. Into a tube containing either heparin or EDTA as an anticoagulant 8 to 10 ml of blood is taken.

Preparation of hemolysate – The blood is centrifuged for 30 min at 3,200 rpm. The plasma and the white cell platelet buffy layer are removed gently with a Pasteur pipette. A volume of 0.85% saline roughly equal to that of the packed erythrocytes is added to the cells, which are then resuspended and centrifuged again. The saline is removed, together with any remaining buffy coat debris, then 1 ml of erythrocytes is added to 9 ml water to make a 1:10 hemolysate (for studies of vitamin B_6 deficient subjects with very low enzyme levels a 1:5 hemolysate is used).

TABLE 5

Assay Mixtures for Erythrocyte Alanine Aminotransferase Determinations

Component	Volume additions (ml)			
	Blank*	Test	Stimulated	Standard
Phosphate buffer	1.0	1.0	0.8	1.0
Hemolysate	1.0	1.0	1.0	–
Substrate	–	1.0	1.0	1.0
Pyridoxal phosphate	–	–	0.2	–
Working standard	–	–	–	0.0 to 1.0
Water	–	–	–	1.0 to 0.0

***1.0 ml substrate is added to the blank after 1 drop of trichloroacetic acid so that the enzyme is inhibited and protein precipitated.**

Assay procedure – The individual assays each require six 15-ml conical centrifuge tubes because the blank, test, and pyridoxal phosphate-stimulated test are all determined in duplicate. Mixtures are prepared according to the protocol in Table 5.

The tubes are mixed immediately and incubated in a water bath for 1 hr at 37°C. The blanks are prepared by adding one drop of trichloroacetic acid to inactivate the enzyme before addition of the substrate. At the end of the incubation period, the other tubes are removed from the water bath, one drop of trichloroacetic acid is added to each, and all tubes are immediately mixed thoroughly with a vortex mixer.

Standard curve – This is prepared for each assay run. A working standard is made by diluting the stock pyruvate solution 1:10 (so 1 ml is equivalent to 100 μg sodium pyruvate) and further dilutions are made to give 20, 40, 60, 80, and 100 μg pyruvate in 1 ml water. Tubes for the standards are set up as in Table 5.

Colorimetry – The timing and mixing are critical; a bench timer and a vortex mixer must be used.

While mixing, 1 ml of 2,4-dinitrophenylhydrazine solution is added to the tubes in turn. The first addition is taken as zero time. After 5 min 2 ml of toluene is added to each tube and these are then mixed vigorously for timed 10-sec periods. The tubes are then centrifuged gently for 2 to 3 min to obtain complete separation of the toluene and aqueous layers. From each tube 1 ml of the toluene layer is removed and added to test tubes containing 2 ml of alcoholic KOH. After mixing, the tubes are left to stand for 10 to 20 min to allow the color to develop and 0.2 ml of water is then added to redissolve any cloudiness due to potassium carbonate precipitating out of the solution. The tubes for the standard curve are treated in an identical manner.

The absorbances are read at a wavelength of 490 nm with a 1-cm light path and zero set with distilled water.

5. Calculation

The absorbances of the duplicate blanks usually agree to within 0.01 units and those of the tests and stimulated tests to within 0.02. The mean values are determined and the blank reading is subtracted from the test and the stimulated test values.

The blank reading of the standards is subtracted from each of the standard readings and a calibration curve is plotted (Figure 8). The

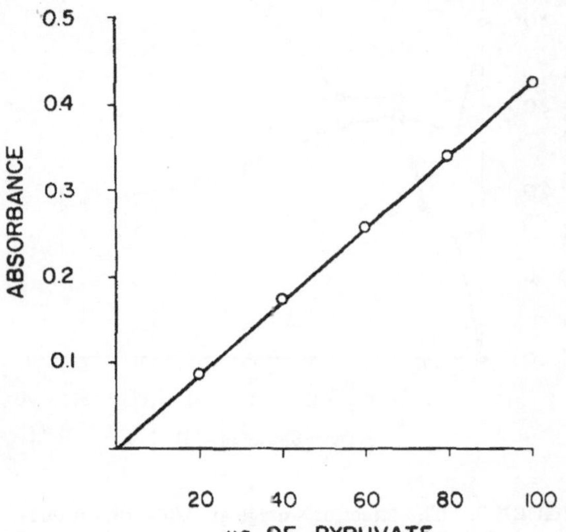

FIGURE 8. Calibration curve for erythrocyte alanine aminotransferase activity.

corrected test and stimulated test absorbances are referred to the standard curve and the amount of pyruvate formed in each case is obtained.

For the investigation of oral contraceptive users it has been convenient to express the enzyme activity in terms of the hemoglobin content of the hemolysate,[16,26] determined by the cyanmethemoglobin method,[69] but if this is done, the whole blood hemoglobin concentration should also be determined because this mode of reference may produce artificially low results in anemic subjects. Alternatives are to relate enzyme activity to the whole blood erythrocyte count[70] or the packed red cell volume.[24] Alanine aminotransferase activity expressed as μg sodium pyruvate produced/hr/ mg Hb is given by:

$$\frac{\text{amount of pyruvate formed in tube}}{\text{Hb concentration (mg/ml)}}$$

% Stimulation by pyridoxal phosphate in vitro is calculated from:

$$\frac{\text{pyruvate formed by stimulated test-pyruvate formed by test}}{\text{pyruvate formed by test}} \times 100$$

6. *Interpretation of Results*

a. Stability of the Hemolysate

It has been reported that alanine aminotransferase in hemolysates stored at -20°C is stable for several weeks,[71] but others have noted some loss of activity during freezing.[24] Figure 9 shows the results of an experiment performed in the author's laboratory in which a hemolysate was prepared and divided into seven aliquots. One of these was assayed immediately and the rest were stored at -20°C for varying intervals of time, after which they too were assayed for enzyme activity. There was a 15% decrease in activity in the first 24 hr and further significant losses occurred over the next 9 days. These reductions in aminotransferase were due to denaturation of the apoenzyme on storage, and not to a loss of coenzyme, because they were not reversed by adding pyridoxal phosphate to the assay medium (i.e., there was no corresponding increase of the stimulation in vitro, Figure 9). In view of these observations, it is considered that erythrocyte alanine aminotransferase has to be assayed on the day of sampling in order to obtain reproducible results.

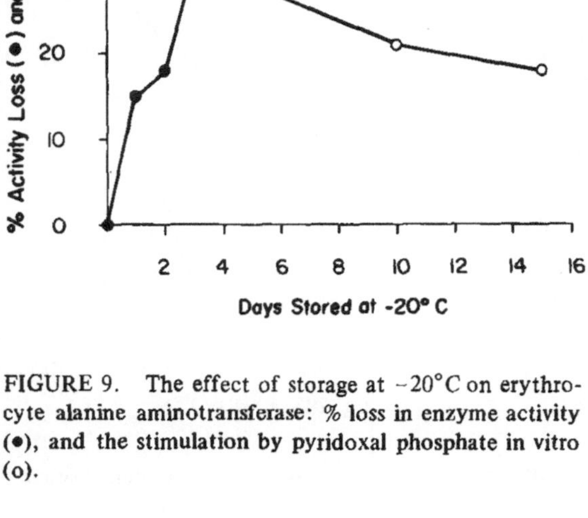

FIGURE 9. The effect of storage at −20°C on erythrocyte alanine aminotransferase: % loss in enzyme activity (•), and the stimulation by pyridoxal phosphate in vitro (o).

b. Within-subject Variation in Assayed Activity

The variation in activity determined in the same healthy adult male subject eating an adequate, self-selected diet has been assayed by performing the erythrocyte alanine aminotransferase assay on ten separate occasions over a period of several months. The results showed that a steady level of both enzyme activity and stimulation in vitro by pyridoxal phosphate are maintained in an individual subject (Table 6).

c. Normal Range and Effect of Oral Contraceptives

The normal ranges for erythrocyte alanine aminotransferase activity and the stimulation in vitro by pyridoxal phosphate that have been obtained in the author's laboratory are given in Table 7. The males had a significantly higher enzyme activity than females, but because there was no difference in the stimulation in vitro, this appears to result from a low apoenzyme level in the women rather than a relative lack of pyridoxal phosphate coenzyme.

Table 7 also summarizes the author's experience of these assays in women taking estrogen-

TABLE 6

Variation in Erythrocyte Alanine Aminotransferase Activity Determined for a Male Adult Subject on 10 Separate Occasions

*Enzyme activity**	% Stimulation
2.03	5
2.38	8
2.33	6
2.25	6
2.00	10
1.80	5
1.60	6
1.80	3
1.92	4
2.13	10
Mean value ± S.D. = 2.02 ± 0.25	**6± 2**

*μg pyruvate/hr/mg hemoglobin.

containing oral contraceptives. Overall, there was no change in the basal enzyme activity for the group of 80 subjects, but a significant increase in the stimulation in vitro compared to controls did suggest the presence of vitamin B_6 deficiency in some of these women.[26]

TABLE 7

Eythrocyte Alanine Aminotransferase Activity in Male and Female Controls and Oral Contraceptive Users

			Enzyme activity*		% Stimulation in vitro	
Group	No.	Age range (yr)	range	mean ± S.D.	range	mean ± S.D.
Male controls	11	21–35	0.72–2.25	1.41 ± 0.49	4–30	14 ± 12
Female controls	50	21–35	0.41–1.80	0.91 ± 0.36†	0–56	18 ± 14
Oral contraceptive users	80	17–28	0.29–2.50	1.04 ± 0.48	1–118	25 ± 21††

*μg pyruvate/hr/mg hemoglobin.
†Significantly different from male controls, $p < 0.001$.
††Significantly different from female controls, $0.01 > p > 0.001$.

IX. THE CLINICAL SIGNIFICANCE OF TRYPTOPHAN METABOLIC STUDIES AND TESTS OF VITAMIN B_6 NUTRITIONAL STATUS IN PREGNANCY AND ORAL CONTRACEPTIVE USERS

A. Pregnancy

Normal pregnancy is associated with a similar disturbance of tryptophan metabolism to that occurring in women taking estrogen-progestogen preparations. Elevated excretions of xanthurenic acid, kynurenine, 3-hydroxykynurenine and kynurenic acid,[10] and 3-hydroxyanthranilic acid[72] are all present in urine collected after a tryptophan load; these abnormalities are reversed by treatment with pyridoxine. In addition, pregnant women excrete increased amounts of N^1-methylnicotinamide[72] and *N*-methyl-2-pyridone-5-carboxamide[10] and have an enhanced yield of these nicotinic acid metabolites after tryptophan administration. On the basis of these data, Brown et al.[10] concluded that two factors are responsible for the altered tryptophan metabolism of pregnancy: vitamin B_6 deficiency brought about by the additional burden of fetal nutritional demands and a direct influence of the hormonal environment of pregnancy on the tryptophan-nicotinic acid ribonucleotide pathway.

Some of the compelling evidence that vitamin B_6 deficiency develops in some women during the latter stages of normal pregnancy is summarized in Table 8. It appears from the published data that a

TABLE 8

Evidence for the Development of Vitamin B_6 Deficiency During Normal Pregnancy

Author(s)	Observation
Wachstein and Gudaitis.[21]	Low 4-pyridoxic acid excretion after test dose of pyridoxine HCL
Wachstein et al.[73]	Low leukocyte pyridoxal phosphate in women at term
Wachstein et al.[74]	Low leukocyte and plasma pyridoxal phosphate in women at term with subnormal rise in plasma level after a pyridoxine load
Hamfelt and Hahn[22]	Gradual decrease in plasma pyridoxal phosphate throughout pregnancy
Brin[75]	Low plasma B_6-vitamers and erythrocyte aminotrasferase activities in maternal blood at term compared to cord blood levels

TABLE 9

Evidence for Vitamin B_6 Deficiency in Oral Contraceptive Users

Author(s)	Test	Incidence of abnormal results
Doberenz et al.[25]	Erythrocyte alanine aminotransferase	5 of 13 (39%)
Rose et al.[16] and unpublished data (Figure 10)	HK/HA ratio and urinary 4-pyridoxic acid excretion	11 of 40 (27%)
Rose et al.[26]	Erythrocyte alanine aminotransferase in vitro stimulation	12 of 80 (15%)
Lumeng et al.[67]	Plasma pyridoxal phosphate	9 of 52 (17%)

true deficiency of vitamin B_6 develops toward term which is due to transplacental uptake by the fetus. Thus, at term the plasma levels of pyridoxal, pyridoxamine, and pyridoxine and the erythrocyte alanine and aspartate aminotransferase activities are considerably higher in cord blood than in maternal blood.[75]

A possible role for vitamin B_6 deficiency in the pathogenesis of toxemia of pregnancy has been explored by a number of workers, with results which remain inconclusive. Sprince and colleagues[76] originally reported that after a 10-g oral dose of DL-tryptophan, the xanthurenic acid excretion by patients with pre-eclampsia was higher than in normal pregnancy. They postulated that toxemia is associated with an abnormal requirement for vitamin B_6 due to disturbed protein metabolism or excessive fibrinolytic activity. Very high xanthurenic acid excretions after DL-tryptophan loading were also obtained in pre-eclamptic patients by Wachstein and Graffeo,[77] who also reported a lower incidence of toxemia in pregnant women treated with 10 mg pyridoxine hydrochloride daily than in a group of untreated controls. Unfortunately, this observation was not confirmed by a double-blind clinical trial in which the occurrence of obstetrical complications was assessed in 956 pregnant women treated with 20 mg pyridoxine hydrochloride daily and 576 who did not receive the vitamin.[78]

The clinical significance of vitamin B_6 deficiency in pregnancy remains uncertain and at the present time there is no good basis on which to recommend routine administration of pyridoxine to pregnant women.

B. Oral Contraceptive Users

The evidence currently available that vitamin B_6 deficiency may be precipitated by the use of oral contraceptives is summarized in Table 9. The most recent data obtained from studies of the ratio of 3-hydroxykynurenine to 3-hydroxyanthranilic acid (HK/HA) excreted and the urinary 4-pyridoxic acid in oral contraceptive users are illustrated in Figure 10. They confirm the previous finding[16] of elevated HK/HA ratios and subnormal 4-pyridoxic acid levels in some of these women.

This is an extremely active area of research at the present time and no doubt more detailed information will be available in the near future. The incidence of deficiency has not yet been defined with any certainty and any figure quoted will depend on the criteria employed, but the available data suggest that reduced tissue pyridoxal phosphate coenzyme levels occur in 15 to 20% of

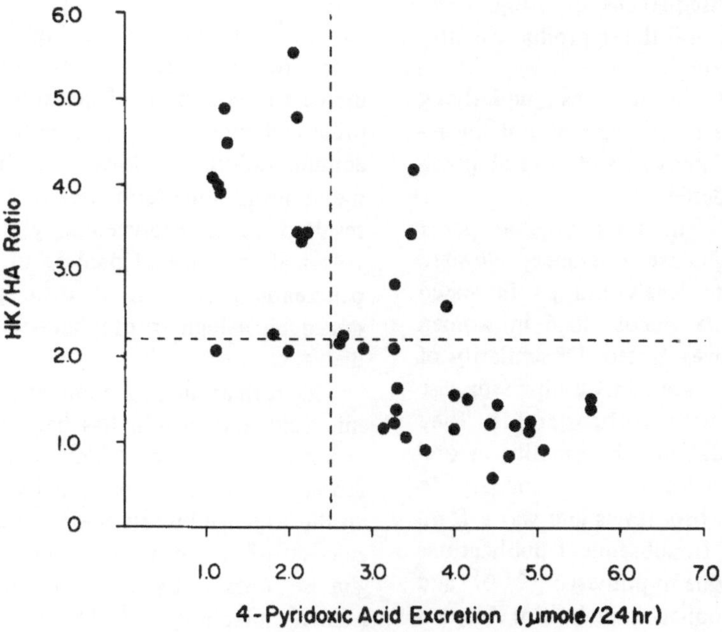

FIGURE 10. HK/HA ratios and 4-pyridoxic acid excretions by 40 oral contraceptive users. The upper limit of normal for the HK/HA ratio is indicated by the horizontal dotted line, and the lower limit of normal for 4-pyridoxic acid by the vertical dotted line. Eleven subjects had biochemical evidence of vitamin B_6 deficiency, as judged by the two tests in combination.

women taking estrogen-containing oral contraceptives.

Altered carbohydrate metabolism with impaired glucose tolerance, an excessive plasma insulin response to a glucose load, and elevated plasma pyruvate levels is a well-recognized biochemical side effect of oral contraceptives.[79,80] In one longitudinal study,[81] a deterioration in glucose tolerance after taking an estrogen-containing oral contraceptive compared with the pretreatment results occurred in 78% and 70% of oral and intravenous glucose tolerance tests, respectively. In 13% of the subjects the abnormalities were sufficiently severe to be classified as "chemical" diabetes mellitus.

Cortisone or prednisone-provocative glucose tolerance tests have been employed in several studies; distinct abnormalities were obtained in a high proportion of women taking estrogen-containing oral contraceptives,[82,83] including impairment of insulin secretion.[84]

Estrogens alone may impair glucose tolerance[82,85] and the changes in carbohydrate metabolism which occur during oral contraceptive administration have been ascribed generally to the estrogenic component. Megestrol acetate does not affect carbohydrate metabolism,[86] but Spellacy et al.[87] have recently reported that another progestogen, ethynodiol diacetate, may increase the fasting blood glucose and diminish glucose tolerance in about 13% of women when used alone for contraceptive purposes.

The clinical importance of altered carbohydrate metabolism and impaired glucose tolerance in oral contraceptive users remains to be established. Wynn and Doar[80] have pointed out that it will require an extensive study of several hundred women over many years before the risk of pancreatic islet exhaustion and the onset of clinically overt diabetes mellitus can be properly assessed. The dangers of permanent damage to pancreatic endocrine function have also been discussed by Kalkhoff and co-workers[84] and Spellacy.[88]

Studies in which the reversal of impaired glucose tolerance has been assessed when oral contraceptive use is discontinued indicate that a return to normal carbohydrate metabolism does take place, although it may take several months.[80] But, it must be pointed out that these women had been using an oral contraceptive for a maximum period of 72 months and a mean duration of only

23.8 months. The consequences of using these preparations over most of the reproductive life span may be very different.

The biochemical mechanisms underlying abnormal carbohydrate metabolism in oral contraceptive users are not known, but several possibilites have been considered.

Wynn and Doar,[89] in their original paper describing impaired glucose tolerance, elevated fasting blood pyruvate levels and an increased pyruvate response to a glucose load in women taking oral contraceptives, noted the similarity of these changes to those seen in Cushing's disease and patients treated with corticosteroids. They suggested that the metabolic abnormality in oral contraceptive-treated women was secondary to elevated circulating cortisol levels and was a form of "steroid diabetes." In subsequent publications they have developed this hypothesis[80,81,90] and the studies of tryptophan metabolism in oral contraceptive users have provided indirect support for a cortisol-mediated, estrogen-induced stimulation of liver enzymes.

Fasting plasma growth hormone levels are elevated both in women taking the combined type of oral contraceptives and in subjects treated with estrogens alone.[91-94] Spellacy et al.[91] also measured the plasma growth hormone levels during insulin-induced hypoglycemia and observed larger increases in the oral contraceptive group as compared with controls.

Because growth hormone exerts a diabetogenic effect, it has been postulated that high levels of this hormone are responsible for the altered carbohydrate metabolism brought about by contraceptive steroids.[91,95] Maw and Wynn,[93] on the other hand, although observing elevated fasting growth hormone levels, could find no correlation between these changes and the plasma glucose levels and noted that deterioration occurred in individuals with widely differing growth hormone responses. It appears from their study that increased growth hormone activity is not of primary etiological significance in abnormal carbohydrate metabolism due to oral contraceptive usage.

Some recent studies have suggested that abnormal tryptophan metabolism or perhaps vitamin B_6 deficiency per se may be concerned in the alteration in carbohydrate metabolism by women taking oral contraceptives.

There are clear indications that tryptophan metabolites are involved in the etiology of some forms of experimental diabetes. Twenty years ago Kotake[96] found that a high tryptophan diet supplemented with sodium butyrate causes an increased excretion of xanthurenic acid in rat urine and that this effect is reversed by pyridoxine administration. In addition to altered tryptophan metabolism, long-term treatment with this diet resulted in hyperglycemia, glycosuria, and histological evidence of damage to the β-cells of the pancreatic islets. An identical picture was produced by a high tryptophan-vitamin B_6-deficient diet.

The formation of a complex between xanthurenic acid and insulin has been demonstrated by experiments in vitro.[97] This complex is stable and shows a loss of hormonal activity compared to free insulin as judged by its ability to depress the blood glucose of experimental animals and stimulate glucose uptake by rat epididymal fat pads or isolated diaphragm.[98] On the basis of these experiments, Kotake et al.[98] postulated that any condition that causes a high circulating level of xanthurenic acid may impair insulin function because of complex formation between the hormone and xanthurenic acid.

One such condition is oral contraceptive administration which raised the possibility that only a proportion of the elevated immunoreactive insulin present in the plasma of oral contraceptive users during a glucose tolerance test has full biological activity, the rest being complexed with xanthurenic acid. Impaired glucose tolerance in these women might then be related directly to their abnormal tryptophan metabolism.[19]

Spellacy et al.[99] reported that vitamin B_6 will improve the glucose tolerance of women whose tests have deteriorated previously during the use of oral contraceptives. As a group, the 12 women showed an improvement in oral glucose tolerance after treatment with pyridoxine hydrochloride 25 mg daily for 1 month, although in 4 there was no benefit. There was no attempt to assess vitamin B_6 nutritional status in this small pilot study.

Adams et al. (to be published) determined oral glucose tolerance, urinary xanthurenic acid excretion and the HK/HA ratio after a 2-g L-tryptophan load, 4-pyridoxic acid excretion, and erythrocyte alanine and aspartate aminotransferase activities in 31 women taking an estrogen-containing oral contraceptive before and after receiving 40 mg pyridoxine daily for 1 month. Twelve had biochemical evidence of vitamin B_6 deficiency and their glucose tolerance was improved by pyridoxine administration. Treatment with the vitamin did not alter glucose tolerance in the nondeficient group. There was no correlation between the initial level of the xanthurenic acid excretion and glucose tolerance.

Evidence has been obtained recently which again suggests that vitamin B_6 deficiency, rather than elevated concentrations of tryptophan metabolites, may be concerned in abnormal carbohydrate metabolism by some oral contraceptive users. Rose et al.[100] have examined the effect of experimentally produced dietary vitamin B_6 deficiency on oral glucose tolerance in groups of oral contraceptive users and control subjects. After 4 weeks on the diet, the women taking oral contraceptives showed a marked deterioration in glucose tolerance, which was reversed completely by supplementation with pyridoxine. There was no correlation between the urinary xanthurenic acid and altered carbohydrate metabolism and the controls, who had excreted large quantities of xanthurenic acid during the vitamin B_6 deficiency period, did not suffer any deterioration in glucose tolerance.

The results of these studies indicate that the presence of vitamin B_6 deficiency in women taking contraceptive steroid preparations may lead to altered carbohydrate metabolism. Further work is required to elucidate the mechanisms involved, but at present there is no support for a role of the xanthurenic acid-insulin complex described by Kotake and co-workers in the production of this metabolic disturbance.

A second complication of oral contraceptive usage that may be related to abnormal tryptophan metabolism is mental depression. There have been a number of reports describing an increased incidence of depression in women taking estrogen-progestogen preparations, usually for contraceptive purposes. The frequency with which this occurs is uncertain, but a figure of 6.6% was obtained in one study.[101]

Various clinical and experimental investigations have evolved a body of evidence which implicates diminished 5-hydroxytryptamine synthesis in the etiology of depressive illness. The metabolism of L-tryptophan to 5-hydroxytryptamine involves two enzymatic steps, the first a hydroxylation reaction to form 5-hydroxytryptophan and the second a pyridoxal phosphate-dependent decarboxylation to yield the amine. Although 5-hydroxytryptamine is synthesized in several organs, it cannot cross the blood-brain barrier, so that functioning as a cerebral transmitter it has to be formed *in situ.*

Curzon[102] has postulated that low brain 5-hydroxytryptamine levels occur in depression because elevated plasma corticosteroids are present which induce a high activity of tryptophan oxygenase and, hence, increase the metabolism of substrate along the tryptophan-nicotinic acid ribonucleotide pathway. This may result in an inadequate supply of free tryptophan for brain 5-hydroxytryptamine production. An alternative mechanism is suggested by the finding that kynurenine and 3-hydroxykynurenine decrease the uptake of L-tryptophan into brain slices and that in the intact rat administration of these metabolites reduces cerebral 5-hydroxytryptamine concentration.[103] These experimental observations may be relevant to depression in oral contraceptive users because here too there are abnormal tryptophan metabolism and probably elevated tryptophan oxygenase activity.[15,105]

Furthermore, in man, unlike the rat, decarboxylation of 5-hydroxytryptophan may be rate-limiting for 5-hydroxytryptamine synthesis.[106] If this is so, oral contraceptives may decrease amine formation by impairing the coenzymic function of pyridoxal phosphate in this reaction.

These possibilities were recognized by Winston[18] who suggested that treatment with pyridoxine might alleviate depressive symptoms in oral contraceptive users and subsequently reported that this was the case.[106]

Although the original study was criticized because it lacked adequate controls and safeguards against a "placebo effect," a formal double-blind cross-over clinical trial has now been completed.[107] An assessment of vitamin B_6 nutrition was made by determination of erythrocyte alanine and aspartate aminotransferase activities, 4-pyridoxic acid excretion, and the urinary HK/HA ratio before the commencement of the therapeutic trial. Evidence of deficiency was found in 11 of 22 women who had developed depression during treatment with contraceptive steroids. These women, but not the other 11, were improved by pyridoxine administration; placebo tablets of identical appearance were without effect.

While it is hoped that other centers will attempt to confirm the findings of this investigation, these first results do indicate that vitamin B_6 deficiency is one of the causes of mental depression in oral contraceptive users. In other cases a lack of available substrate for 5-hydroxytryptamine synthesis may be the principal factor and these women may benefit from supplementing the diet with additional tryptophan as do some patients with endogenous depression.[108]

ACKNOWLEDGMENTS

I wish to thank Dr. Raymond R. Brown, University of Wisconsin Medical School, for allowing me to include the details of his procedure for the assay of plasma pyridoxal phosphate in this chapter. My own work is supported by USPHS, Contract NIH-NICHD-72-2782, Grant CA-13302, and by Grant CA-14520, awarded to the Wisconsin Clinical Cancer Center by the National Cancer Institute.

REFERENCES

1. **Salhanick, H. A., Kipnis, D. M., and Vande Wiele, R. L.**, Eds., *Metabolic Effects of Gonadal Hormones and Contraceptive Steroids,* Plenum Press, New York, 1969.
2. **Altman, K. and Greengard, O.**, Tryptophan pyrrolase induced in human livers by hydrocortisone: effect on excretion of kynurenine, *Science,* 151, 332, 1966.
3. **Yess, N., Price, J. M., Brown, R. R., Swan, P. B., and Linkswiler, H.**, Vitamin B_6 depletion in man: urinary excretion of tryptophan metabolites, *J. Nutr.,* 84, 229, 1964.
4. **Price, S. A., Rose, D. P., and Toseland, P. A.**, Effects of dietary vitamin B_6 deficiency and oral contraceptives on the spontaneous urinary excretion of 3-hydroxyanthranilic acid, *Am. J. Clin. Nutr.,* 25, 494, 1972.
5. **Brown, R. R., Yess, N., Price, J. M., Linkswiler, H., Swan, P., and Hankes, L. V.**, Vitamin B_6 depletion in man: urinary excretion of quinolinic acid and niacin metabolites, *J. Nutr.,* 87, 419, 1965.
6. **Rose, D. P. and Toseland, P. A.**, Urinary excretion of quinolinic acid and other tryptophan metabolites after deoxypyridoxine or oral contraceptive administration, *Metabolism,* 22, 165, 1973.
7. **Bryan, G. T.**, The role of urinary tryptophan metabolites in the etiology of bladder cancer, *Am. J. Clin. Nutr.,* 24, 841, 1971.
8. **Bell, E. M., Mainwaring, W. I. P., Bulbrook, R. D., Tong, D., and Hayward, J. L.**, Relationships between excretion of steroid hormones and tryptophan metabolites in patients with breast cancer, *Am. J. Clin. Nutr.,* 24, 694, 1971.
9. **Sauberlich, H. E.**, Human requirements for vitamin B_6, *Vitam. Horm.,* 22, 807, 1964.
10. **Brown, R. R., Thornton, M. J., and Price, J. M.**, The effect of vitamin supplementation on the urinary excretion of tryptophan metabolites by pregnant women, *J. Clin. Invest.,* 40, 617, 1961.
11. **Rose, D. P.**, Excretion of xanthurenic acid in the urine of women taking progestogen-oestrogen preparations, *Nature,* 210, 196, 1966.
12. **Rose, D. P.**, The influence of oestrogens on tryptophan metabolism in man, *Clin. Sci.,* 31, 265, 1966.
13. **Price, J. M., Thornton, M. J., and Mueller, L. M.**, Tryptophan metabolism in women using steroid hormones for ovulation control, *Am. J. Clin. Nutr.,* 20, 452, 1967.
14. **Rose, D. P., Adams, P. W., and Strong, R.**, Influence of the progestogenic component of oral contraceptives on tryptophan metabolism, *J. Obstet. Gynaecol. Br. Commonw.,* 80, 82, 1973.
15. **Rose, D. P. and McGinty, F.**, The effect of steroid hormones on tryptophan metabolism, in *Advances in Steroid Biochemistry and Pharmacology,* Vol. 1, Briggs, M. H., Ed., Academic Press, London, 1970, 97.
16. **Rose, D. P., Strong, R., Adams, P. W., and Harding, P. E.**, Experimental vitamin B_6 deficiency and the effect of oestrogen-containing oral contraceptives on tryptophan metabolism and vitamin B_6 requirements, *Clin. Sci.,* 42, 465, 1972.
17. **Rose, D. P., Brown, R. R., and Price, J. M.**, Metabolism of tryptophan to nicotinic acid derivatives by women taking oestrogen-progestogen preparations, *Nature,* 219, 1259, 1968.
18. **Winston, F.**, Oral contraceptives and depression, *Lancet,* 1, 1209, 1969.
19. **Rose, D. P. and Adams, P. W.**, Oral contraceptives and tryptophan metabolism: effects of oestrogen in low dose combined with a progestagen and of a low-dose progestagen (megestrol acetate) given alone, *J. Clin. Pathol.,* 25, 252, 1972.
20. **Baysal, A., Johnson, B. A., and Linkswiler, H.**, Vitamin B_6 depletion in man: blood vitamin B_6, plasma pyridoxal-phosphate, serum cholesterol, serum transaminases and urinary vitamin B_6 and 4-pyridoxic acid, *J. Nutr.,* 89, 19, 1966.

21. Wachstein, M. and Gudaitis, A., Disturbance of vitamin B_6 metabolism in pregnancy. III. Abnormal vitamin B_6 load test, *Am. J. Obstet. Gynecol.*, 66, 1207, 1953.
22. Hamfelt, A. and Hahn, L., Pyridoxal phosphate concentration in plasma and tryptophan load test during pregnancy, *Clin. Chim. Acta*, 25, 91, 1969.
23. Cinnamon, A. D. and Beaton, J. R., Biochemical assessment of vitamin B_6 status in man, *Am. J. Clin. Nutr.*, 23, 696, 1970.
24. Woodring, M. J. and Storvick, C. A., Effect of pyridoxine supplementation on glutamic pyruvic transaminase and *in vitro* stimulation in erythrocytes of normal women, *Am. J. Clin. Nutr.*, 23, 1385, 1970.
25. Doberenz, A. R., Van Miller, J. P., Green, J. R., and Beaton, J. R., Vitamin B_6 depletion in women using oral contraceptives as determined by erythrocyte glutamic-pyruvic transaminase activities, *Proc. Soc. Exp. Biol. Med.*, 137, 1100, 1971.
26. Rose, D. P., Strong, R., Folkard, J., and Adams, P. W., Erythrocyte aminotransferase activities in women using oral contraceptives and the effect of vitamin B_6 supplementation, *Am. J. Clin. Nutr.*, 26, 48, 1973.
27. Raica, N. and Sauberlich, H. E., Blood cell transaminase activity in human vitamin B_6 deficiency, *Am. J. Clin. Nutr.*, 15, 67, 1964.
28. Hamfelt, A., Pyridoxal phosphate concentration and aminotransferase activity in human blood cells, *Clin. Chim. Acta*, 16, 19, 1967.
29. Aly, H. E., Donald, E. A., and Simpson, M. H. W., Oral contraceptives and vitamin B_6 metabolism, *Am. J. Clin. Nutr.*, 24, 297, 1971.
30. Cheney, M. C. and Beaton, G. H., Assessment of vitamin B_6 nutriture in pregnancy in the rat, *Proc. Soc. Exp. Biol. Med.*, 120, 692, 1965.
31. Toseland, P. A., Michelin, M. J., and Price, S. A., The determination of 3-hydroxyanthranilic acid in human urine by thin-layer electrophoresis and fluorimetry, *Clin. Chim. Acta*, 37, 477, 1972.
32. Price, J. M., Disorders of tryptophan metabolism, *Univ. Mich. Med. Bull.*, 24, 461, 1958.
33. Coursin, D. B., Recommendations for standardization of the tryptophan load test, *Am. J. Clin. Nutr.*, 14, 56, 1964.
34. Price, J. M., Brown, R. R., and Yess, N., Testing the functional capacity of the tryptophan-niacin pathway in man by analysis of urinary metabolites, *Adv. Metab. Disord.*, 2, 159, 1965.
35. Rose, D. P. and Randall, Z. C., Influence of the loading dose on the demonstration of abnormal tryptophan metabolism by cancer patients, *Clin. Chim. Acta*, 47, 45, 1973.
36. O'Brien, D. and Jensen, C. B., Pyridoxine dependency in two mentally retarded subjects, *Clin. Sci.*, 24, 179, 1963.
37. Rapoport, M. I. and Beisel, W. R., Circadian periodicity of tryptophan metabolism, *J. Clin. Invest.*, 47, 934, 1968.
38. Kowlessar, O. D., Haeffner, L. J., and Benson, G. D., Abnormal tryptophan metabolism in patients with adult celiac disease, with evidence for deficiency of vitamin B_6, *J. Clin. Invest.*, 43, 894, 1964.
39. Rose, D. P. and McGinty, F., The influence of adrenocortical hormones and vitamins upon tryptophan metabolism in man, *Clin. Sci.*, 35, 1, 1968.
40. Kelsall, M. A., Ed., Vitamin B_6 metabolism of the nervous system, *Ann. N. Y. Acad. Sci.*, 166, 1, 1969.
41. Tompsett, S. L., The determination in urine of some metabolites of tryptophan – kynurenine, anthranilic acid and 3-hydroxyanthranilic acid – and reference to the presence of o-aminophenol in urine, *Clin. Chim. Acta*, 4, 411, 1959.
42. Coppini, D., Benassi, C. A., and Montorsi, M., Quantitative determination of tryptophan metabolites (via kynurenine) in biologic fluids, *Clin. Chem.*, 5, 391, 1959.
43. Walsh, M. P., Separation and estimation of tryptophan-nicotinic acid metabolites in urine by thin-layer chromatography, *Clin. Chim. Acta*, 11, 263, 1965.
44. Bell, E. D. and Mainwaring, W. I. P., A method for the determination of urinary metabolites of tryptophan, *Clin. Chim. Acta*, 35, 83, 1971.
45. Watanabe, M. and Hayashi, K., A fluorometric method for 3-hydroxyanthranilic acid in human urine, *Clin. Chim. Acta*, 37, 417, 1972.
46. Hill, H. D., Summer, G. K., and Roszel, N. O., Determination of tryptophan derivatives by thin-layer chromatography, *Anal. Biochem.*, 16, 84, 1966.
47. Chen, N. C. and Gholson, R. K., An improved column chromatographic method for isolation of tryptophan metabolites, *Anal. Biochem.*, 47, 139, 1972.
48. Rose, D. P. and Toseland, P. A., The determination of 3-hydroxyanthranilic acid in urine by gas-liquid chromatography, *Clin. Chim. Acta*, 17, 235, 1967.
49. Hirano, K., Mori, K., Tsuboi, N., Kawai, S., and Ohno, T., Gas chromatography of urinary anthranilic acid and 3-hydroxyanthranilic acid by solvent extraction method, *Chem. Pharm. Bull.*, 20, 1412, 1972.
50. Toseland, P. A., The determination of urinary quinolinic acid by gas-liquid chromatography, *Clin. Chim. Acta*, 25, 185, 1969.
51. Arend, R. A., Leklem, J. E., and Brown, R. R., Direct and steam distillation autoanalyzer methods for assay of diazotizable aromatic amine metabolites of tryptophan in urine and in serum, *Biochem. Med.*, 4, 457, 1970.
52. Rose, D. P., in discussion of Bell, E. M., Mainwaring, W. I. P., Bulbrook, R. D., Tong, D., and Hayward, J. L., Relationships between excretion of steroid hormones and tryptophan metabolites in patients with breast cancer, *Am. J. Clin. Nutr.*, 24, 694, 1971.

53. **Rose, D. P.**, The influence of sex, age, and breast cancer on tryptophan metabolism, *Clin. Chim. Acta*, 18, 221, 1967.
54. **Rose, D. P.**, Tryptophan metabolism in carcinoma of the breast, *Lancet*, 1, 239, 1967.
55. **Rosen, F., Huff, J. W., and Perlzweig, W. A.**, The role of B_6-deficiency in the tryptophan-niacin relationships in rats, *J. Nutr.*, 33, 561, 1947.
56. **Glazer, H. S., Mueller, J. F., Thompson, C., Hawkins, V. R., and Vilter, R. W.**, A study of urinary excretion of xanthurenic acid and other tryptophan metabolites in human beings with pyridoxine deficiency induced by desoxypyridoxine, *Arch. Biochem. Biophys.*, 33, 243, 1951.
57. **Rosen, F., Lowy, R. S., and Sprince, H.**, A rapid assay for xanthurenic acid in urine, *Proc. Soc. Exp. Biol. Med.*, 77, 399, 1951.
58. **Satoh, K. and Price, J. M.**, Fluorometric determination of kynurenic acid and xanthurenic acid in human urine, *J. Biol. Chem.*, 230, 781, 1958.
59. **Heeley, A. F.**, The effect of pyridoxine on tryptophan metabolism in phenylketonuria, *Clin. Sci.*, 29, 465, 1965.
60. **Rose, D. P. and Randall, Z. C.**, Reassessment of tryptophan metabolism in breast cancer five years after an initial study, *Clin. Chim. Acta*, 45, 33, 1973.
61. **Mainardi, L. and Tenconi, L. T.**, Contributo allo studio dei metaboliti della linea triptofano → acido nicotinico in rapporto al sesso, *Acta Vitaminol.*, 18, 249, 1964.
62. **Reddy, S. K., Reynolds, M. S., and Price, J. M.**, The determination of 4-pyridoxic acid in human urine, *J. Biol. Chem.*, 233, 691, 1958.
63. **Huff, J. W. and Perlzweig, W. A.**, A product of oxidative metabolism of pyridoxine, 2-methyl-3-hydroxy-4-carboxy-5-hydroxy-methylpyridine (4-pyridoxic acid). I. Isolation from urine, structure, and synthesis, *J. Biol. Chem.*, 155, 345, 1944.
64. **Sarett, H. P.**, A study of the measurement of 4-pyridoxic acid in urine, *J. Biol. Chem.*, 189, 769, 1951.
65. **Sundaresan, P. R. and Coursin, D. B.**, Microassay of pyridoxal phosphate using L-tyrosine-1-^{14}C and tyrosine apodecarboxylase, *Meth. Enzymol.*, 18, 509, 1970.
66. **Chabner, B. and Livingston, D.**, A simple enzymic assay for pyridoxal phosphate, *Anal. Biochem.*, 34, 413, 1970.
67. **Lumeng, L., Cleary, R. E., and Li, T. K.**, Plasma pyridoxal-5′-phosphate in oral contraceptive users and in chronic alcohol abuse, *Fed. Proc.*, 32, 891, 1973.
68. **Giusti, G., Ruggioro, G., and Cacciatore, L.**, A comparative study of some spectrophotometric and colorimetric procedures for the determination of serum glutamic-oxaloacetic and glutamic-pyruvic transaminase in hepatic diseases, *Enzymol. Biol. Clin.*, 10, 17, 1969.
69. **Dacie, J. V.**, *Practical Haematology*, 2nd ed., Churchill, London, 1958, 31.
70. **Cavill, I. A. J. and Jacobs, A.**, Erythrocyte transaminase activity in iron deficiency anaemia, *Scand. J. Haematol.*, 4, 249, 1967.
71. **Cheney, M., Sabry, Z. I., and Beaton, G. H.**, Erythrocyte glutamic-pyruvic transaminase activity in man, *Am. J. Clin. Nutr.*, 16, 337, 1965.
72. **Hernandez, T.**, Tryptophan metabolite excretion in pregnancy after a tryptophan load test, *Fed. Proc.*, 23, 136, 1964.
73. **Wachstein, M., Moore, C., and Graffeo, L. W.**, Pyridoxal phosphate (B_6-al-PO_4) levels of circulating leukocytes in maternal and cord blood, *Proc. Soc. Exp. Biol. Med.*, 96, 326, 1957.
74. **Wachstein, M., Kellner, J. D., and Ortiz, J. M.**, Pyridoxal phosphate in plasma and leukocytes of normal and pregnant subjects following B_6 load tests, *Proc. Soc. Exp. Biol. Med.*, 103, 350, 1960.
75. **Brin, M.**, Abnormal tryptophan metabolism in pregnancy and with the oral contraceptive pill. II. Relative levels of vitamin B_6 vitamers in cord and maternal blood, *Am. J. Clin. Nutr.*, 24, 704, 1971.
76. **Sprince, H., Lowy, R. S., Folsome, C. E., and Behrman, J. S.**, Studies on the urinary excretion of "xanthurenic acid" during normal and abnormal pregnancy: a survey of the excretion of "xanthurenic acid" in normal nonpregnant, pre-eclamptic, and eclamptic women, *Am. J. Obstet. Gynecol.*, 62, 84, 1951.
77. **Wachstein, M. and Graffeo, L. W.**, Influence of vitamin B_6 on the incidence of pre-eclampsia, *Obstet. Gynecol.*, 8, 177, 1956.
78. **Hillman, R. W., Cabaud, P. G., Nilsson, D. E., Arpin, P. D., and Tufano, R. J.**, Pyridoxine supplementation during pregnancy, *Am. J. Clin. Nutr.*, 12, 427, 1963.
79. **Wynn, V. and Doar, J. W. H.**, Longitudinal studies of the effects of oral contraceptive therapy on plasma glucose, non-esterified fatty acid, insulin and blood pyruvate levels during oral and intravenous glucose tolerance tests, in *Metabolic Effects of Gonadal Hormones and Contraceptive Steroids*, Salhanick, H. A., Kipnis, D. M., and Vande Wiele, R. L., Eds., Plenum Press, New York, 1969, 157.
80. **Wynn, V. and Doar, J. W. H.**, Effects of oral contraceptives on carbohydrate metabolism, *J. Clin. Pathol.*, Suppl. 23, 3, 19, 1970.
81. **Wynn, V. and Doar, J. W. H.**, Some effects of oral contraceptives on carbohydrate metabolism, *Lancet*, ii, 761, 1969.
82. **Javier, Z., Gershberg, H., and Hulse, M.**, Ovulatory suppressants, estrogens, and carbohydrate metabolism, *Metabolism*, 17, 443, 1968.
83. **di Paola, G., Puchulu, F., Robin, M., Nicholson, R., and Marti, M.**, Oral contraceptives and carbohydrate metabolism, *Am. J. Obstet. Gynecol.*, 101, 206, 1968.

84. **Kalkhoff, R. K., Kim, H. J., and Stoddard, F. J.**, Acquired subclinical diabetes mellitus in women receiving oral contraceptive agents, in *Metabolic Effects of Gonadal Hormones and Contraceptive Steroids*, Salhanick, H. A., Kipnis, D. M., and Vande Wiele, R. L., Eds., Plenum Press, New York, 1969, 193.
85. **Goldman, J. A. and Ovadia, J. L.**, The effect of estrogen on intravenous glucose tolerance in women, *Am. J. Obstet. Gynecol.*, 103, 172, 1969.
86. **Adams, P. W. and Wynn, V.**, The effects of a progestogen, megestrol acetate, on carbohydrate and lipid metabolism, *J. Obstet. Gynaecol. Br. Commonw.*, 79, 744, 1972.
87. **Spellacy, W. N., Buhi, W. C., and Birk, S. A.**, The effect of the progestogen ethynodiol diacetate on glucose, insulin, and growth hormone after six months treatment, *Acta Endocrinol.*, 70, 373, 1972.
88. **Spellacy, W. N.**, A review of carbohydrate metabolism and the oral contraceptives, *Am. J. Obstet. Gynecol.*, 104, 448, 1969.
89. **Wynn, V. and Doar, J. W. H.**, Some effects of oral contraceptives on carbohydrate metabolism, *Lancet*, ii, 715, 1966.
90. **Doar, J. W. H. and Wynn, V.**, Effects of obesity, glucocorticoid and oral contraceptive therapy on plasma glucose and blood pyruvate levels, *Br. Med. J.*, 1, 149, 1970.
91. **Spellacy, W. N., Carlson, K. L., and Schade, S. L.**, Human growth hormone levels in normal subjects receiving an oral contraceptive, *J.A.M.A.*, 202, 451, 1967.
92. **Yen, S. S. C. and Vela, P.**, Effects of contraceptive steroids on carbohydrate metabolism, *J. Clin. Endocrinol.*, 28, 1564, 1968.
93. **Maw, D. S. J. and Wynn, V.**, The relation of growth hormone to altered carbohydrate metabolism in women taking oral contraceptives, *J. Clin. Pathol.*, 25, 354, 1972.
94. **Frantz, A. G. and Rabkin, M. T.**, Effects of estrogen and sex difference on secretion of human growth hormone, *J. Clin. Endocrinol.*, 25, 1470, 1965.
95. **Davidson, M. B. and Holzman, G. B.**, Role of growth hormone in the alteration of carbohydrate metabolism induced by oral contraceptive agents, *J. Clin. Endocrinol.*, 36, 246, 1973.
96. **Kotake, Y.**, Xanthurenic acid, an abnormal metabolite of tryptophan and the diabetic symptoms caused in albino rats by its production, *J. Vitaminol.*, 1, 157, 1955.
97. **Murakami, E.**, Studies on the xanthurenic acid-insulin complex. 1. Preparation and properties, *J. Biochem.*, 63, 573, 1968.
98. **Kotake, Y., Sotokawa, T., Murakami, E., Hisatake, A., Abe, M., and Ikeda, Y.**, Studies on the xanthurenic acid-insulin complex. II. Physiological activities, *J. Biochem.*, 63, 578, 1968.
99. **Spellacy, W. N., Buhi, W. C., and Birk, S. A.**, The effects of vitamin B_6 on carbohydrate metabolism in women taking steroid contraceptives: preliminary report, *Contraception*, 6, 265, 1972.
100. **Rose, D. P., Leklem, J. E., Brown, R. R., and Linkswiler, H. M.**, Impairment of glucose tolerance by dietary vitamin B_6 deficiency in oral contraceptive users, *J. Nutr.*, 103, xviii, 1973.
101. **Herzberg, B. N., Johnson, A. L., and Brown, S.**, Depressive symptoms and oral contraceptives, *Br. Med. J.*, 4, 142, 1970.
102. **Curzon, G.**, Tryptophan pyrrolase – a biochemical factor in depressive illness? *Br. J. Psychiatr.*, 115, 1367, 1969.
103. **Green, A. R. and Curzon, G.**, The effect of tryptophan metabolites on brain 5-hydroxytryptamine metabolism, *Biochem. Pharmacol.*, 19, 2061, 1970.
104. **Rose, D. P. and Braidman, I. P.**, Excretion of tryptophan metabolites as affected by pregnancy, contraceptive steroids and steroid hormones, *Am. J. Clin. Nutr.*, 24, 673, 1971.
105. **Robins, E., Robins, J. M., Croninger, A. B., Moses, S. G., Spencer, S. J., and Hudgens, R. W.**, The low level of 5-hydroxytryptophan decarboxylase in human brain, *Biochem. Med.*, 1, 240, 1967.
106. **Baumblatt, M. J. and Winston, F.**, Pyridoxine and the pill, *Lancet*, 1, 832, 1970.
107. **Adams, P. W., Wynn, V., Rose, D. P., Seed, M., Folkard, J., and Strong, R.**, The effect of pyridoxine hydrochloride (vitamin B_6) upon depression associated with oral contraceptive administration, *Lancet*, 1, 897, 1973.
108. **Coppen, A., Shaw, D. M., Herzberg, B., and Maggs, R.**, Tryptophan in the treatment of depression, *Lancet*, ii, 1178, 1967.

INDEX

A

Acid-base balance, in the fetus, 310
Addison's disease and neutral steroids, 146
Adenosine deaminase, and carcinoma, 219
Adenosine deaminase, serum, method, 220
Adrenal activity – plasma or urine?, 133
Adrenal steroidogenesis, 49
Adrenocorticol function, 132
Alanine aminotransferase, 320
Aldosterone, 129, 147
Alkaline phosphatase, and carcinoma, 219
Alkaline phosphatase, in pregnancy, 238
Alkaline phosphatase isoenzymes, 219
Alkaline phosphatase, isoenzymes, serum, method, 223
Alkaline phosphatase, placental, in serum, method, 221–238
Amniotic fluid lecithin levels in R.D.S., 304
Amniotic fluid lecithin, method, 300
Amniotic fluid phospholipids, method, 300
Androgen-estrogen, conversion in the placenta, 23
Androsterone, gas chromatography, 151
Androsterone, in urine, method, 163, 196
Androsterone sulfate in plasma, method, 202
Arylamidase, in serum method, 241

B

BEI, butanol extractable iodine, 265
Bioassays, reliability, 107
Blood pH in the fetus, 311
BMR, basal metabolic rate, 274
BMR, clinical interpretation, 293
BMR, in nonpregnant women, 288
BMR, method, 285
Breast cancer, and gonadotrophins, 210
Breast cancer, and hormones, 191–217
Breast cancer, and thyroid hormones, 212
Breast cancer, plasma 11-deoxy-17-ketosteroids and etiology, 200
Breast cancer, plasma levels of estrone, estradiol, and progesterone, 206

C

Cancer, and enzymes, 219
Cancer, and vaginal fluid, 233
Cancer, estrogens and progesterone, 204
Carbohydrate metabolism, in oral contraceptive users, 343
CBG, corticosteroid binding globulin, 132
CBG, method for progesterone, 208
Cholesterol, and thyroid function, 274
Clomid® *see* Clomiphene
Clomiphene, treatment in infertility, 10
Congenital adrenal hyperplasia, 76
Contraceptives, and depression, 345
Contraceptives, and enzymes, 251
Contraceptives, and thyroid function, 262
Contraceptives, metabolic side effects, 318
Contraceptives, vitamin B_6 and tryptophan, 317
Copper resistant serum acid phosphatase, method, 224
Cortisol, 129
Cortisol, in plasma, 132, 139, 146
Cortisol, in urine, 140, 146
Cortisol in urine, competitive protein binding method, 141
Cortisol, in urine, fluorescence method, 140
Cortisol metabolites, 134
Cortisol, progesterone pathways, 49
Cortisol, relationships between production, effect, and measurement, 134
CPB, competitive protein binding for estrogens in plasma of pregnant women, 27
CPB, for progesterone, method, 50
CPR, cortisol production rate, 134
Creatinine, in neutral steroids assays, 143
C_{18} steroids, structure, 1, 2
Cushing's syndrome, and neutral steroids, 145

D

Dehydroepiandrosterone, 23
Dehydroepiandrosterone, gas chromatography, 151
Dehydroepiandrosterone, in urine, method, 169, 196
Dehydroepiandrosterone sulfate, in plasma, method, 202
11-Deoxy-17-ketosteroids, and breast cancer, 193
11-Deoxyketosteroids, in plasma, 199
11-Deoxy-17-ketosteroids in urine, method, 196
Depression, and oral contraceptives, 345
Dexamethasone, suppression, 147, 168
Diabetes, in pregnancy, 19
Dihydrotestosterone, radioimmunoassay method, 182
Drug interference, in 17-oxosteroid and 17-oxogenic steroid assays, 144

E

Enzymes, and oral contraceptives, 251
Enzymes, and pregnancy, 237
Enzymes, in carcinoma, 219
Erythrocyte alanine aminotransferase, in blood, levels, 340, 341
Erythrocyte alanine aminotransferase, in blood, method, 338
Erythrocyte aspartate aminotransferase, 320
Estradiol, and breast cancer, 206
Estradiol, by gas chromatography, 3
Estradiol, gas chromatography, recommended method, in nonpregnancy urine, 6
Estradiol, in plasma, method, 8, 208
Estradiol, plasma levels in menstrual cycle, 12
Estradiol, recommended methods, 9
Estriol, by gas chromatography, 3
Estriol, gas chromatography, recommended method, in nonpregnancy urine, 6
Estriol, methods, in plasma, 8

Estrogen, biosynthesis, 23
Estrogen excretion, and renal function, 35
Estrogen excretion, compounds that suppress, 35
Estrogens, and cancer, 204
Estrogens, clinical applications in nonpregnant women, 9, 12
Estrogens, colorimetric methods, 2
Estrogens, excretion in normal menstrual cycle, 10
Estrogens, gas chromatography, recommended method, in nonpregnancy urine, 6
Estrogens, in amniotic fluid, 27
Estrogens in blood, review, 7
Estrogens, in maternal plasma, review, 21
Estrogens, in maternal urine, review, 20
Estrogens, in pregnancy, 19
Estrogens, in pregnancy plasma, review, 27
Estrogens, in pregnancy urine, review, 25
Estrogens, interpretation of results, 36
Estrogens, in urine, fluorimetry, recommended methods, 4
Estrogens, in urine, recommended methods, 28
Estrogen, total, in urine, overnight and 24 hr, 13
Estrone, and breast cancer, 206
Estrone, by gas chromatography, 3
Estrone, gas chromatography, recommended method, in nonpregnancy urine, 6
Estrone, in plasma, method, 208
Estrone, method, 2
Etiocholanolone, and breast cancer, 193, 194
Etiocholanolone, gas chromatography, 151
Etiocholanolone, in urine, method, 163, 196

F

Fetal acid-base balance, 310
Fetal monitoring, 299
Fetal scalp blood sampling, 310
Fetus, PCO_2 levels, 313
Fetus, pH of whole blood in the fetus, 311
FSH, 79
FSH, bioassays, 85, 88
FSH, excretion in normal menstrual cycle, 10, 12
FSH, in serum, 115
FSH, in urine, 115

G

Gas chromatography, androsterone, 151, 203
Gas chromatography, dehydroepiandrosterone, 151, 203
Gas chromatography, etiocholanolone, 151
Gas chromatography, for estrogens, 3
Gas chromatography, pregnanediol, 65, 151
Gas chromatography, pregnanetriol, 65
Gas chromatography, pregnanetriolone, 65
GGTP, gamma-glutamyl transpeptidase, 219–226
Glucose tolerance and oral contraceptives, 343
Gonadotrophins, 79
Gonadotrophins, and breast cancer, 210
Gonadotrophins, extraction from urine, 93
Gonadotrophins, immunological properties, 94
Gonadotrophins, interpretation of results, 112
GOT, and pregnancy, 244
GPT, and pregnancy, 244
Graafian follicle, maturation, 45
Growth hormone, 211
Growth hormone, and oral contraceptives, 344

H

HA *see* 3-Hydroxyanthranilic acid, 59
HCG, in serum of fetus and mother, 112
HCG, radioimmunoassay, 96
HFSH, in serum of fetus and mother, 112
HFSH, radioimmunoassay, 96
Hirsutism, 168
HK *see* 3-Hydroxykynurenine
HK/HA ratios, in oral contraceptive users, 343
HLH, in serum of fetus and mother, 112
HLH, radioimmunoassay, 96
HMG, human menopausal gonadotrophin, 11
Hormones, and breast cancer, 191
Hormones, binding to plasma proteins, 209
5-HT, 5-Hydroxytryptamine biosynthesis, from tryptophan, 319
Hydatidiform mole, 290
3-Hydroxyanthranilic acid, by ion exchange and colorimetry, 328
3-Hydroxyanthranilic acid excretion, 323
3-Hydroxyanthranilic acid, in urine, method, 327
17-Hydroxycorticoids, and breast cancer, 193
17-Hydroxycorticoids, in urine, method, 198
17-Hydroxycorticosteroids, in plasma, 203
3-Hydroxykynurenine, by ion exchange and colorimetry, 328
3-Hydroxykynurenine, excretion, 323
3-Hydroxykynurenine, in urine, method, 327
21-Hydroxylase deficiency, 49
17α-Hydroxyprogesterone, metabolism, 47
Hypertension, in pregnancy, 19
Hypopituitarism, and neutral steroids, 146

I

Iodine metabolism, 258

K

Kober reaction, 2, 3

L

Lactic acid dehydrogenase, 219
Lactic acid dehydrogenase, in serum, method, 228
Lactic acid dehydrogenase, isoenzymes, 219
Lactic acid dehydrogenase, isoenzymes, in serum, method, 229
LATS, long acting thyroid stimulator, 275, 290

Lecithin, and phospholipids, chromatographic separation, 303
Lecithin, in amniotic fluid, method, 300
Lecithin, in R.D.S., levels, 304
Lecithin, to sphingomyelin, ratios in amniotic fluid, 304
LH, 79
LH, bioassays, in vivo, 82
LH, excretion, in normal menstrual cycle, 10, 12
LH, plasma levels in normal menstrual cycle, 12

M

Menopause, and gonadotrophins, 117
Metyrapone stimulation test, 146

N

Neutral steroids, 129
Nicotinic acid ribonucleotide and 5-hydroxytryptamine, biosynthesis, 319
Nicotinic acid, urinary metabolites, 318
5′-Nucleotidase, 248

O

Oestradiol *see* Estradiol
Oestriol *see* Estriol
Ovarian ascorbic acid depletion, bioassay for LH, 85
Ovarian cholesterol depletion, bioassay for LH, 86
Ovary-pituitary relationship, 11
17-Oxogenic steroids, in urine, 145
17-Oxogenic steroids, in urine, recommended method, 135
17-Oxosteroids, 129
17-Oxosteroids, in urine, 144
17-Oxosteroids, in urine, method, 130
Oxytocinase, in serum, method, 241

P

PBI, in serum, method, 277
PBI, serum protein bound iodine, 264
PCO_2, in the fetus, 313
Pergonal®, 10
Pertechnetate, 264
PHI *see* Phosphohexoseisomerase
Phosphohexoseisomerase, 219
Phosphohexoseisomerase, in serum, method, 231
Phospholipids, and lecithin, chromatographic separation, 303
Phospholipids, in amniotic fluid, method, 300
Pituitary-ovary relationship, 11
Placental alkaline phosphatase in serum, method, 238
Pregnancy, and alkaline phosphatase, 238
Pregnancy, and enzymes, 237
Pregnancy, and oxytocinase, 241
Pregnancy, and tryptophan metabolism, 341
Pregnancy, choice of thyroid function tests, 276
Pregnancy, complications and 5′-nucleotidase, 250
Pregnancy, thyroid involvement, 257
Pregnanediol, as a metabolite of progesterone, 47
Pregnanediol, excretion in normal menstrual cycle, 10
Pregnanediol, gas chromatography, 65, 151
Pregnanediol, in urine, and breast cancer, 205
Pregnanediol, in urine, method, 60, 155
Pregnanediol, urine levels, 70
Pregnanetetrol, by TLC, 70
Pregnanetriol, 47
Pregnanetriol, by TLC, 70
Pregnanetriol, gas chromatography, 65
Pregnanetriol, in urine, method, 60
Pregnanetriolone, by TLC, 70
Pregnanetriolone, gas chromatography, 65
Pregnanetriol, urine levels, 70
Pregnanolone, urine levels, 70
Progesterone, 45
Progesterone, and breast cancer, 206
Progesterone, and cancer, 204
Progesterone, and the placenta, 75
Progesterone, by radioimmunoassay, in plasma, method, 56
Progesterone, in plasma, method, 51, 208
Progesterone, levels in plasma in women with regular menstrual cycles, 73
Progesterone, metabolism, 47, 48
Progesterone, metabolites, 45
Progesterone, plasma levels, 45
Progesterone, plasma levels in men, 73
Progesterone, plasma levels in normal menstrual cycle, 12
Progesterone, plasma levels in pregnancy, 46, 47
Prolactin, 79
Prolactin, and cancer, 211
Prolactin, bioassays, 88
Prolactin, extraction from urine, 94
Prolactin, in serum in pregnancy, 118
Prolactin, interpretation of results, 112
Proteins, binding of steroids, 209
Pyridoxal phosphate, 318
Pyridoxal phosphate, in plasma, levels, 320
Pyridoxal phosphate, in plasma, method, 334
Pyridoxal phosphate, in plasma, normal levels, 337
4-Pyridoxic acid, in urine, method, 332
4-Pyridoxic acid, metabolite of vitamin B_6, 320
4-Pyridoxic acid, urinary excretion, 335

Q

Quality control, enzymes, 254
Quality control, estrogens, 38
Quality control, pregnanediol and pregnanetriol, 64

R

Radioimmunoassay, dihydrotestosterone, method, 182
Radioimmunoassay, for estrogens, 4
Radioimmunoassay, for progesterone in plasma, method, 56
Radioimmunoassay, HCG, HLH, HFSH, 96

Radioimmunoassay, of estrogens in plasma from pregnant women, 27
Radioimmunoassay, reliability, 107
Radioimmunoassay, standards, 106
Radioimmunoassay, T_3 and T_4, 267
Radioimmunoassay, testosterone and metabolites, 181
Radioligand-receptor assay for LH, 86
Radioligand-receptor assay for prolactin, 90
R.D.S., respiratory distress syndrome, 300
Renal function, and estrogen excretion, 35

S

Sodium, and thyrotoxicosis, 274
Sphingomyelin, to lecithin ratios, in amniotic fluid, 304
Steroids, in urine, 59
Surfactant, in amniotic fluid, method, 308
Surfactant, in lungs, method, 300
Synacthen®, 147

T

T_3, in plasma, 260
T_3, in serum, 266
T_3, resin uptake, interpretation, 291
T_4, free factor, 284
T_4, free factor, interpretation, 292
T_4, in plasma, 260
T_4, in serum, 265
Testosterone, assay method for LH, 87
Testosterone, metabolites, radioimmunoassay techniques, 181
Testosterone, radioimmunoassay techniques, 181
Tetracosactrin, 147
Thyroid function tests, choice, in pregnancy, 276
Thyroid function tests, interpretation, 288
Thyroid hormones, and breast cancer, 212
Thyroid problems in pregnancy, 257
Thyroid status in pregnancy, 262
Thyrotoxicosis, and pregnancy, 289
Thyroxine, in serum, method, 281
TLC of pregnanetetrol, 70
TLC of pregnanetriol, 70
TLC of pregnanetriolone, 70
Toxemia, of pregnancy, 20
Triglyceride, level and thyrotoxicosis, 274
L-Tryptophan, dose in the load test, 321
Tryptophan load test, 319, 320
Tryptophan load test, interference by drugs, 322
Tryptophan metabolism, in pregnancy and oral contraceptive users, 317
Tryptophan metabolites, in urine, after loading test, levels, 330
Tryptophan metabolites, thin-layer chromatography, 327
Tryptophan-nicotinic acid, ribonucleotide pathway, 318
TSH, 275
Tyrosine, and thyrotoxicosis, 274

V

Vaginal fluid, investigations in neoplasia, 233
Ventral prostate weight, 82
Vitamin B_6 deficiency, in normal pregnancy, 342
Vitamin B_6 deficiency, in oral contraceptive users, 318, 342
Vitamin B_6 nutrition, biochemical assessment, 320
Vitamin B_6 nutrition, in pregnancy and oral contraceptive users, 317
Vitamin B_6, urinary metabolites, 320

X

Xanthurenic acid, by ion exchange chromatography, 327
Xanthurenic acid, excretion after tryptophan load test, 323
Xanthurenic acid excretion, and oral contraceptives, 319
Xanthurenic acid, method, 324

Z

Zimmerman reaction, 130
Zimmerman reaction, for 17-oxogenic steroids, 138